Linus Pauling

The Nature of the Chemical Bond

and the Structure of Molecules and Crystals

An Introduction to Modern Structural Chemistry

Third Edition

化学键的本质

第 3 版

世界图书出版公司

北京·广州·上海·西安

图书在版编目（CIP）数据

化学键的本质 : 第 3 版 = The Nature of the Chemical Bond and the Structure of Molecules and Crystals: An Introduction to Modern Structural Chemistry, Third Edition : 英文 /（美）莱纳斯·鲍林（Linus Pauling）著 . — 北京 : 世界图书出版有限公司北京分公司 , 2023.8
ISBN 978-7-5192-9702-2

Ⅰ . ①化… Ⅱ . ①莱… Ⅲ . ①化学键—英文 Ⅳ . ① O641.1

中国版本图书馆 CIP 数据核字（2022）第 131023 号

Copyright © 1939 and 1940, third edition © 1960 by Cornell University
Published by arrangement with the original publisher, Cornell University Press, for sale in China only (excluding the territories of Hong Kong SAR, Macau SAR and Taiwan Province).

本书由康奈尔大学出版社授权世界图书出版有限公司北京分公司出版。此版本仅限在中华人民共和国境内（不包括香港、澳门特别行政区及台湾省）销售发行。

中文书名	化学键的本质（第 3 版）
英文书名	The Nature of the Chemical Bond and the Structure of Molecules and Crystals: An Introduction to Modern Structural Chemistry *3rd edition*
著　　者	［美］莱纳斯·鲍林（Linus Pauling）
策划编辑	陈　亮
责任编辑	陈　亮　刘叶青
出版发行	世界图书出版有限公司北京分公司
地　　址	北京市东城区朝内大街 137 号
邮　　编	100010
电　　话	010-64038355（发行）　64033507（总编室）
网　　址	http://www.wpcbj.com.cn
邮　　箱	wpcbjst@vip.163.com
销　　售	新华书店
印　　刷	北京建宏印刷有限公司
开　　本	710mm×1000mm　1/16
印　　张	41.5
字　　数	644 千字
版　　次	2023 年 8 月第 1 版
印　　次	2023 年 8 月第 1 次印刷
版权登记	01-2019-5194
国际书号	ISBN 978-7-5192-9702-2
定　　价	219.00 元

THE NATURE OF

THE CHEMICAL BOND

AND THE STRUCTURE OF
MOLECULES AND CRYSTALS:

An Introduction to Modern Structural Chemistry

By LINUS PAULING

THIRD EDITION

CORNELL UNIVERSITY PRESS

Ithaca, New York

Third edition published 1960 by Cornell University Press

ISBN 978-0-8014-0333-0 (cloth : alk. paper)

Cornell University Press strives to use environmentally responsible suppliers and materials to the fullest extent possible in the publishing of its books. Such materials include vegetable-based, low-VOC inks and acid-free papers that are recycled, totally chlorine-free, or partly composed of nonwood fibers. For further information, visit our website at www.cornellpress .cornell.edu.

Cloth printing   20   19

TO

GILBERT NEWTON LEWIS

Preface to the Third Edition

DURING the period of a score of years that has passed since the first and second editions of this book were written much progress has been made in the determination of the structures of molecules and crystals and in the development of the theory of the chemical bond. It is no longer possible to discuss in a short book all of the present body of knowledge about the structures of molecules and crystals. I have contented myself in this third edition with the presentation of general principles and the discussion of a rather small number of substances as examples. In some cases the old examples have been retained, and in others they have been replaced by new ones.

The principal innovations that have been made in the discussion of the theory of the chemical bond in this edition are the wide application of the electroneutrality principle and the use of an empirical equation (Sec. 7-10) for the evaluation of the bond numbers of fractional bonds from the observed bond lengths. A new theory of the structure of electron-deficient substances, the resonating-valence-bond theory, is described and used in the discussion of the boranes, ferrocene, and other substances. A detailed discussion of the valence-bond theory of the electronic structure of metals and intermetallic compounds is also presented.

Recognition has been made of some rather strongly worded criticism, from various sides, of the treatment of resonance of molecules among alternative valence-bond structures, as presented in earlier editions of this book, on the basis of its idealistic and arbitrary character, by the introduction of a section (Sec. 6-5) in which it is pointed out that the theory of resonance involves only the same amounts of idealization and arbitrariness as the classical valence-bond theory.

It is my opinion that the student of chemistry may well benefit from the study of modern structural chemistry early in his career; for example, as an undergraduate.   I have thought it wise to change the character of this book somewhat in order to increase its value to such a student.   The principal change is the introduction, in Chapter 2 and Appendix IV, of a moderately detailed treatment of the electronic structure of atoms, atomic energy levels, electron spin, Russell-Saunders coupling, the Pauli exclusion principle, and the magnetic moments of atoms.   Some other aspects of the theory of atomic and molecular structure and of experimental methods of structure determination are presented in other appendices.

The theory of the chemical bond, as presented in this book, is still far from perfect.   Most of the principles that have been developed are crude, and only rarely can they be used in making an accurate quantitative prediction.   However, they are the best that we have, as yet, and I agree with Poincaré that "it is far better to foresee even without certainty than not to foresee at all."

I am grateful to many friends for advice and assistance in the preparation of the revised edition of this book—especially to my colleagues in the California Institute of Technology.   I thank also Mrs. Beatrice Wulf, Mrs. Joan Harris, Mrs. Ruth Hughes, and Mr. Crellin Pauling for their help.

LINUS PAULING

Deer Flat Ranch
San Simeon, California
April 6, 1959

Preface to the Second Edition

THE progress made in the field of modern structural chemistry during the past year has consisted in the main in the determination of the structures of a number of especially interesting molecules and crystals. I have been glad to have the opportunity provided by the exhaustion of the first edition of this book to revise it by the inclusion of references to these researches and of discussion of the new structures. A few corrections have been made, and the argument in some places has been expanded in an effort to improve the clarity of its presentation. Two new sections have been added, dealing with restricted rotation about single bonds (Sec. 14d) and the conditions for equivalence or non-equivalence of bonds (Sec. 22a).

I have again to thank many friends for their advice and assistance; I am grateful especially for their aid to Dr. E. W. Hughes, Research Fellow in Chemistry in the California Institute of Technology, and to Mr. W. S. Schaefer of the Cornell University Press.

LINUS PAULING

Pasadena, California
February 28, 1940

Preface to the First Edition

FOR a long time I have been planning to write a book on the structure of molecules and crystals and the nature of the chemical bond. With the development of the theory of quantum mechanics and its application to chemical problems it became evident that a decision would have to be made regarding the extent to which the mathematical methods of the theory would be incorporated in this book. I formed the opinion that, even though much of the recent progress in structural chemistry has been due to quantum mechanics, it should be possible to describe the new developments in a thorough-going and satisfactory manner without the use of advanced mathematics. A small part only of the body of contributions of quantum mechanics to chemistry has been purely quantum-mechanical in character; only in a few cases, for example, have results of direct chemical interest been obtained by the accurate solution of the Schrödinger wave equation. The advances which have been made have been in the main the result of essentially chemical arguments—the assumption of a simple postulate, which is then tested by empirical comparison with available chemical information, and used in the prediction of new phenomena. The principal contribution of quantum mechanics to chemistry has been the suggestion of new ideas, such as the resonance of molecules among several electronic structures with an accompanying increase in stability.

The ideas involved in modern structural chemistry are no more difficult and require for their understanding no more, or little more, mathematical preparation than the familiar concepts of chemistry. Some of them may seem strange at first, but with practice there can be developed an extended chemical intuition which permits the new concepts to be used just as confidently as the older ones of the valence bond, the tetrahedral carbon atom, etc., which form the basis of classical structural chemistry.

The foundation of the modern theory of valence was laid by G. N. Lewis in his 1916 paper.[1]   The theory was extended in his book *Valence and the Structure of Atoms and Molecules* (Chemical Catalog Co., New York, 1923), in N. V. Sidgwick's volumes *The Electronic Theory of Valency* (Clarendon Press, Oxford, 1927) and *The Covalent Link in Chemistry* (Cornell University Press, 1933), and in numerous publications by Irving Langmuir, W. M. Latimer, W. H. Rodebush, M. L. Huggins, W. A. Noyes, A. Lapworth, Robert Robinson, C. K. Ingold, and many other investigators.   The detailed discussion in the following chapters is based to a large extent on seven papers with the general title "The Nature of the Chemical Bond," published between 1931 and 1933 in the *Journal of the American Chemical Society* and the *Journal of Chemical Physics*, and on other papers by my collaborators and myself.

I have felt that in writing on this complex subject my primary duty should be to present the theory of the chemical bond (from my point of view) in as straightforward a way as possible, relegating the historical development of the subject to a secondary place.   Many references are included to early work in this field; the papers on the electronic theory of valence published during the last twenty years are so numerous, however, and often represent such small differences of opinion as to make the discussion of all of them unnecessary and even undesirable.

The opportunity and incentive to prepare this work for publication have been provided by my tenure of the George Fisher Baker Nonresident Professorship of Chemistry at Cornell University during the Fall Semester of 1937–38.   I wish to express my sincere thanks to Professor Papish and his colleagues in the Department of Chemistry of the University for their kindness in extending to me the invitation to present the Baker Lectures and in making available the facilities of the Baker Laboratory of Chemistry during my period of residence in Ithaca.   I am grateful for advice and assistance in the preparation of the manuscript to many friends, including Dr. E. W. Hughes, Dr. C. D. Coryell, Dr. H. D. Springall, Dr. G. Schwarzenbach, Dr. J. H. Sturdivant, Dr. G. C. Hampson, Mr. P. A. Shaffer, Jr., Dr. E. R. Buchman, Dr. S. Weinbaum, Dr. Fred Stitt, Dr. J. Sherman, and Dr. F. J. Ewing.   My wife joins me in expressing our appreciation to the young men of the Telluride House at Cornell University, who were our hosts during our stay in Ithaca.

Linus Pauling

Gates and Crellin Laboratories of Chemistry
California Institute of Technology, Pasadena, California
June 1938

[1] G. N. Lewis, *J.A.C.S.* **38**, 762 (1916).

Contents

xiii

CHAPTER 3

The Partial Ionic Character of Covalent Bonds and the Relative Electronegativity of Atoms

CHAPTER 4

The Directed Covalent Bond; Bond Strengths and Bond Angles

CHAPTER 5

Complex Bond Orbitals; The Magnetic Criterion for Bond Type

CHAPTER 6

The Resonance of Molecules among Several Valence-Bond Structures

CHAPTER 7

*Interatomic Distances and Their Relation to the Structure
of Molecules and Crystals*

CHAPTER 8

Types of Resonance in Molecules

CHAPTER 9

The Structure of Molecules and Complex Ions Involving Bonds with Partial Double-Bond Character

CHAPTER 10

The One-Electron Bond and the Three-Electron Bond; Electron-deficient Substances

Contents

CHAPTER 11
The Metallic Bond

CHAPTER 12
The Hydrogen Bond

CHAPTER 13

The Sizes of Ions and the Structure of Ionic Crystals

CHAPTER 14

A Summarizing Discussion of Resonance and
Its Significance for Chemistry

Appendices and Indices

THE NATURE OF
THE CHEMICAL BOND

CHAPTER 1

Resonance and the Chemical Bond

MOST of the general principles of molecular structure and the nature of the chemical bond were formulated long ago by chemists by induction from the great body of chemical facts. During recent decades these principles have been made more precise and more useful through the application of the powerful experimental methods and theories of modern physics, and some new principles of structural chemistry have also been discovered. As a result structural chemistry has now become significant not only to the various branches of chemistry but also to biology and medicine.

The amount of knowledge of molecular structure and the nature of the chemical bond is now very great. In this book I shall attempt to present only an introduction to the subject, with emphasis on the most important general principles.

1-1. THE DEVELOPMENT OF THE THEORY OF VALENCE

The study of the structure of molecules was originally carried on by chemists using methods of investigation that were essentially chemical in nature, relating to the chemical composition of substances, the existence of isomers, the nature of the chemical reactions in which a substance takes part, and so on. From the consideration of facts of this kind Frankland, Kekulé, Couper, and Butlerov[1] were led a century ago to formulate the theory of valence and to write the first structural

---

[1] E. Frankland, *Phil. Trans. Roy. Soc. London* 142, 417 (1852), proposed the concept of valence in 1852, stating that each element forms compounds by uniting with a definite number of what we now call equivalents of other elements. F. A. Kekulé, *Ann. Chem.* 104, 129 (1857), and A. W. H. Kolbe, *ibid.* 101, 257 (1857), then extended the concept of valence to carbon and said that carbon usually has the valence 4. In the following year Kekulé, *ibid.* 106, 129 (1858), suggested that carbon atoms can unite with an indefinite number of other carbon atoms into long chains. A. S. Couper, a Scottish chemist, independently discussed the quadrivalence of carbon and the ability of carbon atoms to form chains (*Compt. rend.* 46, 1157 [1858]; *Ann. chim. phys.* 53, 469 [1858]). Couper's

formulas for molecules, van't Hoff and le Bel[2] were led to bring classical organic stereochemistry into its final form by their brilliant postulate of the tetrahedral orientation of the four valence bonds of the carbon atom, and Werner[3] was led to his development of the theory of the stereochemistry of complex inorganic substances.

Modern structural chemistry differs from classical structural chemistry with respect to the detailed picture of molecules and crystals that it presents. By various physical methods, including the study of the structure of crystals by the diffraction of x-rays and of gas molecules by the diffraction of electron waves, the measurement of electric and magnetic dipole moments, the interpretation of band spectra, Raman spectra, microwave spectra, and nuclear magnetic resonance spectra, and the determination of entropy values, a great amount of information has been obtained about the atomic configurations of molecules and crystals and even their electronic structures; a discussion of valence and the chemical bond now must take into account this information as well as the facts of chemistry.

In the nineteenth century the valence bond was represented by a line drawn between the symbols of two chemical elements, which expressed in a concise way many chemical facts, but which had only qualitative significance with regard to molecular structure. The nature of the bond was completely unknown. After the discovery of the electron numerous attempts were made to develop an electronic theory of the chemical bond. These culminated in the work of Lewis, who in

---

chemical formulas were much like the modern ones; he was the first chemist to use a line between symbols to represent the valence bond.

In 1861 the Russian chemist A. M. Butlerov, *Z. Chem. Pharm.* **4**, 549 (1861), used the term "chemical structure" for the first time and stated that it is essential to express the structure by a single formula, which should show how each atom is linked to other atoms in the molecule of the substance. He stated clearly that all properties of a compound are determined by the molecular structure of the substance and suggested that it should be possible to find the correct structural formula of a substance by studying the ways in which it can be synthesized.

None of these chemists stated that the chemical formulas were to be interpreted as showing the way in which the atoms are bonded together in space; the formulas were used to indicate something about the ways in which the substances take part in chemical reactions. The next step, that of assigning structures in three-dimensional space to the molecules, was then taken by van't Hoff and le Bel. Excerpts from some of the papers mentioned above are to be found in H. M. Leicester and H. S. Klickstein, *A Source Book in Chemistry*, McGraw-Hill Book Co., New York, 1952.

[2] J. H. van't Hoff, *Arch. neerland. sci.* **9**, 445 (1874); J. A. le Bel, *Bull. soc. chim. France* **22**, 337 (1874).

[3] A. Werner, *Z. anorg. Chem.* **3**, 267 (1893).

his 1916 paper,[4] which forms the basis of the modern electronic theory of valence, discussed not only the formation of ions by the completion of stable shells of electrons[5] but also the formation of a chemical bond, now called the covalent bond, by the sharing of two electrons between two atoms.[6] Lewis further emphasized the importance of the phenomena of the pairing of unshared as well as of shared electrons and of the stability of the group of eight electrons (shared or unshared) about the lighter atoms.   These ideas were then further developed by many investigators; the work of Langmuir[7] was especially valuable in showing the great extent to which the facts of chemistry could be coordinated and clarified by the application of the new ideas.   Many of the features of the detailed theory that is discussed in this book were suggested in the papers of Langmuir and others written in the decade following 1916, or in the book *Valence and the Structure of Atoms and Molecules* written by Lewis in 1923.

All of these early studies, however, contained, in addition to suggestions that have since been incorporated into the present theory, many others that have been discarded.   The refinement of the electronic theory of valence into its present form has been due almost entirely to the development of the theory of quantum mechanics, which has not only provided a method for the calculation of the properties of simple molecules, leading to the complete elucidation of the phenomena involved in the formation of a covalent bond between two atoms and dispersing the veil of mystery that had shrouded the bond during the decades since its existence was first assumed, but has also introduced into chemical theory a new concept, that of *resonance*, which, if not entirely unanticipated in its applications to chemistry, nevertheless had not before been clearly recognized and understood.

In the following sections of this chapter there are given, after an introductory survey of the types of chemical bonds, discussions of the concept of resonance and of the nature of the one-electron bond and the electron-pair bond.

1-2. TYPES OF CHEMICAL BONDS

It is convenient to consider three general extreme types of chemical bonds: *electrostatic bonds*, *covalent bonds*, and *metallic bonds*.   This classification is not a rigorous one; for, although the bonds of each extreme

---

[4] G. N. Lewis, "The Atom and the Molecule," *J.A.C.S.* **38**, 762 (1916).

[5] This was treated independently at about the same time by W. Kossel, *Ann. Physik* **49**, 229 (1916).

[6] Earlier attempts to develop a theory of valence involving the sharing of electrons by atoms were made by W. Ramsay, J. J. Thomson, J. Stark, A. L. Parson, and others.

[7] I. Langmuir, *J.A.C.S.* **41**, 868, 1543 (1919).

type have well-defined properties, the transition from one extreme type to another may be gradual, permitting the existence of bonds of intermediate type (see Chap. 3 and later chapters).

The Chemical Bond Defined.—We shall say that *there is a chemical bond between two atoms or groups of atoms in case that the forces acting between them are such as to lead to the formation of an aggregate with sufficient stability to make it convenient for the chemist to consider it as an independent molecular species.*

With this definition we accept in the category of chemical bonds not only the directed valence bond of the organic chemist but also, for example, the bonds between sodium cations and chloride anions in the sodium chloride crystal, those between the aluminum ion and the six surrounding water molecules in the hydrated aluminum ion in solution or in crystals, and even the weak bond that holds together the two O_2 molecules in O_4. In general we do not consider the weak van der Waals forces between molecules as leading to chemical-bond formation; but in exceptional cases, such as that of the O_4 molecule mentioned above, it may happen that these forces are strong enough to make it convenient to describe the corresponding intermolecular interaction as bond formation.

The Ionic Bond and Other Electrostatic Bonds.—In case that there can be assigned to each of two atoms or groups of atoms a definite electronic structure essentially independent of the presence of the other atom or group and such that electrostatic interactions are set up that lead to strong attraction and the formation of a chemical bond, we say that the bond is an *electrostatic bond*.

The most important electrostatic bond is the *ionic bond*, resulting from the Coulomb attraction of the excess electric charges of oppositely charged ions. The atoms of metallic elements lose their outer electrons easily, whereas those of nonmetallic elements tend to add additional electrons; in this way stable cations and anions may be formed, which may essentially retain their electronic structures as they approach one another to form a stable molecule or crystal. In the sodium chloride crystal, with the atomic arrangement shown in Figure 1-1, there exist no discrete NaCl molecules. The crystal is instead composed of sodium cations, Na^+, and chloride anions, Cl^-, each of which is strongly attracted to and held by the six oppositely charged ions that surround it octahedrally. We describe the interactions in this crystal by saying that each ion forms ionic bonds with its six neighbors, these bonds combining all of the ions in the crystal into one giant molecule. A detailed treatment of ionic crystals is given in Chapter 13.

In $[Fe(H_2O)_6]^{+++}$, $[Ni(H_2O)_6]^{++}$, and many other complexes the bonds between the central ion and the surrounding molecules result in

FIG. 1-1.—The atomic arrangement in the sodium chloride crystal. (This figure is from the paper by W. Barlow, *Z. Krist.* **29,** 433 (1898), referred to in Sec. 11-5.)

considerable part from the electrostatic attraction of the excess charge of the central ion for the permanent electric dipoles of the molecules.[8] Electrostatic bonds of this type may be called ion-dipole bonds. Electrostatic bonds might also result from the attraction of an ion for the induced dipole of a polarizable molecule or from the mutual interaction of the permanent electric dipoles of two molecules.

The Covalent Bond.[9]—Following Lewis, we interpret the ordinary

valence bond, as in the formulas H—H,   Cl—Cl,   H—Cl,   H—C—H, with H above and H below the C.

[8] I. Langmuir, *loc. cit.* (7), 868, especially pp. 930–931.

[9] The convenient name covalent bond, which we shall often use in this book in place of the more cumbersome expressions shared-electron-pair bond or electron-pair bond, was introduced by Langmuir (*loc. cit.* [7], 868). Lewis preferred to include under the name chemical bond a more restricted class of interatomic interactions than that given by our definition ("the chemical bond is at all times and in all molecules merely a pair of electrons held jointly by two atoms" —Lewis, *op. cit.* p. 78).

etc., as involving the sharing of a pair of electrons by the two bonded atoms, and we write corresponding electronic structures, such as

$$H:H, \quad :\ddot{C}l:\ddot{C}l:, \quad H:\ddot{C}l:, \quad H:\overset{\displaystyle H}{\underset{\displaystyle H}{\ddot{C}}}:H, \quad \text{etc. In these Lewis electronic for-}$$

mulas the symbol of the element represents the *kernel* of the atom, consisting of the nucleus and the inner electrons, but not those in the valence shell, which are shown by dots.   A pair of electrons held jointly by two atoms is considered for some purposes to do double duty, and to be effective in completing a stable electronic configuration for each atom.   It is seen that in methane the carbon atom, with its two inner electrons and its outer shell of eight shared electrons, has assumed the stable ten-electron configuration of neon, and that each of the other atoms in the structures shown has achieved a noble-gas configuration.

A double bond and a triple bond between two atoms can be represented respectively by four and six shared electrons, as in the following examples:

$$\overset{\displaystyle H}{\underset{\displaystyle H}{>}}C\!=\!C\overset{\displaystyle H}{\underset{\displaystyle H}{<}} \qquad\qquad \overset{\displaystyle H}{\underset{\displaystyle H}{:}}\ddot{C}::\ddot{C}\overset{\displaystyle H}{\underset{\displaystyle H}{:}}$$

$$H\!-\!C\!\equiv\!C\!-\!H \qquad\qquad H:C:::C:H$$

$$N\!\equiv\!N \qquad\qquad :N:::N:$$

In order that the nitrogen atom in trimethylamine oxide, $(CH_3)_3NO$, might be assigned the neon structure with a completed octet of valence

electrons, Lewis wrote for it the electronic structure $R:\overset{\displaystyle R}{\underset{\displaystyle R}{\ddot{N}}}:\ddot{O}:$ (with

$R = CH_3$), in which the nitrogen atom forms four single covalent bonds and the oxygen atom one.   If it is assumed that the electrons of a shared pair are divided between the two atoms which they connect, it is found on counting electrons for this formula that the nitrogen atom has the electric charge $+1$ (in units equal in magnitude to the electronic charge, with changed sign) and the oxygen atom the charge -1. We shall call these charges, calculated with use of an electronic structure by dividing shared electrons equally between the bonded atoms, the *formal charges* of the atoms for the corresponding structure,[10] and we shall often represent them by signs near the symbols of the atoms, as in the following examples:

Trimethylamine oxide,

$$\begin{array}{c} \text{R} \\ | \\ \text{R}\!-\!\overset{+}{\text{N}}\!-\!\overset{..}{\underset{..}{\text{O}}}\!:^{-} \\ | \\ \text{R} \end{array}$$

Sulfate ion,

$$\left[\begin{array}{c} :\!\overset{..}{\underset{..}{\text{O}}}\!:^{-} \\ | \\ :\!\overset{..}{\underset{..}{\text{O}}}\!-\!\overset{++}{\underset{|}{\text{S}}}\!-\!\overset{..}{\underset{..}{\text{O}}}\!:^{-} \\ :\!\overset{..}{\underset{..}{\text{O}}}\!:^{-} \end{array} \right]^{--}$$

Ammonium ion,

$$\left[\begin{array}{c} \text{H} \\ | \\ \text{H}\!-\!\overset{+}{\text{N}}\!-\!\text{H} \\ | \\ \text{H} \end{array} \right]^{+}$$

These formal charges are, as indicated by their name, to be considered as conventional in significance; they do not show in general the actual distribution of electric charges among the atoms in a molecule or complex ion. Thus in the ammonium ion the unit positive charge of the complex is not to be considered as residing exclusively on the nitrogen atom; as a consequence of the partial ionic character of the N—H bonds, discussed in Chapter 3, part of the excess positive charge can be considered to be transferred to each of the hydrogen atoms.

We see from the electronic formula that we have just written that the bond between nitrogen and oxygen in trimethylamine oxide may be considered as a sort of double bond, consisting of one single covalent bond and one ionic bond of unit strength. A bond of this type has sometimes been called[11] a semipolar double bond. The name coordinate link has also been used,[12] together with a special symbol, →, to indicate the transfer of electric charge from one atom to another. Electronic formulas have also been used in which the presumable original attachment of electrons to one atom or another is indicated by the use of different symbols (dots and crosses) for different electrons. We

[10] The formal charge (which he called the residual atomic charge) was first discussed by I. Langmuir, *Science* **54**, 59 (1921).

[11] T. M. Lowry, *Trans. Faraday Soc.* **18**, 285 (1923); *J. Chem. Soc.* **123**, 822 (1923).

[12] N. V. Sidgwick, *The Electronic Theory of Valency*, Clarendon Press, Oxford, 1927.

shall not find it convenient to make use of these names or of these symbols.

In a few molecules there occur covalent bonds involving one electron or three electrons, instead of a shared pair. These *one-electron* and *three-electron bonds* are discussed in Section 1-4 and Chapter 10.

The Metallic Bond; Fractional Bonds.—The most striking characteristic of the bond that holds atoms together in a metallic aggregate is the mobility of the bonding electrons, which gives rise to the high electric and thermal conductivity of metals. A discussion of the metallic bond and its relation to the covalent bond is given in Chapter 11. The bonds in metals may be described as fractional bonds. A discussion of other substances involving fractional bonds, called electron-deficient compounds, is given in Chapter 10.

1-3. THE CONCEPT OF RESONANCE[13]

There is one fundamental principle of quantum mechanics that finds expression in most of the chemical applications of the theory to problems dealing with the normal states of molecules. This is the principle that underlies the concept of *resonance*.

A structure for a system is represented in quantum mechanics by a wave function, usually called ψ, a function of the coordinates that in classical theory would be used (with their conjugate momenta) in describing the system. The methods for finding the wave function for a system in a particular state are described in treatises on quantum mechanics. In our discussion of the nature of the chemical bond we shall restrict our interest in the main to the normal states of molecules. The stationary quantum states of a molecule or other system are states that are characterized by definite values of the total energy of the system. These states are designated by a quantum number, repre-

---

[13] In preparing this discussion of a phenomenon that is essentially quantum-mechanical in nature I have introduced concepts and principles that are based on the theory of quantum mechanics whenever they are necessary for the argument, without attempting to place the discussion on a postulatory basis or to make the development of the argument logically complete.

The discussion in this book may be complemented by that in Linus Pauling and E. Bright Wilson, Jr., *Introduction to Quantum Mechanics with Applications to Chemistry*, McGraw-Hill Book Co., New York and London, 1935, to which later reference will be made under the title *Introduction to Quantum Mechanics*.

A thorough and penetrating discussion of the theory of resonance is given in the book by George Willard Wheland, *Resonance in Organic Chemistry*, John Wiley and Sons, New York, 1955. Other valuable reference books are Y. K. Syrkin and M. E. Dyatkina, *Structure of Molecules and the Chemical Bond*, Interscience Publishers, New York, 1950, and C. A. Coulson, *Valence*, Clarendon Press, Oxford, 1952.

sented by the letter n, say, or by a set of two or more quantum numbers, each of which can assume any one of certain integral values. The system in the nth stationary quantum state has the definite energy value W_n and is represented by the wave function ψ_n. Predictions can be made about the behavior of the system known to be in the nth quantum state by use of the wave function. These predictions, which relate to the expected results of experiments to be carried out on the system, are in general not unique, but instead statistical in nature. For example, it is not possible to make a definite prediction of the position of the electron relative to the nucleus in a normal hydrogen atom; instead, a corresponding probability distribution function can be found.

The stationary quantum state that has the lowest value of the total energy of the system, corresponding to maximum stability, is called the normal state. The quantum numbers are usually assigned values 1 or 0 for this state.

Let ψ_0 be the correct wave function for the normal state of the system under discussion. The fundamental principle of quantum mechanics in which we are interested states that *the energy value W_0 calculated by the equations of quantum mechanics with use of the correct wave function ψ_0 for the normal state of the system is less than that calculated with any other wave function ψ that might be proposed;*[14] in consequence, *the actual structure of the normal state of a system is that one, of all conceivable structures, that gives the system the maximum stability.*

Now let us consider two structures, I and II, that might reasonably or conceivably represent the normal state of the system under consideration. The methods of the theory are such that the more general function

$$\psi = a\psi_{\mathrm{I}} + b\psi_{\mathrm{II}} \tag{1-1}$$

formed by multiplying ψ_{I} and ψ_{II} by arbitrary numerical coefficients and adding is also a possible wave function for the system. Only the ratio b/a is significant, the nature of the function not being changed by multiplication by a constant. By calculating the energy corresponding to ψ as a function of the ratio b/a, the value of b/a that gives the energy its minimum value can be found. The corresponding wave function is then the best approximation to the correct wave function for the normal state of the system that can be constructed in this way. If the best value of b/a turns out to be very small, then the best wave function ψ will be essentially equal to ψ_{I} and the normal state will be represented more closely by structure I than by any other structure of

[14] For a detailed discussion of this principle see *Introduction to Quantum Mechanics.*

those considered. It may well happen, however, that the best value of b/a is neither very small nor very large (in the latter case the best ψ would differ little from ψ_{II}), but is of the order of magnitude of unity. In this case the best wave function ψ would be formed in part from ψ_I and in part from ψ_{II} and the normal state of the system would be described correspondingly as involving both structure I and structure II. It has become conventional to speak of such a system as *resonating* between structures I and II, or as being a *resonance hybrid* of structures I and II.

The structure of such a system is not, however, exactly intermediate in character between structures I and II, because as a consequence of the resonance it is stabilized by a certain amount of energy, the *resonance energy*. The best value of b/a is that which gives the total energy of the system its minimum value, this value lying below that for either ψ_I or ψ_{II} by an amount that depends on the magnitude of the interaction between structures I and II and on their energy difference (see Sec. 1-4). It is this extra stability of the system, relative to structure I or structure II (whichever is the more stable), that is called the resonance energy.[15]

The structures considered in the discussion of the normal state of a system need not be restricted in number to two. In general a wave function

$$\psi = a\psi_I + b\psi_{II} + c\psi_{III} + d\psi_{IV} + \cdots \qquad (1\text{-}2)$$

may be formed by linear combination of the wave functions ψ_I, ψ_{II}, ψ_{III}, ψ_{IV}, $\cdots$ corresponding to the structures I, II, III, IV, $\cdots$ that suggest themselves for consideration. In this wave function the best relative values of the numerical coefficients a, b, c, d, $\cdots$ are to be found by minimizing the energy.

The concept of resonance was introduced into quantum mechanics by Heisenberg[16] in connection with the discussion of the quantum states of the helium atom. He pointed out that a quantum-mechanical treatment somewhat analogous to the classical treatment of a system of resonating coupled harmonic oscillators can be applied to many systems. The resonance phenomenon of classical mechanics is observed, for example, for a system of two tuning forks with the same characteristic frequency of oscillation and attached to a common base, which

---

[15] Because the resonating system does not have a structure intermediate between those involved in the resonance, but instead a structure that is further changed by the resonance stabilization, I prefer not to use the word mesomerism, suggested by Ingold in 1933 for the resonance phenomenon (C. K. Ingold, *J. Chem. Soc.*, **1933**, 1120).

[16] W. Heisenberg, *Z. Physik* **39**, 499 (1926).

provides an interaction between them. When one fork is struck, it gradually ceases to oscillate, transferring its energy to the other, which begins its oscillation; the process is then reversed, and the energy resonates back and forth between the two forks until it is dissipated by frictional and other losses. The same phenomenon is shown by two similar pendulums connected by a weak spring. The qualitative analogy between this classical resonance phenomenon and the quantum-mechanical resonance phenomenon described in the first part of this section is obvious; the analogy does not, however, provide a simple nonmathematical explanation of a most important feature of quantum-mechanical resonance in its chemical applications, that of stabilization of the system by the resonance energy, and we shall accordingly not pursue it further. The student of chemistry will, I believe, be able to develop a reliable and useful intuitive understanding of the concept of resonance by the study of its applications to various problems as described throughout this book.

It must be pointed out that there is an element of arbitrariness in the use of the concept of resonance, introduced by the choice of the initial structures I, II, III, IV, etc. as the basis for discussion of the normal state of a system. It is found, however, that for many systems certain structures suggest themselves strongly as most appropriate for this purpose and that great progress can be made in the discussion of complex systems such as molecules by using the structures of related simpler systems as a starting point. A striking example of this is given by the most important chemical application of resonance that has been discovered, the resonance of a molecule among several valence-bond structures: it is found that there are many substances whose properties cannot be accounted for by means of a single electronic structure of the valence-bond type, but which can be fitted into the scheme of classical valence theory by the consideration of resonance among two or more such structures.

The convenience and value of the concept of resonance in discussing the problems of chemistry are so great as to make the disadvantage of the element of arbitrariness of little significance. This element occurs in the classical resonance phenomenon also—it is arbitrary to discuss the behavior of a system of pendulums with connecting springs in terms of the motion of independent pendulums, since the motion can be described in a way that is mathematically simpler by use of the normal coordinates of the system—but the convenience and usefulness of the concept have nevertheless caused it to be widely applied.

Moreover, it must not be forgotten that the element of arbitrariness occurs in essentially the same way in the simple structure theory of organic chemistry as in the theory of resonance: there is the same use of idealized, hypothetical structural elements. For example, the propane molecule, C_3H_8, has its own structure, which cannot be described precisely in terms of structural elements from other molecules; it is not possible to isolate a portion of the propane molecule, involving parts of two carbon atoms and perhaps two electrons between them, and to say that this portion of the propane molecule is the carbon-carbon single bond, identical with a portion of the ethane molecule. The description of the propane molecule as involving carbon-carbon single bonds and carbon-hydrogen single bonds is arbitrary; the concepts themselves are idealizations. Chemists have, however, found that the simple structure theory of organic chemistry and also the resonance theory are valuable, despite their use of idealizations and their arbitrary character.[17]

1-4. THE HYDROGEN MOLECULE-ION AND THE ONE-ELECTRON BOND

In this section we make the first chemical application of the idea of resonance, in connection with the structure of the simplest of all molecules, the hydrogen molecule-ion, H_2^+, and the simplest of all chemical bonds, the one-electron bond, which involves one electron shared by two atoms.

The Normal Hydrogen Atom.—According to the Bohr theory,[18] the electron in the normal hydrogen atom moved about the nucleus in a circular orbit with radius $a_0 = 0.530$ Å and the constant speed $v_0 = 2.182 \times 10^8$ cm/sec. The quantum-mechanical picture is similar but less definite. The wave function ψ_{1s} that represents the orbital motion of the electron in this atom, shown in Figure 1-2, is large in magnitude only within a region close to the nucleus; beyond 1 or 2 Å it falls off rapidly toward zero. The square of ψ represents the *probability distribution function* for the position of the electron, such that $\psi^2 dV$ is the probability that the electron be in the volume element dV, and $4\pi r^2 \psi^2 dr$ is the probability that it be between the distances r and $r + dr$ from the nucleus. It is seen from the figure that this last function has its maximum value at $r = a_0$. The most probable distance of the electron from the nucleus is thus just the Bohr radius a_0; the electron is, however, not restricted to this one distance. The speed of the electron also is not constant, but can be represented by a distribution function, such that the root-mean-square speed has just the Bohr value v_0. We can accordingly describe the normal hydrogen atom by saying that the electron moves in and out about the nucleus, remaining

---

[17] See the more detailed discussion of these points in Sec. 6-5.

[18] A more detailed account of the Bohr theory of the hydrogen atom is given in Chap. 2 and Apps. II and III.

usually within a distance of about 0.5 Å, with a speed that is variable but is of the order of magnitude of v_0. Over a period of time long enough to permit many cycles of motion of the electron the atom can be described as consisting of the nucleus surrounded by a spherically

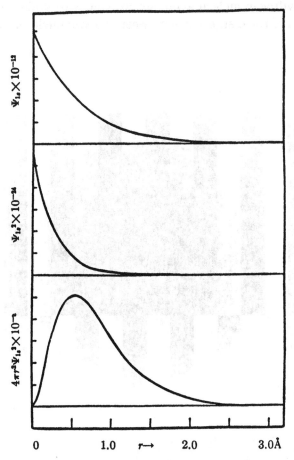

FIG. 1-2.—The wave function ψ_{1s}, its square, and the radial probability distribution function $4\pi r^2\psi_{1s}^2$ for the normal hydrogen atom.

symmetrical ball of negative electricity (the electron blurred by a time-exposure of its rapid motion), as indicated in Figure 1-3.

The Hydrogen Molecule-Ion.—The structure of the hydrogen molecule-ion, H_2^+, as of any molecule, is discussed theoretically by first considering the motion of the electron (or of all the electrons in case that there are several) in the field of the atomic nuclei considered to be fixed in a definite configuration.[19] The electronic energy of the molecule

---

[19] M. Born and J. R. Oppenheimer, *Ann. Physik* **84**, 457 (1927).

is thus obtained as a function of the nuclear configuration.   The configuration for the normal state of the molecule is that corresponding to the minimum value of this energy function, and thus giving the molecule the maximum stability.

For the hydrogen molecule-ion our problem is to evaluate the energy as a function of the distance r_{AB} between the two nuclei A and B.   For

Fɪɢ. 1-3.—A drawing illustrating the decrease in electron density with increasing distance from the nucleus in the normal hydrogen atom.

large values of r_{AB} the system in its normal state consists of a normal hydrogen atom (the electron and nucleus A, say) and a hydrogen ion (nucleus B), which interact with one another only weakly.   If we assume the same structure H + H⁺ to hold as the nuclei approach one another, we find on calculation that the interaction energy has the form shown by the dashed curve in Figure 1-4, with no minimum. From this calculation we would say that a hydrogen atom and a hydrogen ion repel one another, rather than attract one another to form a stable molecule-ion.

However, the structure assumed is too simple to represent the system satisfactorily. We have assumed that the electron forms a normal hydrogen atom with nucleus A:

Structure I $H_A \cdot$ H_B^+

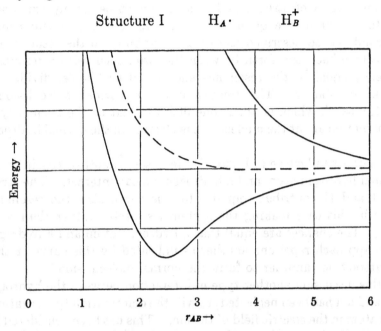

Fig. 1-4.—Curves showing the energy of interaction of a hydrogen atom and a proton. The lower curve corresponds to the formation of a hydrogen molecule-ion in its stable normal state. The scale for the internuclear distance r_{AB} is based on the unit $a_0 = 0.530$ Å.

The structure in which the electron forms a normal hydrogen atom with nucleus B, which then interacts with nucleus A, is just as stable a structure as the first:

Structure II H_A^+ $\cdot H_B$

and we must consider the possibility of resonance between these two structures. These structures are equivalent, and correspond separately to exactly the same energy; the principles of quantum mechanics require that in such a case *the two structures make equal contributions to the normal state of the system.* On repetition of the calculation of the energy curve with use of the corresponding wave function, formed by adding the wave functions for structures I and II, the lower full curve

shown in Figure 1-4 is obtained.[20]   This curve has a pronounced minimum at about $r_{AB} = 1.06$ Å, showing that *as a result of the resonance of
the electron between the two nuclei a stable one-electron bond is formed*,
the energy of the bond being about 50 kcal/mole.   The way in which
this extra stability and the consequent formation of a bond result from
the combination of structures I and II cannot be simply explained;
it is the result of the quantum-mechanical resonance phenomenon.
The bond can be described as owing its stability to the resonance of
the electron back and forth between the two nuclei, with a resonance
frequency equal to the resonance energy, 50 kcal/mole, divided by
Planck's constant h.   This frequency for the normal molecule-ion is
7×10^{14} sec^{-1}, which is about one-fifth as great as the frequency of
orbital motion about the nucleus of the electron in the normal hydrogen
atom.

(The upper full curve in Figure 1-4 represents another way in which
a normal hydrogen atom and a hydrogen ion can interact.   The structures I and II contribute equally to this curve also, the resonance
energy in this case making the system less stable rather than more
stable.   The chances are equal that a hydrogen atom and a hydrogen
ion on approach repel one another as indicated by this curve or that
they attract one another to form the normal molecule-ion.)

In this discussion another type of interaction between the hydrogen
atom and ion has been neglected; to wit, the deformation (polarization)
of the atom in the electric field of the ion.   This has been considered by
Dickinson,[21] who has shown that it contributes an additional 10 kcal
/mole to the energy of the bond.   We may accordingly say that of the
total energy of the one-electron bond in H_2^+ (61 kcal/mole) about 80
percent (50 kcal/mole) is due to the resonance of the electron between
the two nuclei, and the remainder is due to deformation.

Very accurate calculations[22] have led to the value

$$D_0(H_2^+) = 60.95 \pm 0.10 \text{ kcal/mole}$$

for the energy of formation of the normal hydrogen molecule-ion from
a hydrogen atom and a hydrogen ion.   This is in agreement with the
experimental value, which is less accurately known.   The calculated
values of the equilibrium internuclear distance, 1.06 Å, and the vibrational frequency, 2250 cm^{-1}, also agree with the experimental values to

[20] L. Pauling, *Chem. Revs.* **5**, 173 (1928); B. N. Finkelstein and G. E. Horo-
witz, *Z. Physik* **48**, 118 (1928).

[21] B. N. Dickinson, *J. Chem. Phys.* **1**, 317 (1933)

[22] Ø. Burrau, *Kgl. Danske Videnskab. Selskab.* **7**, 1 (1927); E. A. Hylleraas,
Z. Physik **71**, 739 (1931); G. Jaffé, *ibid.* **87**, 535 (1934), and later investigators.

within the accuracy of their calculation and experimental determination.[23]

The electron distribution function for the molecule-ion is shown in Figure 1-5. It is seen that the electron remains for most of the time in the small region just between the nuclei, only rarely getting on the far side of one of them; and we may feel that the presence of the electron between the two nuclei, where it can draw them together, provides some explanation of the stability of the bond. The electron dis-

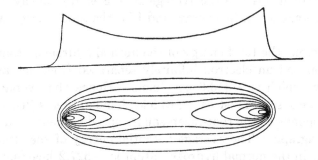

Fig. 1-5.—The electron distribution function for the hydrogen molecule-ion. The upper curve shows the value of the function along the line through the two nuclei, and the lower figure shows contour lines, increasing from 0.1 for the outermost to 1 at the nuclei.

tribution function is concentrated relative to that for the hydrogen atom, the volume within the outermost contour surface shown (with one-tenth the maximum value) being only 70 percent as great as for the atom.

For convenience we may represent the one-electron bond by a dot midway between the symbols of the bonded atoms, the hydrogen molecule-ion then having the structural formula $(H \cdot H)^+$.

The Virial Theorem.—There is another way of discussing the hydrogen molecule-ion that throws added light on the question of the nature of the one-electron bond. This discussion involves use of the virial theorem. The virial theorem, which is valid for quantum mechanics as well as for classical mechanics, states that in any system built up of atomic nuclei and electrons—any atom, molecule, crystal—that is in a steady state (the normal state or one of the excited states) the average kinetic energy is equal to minus one-half of the average potential energy; and, inasmuch as the total energy is the sum of the kinetic

---

[23] For further discussion of wave functions for the hydrogen molecule-ion see *Introduction to Quantum Mechanics.*

energy and the potential energy, the average kinetic energy is equal to the total energy with changed sign, and the average potential energy is equal to twice the total energy:

$$\overline{V} = -2\overline{K}$$
$$W = -\overline{K}$$
$$\overline{V} = 2W$$

In these equations $\overline{K}$ is the average kinetic energy (always positive), $\overline{V}$ is the average potential energy, and W is the total energy, which is a constant.

For example, the total energy of the normal hydrogen atom relative to a proton and an electron infinitely separated from one another is -13.60 ev, which is -313.6 kcal/mole. Hence the average kinetic energy of the system must be $+313.6$ kcal/mole; this is just the value that corresponds to the root-mean-square speed v_0 mentioned in a preceding paragraph. The average potential energy of the electron and the nucleus in the normal hydrogen atom is -627.2 kcal/mole, which corresponds to the Coulomb energy expression $-e^2/r$, with $r = 0.530$ Å, the Bohr radius. For the hydrogen molecule-ion in its normal state the energy (relative to two protons and one electron) is -313.6 $-60.9 = -374.5$ kcal/mole. Accordingly the average kinetic energy of the electron in this molecule-ion is about 374.5 kcal/mole (the two nuclei are nearly stationary; hence most of the kinetic energy of the system is the kinetic energy of the electron); the electron in this molecule-ion is moving faster than in the normal hydrogen atom.

The average potential energy of the hydrogen molecule-ion has the value -749 kcal/mole. This average potential energy consists of three terms: the average potential energy of the two protons, the average potential energy of the electron and the first proton, and the average potential energy of the electron and the second proton, the last two being equal to one another. We may obtain the sum of these two by correcting the total average potential energy by the known potential energy of interaction of two protons, at distance 1.06 Å from one another, the equilibrium internuclear distance for the molecule-ion. The value of the proton-proton potential energy, which is positive because the Coulomb interaction is repulsion, is e^2/r, with $r = 1.06$ Å; this value is 314 kcal/mole. Accordingly the average potential energy of the electron with respect to the two protons is $-749 - 314$ $= -1063$, and the average potential energy of interaction of the electron with each of the protons is one-half of this quantity, being equal to -532 kcal/mole, as compared with -627.2 for the hydrogen atom.

We may say that the stability of the hydrogen molecule-ion relative to a hydrogen atom plus a hydrogen ion is the result of the great concentration of the electron distribution function into the region between the two protons.   This concentration is such as to permit the stabilizing Coulomb interaction ($-e^2/r$) of the electron with each of the two protons to be nearly as great as it is with the single proton in the normal hydrogen atom.   The stability of the one-electron bond in the hydrogen molecule-ion may be attributed to the concentration of the electron into this region.   A study of the wave function for the hydrogen molecule-ion shows that the concentration of the distribution function into the region between the two nuclei can be explained in large part as resulting from the addition of the two wave functions, corresponding respectively to the structures I, H· H⁺, and II, H⁺ ·H.   Accordingly we may say that the resonance phenomenon causes the electron to be concentrated in the region where it can interact most strongly with both of the nuclei and in this way give rise to the bond energy.

The Hellmann-Feynman Theorem.—An interesting quantum-mechanical theorem was discovered independently by Hellmann[24] and Feynman.[25]   This theorem states that the force acting on each nucleus in a molecule is exactly that calculated by the principles of classical electrostatic theory from the charges and positions of the other nuclei and of the electrons; the electrons are taken to have the spatial distribution given by the square of the electronic wave function.   At the equilibrium configuration of a molecule the resultant force acting on each nucleus vanishes; hence for this configuration the repulsion of a nucleus by the other nuclei is just balanced by its attraction by the electrons.

For example, in the hydrogen molecule-ion in its equilibrium configuration we may say that the electron is distributed in a way equivalent to having 3/7 of the electron spherically distributed about each nucleus and 1/7 at the midpoint between the two nuclei, giving a force of attraction by the electron for each nucleus that balances the force of repulsion by the other nucleus.

The Conditions for the Formation of a One-Electron Bond.—The magnitude of the resonance energy of the one-electron bond in the hydrogen molecule-ion is determined by the amount of interaction of the two structures I and II (H· H⁺ and H⁺ ·H), as calculated by the methods of quantum mechanics.   Because the two structures correspond to the same energy, the interaction energy is completely manifested as resonance energy; there is complete resonance.   If, however,

[24] H. Hellmann, *Einführung in die Quantenchemie*, Franz Deuticke, Leipzig, 1937, p. 285.

[25] R. P. Feynman, *Phys. Rev.* **56**, 340 (1939).

the two nuclei A and B were unlike, so that the two structures

$$\text{I} \quad A\cdot \qquad\qquad B^+$$

and $\qquad\qquad \text{II} \quad A^+ \qquad\qquad \cdot B$

corresponded to different energy values, the conditions for complete resonance would not be satisfied. The more stable of the two structures (structure I, say) would contribute more to the normal state of the system than the other, and the system would be stabilized (relative to structure I) by an amount of resonance energy less than the inter-

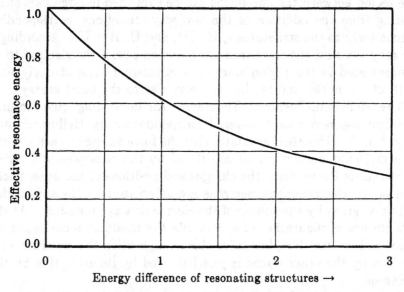

Fig. 1-6.—Curve showing the amount of energy stabilizing the normal state of a system with two resonating structures, relative to the more stable of the two resonating structures, as a function of the energy difference of the two structures. The unit of energy used in the graph is the interaction energy of the two structures (the resonance integral).

action energy. A curve showing the effect of difference in energy of two resonating structures in inhibiting resonance is shown in Figure 1-6. The way in which this curve is calculated is described in Appendix V. As structure I becomes more and more stable relative to structure II it makes up a larger and larger part of the normal state of the system, and resonance with structure II stabilizes the system by a smaller and smaller amount. For this reason we expect the one-electron bond to be formed only between like atoms or occasionally between unlike atoms which happen to be of such a nature (similarity in electronegativity) as to make structures I and II approximately equal in energy.

1-5. THE HYDROGEN MOLECULE AND THE ELECTRON-PAIR BOND

Before 1927 there was no satisfactory theory of the covalent bond. The chemist had postulated the existence of the valence bond between atoms and had built up a body of empirical information about it, but his inquiries into its structure had been futile. The step taken by Lewis of associating two electrons with a bond can hardly be called the development of a theory, since it left unanswered the fundamental questions as to the nature of the interactions involved and the source of the energy of the bond. Only in 1927 was the development of the theory of the covalent bond initiated by the work of Condon[26] and of Heitler and London[27] on the hydrogen molecule, described in the following paragraphs.

Condon's Treatment of the Hydrogen Molecule.—Condon gave a discussion of the hydrogen molecule based upon Burrau's treatment of the hydrogen molecule-ion. He introduced two electrons into the normal-state orbital described by Burrau for the one electron of H_2^+. The total energy of the H_2 molecule for this structure consists of four parts: the energy of repulsion of the two nuclei, the energy of the first electron moving in the field of the two nuclei (as evaluated by Burrau), the equal energy of the second electron, and the energy of mutual electrostatic repulsion of the two electrons. Condon did not attempt to evaluate the last of these terms by integration, but instead assumed it to be the same fraction of the interaction energy of the two electrons and the nuclei as for the normal helium atom, which corresponds to the limiting case of the hydrogen molecule when the two protons are fused into a single nucleus.

In this way he obtained an energy curve for H_2 with its minimum at $r_{AB} = 0.73$ Å and with bond energy 100 kcal/mole, in excellent agreement with experiment. This agreement could not be assigned great significance, however, because of uncertainty as to the accuracy of the estimate of the electron repulsion energy.

Condon's treatment is the prototype of the *molecular-orbital method* of discussing the electronic structure of molecules. In this method a wave function is formulated that involves the introduction of a pair of electrons in an electron orbital that extends about two or more atomic nuclei.

[26] E. U. Condon, *Proc. Nat. Acad. Sci. U. S.* **13**, 466 (1927).
[27] W. Heitler and F. London, *Z. Physik* **44**, 455 (1927). A mathematical improvement of this work was made by Y. Sugiura, *ibid.* **45**, 484 (1927).

The second method of discussing the electronic structure of molecules, usually called the *valence-bond method*, involves the use of a wave function of such a nature that the two electrons of the electron-pair bond between two atoms tend to remain on the two different atoms. The prototype of this method is the Heitler-London treatment of the hydrogen molecule, which we shall now discuss.

The Heitler-London Treatment of the Hydrogen Molecule.—The hydrogen molecule consists of two nuclei, which may be designated A

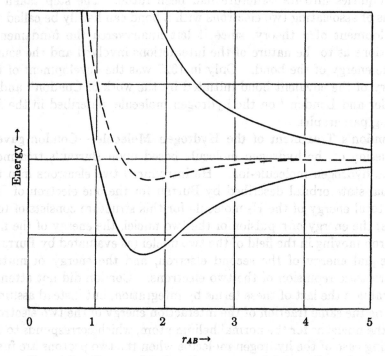

Fig. 1-7.—Curves showing the energy of interaction of two normal hydrogen atoms. The scale for the internuclear distance r_{AB} is based on the unit $a_0 = 0.530$ Å.

and B, and two electrons, 1 and 2. As in the treatment of the hydrogen molecule-ion, we calculate the interaction energy for various values of the internuclear distance r_{AB}. When the nuclei are far apart the normal state of the system involves two normal hydrogen atoms. Let us assume that one of the electrons, electron 1, say, is associated with nucleus A, and the other, electron 2, with nucleus B. On calculating the interaction energy as a function of the internuclear distance, we find that at large distances there is a weak attraction, which soon turns into strong repulsion as r_{AB} is further diminished (dashed curve of Figure 1-7). According to this calculation the two atoms would not combine to form a stable molecule.

Here again, however, we have neglected the resonance phenomenon; for the structure with electron 2 attached to nucleus A and electron 1 to nucleus B is just as stable as the equivalent structure assumed above, and in accordance with quantum-mechanical principles we must consider as a representation of the normal state of the system neither one structure nor the other, but rather a combination to which the two contribute equally; that is, we must make the calculation in such a way as to take into consideration the possibility of the exchange of places of the two electrons:

$$\text{Structure I} \qquad H_A \cdot 1 \qquad 2 \cdot H_B$$
$$\text{Structure II} \qquad H_A \cdot 2 \qquad 1 \cdot H_B$$

In this way there is obtained an interaction-energy curve (the lower full curve in Figure 1-7) that shows a pronounced minimum, corresponding to the formation of a stable molecule. The energy of formation of the molecule from separated atoms as calculated by Heitler, London, and Sugiura is about 67 percent of the experimental value of 102.6 kcal/mole, and the calculated equilibrium distance between the nuclei is 0.05 Å larger than the observed value 0.74 Å.

Moreover, the Heitler-London wave function does not correspond to the virial theorem (it does not make the average value of the potential energy equal to minus twice the average value of the kinetic energy), and it is accordingly a poor approximation to the correct wave function for the molecule.

A simple improvement in the wave function was then made by Wang.[28] In place of the normal 1s hydrogen-atom wave functions about nucleus A and nucleus B, with the radial dependence corresponding to unit nuclear charge, he used similar 1s functions with an effective nuclear charge Z' that was allowed to vary in such a way as to minimize the energy. This treatment gives agreement with the virial theorem. The calculated equilibrium internuclear distance with this wave function is 0.75 Å, in agreement with experiment, and the calculated bond energy is 80 percent of the correct value. The effective nuclear charge Z' is 1.17, which corresponds to a significant shrinkage of the electron distribution function into the region close to the two nuclei.

Hence we see that a very simple treatment of the system of two hydrogen atoms leads to an explanation of the formation of a stable molecule, *the energy of the electron-pair bond being in the main the resonance energy corresponding to the interchange of the two electrons between the two atomic orbitals.*

Partial Ionic Character and Deformation.—We have so far considered only structures for the hydrogen molecule in which the two

[28] S. C. Wang, *Phys. Rev.* **31**, 579 (1928).

electrons remain near different nuclei. The two ionic structures III and IV

$$\text{Structure III} \qquad \text{H}_A^- : \qquad \text{H}_B^+$$
$$\text{Structure IV} \qquad \text{H}_A^+ \qquad : \text{H}_B^-$$

in which both electrons are attached to the same nucleus must also be considered. These structures involve a positive hydrogen ion H^+ and a negative hydrogen ion $\text{H}:^-$ with the helium structure (K shell completed).

One of the most important rules about resonance is that *resonance can occur only among structures with the same number of unpaired electrons.* Since the electrons in the negative hydrogen ion, occupying the same orbital, are paired and the electrons involved in a bond formed by structures I and II are paired, this condition is satisfied, and we expect structures III and IV as well as I and II to be of significance for the normal hydrogen molecule.

At large internuclear distances the ionic structures III and IV are not of importance. The energy of the reaction

$$\text{H} + \text{H} \rightarrow \text{H}^+ + \text{H}^-$$

is $- 295.6$ kcal/mole, the difference between the electron affinity of hydrogen

$$\text{H} + e^- \rightarrow \text{H}^- + 16.4 \text{ kcal/mole}$$

and the ionization energy of hydrogen

$$\text{H}^+ + e^- \rightarrow \text{H} + 312.0 \text{ kcal/mole}$$

and this makes the structures III and IV so unstable relative to I and II that they make no contribution. As r_{AB} is decreased, however, the Coulomb attraction of H^+ and H^- stabilizes structures III and IV; at the equilibrium distance $r_{AB} = 0.74$ Å each of these two ionic structures makes a contribution of about 2 percent to the normal state of the molecule. The corresponding extra ionic resonance energy is about 5.5 kcal/mole, or 5 percent of the total bond energy.[29]

The remaining 15 percent of the observed bond energy may be attributed to deformation, this term being used to cover all the complicated interactions neglected in the foregoing simple treatments. As the culmination of several attacks on the problem, a thoroughly satisfactory and accurate theoretical treatment of the normal hydrogen molecule was made by James and Coolidge.[30] Their careful and labo-

[29] S. Weinbaum, *J. Chem. Phys.* 1, 593 (1933).

[30] H. M. James and A. S. Coolidge, *J. Chem. Phys.* 1, 825 (1933). Highly accurate calculations for the ground states of two-electron atoms from H^- to Ne^{8+} have been made by C. L. Pekeris, *Phys. Rev.* 112, 1649 (1958).

rious investigation led to a value for the bond energy of the molecule

$$D_0(H_2) = 103.2 \text{ kcal/mole}$$

in complete agreement with experiment, with similar agreement for the equilibrium internuclear distance and the vibrational frequency. Theoretical calculations of other properties of the normal hydrogen molecule—the diamagnetic susceptibility, the electric polarizability, the anisotropy of the latter quantity, van der Waals forces, and so on—have also been made, with satisfactory results, so that the structure of this simple covalent molecule is now well understood.

As a summary of the results reported above, the bond in the hydrogen molecule may be described as resulting in the main from the resonance of the two electrons between the two nuclei, this phenomenon contributing 80 percent of the total bond energy. An additional 5 percent is contributed by the ionic structures H^-H^+ and H^+H^-, which are of equal importance. The remaining 15 percent of the energy of the bond can be ascribed to complex interactions included under the term deformation.[31]

The Conditions for the Formation of an Electron-Pair Bond.—In Section 1-4 it was pointed out that the resonance that leads to the formation of a stable one-electron bond between two atoms is in general largely inhibited in case that the two atoms are unlike, and that in consequence such a bond occurs only rarely. We see that for the electron-pair bond no such restriction exists; the two structures I and II that differ only in interchange of the two electrons 1 and 2 by two atoms A and B are equivalent even though the two atoms involved are unlike, and accordingly complete resonance occurs for unlike atoms as well as for like atoms, the resonance energy of the bond being equal to the interaction energy of the two structures. Thus there is no special condition as to the nature of the atoms that must be satisfied in order for an electron-pair bond to be formed, and we need not be surprised by the widespread occurrence and great importance of this bond.

Resonance with the ionic structures A^+B^- and A^-B^+ also occurs for unlike atoms as well as for like atoms, and is, indeed, of great importance in case that the atoms A and B differ greatly in electronegativity, the contribution of the favored ionic structure then being large. This aspect of the covalent bond is treated in Chapter 3.

A detailed discussion of the electronic structure of atoms is given in the following chapter, preliminary to the statement of the formal rules for covalent-bond formation at the end of the chapter.

[31] For further discussion of wave functions for the hydrogen molecule see *Introduction to Quantum Mechanics* and the paper by H. Shull, *J.A.C.S.*, in press.

CHAPTER 2

The Electronic Structure of Atoms and
the Formal Rules for the Formation
of Covalent Bonds

AN understanding of the electronic structure of atoms is necessary for the study of the electronic structure of molecules and the nature of the chemical bond. Our knowledge of the electronic structure of atoms has been obtained almost entirely from the analysis of the spectra of gases. In this chapter we shall discuss the nature of spectra and the information about the electronic structure of atoms that has been derived from this information, in preparation for the later chapters of the book. The chapter ends with the statement of the formal rules for the formation of covalent bonds.

2-1. THE INTERPRETATION OF LINE SPECTRA

When the radiation emitted from a source of light is resolved into a spectrum by use of a prism or a grating it is found that the distribution of intensity with wavelength depends on the nature of the source. The intensity of light from a glowing solid body varies gradually from place to place in the spectrum and is a function principally of the temperature of the body. A hot gas or a gas excited to the emission of light by an electric discharge or in some other way may emit an *emission spectrum* that consists of sharp lines, each line having a well-defined wavelength. Such a spectrum is called a *line spectrum*. Sometimes many lines occur close together and separated by approximately equal intervals; they are then said to compose a band, and the spectrum is called a *band spectrum*. Lines and bands are also observed to be absorbed when continuous radiation is passed through a gas. Such a spectrum of dark lines or bands on a light background is called an *absorption spectrum*.

Band spectra are produced in emission or absorption by molecules containing two or more atoms, and line spectra by single atoms or monatomic ions.   The structure of bands is related to the vibration of the nuclei of the atoms within the molecule and to the rotation of the molecule.

The intensities and wavelengths of spectral lines are characteristic of the emitting atoms or molecules.   A representative spectrum is shown in Figure 2-1; this is the emission spectrum of atomic hydrogen,

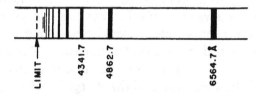

Fig. 2-1.—The Balmer series of spectral lines of atomic hydrogen.   The line at the right, with the longest wavelength, is Hα.   It corresponds to the transition from the state with $n = 3$ to the state $n = 2$.   The other lines correspond to the transitions from the states with $n = 4$, 5, 6, · · · to the state with $n = 2$.

obtained by passing an electric spark through a tube containing hydrogen.   A diagram representing the positions of the spectral lines is given below the spectral photograph.   The position of a line in the spectrum is indicated by giving either its wavelength λ, usually measured in Ångström units, its frequency $\nu = c/\lambda$ (with c the velocity of light), measured in sec⁻¹, or its wave number or reciprocal wavelength $\nu = 1/\lambda$, measured in cm⁻¹.   (Note that the symbol ν is often used both for frequency and wave number, its significance being evident from the context; sometimes ν is used for frequency and ω for wave number.)   The visible region of the spectrum extends from about $\lambda = 7700$ Å (red) to about $\lambda = 3800$ Å (violet).   It is customary to write for a line at, say, $\lambda = 2536$ Å the symbol λ2536.

A characteristic feature of simple line spectra is that the lines can be grouped in *series*.   The separation between succeeding lines in a series decreases gradually toward the violet[1] (Fig. 2-1), the sequence of wavelengths being such that a series limit can be found by extrapolation.

[1] The expression "toward the violet" is often used to mean toward higher frequency (shorter wavelength) of the light and "toward the red" to mean toward lower frequency.

The concept of the atom as a system of one nucleus and one or more electrons was developed to explain the experiments of Lenard and of Rutherford on the passage of anode rays (rapidly moving positive ions) and of alpha particles (helium nuclei emitted by radioactive materials) through matter.   The electrons and the nuclei are very small in comparison with atoms—their diameters are between 10^{-13} and 10^{-12} cm, that is, between 10^{-5} and 10^{-4} Å, whereas atoms have diameters of the order of 2 to 5 Å.   The magnitude of the charge of a nucleus is always an integral multiple of that of the electron, with positive sign; it is expressed as Ze, Z being the atomic number of the element.   An electrically neutral atom has Z electrons about the nucleus.

According to the laws of classical mechanics the system of electrons and nucleus comprising an atom would reach final equilibrium only when the electrons had fallen into the nucleus.   The electrons would be expected, according to classical mechanics, to describe orbits about the nucleus, and the acceleration of the charged particles, the electrons, in the orbits would give rise to the emission of energy as radiation. The frequencies involved in the motion of the electrons would then gradually change during the emission of light.   This sort of structure for the atom is incompatible with the observed sharply defined frequencies of spectral lines.   Moreover, the spectral lines do not show the overtones, with frequencies double, triple, and so forth that of the fundamental frequency, that would be expected classically.   The existence of nonradiating normal states of atoms in which the electrons have certainly not fallen into the nucleus is a further point of disagreement with classical theory, indicating the necessity for the development of a new atomic mechanics, differing from the classical mechanics of macroscopic systems.   This new atomic mechanics is called quantum mechanics.

Two postulates that are fundamental to the interpretation of spectra are the *existence of stationary states* and the *Bohr frequency rule*.   They were enunciated by Bohr in 1913 in the famous paper[2] that led in a few years to the complete elucidation of spectral phenomena.   Planck[3] had previously announced (in 1900) that the amount of energy dW in unit volume (1 cm^3) and contained between the frequencies ν and $\nu + d\nu$ in empty space in equilibrium with matter at temperature T, as measured experimentally, could be represented by the equation

$$dW = \frac{8\pi h\nu^3}{c^\cdot(e^{h\nu/kT} - 1)}\, d\nu \qquad (2\text{-}1)$$

in which ν is the frequency of the light, k is Boltzmann' constant, T

[2] N. Bohr, *Phil. Mag.* 26, 1, 476, 857 (1913).
[3] M. Planck, *Ann. Physik* 4, 553 (1901).

is the absolute temperature, and h is a constant of nature that has been given the name Planck's constant. This equation is not the one that would be obtained from classical statistical mechanics; Planck showed that it could be derived if the assumption were made that radiant energy (light) is not emitted continuously by atoms or molecules, but only in discrete portions, each portion carrying the quantity of energy $h\nu$. Einstein[4] suggested that one of these energy quantities is not emitted uniformly in all directions by a radiating atom, but instead in one direction, like a particle. These portions of radiant energy are called *photons* or *light quanta*.

The next phenomenon explained in terms of quanta was the photoelectric effect, which was interpreted by Einstein in 1908. When light falls on a metal plate electrons are emitted from the surface of the plate, but not with velocities related to the intensity of the light, as would be expected from classical theory. Instead, the maximum velocity of the ejected electrons (the photoelectrons) depends on the frequency of the light: it corresponds to the conversion of just the energy $h\nu$ of one light quantum into the energy of removal of the electron from the metal plate plus the kinetic energy of the liberated electron. Einstein also announced at the same time his law of photochemical equivalence, according to which the absorption of one light quantum of energy $h\nu$ may activate one molecule to chemical reaction. In all of these cases the system (atom, molecule, or crystal) emitting or absorbing radiation in quanta changes discontinuously from a state with a given amount of energy to one with energy $h\nu$ less or greater.

2-2. STATIONARY STATES; THE BOHR FREQUENCY PRINCIPLE

These facts and some observations about the frequencies of spectral lines were the inspiration for Bohr's two postulates, which may be expressed in the following way:

I. The existence of stationary states. *An atomic system can exist in certain stationary states, each one corresponding to a definite value of the energy W of the system, and transition from one state to another is accompanied by the emission or absorption as radiation, or the transfer to or from another system of atoms or molecules, of an amount of energy equal to the energy difference of the two states.*

II. The Bohr frequency rule. *The frequency of the radiation absorbed by a system and associated with the transition from an initial state with energy W_1 to a final state with energy W_2 is*

$$\nu = \frac{W_2 - W_1}{h} \qquad (2\text{-}2)$$

(negative values of ν correspond to emission).

[4] A. Einstein, *Ann. Physik* **22**, 180 (1907).

The two postulates are compatible with the observation that the frequencies of the spectral lines emitted by an atom can be represented as the differences between pairs of a set of frequency values, called the *term values* or *spectral terms* of the atom.   These term values are now seen to be the values of the energy of the stationary states divided by h (to give frequency, in sec^{-1}), or by hc (to give wave number, in cm^{-1}, as is customarily given in tables of term values).

It is pointed out in the following section that Balmer discovered in 1885 that the frequencies of some lines of the hydrogen spectrum could be represented as the differences of term values.   Rydberg, a Swedish spectroscopist, gave a similar representation[5] of lines of sodium in 1889, and the concept of spectral term values was generalized in 1908 by W. Ritz.   In 1901 the American investigator C. P. Snyder published an analysis of a complex spectrum, that of rhodium, which accounted for 476 spectral lines by a set of term values.[6]   During the following 25 years, and especially after Bohr's formulation of his postulates, rapid progress was made in the analysis of spectra and the associated development of the modern theory of atomic structure.

2-3. STATIONARY STATES OF THE HYDROGEN ATOM

An *energy-level diagram* for the hydrogen atom is given in Figure 2-2. The reference state, with zero energy, is that of a proton and an electron infinitely separated from one another, that is, the ionized hydrogen atom.   The stationary states of the hydrogen atom have negative energy values with reference to the ionized state.   The values of the energy for the various stationary states are given by the Bohr equation

$$W_n = -\frac{R_H hc}{n^2} \qquad (2\text{-}3)$$

In this equation R_H is called the Rydberg constant for hydrogen; the value for the Rydberg constant for hydrogen is 109,677.76 cm^{-1}.   The letter h is Planck's constant, and c the velocity of light.   The letter n is the *total quantum number;* it may have integral values 1, 2, 3, 4, $\cdots$.

The frequencies of the spectral lines emitted by a hydrogen atom when it undergoes transition from one stationary state to a lower stationary state can be calculated by the Bohr frequency rule, with use of this expression for the energy values of the stationary state.   It is seen, for example, that the frequencies for the lines corresponding to the transitions indicated by arrows in Figure 2-2, corresponding to transitions from states with $n = 3, 4, 5, \cdots$ to the state with $n = 2$,

[5] J. R. Rydberg, *K. Svenska Akad. Handl.* **1889,** 23.

[6] C. P. Snyder, *Astrophys. J.* **14,** 179 (1901).   This paper is Snyder's only publication

are given by the equation

$$\nu = R_{\text{H}}c\left(\frac{1}{2^2} - \frac{1}{n^2}\right) \qquad (2\text{-}4)$$

This equation was discovered by Balmer in 1885.[7] These spectral lines constitute the Balmer series. Other series of lines for hydrogen correspond to transitions from upper states to the state with $n = 1$ (the Lyman series), to the state with $n = 3$ (the Paschen series), and so on.

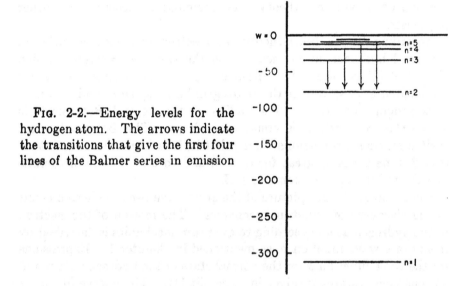

FIG. 2-2.—Energy levels for the hydrogen atom. The arrows indicate the transitions that give the first four lines of the Balmer series in emission

The value of the Rydberg constant R_{H}, as determined from the measured wavelength of the lines of the Lyman series and the other series for atomic hydrogen, is such that the energy of the normal state of the hydrogen atom, with $n = 1$, is -313.6 kcal/mole (-13.60 ev). The amount of energy required to ionize the normal hydrogen atom is accordingly 313.6 kcal/mole; this is called the *ionization energy* of hydrogen. Spectroscopic studies have provided values for the ionization energy of atoms of most of the elements.

In his 1913 papers Bohr developed a theory of the stationary states of the hydrogen atom. According to his theory the electron was to be considered as moving in a circular orbit about the proton. The amount of angular momentum for a stationary state was assumed by Bohr to be equal to $nh/2\pi$, with $n = 1, 2, 3, \cdots$. In Appendix II there is given the derivation of the energy values for the Bohr circular orbits. For

[7] J. J. Balmer, *Wied. Ann.* **25**, 80 (1885).

an electron moving about a nucleus with charge Ze (a hydrogen atom for $Z = 1$, helium ion He$^+$ for $Z = 2$, etc.), the Bohr theory leads to the following expression for the energy of the stationary states:

$$W = -\frac{2\pi^2 m_0 Z^2 e^4}{n^2 h^2} \tag{2-5}$$

Bohr showed that the known values of the mass of the electron, m_0, the electronic charge, e, and Planck's constant, h, led to a value of $2\pi^2 m_0 e^4 / ch^3$ equal to the experimental value for the Rydberg constant for hydrogen, and his theory was immediately accepted by other physicists.

According to the Bohr theory the electron in a circular orbit in a normal hydrogen atom moves with the speed $v_0 = 2\pi e^2/h$, which is 2.18×10^8 cm/sec. The speed changes in proportion to $1/n$ for the excited states, and in the hydrogenlike ions, He$^+$, and so forth, it is proportional to Z. The radius of the normal Bohr orbit is $a_0 = h^2/4\pi^2 m_0 e^2$, which is equal to 0.530 Å. The Bohr radius for excited states is proportional to n^2; that is, it is four times as great for $n = 2$, nine times as great for $n = 3$, and so on. For hydrogenlike ions the radius is proportional to $1/Z$.

Some changes in this picture of the atom have been made as a result of the discovery of quantum mechanics. The motion of the electron in the hydrogen atom according to quantum mechanics is described by means of a wave function ψ, as mentioned in Chapter 1. Expressions for the wave function ψ for the normal state of the hydrogen atom and various excited states are given in Appendix III. These wave functions are designated by three quantum numbers: the total quantum number n, with values $1, 2, \cdots$; the azimuthal quantum number l, with values $0, 1, 2, \cdots, n - 1$; and the magnetic quantum number m_l, with values $-l, -l + 1, \cdots, 0, \cdots, +l$. For the hydrogen atom and hydrogenlike ions the energy depends only on the total quantum number n (except for very small changes in energy determined by the other quantum numbers). The normal state of the hydrogen atom, with $n = 1$, is represented by a single set of quantum numbers: $n = 1$, $l = 0$, and $m_l = 0$.

The azimuthal quantum number l is a measure of the angular momentum of the electron in its orbit. The orbital angular momentum is equal to $\sqrt{l(l + 1)}\,h/2\pi$. The electron in the hydrogen atom in its normal state (with $l = 0$) does not have any angular momentum, and the picture that we must form of the normal hydrogen atom is accordingly somewhat different from that assumed by Bohr. On the left in Figure 2-3 there is shown the Bohr circular orbit for hydrogen, with the electron moving in a circular orbit with radius a_0. This picture is unsatisfactory because it gives orbital angular momentum to the atom, and it has been found by experiment that the normal hydrogen atom does not have any orbital angular momentum. On the right of Figure 2-3 there is shown the extreme case of an elliptical orbit with zero minor axis, that is, the orbit corresponding to zero angular momentum. This picture represents a type of classical motion of a particle about an attracting center. It corresponds to describing the electron as moving out from the nucleus to the distance $2a_0$ and then

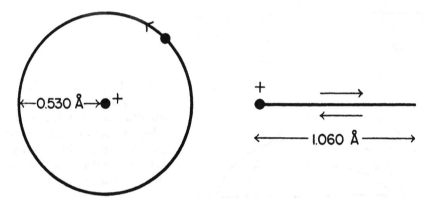

Fig. 2-3.—At the left is represented the circular orbit of the Bohr atom. At the right is shown the very eccentric orbit (line orbit), with no angular momentum, that corresponds somewhat more closely to the description of the hydrogen atom in its normal state given by quantum mechanics.

returning to the nucleus. Comparison with Figure 1-3, representing the electron distribution in the normal hydrogen atom, shows that the electron is to be considered as moving in and out from the nucleus in all directions in space, so as to give spherical symmetry to the atom; moreover, the distance of the electron from the nucleus is not limited rigorously to values less than $2a_0$. The Heisenberg uncertainty principle of quantum mechanics, according to which the position and the momentum of a particle cannot be exactly measured simultaneously, shows that we cannot hope to describe the motion of the electron in the normal hydrogen atom in terms of a definite orbit, such as that shown in Figure 2-3; nevertheless, there is some value in discussing the type of classical motion that corresponds reasonably closely to the quantum-mechanical description of the normal hydrogen atom.

In Figure 2-4 there are shown drawings of the Bohr orbits for hydrogen in the excited states with $n = 2$, $n = 3$, and $n = 4$, with the angular momentum taken equal to $\sqrt{l(l + 1)}h/2\pi$, as required by quantum mechanics. An electron with $l = 0$ is said to be an s electron, one with $l = 1$ a p electron, and so on through the sequence $d, f, g, h, \cdots$. An s electron does not have any angular momentum, whereas $p, d, f, \cdots$ electrons have angular momentum, with increasing magnitude in this sequence.

The electrons with a given value of the total quantum number n constitute an *electron shell*. The shells have been given the designa-

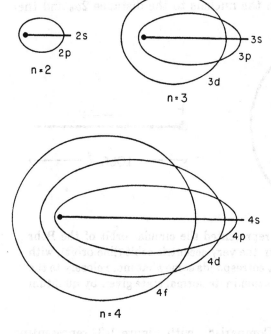

Fig. 2-4.—Bohr orbits for the hydrogen atom, total quantum number 2, 3, and 4. These orbits are represented as having the values of angular momentum given by quantum mechanics.

tions $K, L, M, N, O, \cdots$, corresponding to $n = 1, 2, 3, 4, 5, \cdots$. Another classification of electron shells that is especially useful in chemistry (the helium shell, neon shell, argon shell, etc.) is described in Section 2-7.

There is only one s orbital in each shell (see App. III); it corresponds to the values $l = 0$, $m_l = 0$ for these quantum numbers. There are three p orbitals (with $l = 1$) in each shell (beginning with the L shell) corresponding to the values $-1, 0$, and $+1$ for the magnetic quantum number m_l. Similarly, there are five d orbitals per shell from the M shell on ($m_l = -2, -1, 0, +1, +2$), and seven f orbitals from the N shell on ($m_l = -3, -2, -1, 0, +1, +2, +3$). The orbitals with given values of both n and l are called *subshells*.

The different values of the magnetic quantum number m_l correspond to different orientations in space of the angular momentum vector for the electron. It is customary to represent the angular momentum of a system by a vector; for example, the angular momentum vector for a circular Bohr orbit would extend in a direction perpendicular to the plane of the orbit and would have magnitude proportional to the magnitude of the angular momentum.   The magnetic quantum number m_l represents the component of angular momentum along a designated direction in space, in particular the direction of a magnetic field.   The diagrams in Figure 2-5 show the angles between the angular momentum vector and the field direction for the p orbitals, the d orbitals, and the f orbitals.   The value $m_l = 0$, in each case, corresponds to zero com-

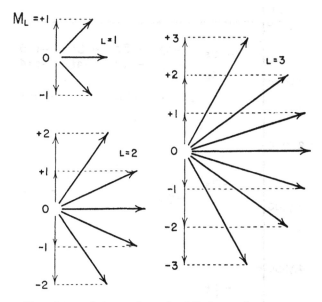

Fɪɢ. 2-5.—Orientation of robital angular momentum vectors for values 1, 2, and 3 of the angular momentum quantum number L.

ponent of angular momentum in the field direction, the value $m_l = +1$ to the component $h/2\pi$, $m_l = +2$ to $2h/2\pi$, and so on.

The electron distribution function ψ^2 as given by quantum mechanics for the normal hydrogen atom has been discussed briefly in Chapter 1. The corresponding electron distribution functions for other orbitals will be discussed in the following chapter.

2-4. THE ELECTRONIC STRUCTURE OF ALKALI ATOMS

The normal lithium atom has two electrons in the K shell, with $n = 1$, and one electron in the $2s$ orbital of the L shell.   The electronic configurations of all of the alkali atoms are given in Table 2-1; in each case there is a single electron in the outermost shell.

TABLE 2-1.—ELECTRON CONFIGURATIONS OF ALKALI ATOMS

Atom	Z	Configuration
Li	3	$1s^2 2s$
Na	11	$1s^2 2s^2 2p^6 3s$
K	19	$1s^2 2s^2 2p^6 3s^2 3p^6 4s$
Rb	37	$1s^2 2s^2 2p^6 3s^2 3p^6 3d^{10} 4s^2 4p^6 5s$
Cs	55	$1s^2 2s^2 2p^6 3s^2 3p^6 3d^{10} 4s^2 4p^6 4d^{10} 5s^2 5p^6 6s$
Fr	87	$1s^2 2s^2 2p^6 3s^2 3p^6 3d^{10} 4s^2 4p^6 4d^{10} 4f^{14} 5s^2 5p^6 5d^{10} 6s^2 6p^6 7s$

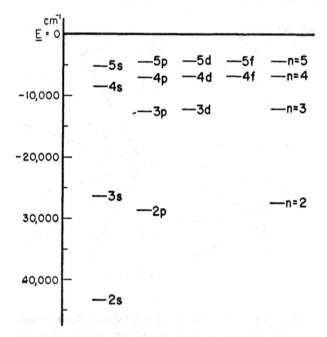

FIG. 2-6.—Energy levels for the lithium atom. The symbols 2s and so forth give the quantum number for one electron; the other two electrons are in the 1s orbital. The levels at the right are those for hydrogen.

Some of the energy levels for the lithium atom, as found by the analysis of the line spectrum of lithium, are shown in Figure 2-6. It is seen that there is a significant difference from the energy-level diagram for hydrogen: for hydrogen the levels 2s and 2p have the same energy value, as have also 3s, 3p, and 3d, and so on, whereas for lithium the levels are split—the energy depends on the azimuthal quantum number l as well as on the total quantum number.

The energy values 4f, 5f, and 6f lie very close to those for hydrogen, which are represented at the right side of the diagram. Those for 3d, 4d, $\cdots$ lie somewhat below the hydrogen values, those for the

p states still lower, and those for the s states much lower.   An explanation of this behavior was suggested by Schrödinger in 1921, before the development of quantum mechanics.[8]   This explanation is indicated by the drawings in Figures 2-7 and 2-8.   Schrödinger suggested that the inner electron shell of lithium might be replaced by an equivalent charge of electricity distributed uniformly over the surface of a sphere of suitable radius, which for lithium would be about 0.28 Å.   The valence electron, outside this shell, would be moving in an electric field due to the nucleus, with charge $+3e$, plus the two K electrons, with

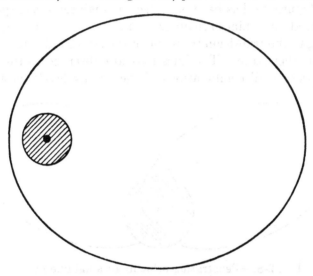

FIG. 2-7.—A nonpenetrating orbit in an alkalilike atom.   The inner electrons are represented by the shaded region about the nucleus.

charge $-2e$, that is, in a field due to a charge $+e$, the same as the charge of the proton.   So long as the electron stayed outside the K shell, it could be expected to behave in a way corresponding to a hydrogenlike electron.   An orbit of this sort, shown in Figure 2-7, is called a *nonpenetrating orbit*.   Reference to Figure 2-4 indicates that an f electron or a d electron in an excited lithium atom would be essentially nonpenetrating, but that surely an s electron, in an orbit that extends to the nucleus, would penetrate the K shell, and probably a p electron would also penetrate the K shell to some extent.   An electron in a *penetrating orbit* (Fig. 2-8) would move into the field of attraction of the nucleus with charge $+3e$ only partially shielded by the K electrons and would accordingly be stabilized by a large amount.

During recent years many detailed quantum-mechanical calculations have been made of the energy levels of the lithium atom and other

[8] E. Schrödinger, *Z. Physik* **4,** 347 (1921).

atoms, giving results in good agreement with experiment. There is no doubt that the Schrödinger wave equation provides a satisfactory theory of the electronic structure of atoms and molecules. The amount of calculation necessary to obtain a reliable energy value for an atom or molecule containing several electrons is, however, so great that most of the information that we have about electronic structure of atoms and molecules has been obtained from experiment, rather than by theoretical calculation.

The Selection Rule for *l*.—The energy-level diagram for lithium as shown in Figure 2-6 has been obtained by analysis of the spectrum of the lithium atom. Lines are observed in the spectrum of lithium corresponding to the transition from one of the states indicated in the diagram to another state. The lines that are observed in the spectrum do not represent all combinations of the energy levels, however, but

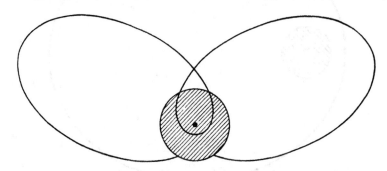

FIG. 2-8.—Penetrating orbit in an alkalilike atom.

only those combinations in which the quantum number l changes by +1 or −1. This rule is called the *selection rule for l*. For example, an atom in a *p* orbital can undergo transition to a lower-lying state with the electron in an *s* orbital or a *d* orbital, with the emission of the corresponding spectral line, but not to an *f* orbital.

If light is passed through lithium vapor containing lithium atoms in the normal state, with the valence electron in the 2*s* orbital, the only

TABLE 2-2.—IONIZATION ENERGIES OF ALKALI ATOMS

Atom	First ionization energy (enthalpy)	
	0°K	298.16°K (15°C)
Li	124.21 kcal/mole	125.79 kcal/mole
Na	118.48	120.04
K	100.08	101.56
Rb	96.29	97.79
Cs	89.75	91.25

transitions that occur with absorption of radiant energy are those to the levels 2p, 3p, 4p, and so on. These transitions are indicated in Figure 2-6; they constitute the lines of the absorption spectrum of lithium.

The frequencies of the lines in such a spectral series can be extrapolated to give the value corresponding to ionization. In this way values of ionization energies for atoms and ions have been determined from spectroscopic information. The ionization energies of the alkali metals are given in Table 2-2.

2-5. THE SPINNING ELECTRON AND THE FINE STRUCTURE OF SPECTRAL LINES

The atomic model that has been discussed in the preceding paragraphs gives a good representation of simple spectra, but not a com-

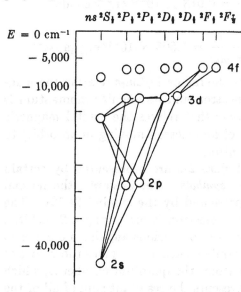

FIG. 2-9.—Energy levels for the lithium atom, showing the separation of the doublet levels and the transitions accompanying absorption and emission of radiant energy.

plete one. For example, the transition from the state 2p to the state 2s for lithium, indicated in Figure 2-6 to be a single line (wavelength 6707.8 Å), is in fact a doublet, consisting of two components with wavelengths differing by 0.15 Å. Similarly, the transition from 3p to 3s for sodium is also a doublet, consisting of two components with wavelengths 5889.95 Å and 5895.92 Å; these are the well-known yellow doublet lines of sodium, seen in sodium vapor lamps.

The splitting of these lines and of other lines that show fine structure can be accounted for by means of the energy-level diagram for lithium shown in Figure 2-9. In this diagram each of the levels 2p, 3p, 3d, and so forth is shown as two levels, only slightly separated from one another, whereas the levels 2s, 3s, 4s, and so forth are not split.

The explanation of this complexity of the energy levels is that the individual electrons have rotatory motion, a spin.[9]   Each electron has the angular momentum $\sqrt{s(s+1)}h/2\pi$ with s, the spin quantum number, always $\frac{1}{2}$. The electron has a magnetic moment associated with this rotation. The experiments that have been carried out on the properties of the electron show that the magnetic moment of the electron is

$$2 \cdot \frac{e}{2m_0c} \cdot \frac{\sqrt{3}}{2} \cdot \frac{h}{2\pi}$$

The electron accordingly has the following properties:

   charge $-e$ = $-$ 4.803 $\times$ 10^{-10} statcoulombs

   mass m_0 = 0.911 $\times$ 10^{-22} g

   angular momentum $\dfrac{\sqrt{3}}{2} \cdot \dfrac{h}{2\pi}$ = 0.913 $\times$ 10^{-27} erg seconds

   magnetic moment $\dfrac{\sqrt{3}}{2} \cdot \dfrac{h}{2\pi} \cdot \dfrac{e}{m_0c}$ = 1.608 $\times$ 10^{-20} erg gauss^{-1}

It is especially interesting that the factor $2e/2m_0c$ relating the magnetic moment of the spinning electron and its angular momentum is twice as great as the factor $e/2m_0c$ that relates the orbital magnetic moment (the magnetic moment of an electron moving in an orbit) to the corresponding angular momentum.

The energy levels shown in Figure 2-9 are represented by certain symbols, called *Russell-Saunders symbols*. For example, the normal state of the lithium atom is represented by the symbol $2s\ ^2S_{\frac{1}{2}}$. The symbol $2s$ means that the valence electron is occupying a $2s$ orbital. The remaining symbol, $^2S_{\frac{1}{2}}$, describes the various angular momenta in the atom.   A Russell-Saunders symbol, such as $^2S_{\frac{1}{2}}$, gives the values of three quantum numbers for the atom: the quantum number S, which is the quantum number that represents the resultant spin of all of the electrons in the atom; the quantum number L, which is the quantum number that represents the resultant orbital angular momentum of all of the electrons in the atom; and the quantum number J, which is the quantum number that represents the total angular momentum of the atom, due to both spin and orbital motion of the electrons, and is the resultant of S and L.[10] S, the spin quantum number for the atom, has the value $\frac{1}{2}$ when there is only one valence electron in the atom. The superscript on the left side of the term symbol is equal to $2S + 1$ (which is equal to 2, for $S = \frac{1}{2}$); it represents the *multiplicity* of the

   [9] G. E. Uhlenbeck and S. Goudsmit, *Naturwissenschaften* 13, 953 (1925); *Nature* 117, 264 (1926).

   [10] Note that the capital letter S is used in two different ways; in general no confusion results from this usage.

energy level and corresponds to the number of ways in which the quantum number S can be oriented in space. The capital letter in the symbol gives the value of the quantum number for the resultant angular momentum; the letters S, P, D, F, G, $\cdots$ correspond respectively to $L = 0, 1, 2, 3, 4, \cdots$. For an atom with a single valence electron, as for the states represented in Figure 2-9, the capital letter is identical with the small letter representing the orbital of the valence electron. The subscript gives the value of the resultant quantum number J, corre-

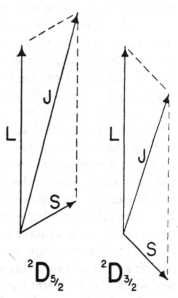

Fig. 2-10.—The interaction of spin angular momentum and orbital angular momentum to form the total angular momentum for the states $^2D_{\frac{5}{2}}$ and $^2D_{\frac{3}{2}}$.

sponding to the resultant of the spin angular momentum and the orbital angular momentum.

For a single valence electron S is equal to $\frac{1}{2}$, and there are only two possible values for J, namely, $L + \frac{1}{2}$ and $L - \frac{1}{2}$. Vector diagrams showing the composition of the spin angular momentum and the orbital angular momentum for the two states $^2D_{\frac{3}{2}}$ and $^2D_{\frac{5}{2}}$ are shown in Figure 2-10.

It has been found by observation that J can change during a quantum jump with emission or absorption of light only by $+1$, 0, or -1. This is a statement of the selection rule for J. The transitions allowed by the selection rules for l and J are shown in Figure 2-9. It is seen that only lines due to transitions involving an S state have two components; all others have three. The name doublets refers not to the number of components of the spectral line but to the multiplicity of the energy levels. The superscript 2 on the left of the term symbols shown in Figure 2-10 is usually read as "doublet." The normal state is thus described as a doublet state, even though it is not split into two levels.

It is seen from the term diagram that the energy of interaction of the spin of the electron and its orbital motion is not very great. This energy of interaction increases rapidly with increase in the atomic number of the element and becomes large for the heavy atoms.

2-6. THE ELECTRONIC STRUCTURE OF ATOMS WITH TWO OR MORE VALENCE ELECTRONS

The energy of an atom containing two or more electrons depends upon many kinds of interaction of the electrons and the nucleus. First of all, there are the interactions of each electron with the nucleus, which in a simple theory give rise to energy terms similar to those described for a single electron in Section 2-4; in general all of the electrons may be described as occupying penetrating orbitals. Other interactions can be correlated with the spin of the electrons and with their orbital angular momenta. The spectroscopists have developed a *vector model* of the atom that provides a simple way of describing the stationary states of the atom. In the following paragraphs we shall discuss the Russell-Saunders vector model,[11] in which, as mentioned in the preceding section, the vectors representing the spin of the individual electrons combine to form a resultant spin vector, represented by the quantum number S, the vectors representing the orbital angular momenta combine to form an orbital angular momentum vector, represented by the quantum number L, and these two resultant vectors combine to form a total angular momentum vector for the atom, represented by the quantum number J. This description of the atom has been found to be a good one for light atoms, with small atomic number; the electronic structure of the heavier atoms is usually more complicated, and, although the Russell-Saunders symbols are commonly used in the description of the stationary states of the heavier atoms, the rules corresponding to the symbols do not in general apply well to the heavier elements.

Let us consider an atom with two *s* electrons, with different total quantum numbers; for example, a beryllium atom with one valence electron in a 2*s* orbital and the other in a 3*s* orbital, in addition to the two electrons in the K shell. The orbital angular momenta of the two valence electrons are zero ($l_1 = 0$, $l_2 = 0$), and accordingly the resultant angular momentum is zero ($L = 0$). Each of the two electrons has spin quantum number $\frac{1}{2}$ ($s_1 = \frac{1}{2}$, $s_2 = \frac{1}{2}$), and each spin angular mo-

mentum vector accordingly has the magnitude $\sqrt{\dfrac{1}{2} \cdot \dfrac{3}{2}} \dfrac{h}{2\pi}$. These

[11] The Russell-Saunders coupling was discovered by H. N. Russell and F. A. Saunders, *Astrophys. J.* **61**, 38 (1925).

two vectors can combine to form a resultant vector corresponding to the values 0 and 1 for the total spin quantum number S, as shown in Figure 2-11. The state $S = 1$ is usually described by saying that the two vectors are parallel (the figure shows that they are not exactly parallel, but are as close to parallel as is allowed by nature) and the state $S = 0$ by saying that they are antiparallel. Inasmuch as L is equal to 0, the total angular momentum quantum number J for the atoms is equal to 0 when S equals 0 and is equal to 1 when S equals 1.

Experience shows that these two states differ greatly in energy. The interaction energy of the magnetic moment of the two electron spins is very small, and the observed energy difference is not due to a direct spin-spin magnetic interaction. It was shown by Heisenberg[12] that

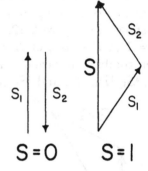

Fig. 2-11.—The interaction of the spin angular momentum vectors of two electrons to form a resultant total spin angular momentum vector, corresponding to the value of the total spin quantum number $S = 0$ or $S = 1$.

the difference between the state with $S = 0$ (which is called a singlet state) and the state with $S = 1$ (which is called a triplet state) is due to the resonance phenomenon, which has been discussed briefly in Chapter 1.

The way in which the resonance energy contributes to the energy of an atom is correlated with the relative orientation of the electron spins. The resonance energy is in fact a result of the electrostatic repulsion of the electrons and not of a direct spin-spin interaction, but it is correlated with the relative orientation of the spins in such a way that it can be discussed as though it were a spin-spin interaction.

Now let us consider the states of the atom of beryllium in which one of the valence electrons occupies a $2p$ orbital and the other occupies a $3p$ orbital. The two electron spins can combine, as shown in Figure

[12] W. Heisenberg, *Z. Physik* **38**, 411 (1926); **39**, 499 (1926); **41**, 239 (1927). The resonance phenomenon was discovered independently by P. A. M. Dirac, *Proc. Roy. Soc. London* A112, 661 (1926). For a detailed discussion of the phenomenon, see books on quantum mechanics, such as *Introduction to Quantum Mechanics*, or G. W. Wheland, *The Theory of Resonance and Its Application to Organic Chemistry*, John Wiley and Sons, New York, 1955.

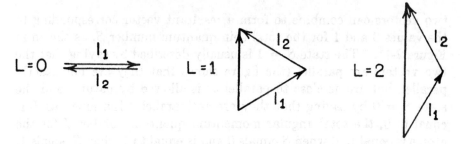

Fig. 2-12.—The interaction of the orbital angular momentum vectors for two *p* electrons ($l_1 = 1$, $l_2 = 1$) to form the resultant total angular momentum vector, with values corresponding to 0, 1, and 2 for the total angular momentum quantum number *L*.

2-11, to form the resultant $S = 0$ or $S = 1$. The two orbital moments, corresponding to $l_1 = 1$ and $l_2 = 1$, can combine in three ways, as shown in Figure 2-12, to give the resultant $L = 0$ (an *S* state), $L = 1$ (a *P* state), or $L = 2$ (a *D* state). The vectors *S* and *L* can then combine in various ways to give the states ${}^3D_1$, ${}^3D_2$, ${}^3D_3$, ${}^3P_0$, ${}^3P_1$, ${}^3P_2$, ${}^3S_1$, ${}^1D_2$, ${}^1P_1$, and ${}^1S_0$ (right to left in Fig. 2-14).

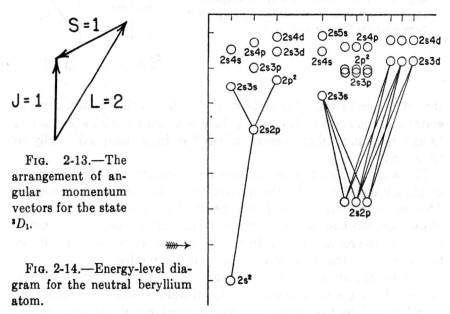

Fig. 2-13.—The arrangement of angular momentum vectors for the state ${}^3D_1$.

Fig. 2-14.—Energy-level diagram for the neutral beryllium atom.

The energy values of all these states can be seen in Figure 2-14, which is the energy-level diagram for beryllium as determined by analysis of the spectrum. In addition, other energy levels are shown, corresponding to other pairs of orbitals for the two valence electrons. A discussion of these energy levels is given in the following section, which deals with the Pauli exclusion principle.

2-7. THE PAULI EXCLUSION PRINCIPLE AND THE PERIODIC SYSTEM OF THE ELEMENTS

A principle of extreme importance to spectroscopy as well as to other phases of physics and chemistry is the *exclusion principle* discovered by Pauli in 1925.[18]

Let us consider an atom in an external magnetic field so strong that the couplings among the various electrons are broken and the electrons orient themselves independently with respect to the field. The state of each electron is then given by fixing the values of a set of quantum numbers: for each electron we may give the values of the total quantum number n of the orbit, the azimuthal quantum number l, the orbital magnetic quantum number m_l (stating the component of orbital angular momentum in the field direction), the spin quantum number s (which has the value $\frac{1}{2}$ for every electron), and the spin magnetic quantum number m_s (which can be equal to either $+\frac{1}{2}$, corresponding to the spin oriented roughly in the direction of the field, or $-\frac{1}{2}$, corresponding to the spin oriented roughly in the opposite direction). The Pauli exclusion principle can be expressed in the following way: *there cannot exist an atom in such a quantum state that two electrons within it have the same set of quantum numbers.*

The Pauli exclusion principle provides an immediate explanation of the principal features of the periodic system of the elements, and also of the energy-level diagrams, such as that for beryllium shown in Figure 2-14.

Let us first discuss the helium atom. The most stable orbital in the helium atom is the $1s$ orbital, with $n = 1$, $l = 0$, $m_l = 0$. There are two electrons in the neutral helium atom, which we place in the $1s$ orbital. In the discussion above of the beryllium atom, it was pointed out that two s electrons give rise to the Russell-Saunders states $^3S_1$ and $^1S_0$. The discussion concerned, however, a $2s$ electron and a $3s$ electron; the two electrons differ in the value of the total quantum number n. For the helium atom with two electrons in the $1s$ orbital, the Pauli exclusion principle requires that the two electrons differ in the value of one quantum number. Their values of n, l, and m_l are the same; moreover, they have the same spin quantum number, $s = \frac{1}{2}$. Accordingly they must differ in the value of m_s, which can have the value $+\frac{1}{2}$ for one electron and $-\frac{1}{2}$ for the other. The resultant spin for the two electrons must accordingly be 0, and only the singlet state $^1S_0$ can exist for two electrons in the $1s$ orbital. The normal state of the helium atom is accordingly $1s^2\,{}^1S_0$, and there is no other state based upon the electron configuration $1s^2$.

An atom of lithium, with three electrons, can have only two electrons

[18] W. Pauli, *Z. Physik* **31**, 765 (1925).

in the 1s orbital, and these two electrons must have their spins opposed. Two electrons constitute a *completed K shell* in any atom. A third electron must occupy an outer orbital. The next most stable orbital is the 2s orbital, which penetrates deeply into the inner electron shell and is hence more stable than 2p, so that lithium in the normal state will have the configuration $1s^2 2s\ ^2S_{\frac{1}{2}}$.

In general two electrons, with opposed spins, may occupy each atomic orbital. There is one s orbital in each electron shell, with a given value of the total quantum number n; three p orbitals, corresponding to $m_l = -1$, 0, and +1, in each shell beginning with the L shell; five d orbitals in each shell beginning with the M shell, and so on. The numbers of electrons in completed subshells and shells of an atom are shown in Table 2-3. Note that there are alternative ways of naming the shells.

TABLE 2-3.—NAMES OF ELECTRON SHELLS

Spectroscopists' names	Chemists' names
$K\ 1s^2$	Helium $1s^2$
$L\ 2s^2 2p^6$	Neon $2s^2 2p^6$
$M\ 3s^2 3p^6 3d^{10}$	Argon $3s^2 3p^6$
$N\ 4s^2 4p^6 4d^{10} 4f^{14}$	Krypton $3d^{10} 4s^2 4p^6$
	Xenon $4d^{10} 5s^2 5p^6$
	Radon $4f^{14} 5d^{10} 6s^2 6p^6$
	Eka-radon $5f^{14} 6d^{10} 7s^2 7p^6$

The nature of the normal states of all atoms can be discussed in terms of the principles that have been mentioned in the preceding paragraphs and sections. The normal state of an atom is the state with lowest energy. The terms that make the principal contribution to the energy of the atom are the energy values of the individual electrons, which depend upon the orbitals assigned to them. The 1s orbital, of the K shell, is the most stable orbital in all atoms. Next come the 2s orbital and then the three 2p orbitals, of the L shell. The following shells overlap one another, in a way determined by the atomic number of the atom and its degree of ionization. The 3s orbital is the next most stable, followed by the three 3p orbitals; but for the lighter elements, such as potassium, the 4s orbital, of the N shell, is more stable than the five 3d orbitals of the M shell. The sequence of stabilities of the orbitals is represented to good approximation in Figure 2-15 This representation is only approximate; for example, for copper, with atomic number 29, the configuration of the normal state is $1s^2 2s^2 2p^6 3s^2 3p^6 3d^{10} 4s$; there are ten 3d electrons and one 4s electron, rather than nine $3d$ electrons and two $4s$ electrons as indicated in Figure 2-15.

The electronic configuration and the Russell-Saunders term symbols for the elements as determined spectroscopically or as predicted are given in Table 2-4. It must be emphasized that these electronic con-

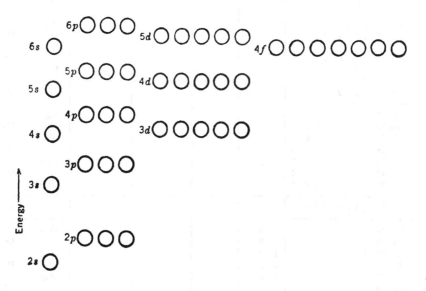

FIG. 2-15.—The approximate sequence of energy values for atomic orbitals, the lowest circle representing the most stable orbital ($1s$). Each circle represents one atomic orbital, which can be occupied either by one electron or by two electrons with opposed spins.

figurations do not have great chemical significance, because for many of the atoms there are excited states differing only by a small amount of energy from the normal state of the atom, and the electronic structure of the atom in a molecule may be more closely represented by one of the excited states than by the normal state, or, as is in fact usually the case, the electronic structure in a molecule or crystal is in general to be described as corresponding to a resonance hybrid of many of the low-lying states of the isolated atom. For copper, for example, the state $1s^2 2s^2 2p^6 3s^2 3p^6 3d^9 4s^2$ $^2D_{\frac{3}{2}}$ lies only 11,202 cm^{-1} (31.9 kcal/mole) above the normal state.

TABLE 2-4.—ELECTRON CONFIGURATION OF ATOMS IN THEIR NORMAL STATES

	He	Neon		Argon		Krypton			Xenon			Radon				Eka-radon				Term symbol
	1s	2s	2p	3s	3p	3d	4s	4p	4d	5s	5p	4f	5d	6s	6p	5f	6d	7s	7p	
H 1	1																			$^2S_{1/2}$
He 2	2																			$^1S_0$
Li 3	2	1																		$^2S_{1/2}$
Be 4	2	2																		$^1S_0$
B 5	2	2	1																	$^2P_{1/2}$
C 6	2	2	2																	$^3P_0$
N 7	2	2	3																	$^4S_{3/2}$
O 8	2	2	4																	$^3P_2$
F 9	2	2	5																	$^2P_{1/2}$
Ne 10	2	2	6																	$^1S_0$
Na 11	10 Neon core			1																$^2S_{1/2}$
Mg 12				2																$^1S_0$
Al 13				2	1															$^2P_{1/2}$
Si 14				2	2															$^3P_0$
P 15				2	3															$^4S_{3/2}$
S 16				2	4															$^3P_2$
Cl 17				2	5															$^2P_{3/2}$
Ar 18	2	2	6	2	6															$^1S_0$
K 19	18 Argon core						1													$^2S_{1/2}$
Ca 20							2													$^1S_0$
Sc 21						1	2													$^2D_{3/2}$
Ti 22						2	2													$^3F_2$
V 23						3	2													$^4F_{3/2}$
Cr 24						5	1													$^7S_3$
Mn 25						5	2													$^6S_{5/2}$
Fe 26						6	2													$^5D_4$
Co 27						7	2													$^4F_{9/2}$
Ni 28						8	2													$^3F_4$
Cu 29						10	1													$^2S_{1/2}$
Zn 30						10	2													$^1S_0$
Ga 31						10	2	1												$^2P_{1/2}$
Ge 32						10	2	2												$^3P_0$
As 33						10	2	3												$^4S_{3/2}$
Se 34						10	2	4												$^3P_2$
Br 35						10	2	5												$^2P_{3/2}$
Kr 36	2	2	6	2	6	10	2	6												$^1S_0$
Rb 37	36 Krypton core									1										$^2S_{1/2}$
Sr 38										2										$^1S_0$
Y 39									1	2										$^2D_{3/2}$
Zr 40									2	2										$^3F_2$
Nb 41									4	1										$^6D_{1/2}$
Mo 42									5	1										$^7S_3$
Tc 43									5	2										$^6S_{5/2}$
Ru 44									7	1										$^5F_5$
Rh 45									8	1										$^4F_{9/2}$
Pd 46									10											$^1S_0$
Ag 47									10	1										$^2S_{1/2}$
Cd 48									10	2										$^1S_0$
In 49									10	2	1									$^2P_{1/2}$
Sn 50									10	2	2									$^3P_0$
Sb 51									10	2	3									$^4S_{3/2}$
Te 52									10	2	4									$^3P_2$
I 53									10	2	5									$^2P_{3/2}$

TABLE 2-4.—(continued)

	He	Neon		Argon		Krypton			Xenon			Radon				Eka-radon				Term symbol
	1s	2s	2p	3s	3p	3d	4s	4p	4d	5s	5p	4f	5d	6s	6p	5f	6d	7s	7p	
Xe 54	2	2	6	2	6	10	2	6	10	2	6									$^1S_0$
Cs 55														1						$^2S_{1/2}$
Ba 56														2						$^1S_0$
La 57													1	2						$^2D_{3/2}$
Ce 58												1	1	2						$^3H_4$
Pr 59												2	1	2						$^4K_{11/2}$
Nd 60												3	1	2						$^5L_6$
Pm 61												4	1	2						$^6L_{9/2}$
Sm 62												5	1	2						$^7K_4$
Eu 63												6	1	2						$^8H_{3/2}$
Gd 64												7	1	2						$^9D_2$
Tb 65												8	1	2						$^8H_{17/2}$
Dy 66												9	1	2						$^7K_{10}$
Ho 67												10	1	2						$^6K_{19/2}$
Er 68							54					11	1	2						$^5L_{10}$
Tm 69					Xenon core							12	1	2						$^4K_{17/2}$
Yb 70												13	1	2						$^3H_6$
Lu 71												14	1	2						$^2D_{3/2}$
Hf 72												14	2	2						$^3F_2$
Ta 73												14	3	2						$^4F_{3/2}$
W 74												14	4	2						$^5D_0$
Re 75												14	5	2						$^6S_{5/2}$
Os 76												14	6	2						$^5D_4$
Ir 77												14	7	2						$^4F_{9/2}$
Pt 78												14	9	1						$^3D_3$
Au 79												14	10	1						$^2S_{1/2}$
Hg 80												14	10	2						$^1S_0$
Tl 81												14	10	2	1					$^2P_{1/2}$
Pb 82												14	10	2	2					$^3P_0$
Bi 83												14	10	2	3					$^4S_{3/2}$
Po 84												14	10	2	4					$^3P_2$
At 85												14	10	2	5					$^2P_{3/2}$
Rn 86	2	2	6	2	6	10	2	6	10	2	6	14	10	2	6					$^1S_0$
Fr 87																		1		$^2S_{1/2}$
Ra 88																		2		$^1S_0$
Ac 89							86										1	2		$^2D_{3/2}$
Th 90					Radon core												2	2		$^3F_2$
Pa 91																	3	2		$^4F_{3/2}$
U 92																	4	2		$^5D_4$
Eka-Rn 118	2	2	6	2	6	10	2	6	10	2	6	14	10	2	6	14	10	2	6	$^1S_0$

It has been pointed out above that two electrons in the $1s$ orbital must have their spins opposed, and hence give rise to the singlet state $^1S_0$, with no spin or orbital angular momentum, and hence with no magnetic moment. Similarly it is found that a completed subshell of electrons, such as six electrons occupying the three $2p$ orbitals, must have $S = 0$ and $L = 0$, corresponding to the Russell-Saunders term symbol $^1S_0$; such a completed subshell has spherical symmetry and zero magnetic moment. The application of the Pauli exclusion principle to electron configurations in which there are several electrons in the same subshell is discussed in Appendix IV.

The stability of the various Russell-Saunders states resulting from the same electron configuration (the same distribution of electrons among orbitals) can be described by means of a set of rules usually called Hund's rules.[14]   These rules are the following:

1. *Of the Russell-Saunders states arising from a given electron configuration those with the largest value of S lie lowest, those with the next largest*

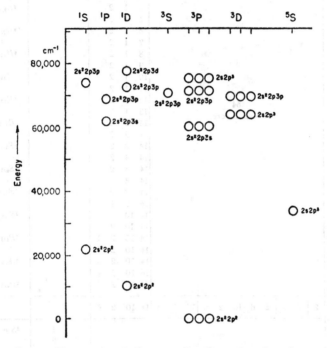

FIG. 2-16.—Energy-level diagram for the neutral carbon atom.

next, and so on; in other words, the states with the largest multiplicity are the most stable.

2. *Of the group of terms with a given value of S, that with the largest value of L lies lowest.*

3. *Of the states with given values of S and L in a configuration consisting of less than half the electrons in a completed subgroup, the state with the smallest value of J is usually the most stable, and for a configuration consisting of more than half the electrons in a subgroup the state with largest J is the most stable.*   The multiplets of the first sort, smallest J most stable, are called normal multiplets, and those of the second sort are called inverted multiplets.

[14] F. Hund, *Z. Physik* **33,** 345 (1925).

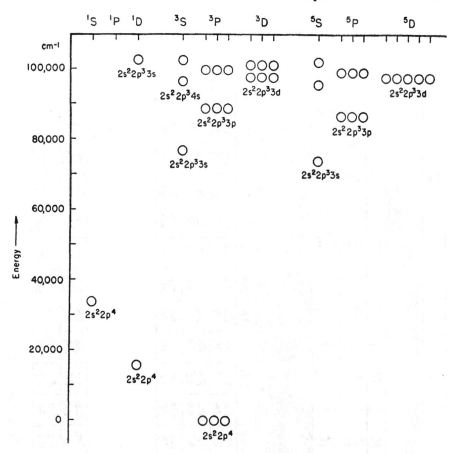

FIG. 2-17.—Energy-level diagram for the neutral oxygen atom.

The application of these rules may be illustrated by the example of the carbon atom and the oxygen atom, for which the most stable spectroscopic states are indicated in the diagrams of Figures 2-16 and 2-17. The stable electron configuration for carbon is $1s^2 2s^2 2p^2$, which gives rise to the Russell-Saunders states ^1S, ^1D, and ^3P. For oxygen the stable configuration is $1s^2 2s^2 2p^4$, which gives rise to the same set of Russell-Saunders states. (Note that the same set of Russell-Saunders states results from x electrons missing from a completed subshell as from x electrons present in the subshell.) As seen from Figure 2-16, the states ^3P are the most stable for each atom, with ^1D next, and ^1S least stable, in accordance with the first two Hund rules. Carbon, with two electrons in the $2p$ subshell (which can hold six electrons) is predicted by the third rule to give a normal multiplet, with the smallest value of J lowest, whereas oxygen, with four $2p$ electrons, should give rise to inverted multiplets. It is seen from the term diagram that the rule corresponds to the spectroscopic observations.

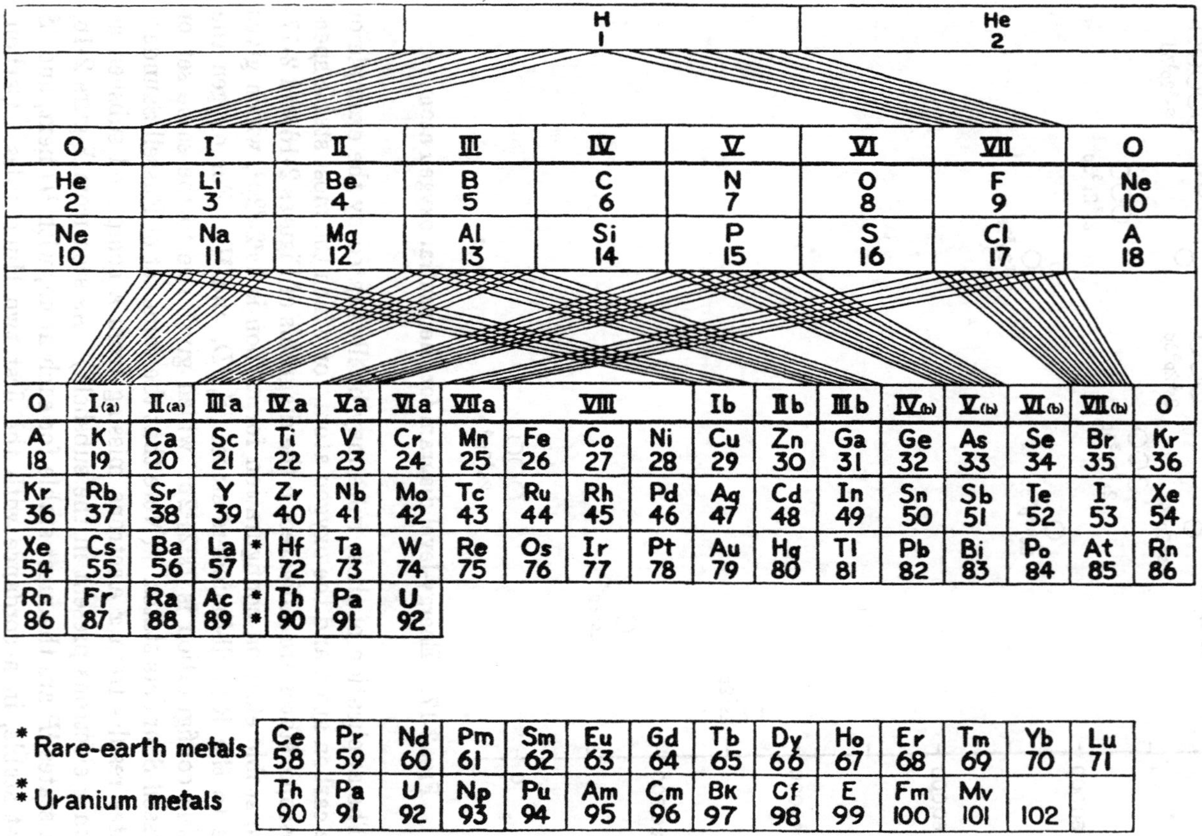

Fig. 2-18.—The periodic system of the elements.

* Rare-earth metals
** Uranium metals

The correlation between the discussion of the electronic structure of atoms and the periodic system of the elements can be seen by comparing the energy-level diagram of Figure 2-15 with the periodic table, shown as Figure 2-18.   It is seen that each of the noble gases has eight electrons in its outer shell, two s electrons and six p electrons.   This electron configuration corresponds to a special stability.   The first long period and the second long period result from the introduction of ten electrons into the five $3d$ orbitals and the five $4d$ orbitals, respectively, as well as the addition of the eight electrons that constitute the outer shell of the corresponding noble-gas atoms.   The first very long period involves the introduction of 14 electrons into the seven $4f$ orbitals, in addition to the 18 in the $5d$, $6s$, and $6p$ orbitals.   The heaviest elements that have been discovered or made are in the second very long period, which involves occupancy of the $5f$ orbitals as well as the $6d$, $7s$, and $7p$ orbitals.

The extent to which the periodic system of the elements can be represented as having its basis in the Schrödinger wave equation can be indicated by the consideration of the values of electron energies obtained by approximate solution of the wave equation by the Thomas-Fermi-Dirac method.[15]   The best method of approximate solution of the wave equation for a many-electron atom at the present time is the Hartree-Fock method of the self-consistent field.[16]   This method is, however, so complex as to have permitted application to only a few atoms as yet, not enough to provide the basis for a discussion of all elements.   The Thomas-Fermi-Dirac method of the statistic atomic potential[17] can be applied in a systematic way to give the curves[18] shown in Figure 2-19.   Each curve represents the energy of an electron in a spherical-field orbital ($1s$, $2s$, $2p$, etc.) as a function of the atomic number of the neutral atom from 1 to 100.

The major features of the periodic system and of the sequence of electron distributions of Table 2-4 are accounted for by the curves. For elements with atomic numbers less than 27 a $3d$ electron is shown in the figure as being less stable than a $4p$ or $4s$ electron; the $3d$ curve crosses the $4p$ curve at $Z = 28$.   In fact, the crossover of $3d$ and $4p$ should occur about $Z = 21$.   Similarly the crossover of the curves for

[15] R. Latter, *Phys. Rev.* 99, 510 (1955).

[16] D. R. Hartree, *Proc. Cambridge Phil. Soc.* 24, 89 (1928); V. Fock, *Z. Physik* 61, 126 (1930).

[17] L. H. Thomas, *Proc. Cambridge Phil. Soc.* 23, 542 (1927); E. Fermi, *Atti accad. nazl. Lincei* 6, 602 (1927); 7, 342, 726 (1928); P. A. M. Dirac, *Proc. Cambridge Phil. Soc.* 26, 376 (1930).

[18] Latter, *loc. cit.* (15).

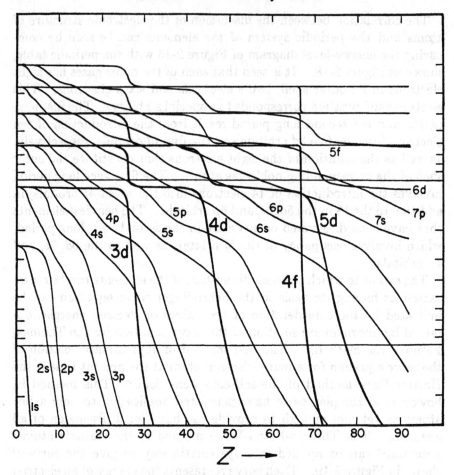

FIG. 2-19.—Curves representing values of electron energies, as a function of atomic number. These curves were obtained by approximate solution of the wave equation by the Thomas-Fermi-Dirac method.

$4d$ and $5p$ at $Z = 45$ should occur at 39 or 40, the beginning of the second sequence of transition metals, and that of the curves for $4f$ and $6p$, at 67, should occur at 57, the beginning of the sequence of rare-earth metals. The shapes of the curves of Figure 2-19 are apparently essentially correct, but the atomic numbers at which the rapid increase in stability of the d and f electrons occurs as given by the approximate solution of the wave equation are too large (by about 6 units for the d orbitals and 10 units for $4f$; probably also for $5f$, which should show the increase in stability at about $Z = 93$).

Values of the first and second ionization energies of the elements, as determined spectroscopically, are given in Table 2-5.

TABLE 2-5.—FIRST AND SECOND IONIZATION ENERGIES
OF THE ELEMENTS[a]

Z		I_1	I_2	Z		I_1	I_2
1	H	313.4[a]		39	Y	147	282
2	He	566.7	1254.2	40	Zr	158	303
3	Li	124.3	1743.2	41	Nb	159	330
4	Be	214.9	419.7	42	Mo	164	372
5	B	191.2	579.8	43	Tc	168	352
6	C	259.5	561.9	44	Ru	169.8	386.4
7	N	335	682.2	45	Rh	172	417
8	O	313.8	809.3	46	Pd	192	448
9	F	401.5	806	47	Ag	174.6	495
10	Ne	497.0	947	48	Cd	207.3	389.7
11	Na	118.4	1090	49	In	133.4	434.8
12	Mg	176.2	346.5	50	Sn	169.3	337.2
13	Al	137.9	433.9	51	Sb	199.2	380.4
14	Si	187.9	377	52	Te	208	429
15	P	241.7	455	53	I	241.0	440.0
16	S	238.8	539	54	Xe	279.6	488.7
17	Cl	300	548.7	55	Cs	89.7	578.6
18	Ar	363.2	636.7	56	Ba	120.1	230.6
19	K	100.0	733.3	57	La	129	263
20	Ca	140.9	273.6				
21	Sc	151	295	72	Hf	160	343
22	Ti	157	313	73	Ta	182	373
23	V	155	338	74	W	184	408
24	Cr	155.9	380.1	75	Re	181	383
25	Mn	171.3	360.5	76	Os	200	390
26	Fe	181	373	77	Ir	210	
27	Co	181	393	78	Pt	210	427.9
28	Ni	176.0	418.4	79	Au	213	473
29	Cu	178.1	467.7	80	Hg	240	432.3
30	Zn	216.5	414.0	81	Tl	140.8	470.7
31	Ga	138	473	82	Pb	170.9	346.4
32	Ge	182	367	83	Bi	168.0	384.5
33	As	226	429	84	Po	194	—
34	Se	225	495.6	85	At	—	—
35	Br	272.9	497.9	86	Rn	247.7	—
36	Kr	322.6	566.2	87	Fr	—	—
37	Rb	96.3	634.0	88	Ra	121.7	233 8
38	Sr	131.2	254.2	89	Ac	160	280

• Values in kcal/mole. The values are obtained from the ionization potentials in Charlotte E. Moore, *Atomic Energy Levels as Derived from the Analyses of Optical Spectra* (Circular of the National Bureau of Standards 467, Government Printing Office, Washington, D. C., 1949–1958, vol. III), by multiplying by the conversion factor from electron volts to kcal/mole, 23.053.

2-8. THE ZEEMAN EFFECT AND THE MAGNETIC PROPERTIES
OF ATOMS AND MONATOMIC IONS

In 1896 the Dutch physicist P. Zeeman found that spectral lines in general are split into a number of components by the application of an external magnetic field to the emitting atoms. In some cases this splitting is of a simple type shown by H. A. Lorentz to be explicable on the basis of classical theory; the phenomenon is then called the normal Zeeman effect. In general, however, the splitting is more complicated: there occurs the anomalous Zeeman effect. The anomalous Zeeman effect results from the presence in the atom of two kinds of angular momenta, with their associated magnetic moments. These two kinds are the angular momentum corresponding to the orbital motion of the electrons about the nucleus, on the one hand, and the angular momentum corresponding to the spin of the electrons. The normal Zeeman effect occurs only when the spin makes no contribution to the angular momentum and the magnetic moment of the atom.

The effect of applying a magnetic field to an atom, neglecting powers of the magnetic field H higher than the first, is to impose an additional rotation about the field direction, in accordance with Larmor's theorem of classical mechanics. This rotation is called the Larmor precession. The angular velocity ω of this precession is equal to the product of the field strength H and the ratio of the magnetic moment to the angular momentum:

$$\omega = Hge/2m_0c$$

In this equation there has been introduced the symbol g, which is called the g-factor, or sometimes the Landé g-factor, after A. Landé, who introduced it. For orbital motion of an electron the g-factor has the value 1, the ratio of magnetic moment to angular momentum being $e/2m_0c$. For the spin of an electron, however, the g-factor has the value 2. This value of the g-factor cannot be explained in any simple way; it has to be accepted as part of the nature of the electron.

When the angular momentum of an atom is due entirely to the orbital motion of the electrons the value of the g-factor is 1, and when it is due entirely to the spin of the electrons the value of the g-factor is 2. For example, the normal state of the nitrogen atom is $^4S_{\frac{3}{2}}$; hence the g-factor for the normal nitrogen atom is 2.

In general the value of the g-factor is neither 1 nor 2, but has some other value. In case that the electronic state of the atom is such as to approximate closely to Russell-Saunders coupling the value of the g-factor can be calculated in a simple way. The total angular momentum vector of the atom is the resultant of the vector corresponding to the orbital angular momentum of all of the electrons in the atom and the vector corresponding to the spin angular momentum of all the electrons. The magnitudes of these three vectors are $\sqrt{J(J+1)}$, $\sqrt{L(L+1)}$, and $\sqrt{S(S+1)}$, respectively. When the cosines of the angles between the orbital and spin angular momentum vectors and the resultant total angular momentum vector are evaluated by the use of these vector magnitudes, and the total magnetic moment is calculated as the sum of the components of the magnetic moments of the orbital motion, with $g = 1$, and the spin, with $g = 2$, along the direction of the total angular momentum, it is found that the value of g is given by the equation

$$g = 1 + \frac{J(J+1) + S(S+1) - L(L+1)}{2J(J+1)}$$

Values of g are given in app. Table IV-3.

The modern unit of magnetic moment is $he/4\pi m_0 c$. This unit is called the Bohr magneton. Its value is 0.9273×10^{-20} erg gauss^{-1}. The magnetic moment of an atom with total angular momentum quantum number J is equal, in Bohr magnetons, to $g\sqrt{J(J+1)}$. When the atom is in a magnetic field its angular momentum vector is oriented with respect to the field in such a way that the component of angular momentum along the field direction is given by the quantum number M, and has the value $Mh/2\pi$. The component of magnetic moment in the direction of the field, in Bohr magnetons, is then equal to Mg, and the field energy of the atom is the product of this component by the strength of the field. Accordingly the energy level is split by the magnetic field into $2J + 1$ equally spaced levels, corresponding to the $2J + 1$ values that can be assumed by M. By analysis of the observed Zeeman splitting of spectral lines it is possible to evaluate g both for the upper state and the lower state of each spectral line.

For example, the g-factor for the normal state of the neutral silver atom, assigned the symbol $4d^{10}5s\ ^2S_{\frac{1}{2}}$, is observed to be 1.998, and the g-factors for the first two excited states, $4d^{10}5p\ ^2P_{\frac{1}{2}}$ and $^2P_{\frac{3}{2}}$, are observed to be 0.666 and 1.330, respectively; the theoretical values for these three states are 2.000, 0.667, and 1.333, so that the agreement is excellent, and one may conclude that the states are correctly assigned.

Hybrid Atomic States.—For many atomic states the observed properties are not those corresponding closely to a single Russell-Saunders structure. For example, the four most stable states of the neutral tin atom are the following:

Configuration	Symbol	J	Energy value	Observed g	Calculated g
$5s^25p^2$	^3P	0	0.0	—	—
		1	1691.8	1.502	1.500
		2	3427.7	1.452	1.500
$5s^25p^2$	^1D	2	8613.0	1.052	1.000

It is seen that there is good agreement between the observed g and the calculated g for the state $^3P_1$, but poor agreement for the two states $^3P_2$ and $^1D_2$. This poor agreement in the g-factors means that these states are not closely similar to the structures described by the Russell-Saunders symbols. Thus a ^3P state, in which the orbital angular momentum vector and the spin angular momentum vector have the same magnitude, must have the g-factor equal to 1.500, the average of the orbital value and the spin value. The fact that the observed g-factor is somewhat smaller can be explained in a simple way. The quantum number J is a rigorous quantum number for the atom, but the quantum numbers S and L are not rigorous quantum numbers; instead, they correspond to a certain type of interaction, with the electron orbital angular momenta combining to form a resultant and the spins combining to form a resultant, which represents only one extreme of the many alternative ways of interaction of the electrons in the atom. We may, however, continue to use the Russell Saunders structures in describing the two states with $J = 2$. The state with $g = 1.452$ may be said to be a hybrid of the structures $^3P_2$ and $^1D_2$, with the first of these structures making a large contribution and the second only a small contribution; perhaps, as an approximation indicated by the value of the g-factor, we can say that the state is a hybrid with about 90 percent $^3P_2$ character and 10 percent $^1D_2$ character. Similarly, the second state with $J = 2$, which has an observed g-factor equal to 1.052, can be described as having a structure that is about 90 percent $^1D_2$ and 10 percent $^3P_2$.

This description of these two states, as resonance hybrids of the two states $^3P_2$ and $^1D_2$, is arbitrary, but it is useful, inasmuch as the Russell-Saunders structures correspond closely to the actual properties for many atomic states, and it is convenient to continue to use these structures in the description of states for which no single Russell-Saunders structure provides a completely satisfactory representation of the observed properties.

Even the electron configuration represents an idealization, which for some atomic states is not satisfactory. For example, the neutral osmium atom is conventionally described as having $5d^66s^2$ as its most stable electron configuration. The lowest states to which this configuration is assigned are given the Russell-Saunders symbol ^5D, with $J = 4, 3, 2, 1$, and 0. For the first four of these states the observed values of the g-factor lie between 1.44 and 1.47, representing a definite deviation from the theoretical value 1.500. However, the rather stable states with the same values of J with which these states might be hybridized are those corresponding to the configuration $5d^76s$. We conclude that the most stable states of the neutral osmium atom are to be described as having a hybrid configuration, to which the configuration $5d^66s^2$ makes a large contribution and the configuration $5d^76s$ makes a small contribution.

Hybrid states of this sort are to be formed from structures with the same value of J, and also with the same parity. The parity of a configuration is even in case that it involves an even number of electrons in orbitals with odd value of l (p, f, etc.) and odd in case that it involves an odd number of electrons in orbitals with odd l. In tables of spectral terms the parity is often indicated by use of a superscript $°$ on the symbols of states with odd parity. In the above example of neutral osmium the two configurations considered have even parity.

2-9. THE FORMAL RULES FOR THE FORMATION OF COVALENT BONDS

The formal results of the quantum-mechanical treatment of valence, developed by Heitler, London, Born, Weyl, Slater, and other investigators, can be given the following simple statement: *an atom can form an electron-pair bond for each stable orbital*, the bond being of the type described for the hydrogen molecule and owing its stability to the same resonance phenomenon. In other words, for the formation of an electron-pair bond two electrons with opposed spins and a stable orbital of each of the two bonded atoms are needed.

The hydrogen atom, with only one stable orbital ($1s$), is thus limited to the formation of one covalent bond; the structures originally drawn for the hydrogen bond (Chap. 12), with bicovalent hydrogen, cannot be accepted.[19]

The carbon atom, nitrogen atom, and other first-row atoms are limited to four covalent bonds using the four orbitals of the L shell. This restriction forms much of the justification of the importance of the octet postulated by Lewis and Langmuir.

The quantum-mechanical treatment also leads to the conclusion that in general each additional electron-pair bond formed within a molecule stabilizes the molecule further, so that the most stable electronic structures of a molecule are those in which all of the stable orbitals of each

[19] L. Pauling, *Proc. Nat. Acad. Sci. U. S.* 14, 359 (1928).

atom are used either in bond formation or for occupancy by an un-shared pair of electrons. Stable electronic structures for a molecule containing first-row atoms would accordingly in general involve use of all four orbitals of the *L* shell; the sharing of electron pairs occurs to as great an extent as is permitted by the number of electrons present.[20] Electronic structures such as :N̈:N̈:, in which each nitrogen atom has only a sextet of electrons in the outer shell, occupying only three *L* orbitals of each atom, are less stable than structures such as :N:::N:, in which use is made of all the *L* orbitals.[21]

For second-row atoms too the octet retains some significance, since the 3*s* and 3*p* orbitals are more stable than the 3*d* orbitals. In a mole-cule such as phosphine, with the structure

$$\begin{array}{c} \text{H} \\ \ddot{\text{:P:H}} \\ \ddot{\text{H}} \end{array}$$

three of the *M* orbitals of phosphorus are used for bond formation and one for an unshared pair, and in the phosphonium ion

$$\left[\begin{array}{c} \text{H} \\ \ddot{\text{H:P:H}} \\ \ddot{\text{H}} \end{array} \right]^{+}$$

four *M* orbitals are used for bond formation, the five 3*d* orbitals in the *M* shell not being called on for bond formation. In phosphorus pentachloride, on the other hand, for which the structure

$$\begin{array}{c} \text{Cl} \quad \text{Cl} \\ \diagdown \quad \diagup \\ \text{Cl—P} \\ \diagup \quad \diagdown \\ \text{Cl} \quad \text{Cl} \end{array}$$

[20] Simple algebraic equations for calculating the number of shared electrons for structures with completed octets and other completed electron shells were given by I. Langmuir (*J.A.C.S.* **41**, 868 [1919]). These equations usually need not be called upon, since electronic formulas of the sort desired can be written easily with a little practice.

[21] The difference in stability of structures :N̈:N̈: and :N:::N: is the differ-ence in energy of a single and a triple bond, which is about 146 kcal/mole in favor of the triple bond. The chemical properties of unsaturated substances might suggest the double bond and triple bond to be weaker than the single bond; however, these properties involve comparison of the energy of the dou-ble bond with that of *two* single bonds, and similarly of the energy of the triple bond with that of *three* single bonds, whereas in the discussion above the com-parison is with only one single bond.

can be written, one of the $3d$ orbitals (or the $4s$ orbital) as well as the $3s$ and three $3p$ orbitals must be called on, and in order to form six covalent bonds in the hexafluophosphate ion

$$\begin{bmatrix} F & & F \\ & \diagdown & \diagup \\ F & \!\!-P-\!\! & F \\ & \diagup & \diagdown \\ F & & F \end{bmatrix}^{-}$$

two additional orbitals would be needed.

A maximum of nine covalent bonds can be formed by use of the orbitals of the M shell. This limitation is, however, not of great significance, inasmuch as other factors, discussed later in the book, provide a more serious limitation with respect to the number of atoms which can be bonded to a central atom.[22]

The octet rule similarly retains some significance for third-row atoms and still heavier atoms, aside from those of the transition elements. Thus we can, for example, assign to arsine and stibine structures analogous to those for phosphine, using the four s and p orbitals of the valence shell of the central atom.

For the transition elements use is often made in covalent bond formation of some of the d orbitals of the shell just inside the valence shell, as well as of the s and p orbitals of the valence shell. We write for the hexachloropalladate ion, for example, the structure

$$\begin{bmatrix} Cl & & Cl \\ & \diagdown & \diagup \\ Cl & \!\!-Pd-\!\! & Cl \\ & \diagup & \diagdown \\ Cl & & Cl \end{bmatrix}^{--}$$

with six covalent bonds from the palladium atom to the six surrounding chlorine atoms. There are in the palladium atom, in addition to the six bonding electron pairs, 42 electrons. These, in pairs, occupy the $1s$, $2s$, three $2p$, $3s$, three $3p$, five $3d$, $4s$, three $4p$, and three of the $4d$ orbitals. The six bonds are formed by use of the remaining two $4d$ orbitals, the $5s$ orbital, and the three $5p$ orbitals. A detailed discussion of the selection and use of atomic orbitals in bond formation is given in later chapters.

[22] A sharp distinction is to be made between the number of atoms bonded to a central atom (the *ligancy* or *coordination number* of the central atom) and the number of covalent bonds formed by the central atom (its *covalence*). These numbers may, and often do, differ as a result of the attachment of some of the surrounding atoms by bonds of types other than single covalent bonds, such as double bonds or electrostatic bonds.

CHAPTER 3

The Partial Ionic Character of Covalent Bonds and the Relative Electro-negativity of Atoms

THE chemical structure theory that has been developed during the past century is neither simple nor precise. It is customary to describe molecules not only in terms of single bonds, each involving a pair of electrons held jointly by two atoms, but also in terms of double bonds and triple bonds (no one has as yet found evidence justifying the assignment to any molecule of a structure involving a quadruple bond between a pair of atoms). Moreover, for some molecules and crystals there is no one valence-bond structure, with single bonds, double bonds, and triple bonds assigned to positions between pairs of atoms, that provides a satisfactory representation of the properties of the substance; and, as we shall see later, it is convenient to introduce new ideas, such as the resonance of molecules among two or more valence-bond structures or the assignment of fractional bonds, in order to extend chemical structure theory to include these substances.

In the present and the following chapter we shall concentrate our attention on the single bond and on substances for which one valence-bond structure involving only single bonds provides a satisfactory representation of the molecule.

The discussion begins with the consideration of diatomic molecules formed by univalent elements—molecules in which there are two atoms held to one another by a single bond. The hydrogen molecule is the only molecule of this kind for which an accurate solution of the Schrödinger wave equation has been obtained. The approximate quantum-mechanical treatment of more complex molecules has provided interesting information about their electronic structure, but work along these lines has not been sufficiently extensive to permit the formulation of precise generalizations about the nature of the single bond from theory alone.

It has, however, been possible to induce from the properties of substances a number of general principles about the nature of the single bond in its dependence on the nature of the two atoms connected by it. These principles are in general qualitative or only roughly quantitative. For example, in the following paragraphs we shall talk about the partial ionic character of single bonds and shall suggest a method of estimating the partial ionic character of the bond between the atoms of two elements, but the estimated value is not held to be accurate. This discussion of the partial ionic character of bonds permits the prediction of values of the heat of formation of substances containing only single bonds, but these predicted heats of formation are only rough values, reliable to a few kcal/mole—there is no chemical theory that permits the prediction of values accurate to 0.01 or 0.001 kcal/mole, the accuracy of some experimental determinations of heats of formation. Nevertheless, despite their approximate nature, the principles described in this chapter and the following ones have been found useful in helping the student to correlate the facts of descriptive chemistry into a system and to make predictions about the properties of substances that have not yet been synthesized.

3-1. THE TRANSITION FROM ONE EXTREME BOND TYPE TO ANOTHER

After the development some decades ago of the modern ideas of the ionic bond and the covalent bond the following question was formulated and vigorously discussed: If it were possible to vary continuously one or more of the parameters determining the nature of a molecule or a crystal, such as the effective nuclear charges of the atoms, then would the transition from one extreme bond type to another take place continuously, or would it show discontinuities? With the extension of our understanding of the nature of the chemical bond we may now answer this question; the pertinent argument, given in the following paragraph, leads to the conclusion that in some cases the transition would take place continuously, whereas in others an effective discontinuity would appear.[1]

[1] Lewis in 1916 and later years supported the idea that the transition would be continuous and that the shared electron pair is in general attracted more strongly by one than by the other of two unlike bonded atoms, the bond having a corresponding amount of ionic or "polar" character. N. V. Sidgwick (*Some Physical Properties of the Covalent Link in Chemistry*, Cornell University Press, 1933, pp. 42 ff.) and F. London (*Naturwissenschaften* **17**, 525 [1929]) expressed the opinion that, although the transition between two extreme bond types might occur without discontinuity, there is an essential difference between the two

Continuous Change in Bond Type.—Let us first consider the case of a molecule involving a single bond between two atoms A and B, which for certain values of the structural parameters for the molecule is a normal covalent bond of the type formed by like atoms and discussed in Sections 1-5 and 3-4 and for other values is an ionic bond A^+B^-, the more electronegative of the two atoms holding both of the electrons as an unshared pair occupying one of the orbitals of its outer shell. For intermediate values of the structural parameters of the molecule the wave function $a\psi_{A:B} + b\psi_{A^+B^-}$, formed by the linear combination of the wave functions corresponding to the normal covalent structure A:B and the ionic structure A^+B^-, with numerical coefficients a and b, can be used to represent the structure of the molecule, the value of the ratio of the coefficients, b/a, for each set of values of the structural parameters being such as to make the bond energy a maximum.[2] As the parameters of the molecule (in particular, the relative electronegativity of A and B) were changed, the ratio b/a would change from zero to infinity, the bond changing in type without discontinuity from the covalent extreme to the ionic extreme by passing through all intermediate stages. In the case under discussion the two extreme structures are of such a nature (each involving only paired electrons and essentially the same configuration of the atomic nuclei) as to permit resonance, and hence the transition from one extreme type of bond to the other would be continuous.

For an intermediate value of the relative electronegativity of A and B such that the coefficients a and b in the wave function $a\psi_{A:B} + b\psi_{A^+B^-}$ are about equal in magnitude, the bond might be described as *resonating between the covalent extreme and the ionic extreme*, the contributions of the two being given[3] by the values of a^2 and b^2. If the extreme covalent structure A:B and the extreme ionic structure A^+B^- correspond separately to the same bond-energy value, then the two structures will contribute equally to the actual state of the molecule, and the actual bond energy will be greater than the bond energy for either structure alone by an amount equal to the interaction energy of the two structures; that is, resonance between the two structures will stabilize the molecule. If one of the two extreme structures corresponds to a greater bond energy than the other, the more stable structure will

---

types of bonds and that only rarely does there occur a molecule containing a bond of intermediate type. The latter opinion is contrary to the one that we shall form as a result of the discussion given in this chapter.

[2] That is, to minimize the energy of the system (Sec. 1-3).

[3] The squares of the coefficients of terms in a composite wave function are interpreted as representing the magnitudes of the contributions of the corresponding structures.

contribute more to the actual state of the molecule than the less stable one, and the actual bond energy will be increased somewhat by resonance over that for the more stable structure. The relation between the extra resonance energy stabilizing the bond, the interaction energy of the two structures, and the bond-energy values of the two structures is the same as that described in an analogous case in Section 1-4.

For a molecule such as hydrogen chloride we write the two reasonable electronic structures H:Cl: and H⁺ :Cl:⁻. (The third structure that suggests itself, H:⁻ Cl:⁺, is not given much importance because hydrogen is recognized as less electronegative than chlorine; a discussion of the extent to which such a structure contributes to the normal state of a molecule is given in Sec. 3-3.) In accordance with the foregoing argument the actual state of the molecule can be described as corresponding to resonance between these two structures. The extent to which each structure contributes its character to the bond is discussed in detail for hydrogen chloride and other molecules in the following sections of this chapter.

Instead of using this description of the bond as involving resonance between an extreme covalent bond H:Cl: and an extreme ionic bond H⁺Cl⁻, we may describe the bond as a *covalent bond with partial ionic character*, and make use of the *valence line*, writing H—Cl (or H—Cl:) in place of {H:Cl:, H⁺Cl⁻} or some similar complex symbol showing resonance between the two extremes. This alternative description is to be recognized as equivalent to the first; whenever a question arises as to the properties expected for a covalent bond with partial ionic character, it is to be answered by consideration of the corresponding resonating structures.

The amount of ionic character of a bond in a molecule must not be confused with the tendency of the molecule to ionize in a suitable solvent. The ionic character of the bond is determined by the importance of the ionic structure (A⁺B⁻) *when the nuclei are at their equilibrium distance* (1.275 Å for HCl, for example), whereas the tendency to ionize in solution is determined by the relative stability of the actual molecules in the solution and the separated ions in the solution. It is reasonable, however, for the tendency toward ionization in solution to accompany large ionic character of bonds in general, since both result from great difference in electronegativity of the bonded atoms.[4]

[4] A discussion of the ionization of the hydrohalogenic acids in aqueous solution is given in App. XI.

Transitions between other extreme types of bonds (covalent to metallic, covalent to ion-dipole, etc.) can also occur without discontinuity, and the bonds of intermediate character can be discussed in terms of resonance between structures of extreme type in the same way as for covalent-ionic bonds.

Discontinuous Change in Bond Type.[5]—In molecules and complex ions of certain types continuous transition from one extreme bond type to another is not possible. In order for continuous transition to be possible between two extreme bond types the conditions for resonance between the corresponding structures must be satisfied. The most important of these conditions is that the two structures must involve the same numbers of unpaired electrons. *If the two structures under consideration involve different numbers of unpaired electrons, then the transition between the two must be discontinuous, the discontinuity being associated with the pairing or unpairing of electrons.*[6]

The most important molecules and complex ions for which this phenomenon occurs are those containing a transition-group atom. Let us discuss as an example the octahedral complexes FeX_6 of ferric iron. In some of these complexes ($[FeF_6]^{---}$, $[Fe(H_2O)_6]^{+++}$) the bonds are of such a nature that the electronic structure about the iron nucleus is the same as for the Fe^{+++} ion; of the 23 electrons of this ion, 18 occupy the $1s$, $2s$, three $2p$, $3s$, and three $3p$ orbitals in pairs and the remaining five occupy the five $3d$ orbitals without pairing, as described in Section 2-7. If, however, covalent Fe—X bonds are formed with use of two of the $3d$ orbitals (as well as some of the other orbitals—see Chap. 5), as in the ferricyanide ion, $[(Fe(CN)_6]^=$, then the five unshared $3d$ electrons of the iron atom must crowd into the remaining three $3d$ orbitals, with formation of two pairs. This complex contains only one unpaired electron, whereas the complexes of the first kind contain five unpaired electrons. Transition between these structures cannot be continuous.

Resonance is possible, of course, between an ionic FeX_6 structure, with ionic bonds between the Fe^{+++} ion and surrounding anions, and a covalent structure in which only the outer orbitals $4s$, $4p$, $4d$, and so on are used in bond formation. This covalent structure with five unpaired electrons would be different in character from that using two $3d$ orbitals, however, and continuous transition to the latter could not occur.

The nature of the discontinuity under discussion is shown in Figure 3-1. Two states of a complex FeX_6 are represented, one with five unpaired electrons and the other with one. For certain atoms or groups

---

[5] L. Pauling, *J.A.C.S.* **53**, 1367 (1931); **54**, 988 (1932).

[6] This statement is rigorously true in case that spin-orbit and spin-spin interactions are negligible (as for all light atoms), and is practically true in general.

X one of the states is the more stable, and represents the normal complex, and for others the other state is the stable one. At the discontinuity in the nature of the normal state of the complex the energy curves of the two states cross. An actual system would contain complexes in both states, with concentrations determined by the energy difference of the two; an appreciable number of complexes in the less stable state would be present, however, only for the region near the intersection of the two curves.

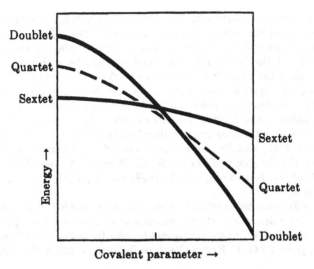

FIG. 3-1.—Energy curves for three states of an iron(III) complex FeX₆. The sextet curve represents a stable structure for extreme ionic bonds and an excited structure for extreme covalent bonds; it has five electrons with unpaired spins. The doublet curve represents an excited structure for extreme ionic bonds and a stable structure for extreme covalent bonds; it has one unpaired electron. The dashed curve represents a quartet state, with three electrons with unpaired spins. The parameter representing the abscissa determines the nature of the bonds.

These complexes and others of similar character are discussed further in Chapter 5, in which a magnetic criterion for bond type applicable to complexes of the transition elements is described.

3-2. BOND TYPE AND ATOMIC ARRANGEMENT

The properties of a substance depend in part upon the type of bonds between the atoms of the substance and in part upon the atomic arrangement and the distribution of the bonds. The atomic arrangement is itself determined to a great extent by the nature of the bonds: the directed character of covalent bonds (as in the tetrahedral carbon atom) plays an especially important part in determining the configurations of molecules and crystals; an important part is also played by the interatomic repulsive forces that give "size" to atoms and ions (Chaps. 7, 13).

Since 1913 a great amount of information about the atomic arrangement in molecules and crystals has been collected.[7] This information can often be interpreted in terms of the nature and distribution of bonds; a detailed discussion of the dependence of interatomic distances and bond angles on bond type will be given in later chapters.

[7] A great amount of information about the structure of crystals has been obtained by use of the x-ray diffraction method. The diffraction of x-rays by crystals was discovered by Max von Laue in 1912. Shortly thereafter W. L. Bragg discovered the Bragg equation, and in 1913 he and his father, W. H. Bragg, published the first structure determinations of crystals.

Thousands of crystals have been subjected to x-ray investigation. The results of the studies have been published in many journals; at present *Acta Crystallographica* is the leading journal in this field. The principal reference books for crystal structures are *Strukturbericht* (vols. I to VII, covering the period 1913 to 1939) and *Structure Reports* (vol. 8 on, covering the later period). Another useful reference book is R. W. G. Wyckoff, *Crystal Structures*, Interscience Publishers, New York, vol. I, 1948; vol. II, 1951; vol. III, 1953 (with later additions).

Neutron diffraction by crystals has been found valuable for locating hydrogen atoms (especially deuterium atoms, which scatter neutrons strongly), for studying the arrangement of magnetic moments, and for other special purposes. A summary is given by G. E. Bacon, *Neutron Diffraction*, Clarendon Press, Oxford, 1955.

Information about the structure of gas molecules has been obtained by several methods. Spectroscopic studies in the infrared, visible, and ultraviolet regions have provided much information about the simplest molecules, especially diatomic molecules, and a few polyatomic molecules. Microwave spectroscopy and molecular-beam studies have yielded very accurate interatomic distances and other structural information about many molecules, including some of moderate complexity. Molecular properties determined by spectroscopic methods are given in the two books by G. Herzberg, *Spectra of Diatomic Molecules*, 1950, and *Infrared and Raman Spectra*, 1945, Van Nostrand Co., New York. The information obtained about molecules by microwave spectroscopy is summarized by C. H. Townes and A. L. Schawlow in their book *Microwave Spectroscopy of Gases*, McGraw-Hill Book Co., New York, 1955.

Most of the structural information about complex gas molecules has been obtained by the electron-diffraction method. Values of interatomic distances and bond angles determined by this method before 1950 are summarized in a review article by P. W. Allen and L. E. Sutton, *Acta Cryst.* **3**, 46 (1950). Values of interatomic distances and bond angles for organic molecules determined by both x-ray diffraction of crystals and electron diffraction of gas molecules are summarized in a 90-page table in G. W. Wheland's book *Resonance in Organic Chemistry*, John Wiley and Sons, New York, 1955, and by L. E. Sutton in *Tables of Interatomic Distances and Configurations in Molecules and Ions*, Chemical Society, London, 1958. (Later references to the latter book will be to Sutton, *Interatomic Distances*.)

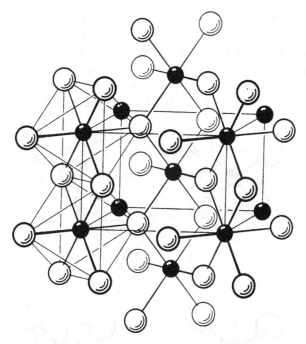

Fig. 3-2.—The atomic arrangement of the tetragonal crystal rutile, TiO_2. Large circles represent oxygen atoms, small circles titanium atoms. Each titanium atom is surrounded by oxygen atoms at the corners of an octahedron. Each octahedron shares two opposite edges with adjacent octahedra, to form long strings of octahedra that extend along the c axis of the crystal (vertically, in the drawing).

An abrupt change in properties in a series of compounds, such as in the melting points or boiling points of metal halogenides, has sometimes been considered to indicate an abrupt change in bond type. Thus of the fluorides of the second-row elements,

	NaF	MgF$_2$	AlF$_3$[8]	SiF$_4$	PF$_5$	SF$_6$
Melting point	995°	1263°	1257°	− 90°	− 94°	− 51°C

those of high melting points have been described as salts, and the others as covalent compounds; and the great change in melting points from aluminum fluoride to silicon fluoride has been interpreted as showing that the bonds change sharply from the extreme ionic type to the extreme covalent type.[9] I consider the bonds in aluminum fluoride to be only slightly different in character from those in silicon fluoride, and I attribute the abrupt change in properties to a change in the nature

[8] Sublimes.
[9] N. V. Sidgwick, *The Electronic Theory of Valency*, Clarendon Press, Oxford, 1927, p. 88; *The Covalent Link in Chemistry*, p. 52.

of the atomic arrangement.[10]   In NaF, MgF_2, and AlF_3 the relative
sizes of the metal and nonmetal atoms are such as to make the stable
ligancy (coordination number) of the metal six; each of the metal atoms
is surrounded by an octahedron of fluorine atoms, and the stoichiometric relations then require that each fluorine atom be held jointly
by six sodium atoms in NaF (which has the sodium chloride structure,

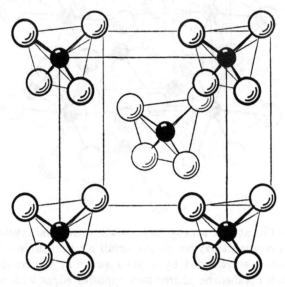

Fig. 3-3.—The atomic arrangement in the cubic crystal SiF_4.   The atoms
form tetrahedral molecules, with four fluorine atoms surrounding a silicon
atom.   The molecules are arranged at the points of a body-centered cubic
lattice.

Fig. 1-1), by three magnesium atoms in MgF_2 (with the rutile structure, Fig. 3-2), or by two aluminum atoms in AlF_3.   In each of these
crystals the molecules are thus combined into giant polymers, and the
processes of fusion and vaporization can take place only by breaking
the strong chemical bonds between metal and nonmetal atoms; in consequence the substances have high melting points and boiling points.
The stable ligancy of silicon relative to fluorine is, on the other hand,
four, so that the SiF_4 molecule has little tendency to form polymers.[11]
The crystal of silicon fluoride consists of SiF_4 molecules piled together
as shown in Figure 3-3, and held together only by weak van der Waals
forces.   Fusion and vaporization of this substance involve no great
change in the molecule; the strong Si—F bonds are not broken, but
only the weak intermolecular bonds, and hence the melting point and

[10] L. Pauling, *J.A.C.S.* **54**, 988 (1932).
[11] Silicon also assumes the ligancy six with fluorine in the **fluorosilicate** ion.

boiling point are low. In phosphorus pentafluoride and sulfur hexa-fluoride too there is no tendency for the ligancies of the central atoms to be increased by polymerization, and the physical properties of these substances are accordingly similar to those of silicon fluoride. It was mentioned many years ago by Kossel that ease of fusion and volatili-zation would be expected for ionic molecules in which a central cation is surrounded by several anions, and is not good evidence for the pres-ence of covalent bonds.[12] Volatility and many other properties such as hardness and cleavability depend mainly not so much on bond type as on the atomic arrangement and the distribution of bonds.

There is, to be sure, some correlation between bond type and type of atomic arrangement. Ionic crystals often possess a coordinated struc-ture such that ionic bonds extend throughout the crystal, leading to low volatility. Another structural feature that leads to high melting points and striking hardness of crystals is the hydrogen bond between molecules (Chap. 12).

3-3. THE NATURE OF THE BONDS IN DIATOMIC HALOGENIDE MOLECULES

In the hydrogen molecule a quantum-mechanical treatment has shown that the two ionic structures H^+H^- and H^-H^+ enter into reso-nance with the extreme covalent structure $H:H$ to only a small extent, each ionic structure contributing only about 2 percent to the normal state of the molecule (Sec. 1-5). The reason for this small contribution by the ionic structures is that these structures are unstable relative to the covalent structure, the large amount of energy (295 kcal/mole) required to transfer an electron from one nucleus to the other and form a positive and a negative ion not being completely counterbalanced by the mutual Coulomb energy of the ions. There is not much evidence as to the amount of ionic character of other single bonds between like atoms, as in the chlorine molecule, Cl_2. Consideration of energy values makes it probable,[13] however, that in this molecule the ionic structures Cl^+Cl^- and Cl^-Cl^+ make a still smaller contribution to the normal

[12] W. Kossel, *Z. Physik* 1, 395 (1920).

[13] The ionization energy of chlorine is 299 kcal/mole and its electron affinity is 86 kcal/mole, making separated ions Cl^+ and Cl^- unstable relative to atoms by 213 kcal/mole (see Chap. 13). The Coulomb energy $-e^2/R$ of two ions at the Cl—Cl equilibrium distance $R = 1.988$ Å is -166 kcal/mole, and the extreme covalent bond energy is about 55 kcal/mole. Hence in the equilibrium configu-ration the ionic structures Cl^+ Cl^- and Cl^- Cl^+ are unstable relative to the co-valent structure by at least 102 kcal/mole. (The characteristic repulsion of the two ions is neglected.) This energy difference is so great as to permit the ionic structures to enter into resonance with the covalent structure to a small extent only.

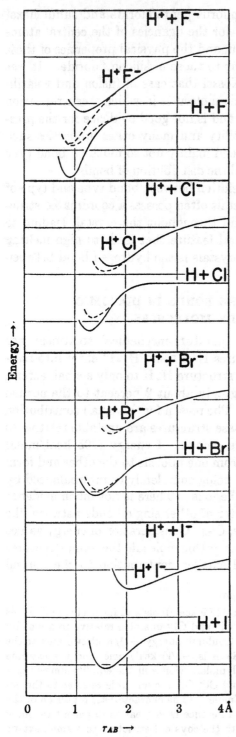

state than the corresponding structures do for the hydrogen molecule. In general we shall use the symbol Cl—Cl or :Cl—Cl: to represent a single covalent bond of the sort existent between the two like atoms, including the small equal contributions of the two ionic structures.

Now let us consider the bond between two unlike atoms that do not differ greatly in electronegativity, such as chlorine and bromine. The energy required to form the ions Br^+ and Cl^- from the atoms Br and Cl is only 186 kcal/mole, whereas that required to form the ions Br^- and Cl^+ is 218 kcal/mole; accordingly the ionic structure Br^-Cl^+ makes only a very small resonance contribution to the normal state of the BrCl molecule (smaller than that made by the ionic structures in Br_2 and Cl_2), and the ionic structure Br^+Cl^- makes a somewhat larger contribution than that

FIG. 3-4.—Calculated energy curves for the hydrogen halogenide molecules. The two dashed curves for each molecule represent extreme ionic and extreme covalent structures, and the two full curves represent the actual structures resulting from resonance between these extreme structures. The dashed curves for HI lie very close to the full curves.

made by the ionic structures in the symmetric molecules.

In hydrogen chloride the energy of formation of the ions H^+ and Cl^- from atoms is 226 kcal/mole, and that of the ions H^- and Cl^+ is 283 kcal/mole. The ionic structure H^-Cl^+ is accordingly very much less important than the structure H^+Cl^-. A qualitative estimate of the extent to which the ionic structures H^+X^- and the extreme covalent structures H:Ẍ: contribute to the normal states of the four hydrogen halides can be made by the consideration of energy curves. In Figure 3-4 calculated energy curves are shown for the structures H^+X^- and H:Ẍ: for each of the molecules HF, HCl, HBr, and HI. It is seen that in the neighborhood of the equilibrium internuclear distances (at the minima of the curves) the covalent curves for HCl, HBr, and HI lie below the ionic curves, the separation increasing from HCl to HI. This shows that the bonds in these molecules are essentially covalent, with, however, a small amount of H^+X^- ionic character, which is presumably greatest in HCl (of the three) and least in HI. (The reasonable assumption is here made that the interaction-energy integral of the two structures H^+X^- and HX is about the same in the three molecules, and that the manifestation of this interaction as resonance energy is decreased by increase in the energy difference of the resonating structures, as discussed in Sec. 1-4.) The effect of resonance on the energy is shown by the full lines in the figure, representing actual states (normal and excited) of the molecules.

In hydrogen fluoride the situation is different. For this molecule the ionic curve and the covalent curve are nearly coincident in the neighborhood of the equilibrium internuclear distance. In consequence of this *the ionic structure* H^+F^- *and the covalent structure* H:F̈: *make nearly equal contributions to the normal state of the molecule;* the hydrogen-fluorine bond has about 50 percent ionic character.[14] Because the two energy curves lie close together, resonance is nearly complete, and almost the entire interaction energy between the two structures is effective as resonance energy.[15]

---

[14] This conclusion follows from the general theorem that two structures with the same energy in resonance make equal contributions to the normal state of the system.

[15] The discussion of hydrogen fluoride in this paragraph is a little different from that in the first two editions of this book. In the first two editions the calculated energy curve for the extreme ionic structure was shown as falling below that for the normal covalent structure, and the conclusion was reached that the bond between the hydrogen atom and the fluorine atom in the molecule

The curves of Figure 3-4 have been drawn in the following way. The extreme covalent curves are Morse curves[16] with equilibrium distances given by the sums of the single-bond covalent radii (Sec. 7-1), curvatures calculated by Badger's rule (Sec. 7-4), and bond energies calculated by the method of the arithmetic mean (Sec. 3-4). The curves for the ionic structures H^+X^- represent the interaction of a proton with negative ions with electron distribution functions calculated by the use of hydrogenlike wave functions with suitable screening constants, polarization being neglected.[17] The curves representing the normal states are Morse curves drawn with the experimentally determined values of the parameters.

The alkali halogenide gas molecules MX present a still more extreme case, the bonds being essentially ionic with only a small amount of covalent character. For cesium chloride, involving the most electropositive of the metals and one of the most electronegative of the nonmetals, the electron affinity of the nonmetal (86 kcal/mole) is about as great as the ionization potential of the metal (89 kcal/mole), so that at large internuclear distances the ionic structure Cs^+Cl^- is about as stable as the covalent structure $Cs\!:\!\overset{\cdot\cdot}{\underset{\cdot\cdot}{Cl}}\!:$. With decreasing internuclear distance the Coulomb energy of the ions causes the ionic structure to be favored relative to the covalent structure, until at the equilibrium distance the energy difference amounts to 100 kcal/mole. This molecule contains a bond that is nearly completely ionic in character, the covalent contribution being very small—only a few percent.

The other alkali halogenide molecules are also largely ionic. The energy curves for a representative molecule, sodium chloride, are shown in Figure 3-5. At very large internuclear distances the covalent structure is more stable than the ionic structure; but at about 10.5 Å the

---

has a little more than 50 percent ionic character. Small changes in the curves and in the discussion have been made because of the discovery that the value of the dissociation energy of the F_2 molecule is 27 kcal/mole less than the value that had been previously accepted. This change leads to a change one half as great in the electron affinity of fluorine and causes a corresponding shift in the ionic curve relative to the covalent curve. The uncertainties of the calculation are such that either curve may lie as much as 10 kcal/mole below the other, and the amount of ionic character may differ considerably (perhaps by as much as 10 percent) from the value 50 percent given above.

[16] See App. VII.

[17] L. Pauling, *Proc. Roy. Soc. London* A114, 181 (1927); *loc. cit.* (10); L. Pauling and J. Sherman, *Z. Krist.* 81, 1 (1932); see also F. T. Wall, *J.A.C.S.* 61, 1051 (1939). A similar discussion of carbon-hydrogen and carbon-halogen bonds has been published by E. C. Baughan, M. G. Evans, and M. Polanyi, *Trans. Faraday Soc.* 37, 377 (1941).

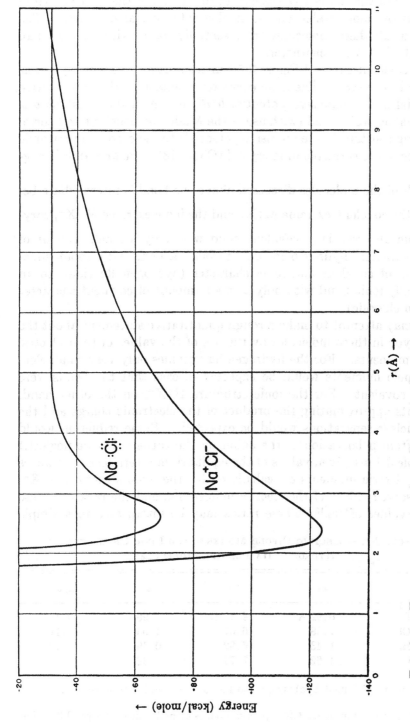

Fig. 3-5.—Energy curves for the sodium chloride molecule. At very large internuclear distances the curve for the ionic structure lies above that for the covalent structure. The curves cross at 10.5 Å, and at smaller internuclear distances the ionic structure is the more stable one.

r(Å) →

← Energy (kcal/mole)

Na :Ö:

Na⁺Cl⁻

Coulomb attraction causes the curves to cross, and the ionic structure remains the more stable one at smaller internuclear distances. The bonds in all these molecules are essentially ionic, with only a small amount of covalent character.

The covalent curve of Figure 3-5 has been drawn in the same way as for the HX curves. The ionic curve represents a Coulomb attractive potential and a repulsive potential b/R^9, in which the constant b is given values which lead (with use of the Madelung constant and corresponding constant in the repulsive potential) to the correct interatomic distance in the corresponding crystal (Chap. 13). Polarization is neglected.

In all of the molecules discussed above the bond is intermediate between the covalent extreme M:X: and the ionic extreme M⁺X⁻, varying from an essentially covalent bond with only a small amount of ionic character (hydrogen iodide), through a bond with about equal amounts of covalent and ionic character (hydrogen fluoride), to an essentially ionic bond with only a small amount of covalent character (cesium chloride).

We may attempt to make a rough quantitative statement about the bond type in these molecules by the use of the values of their electric dipole moments. For the hydrogen halogenides only very small electric dipole moments would be expected in case that the bonds were purely covalent. For the ionic structure H⁺X⁻, on the other hand, moments approximating the product of the electronic charge and the internuclear separations would be expected. (Some reduction would result from polarization of the anion by the cation; this we neglect.) In Table 3-1 are given values of the equilibrium internuclear distances r_0, the electric moments er_0 calculated for the ionic structure H⁺X⁻, the observed values of the electric moments μ, and the ratios of these to the values of er_0.[18] These ratios may be interpreted in a simple

TABLE 3-1.—ELECTRIC DIPOLE MOMENTS AND IONIC CHARACTER OF HYDROGEN HALOGENIDE MOLECULES

	r_0	er_0	μ	μ/er_0
HF	0.92 Å	4.42 Dᵃ	1.98 D	0.45
HCl	1.28	6.07	1.03	.17
HBr	1.43	6.82	0.79	.12
HI	1.62	7.74	.38	.05

ᵃ The unit 1 D (one debye) is equal to 1×10^{-18} statcoulomb centimeters.

[18] For a discussion of electric dipole moments of molecules see App. IX. The value for HF in Table 3-1 involves a special treatment: R. A. Oriani and C. P. Smyth, *J. Chem. Phys* **16**, 1167 (1948).

way as representing approximately the magnitudes of the contributions of the ionic structures to the normal states of the molecules; that is, the amounts of ionic character of the bonds. It is seen that on this basis the bond in hydrogen fluoride is 45 percent ionic, that in hydrogen chloride 17 percent ionic, that in hydrogen bromide 12 percent ionic, and that in hydrogen iodide 5 percent ionic.

The values of the electric dipole moments of the alkali halogenide gas molecules are found to be about 80 percent of er_0; for example, for KCl the value of r_0 is 2.671 Å and that of μ is 10.48 D, which is 82 percent of er_0. The deviation of μ from er_0 is about the magnitude expected to result from the polarization of each ion in the electric field of the other, and because of the uncertainty in the theoretical calculation of the polarization correction it is not possible to say more than that the observed moments agree roughly with those expected for completely ionic structures.

The discussion of the amount of partial ionic character of single bonds will be continued in Section 3-9.

It must be mentioned that the attempt to discuss bond type in this roughly quantitative way without giving a complete quantum-mechanical treatment of the molecules cannot be rigorously justified. We have adopted the procedure of discussing the structure of molecules and the nature of chemical bonds as completely as possible with use of only the most stable of the atomic orbitals; following this procedure, we are led to base our discussion on the simple structures $M:X$, M^+X^-, and M^-X^+. It is possible,[19] on the other hand, to develop (at least in principle) a complete discussion of the structure of a molecule from either the purely ionic point of view (with extreme polarization or deformation of the ions) or the covalent point of view, provided that all the unstable atomic orbitals are used in the discussion. No treatment of either of these types has been carried out for molecules of any complexity, however, whereas the reasonable procedure that forms the basis of our argument has found extensive application to the problems of structural chemistry.

3-4. BOND ENERGIES OF HALOGENIDE MOLECULES; THE ENERGIES OF NORMAL COVALENT BONDS[20]

The wave function representing the single bond in a symmetric molecule A—A can be written in the form

$$a\psi_{A:A} + b\psi_{A^+A^-} + b\psi_{A^-A^+} \tag{3-1}$$

and a similar expression can be written for another molecule B—B.

[19] J. C. Slater, *Phys. Rev.* **41**, 255 (1932).

[20] L. Pauling and D. M. Yost, *Proc. Nat. Acad. Sci. U. S.* **18**, 414 (1932); L. Pauling, *J.A.C.S.* **54**, 3570 (1932).

The ratio b/a, determining the contributions of the ionic structures, is small, and probably about the same for all bonds between like atoms.

Now let us consider a molecule A—B, involving a single bond between two unlike atoms. If the atoms were closely similar in character, the bond in this molecule could be represented by a wave function such as 3-1, an average of those for the symmetric molecules A—A and B—B. Let us describe such a bond as a *normal covalent bond*.

If, now, we consider a molecule A—B in which the atoms A and B are dissimilar, one being more electronegative than the other, we must use a more general wave function,

$$a\psi_{A:B} + c\psi_{A^+B^-} + d\psi_{A^-B^+} \qquad (3\text{-}2)$$

to represent the bond, the best values of c/a and d/a being those that make the bond energy a maximum (minimize the total energy of the molecule). These values will in general be different from b/a of Equation 3-1, one being smaller and one larger. Since they make the bond energy a maximum, we see that *the energy of an actual bond between unlike atoms is greater than (or equal to) the energy of a normal covalent bond between these atoms.* This additional bond energy is due to the *additional ionic character of the bond;* that is, it is the *additional ionic resonance energy* that the bond has as compared with a bond between like atoms. In referring to these quantities later we shall omit the word "additional" and say "ionic character of the bond" and "ionic resonance energy."

To test this conclusion we need values of the energies of normal covalent bonds between unlike atoms. These values might be calculated by quantum-mechanical methods; it is simpler, however, to make a postulate and test it empirically. Since a normal covalent bond A—B is similar in character to the bonds A—A and B—B, we expect the value of the bond energy to be intermediate between the values for A—A and B—B. This result follows from the *postulate of the additivity of normal covalent bonds.* That is, we assume that the arithmetic mean of the two bond-energy values $D(A—A)$ and $D(B—B)$ is the energy of the normal covalent bond between the unlike atoms A and B.

If this postulate were true, actual bond energies $D(A—B)$ between unlike atoms would always be greater than or equal to the arithmetic means of the corresponding symmetrical bond energies; the difference $\triangle$ defined as

$$\Delta = D(A—B) - \tfrac{1}{2}\{D(A—A) + D(B—B)\} \qquad (3\text{-}3)$$

would never be negative. In Table 3-2 values of bond energies and

of Δ for the hydrogen halogenides and halogen halogenides are given.[21]
It is seen that *for each of the eight molecules Δ is positive.* Moreover,
the magnitudes of the Δ values, which measure the resonance energy
due to ionic character of the unsymmetrical bonds, are in agreement
with our previously formed conceptions as to the nature of the bonds
in these molecules. In the series HI, HBr, HCl, HF we have estimated the amounts of ionic character of the bonds to be 5, 12, 17,
and 45 percent, respectively. The corresponding values of Δ, 1.2, 12.3,
22.1, and 64.2 kcal/mole, increase in the same general way and show

TABLE 3-2.—BOND ENERGIES FOR HYDROGEN HALOGENIDE AND HALOGEN
HALOGENIDE MOLECULES (KCAL/MOLE)

	H—H	F—F	Cl—Cl	Br—Br	I—I
Bond energy	104.2	36.6	58.0	46.1	36.1

	H—F	H—Cl	H—Br	H—I
Bond energy	134.6	103.2	87.5	71.4
$\frac{1}{2}\{D(\text{H—H}) + D(\text{X—X})\}$	*70.4*	*81.1*	*75.2*	*70.2*
Δ	64.2	22.1	12.3	1.2

	Cl—F	Br—Cl	I—Cl	I—Br
Bond energy	60.6	52.3	50.3	42.5
$\frac{1}{2}\{D(\text{X—X}) + D(\text{X'—X'})\}$	*47.3*	*52.1*	*47.1*	*41.1*
Δ	13.3	0.2	3.2	1.4

the expected large change from HCl to HF. (The only unexpected
feature is the very small value of Δ for HI.) The molecule BrCl approaches the normal covalent type still more closely, with Δ equal to
only 0.2 kcal/mole. This is the expected result for a bond between
two atoms that resemble one another as closely as chlorine and bromine.
The values of Δ for IBr and ICl are also small, but that for ClF is
about as large as that for HBr, showing that chlorine fluoride is about
as ionic in character as hydrogen bromide. Chlorine, bromine, and
iodine do not differ greatly in electronegativity, chlorine and bromine
being more nearly alike in this respect, as in other respects, than bromine and iodine. But fluorine is very much more electronegative than

[21] In Tables 3-2, 3-3, 3-4, and 3-5 the enthalpies at 25°C are used as the basis
for the calculation of bond energies, which accordingly include not only the
energies of dissociation D_0 of the molecules but also small terms corresponding
to the rotational, oscillational, and translational energy of the molecules and
a pressure-volume term. These small terms are not significant for our arguments. Enthalpies rather than energy values are used to give uniformity with
Sec. 3-5.

the other halogens; it deserves to be classed by itself as a superhalogen.

It is seen that the quantity Δ is just the heat liberated in the reaction

$$\tfrac{1}{2}A_2(g) + \tfrac{1}{2}B_2(g) \rightarrow AB(g) \qquad (3\text{-}4)$$

and our requirement that Δ be greater than or equal to zero is equivalent to the requirement that a reaction of this type not be endothermic.

It will be shown in the following section that the postulate of additivity is valid for a large number of single bonds and that the values of Δ can be used as the basis for the formulation of an extensive scale of electronegativities of the elements. In a few cases, however, the postulate of additivity is found not to hold. The following section is devoted to a discussion of these cases.

TABLE 3-3.—BOND-ENERGY VALUES FOR ALKALI HYDRIDE MOLECULES (KCAL/MOLE)

	H—H	Li—Li	Na—Na	K—K	Rb—Rb	Cs—Cs
Bond energy	104.2	26.5	18.0	13.2	12.4	10.7

	Li—H	Na—H	K—H	Rb—H	Cs—H
Bond energy	58.5	48.2	43.6	40.	41.9
$\tfrac{1}{2}\{D(M\text{—}M) + D(H\text{—}H)\}$	65.4	61.1	58.7	58.3	57.5
Δ	− 6.9	−12.9	−15.1	−18	−15.6
$\{D(M\text{—}M)D(H\text{—}H)\}^{1/2}$	52.6	43.4	37.1	36.0	33.4
Δ'	5.9	4.8	6.5	4	8.5

The Postulate of the Geometric Mean.—The alkali metals form double molecules, M_2, which are present in small concentrations in their vapors. The bonds in these molecules are covalent bonds formed by the valence electrons of the atoms; for example, the $2s$ electron of each lithium atom is used in bond formation in the molecule Li:Li. Because of the large spatial extension of the orbitals and the small binding energy of the valence electrons, the bonds in the alkali metal molecules are weak, with bond energies between 26.5 kcal/mole (in Li_2) and 10.7 (in Cs_2). The alkali metals also form hydride molecules, MH. In crystals of the alkali hydrides, which have the sodium chloride arrangement, hydrogen forms the anions, H^-, and the alkalis the cations. We might accordingly expect the bonds in the alkali hydride gas molecules to have some ionic character M^+H^-, leading to ionic resonance energy and positive values of Δ. It is seen from Table 3-3, however, that the values of Δ are negative.

This result shows that the postulate of the additivity of the energies of normal covalent bonds is not valid for these molecules. A quantum-mechanical treatment of one-electron bonds has been carried out[22] that leads to the conclusion that the postulate of additivity should be replaced by a similar postulate involving the geometric mean of the bond energies $D(A{-}A)$ and $D(B{-}B)$ (that is, the square root of their product) in place of the arithmetic mean. This *postulate of the geometric mean* states that the energy of a normal covalent bond between atoms A and B is equal to $\{D(A{-}A)D(B{-}B)\}^{1/2}$, and that in consequence the quantity Δ', defined as

$$\Delta' = D(A{-}B) - \{D(A{-}A)D(B{-}B)\}^{1/2}, \qquad (3\text{-}5)$$

should always be greater than or equal to zero. With the new postulate Δ' replaces Δ as the ionic resonance energy of the unsymmetrical bond.

If the bond energies $D(A{-}A)$ and $D(B{-}B)$ do not differ greatly in value there is only a small difference between their geometric and arithmetic means (which for 30 and 40, for example, are 34.6 and 35.0, respectively); and for this reason the arguments based on the earlier postulate are in general valid for the new one. For the alkali hydrides, however, the new postulate leads to results much different from those given by the earlier one, since the bond energy of the hydrogen molecule is very much greater than those of the alkali molecules, and the geometric and arithmetic means are correspondingly different. The values of Δ' given in Table 3-3 are seen to be positive, as required by the fundamental resonance theorem in case that the postulate of the geometric mean is valid.

It is probable that in general the postulate of the geometric mean leads to somewhat more satisfactory values for the energy of normal covalent bonds between unlike atoms than does the postulate of additivity. The postulate of the geometric mean is more difficult to apply than the postulate of additivity, however, since values of Δ can be obtained directly from heats of reaction, whereas knowledge of individual bond energies is needed for the calculation of values of Δ', and in the following sections of this chapter we shall sometimes use the postulate of additivity.

3-5. EMPIRICAL VALUES OF SINGLE-BOND ENERGIES

Empirical values of bond energies in diatomic molecules are given directly by the energies of dissociation into atoms, which may be determined by thermochemical or spectroscopic methods. In the case of a polyatomic molecule thermochemical data provide a value for the total energy of dissociation into atoms, that is, for the sum of the bond

[22] L. Pauling and J. Sherman, *J.A.C.S.* **59**, 1450 (1937).

energies in the molecule, but not for the individual bond energies. Thus from the enthalpy of formation of gaseous water from the elements (57.80 kcal/mole) and the enthalpies of dissociation of hydrogen and oxygen (104.18 and 118.32 kcal/mole, respectively), we find that the enthalpy of the reaction

$$2H + O \rightarrow H_2O(g)$$

is 221.14 kcal/mole.   This is the sum of the amounts of energy required to remove first one hydrogen atom from the water molecule, breaking one O—H bond, and then the second hydrogen atom, breaking the other O—H bond.   These two energy quantities are not equal, although they are not much different in value.   It is convenient for us to define their average, 110.6 kcal/mole, as the energy of the O—H bond in the water molecule.   In a similar way values can be obtained for bond energies in polyatomic molecules in which all the bonds are alike.   It is to be emphasized that each of these bond-energy values represents not the amount of energy required to break one bond in the molecule, but instead the average amount required to break all the bonds.[23]

Values for single-bond energies, defined in this way, for many bonds can be found by this process—for the S—S bond from the S_8 molecule (an eight-membered ring containing eight S—S bonds), for N—H, P—H, S—H, and so forth from NH_3, PH_3, H_2S, and so forth.   These values are given in Table 3-4.

There is no allotropic form of oxygen in which the atoms are connected by single O—O bonds. The value of the O—O bond energy given in the table has been obtained from the heat of formation of hydrogen peroxide, with use of the assumption that the H—O bond energy in H_2O_2 is the same as in H_2O. The calculation, typical of those used in evaluating bond energies from thermochemical information, is made in the following way: The enthalpy of formation of $H_2O_2(g)$ from the elements in the standard state is 31.83 kcal/mole.   By adding 104.2 and 118.3, for H_2 and O_2, we obtain 254.3 kcal/mole as the heat of formation of $H_2O_2(g)$ from the atoms 2H and 2O.   Subtraction of 221.1 for two O—H bonds leaves 33.2 kcal/mole as the energy of the O—O bond; this is the value given in the table.

The methods used in obtaining the remaining values in the table are described in the following paragraphs.

The thermochemical values used in this work have been taken for the most part from the compilation *Selected Values of Chemical Thermo-*

---

[23] A discussion of *bond dissociation energy*, the energy required to break one bond in a molecule, is given in App. XII.

TABLE 3-4.—ENERGY VALUES FOR SINGLE BONDS (KCAL/MOLE)[a]

Bond	Bond energy	Bond	Bond energy	Bond	Bond energy
H—H	104.2	P—H	76.4	Si—Cl	85.7
C—C	83.1	As—H	58.6	Si—Br	69.1
Si—Si	42.2	O—H	110.6	Si—I	50.9
Ge—Ge	37.6	S—H	81.1	Ge—Cl	97.5
Sn—Sn	34.2	Se—H	66.1	N—F	64.5
N—N	38.4	Te—H	57.5	N—Cl	47.7
P—P	51.3	H—F	134.6	P—Cl	79.1
As—As	32.1	H—Cl	103.2	P—Br	65.4
Sb—Sb	30.2	H—Br	87.5	P—I	51.4
Bi—Bi	25	H—I	71.4	As—F	111.3
O—O	33.2	C—Si	69.3	As—Cl	68.9
S—S	50.9	C—N	69.7	As—Br	56.5
Se—Se	44.0	C—O	84.0	As—I	41.6
Te—Te	33	C—S	62.0	O—F	44.2
F—F	36.6	C—F	105.4	O—Cl	48.5
Cl—Cl	58.0	C—Cl	78.5	S—Cl	59.7
Br—Br	46.1	C—Br	65.9	S—Br	50.7
I—I	36.1	C—I	57.4	Cl—F	60.6
C—H	98.8	Si—O	88.2	Br—Cl	52.3
Si—H	70.4	Si—S	54.2	I—Cl	50.3
N—H	93.4	Si—F	129.3	I—Br	42.5

[a] Bond-energy values for diatomic molecules of alkali metals and for alkali-metal hydrides are given in Table 3-3.

dynamic Properties, by F. D. Rossini, D. D. Wagman, W. H. Evans, S. Levine, and I. Jaffe (Circular of the National Bureau of Standards 500, Government Printing Office, Washington, D. C., 1952). The bond-energy values are so chosen that their sums represent the enthalpy changes $(-\Delta H)$ at 25°C accompanying the formation of molecules from atoms, all in the gas phase. There is no significance in the inclusion of the vibrational, rotational, and translational energy of the molecules and atoms in the bond energies; it is more convenient to do this than to correct the thermochemical values to 0°K, the information required for this correction being often not available, and there are no appreciable disadvantages involved in this procedure.

The values given in Table 3-5 were used for the enthalpies of the gases of atoms in their normal states (the reference states for the bond energies) relative to the standard states of the elements, to which the enthalpies of formation given in the Bureau of Standards compilation refer. Most of the values in Table 3-5 are taken from the Bureau of Standards compilation; an important exception is the value for nitrogen, which has been shown by recent spectroscopic and thermochemical

TABLE 3-5.—ENTHALPY (IN KCAL/MOLE) OF MONATOMIC GASES OF
ELEMENTS RELATIVE TO THEIR STANDARD STATES

H	52.09								
Li	37.07	C	171.70	N	113.0	O	59.16	F	18.3
Na	25.98	Si	88.04	P	75.18	S	53.25	Cl	29.01
K	21.51	Ge	78.44	As	60.64	Se	48.37	Br	26.71
Rb	20.51	Sn	72	Sb	60.8	Te	47.6	I	25.48
Cs	18.83	Pb	46.34	Bi	49.7				

studies to be the high value given in the table rather than the lower value given in the compilation.

In the discussion of heats of combustion the following values of the enthalpy of reaction at 25°C are useful:

$$H_2(g) + \tfrac{1}{2}O_2(g) \rightarrow H_2O(g) - \Delta H° = 57.7979 \text{ kcal/mole} \quad (3\text{-}6)$$

$$H_2(g) + \tfrac{1}{2}O_2(g) \rightarrow H_2O(l) - \Delta H° = 68.3174 \text{ kcal/mole} \quad (3\text{-}7)$$

$$C \text{ (graphite)} + O_2(g) \rightarrow CO_2(g) - \Delta H° = 94.0518 \text{ kcal/mole} \quad (3\text{-}8)$$

$$C \text{ (graphite)} + \tfrac{1}{2}O_2(g) \rightarrow CO(g) - \Delta H° = 26.4157 \text{ kcal/mole} \quad (3\text{-}9)$$

The bond-energy value given for each of the bonds H—H, F—F, Cl—Cl, Br—Br, I—I, H—F, H—Cl, H—Br, H—I, Cl—F, Br—Cl, I—Cl, and I—Br is the thermochemically or spectroscopically determined value of the enthalpy of dissociation of the corresponding diatomic molecule. The values for the bonds Si—Si and Ge—Ge are half the enthalpies of sublimation of the crystals, which have the diamond structure.

The original set of carbon bond-energy values[24] was referred to gaseous carbon atoms with enthalpy assumed to be 176 kcal/mole greater than that of graphite. In the first edition of this book a change was made to the value 124.3 kcal/mole, which seemed at that time to be the correct value.[25] Since then it has become clear[26] that the correct value of the heat of sublimation of graphite is close to 171.70 kcal/mole, and I have now revised the carbon bond-energy values accordingly; the new values differ only slightly from the original ones.

A fundamental assumption adopted in the formulation and use of the bond-energy values of Table 3-4 is that the energy of a molecule to

[24] Pauling. *loc. cit.* (20).

[25] G. Herzberg, *Chem. Revs.* **20**, 145 (1937).

[26] J. U. White, *J. Chem. Phys.* **8**, 459 (1940); E. C. Baughan, *Nature* **147**, 542 (1941); G. J. Kynch and W. G. Penney, *Proc. Roy. Soc. London* A179, 214 (1941); L. Brewer, P. W. Gilles, and F. A. Jenkins, *J. Chem. Phys.* **16**, 797 (1948); G. B. Kistiakowsky, H. T. Knight, and M. E. Malin, *ibid.*, **20**, 876 (1952); J. M. Hendrie, *ibid.* **22**, 1503 (1954); R. I. Reed and W. Snedden, *Trans Faraday Soc.* **54**, 301 (1958); and others.

which a single valence-bond structure can be confidently assigned can be approximated closely by the sum of constant terms corresponding to the bonds. This assumption is found to be justified empirically to a considerable extent, the enthalpies of formation calculated by summing the bond energies agreeing with the experimental values to within a few kcal/mole for nearly all molecules. As an example selected at random, the enthalpy of formation of $CH_2FCH_2OH(g)$ from elements in their standard states is 95.7 kcal/mole; this leads to 777.0 kcal/mole on addition of the suitable terms from Table 3-5 for the enthalpy of formation from monatomic gases. The sum of the bond energies for four C—H bonds, one C—F bond, one C—C bond, one C—O bond, and one O—H bond from Table 3-4 is 778.3 kcal/mole, the agreement in this case thus being excellent.

The bond-energy values are devised for use only with molecules containing atoms that show their normal covalences (four for carbon, three for nitrogen, etc.). An ammonium salt or a substance such as trimethylamine oxide cannot be treated in this way. The bond energies are also not expected to be valid for a molecule such as phosphorus pentachloride. It is interesting to point out that the enthalpy of the reaction $PCl_3(g) + 2Cl(g) \rightarrow PCl_5(g)$, 80.2 kcal/mole, corresponds to the formation of two new P—Cl bonds with apparent bond energy 40.1 kcal/mole, which is much less than the normal P—Cl bond energy, 79.1 kcal/mole.

Bond energies can be used in the discussion of the structure of molecules. For example, in 1932 it was pointed out[27] that the enthalpy of formation expected for ozone from molecular oxygen would be −77.9 kcal/mole (or less if a correction were made for strain in the three-membered ring), if the molecule had the structure

the observed value −34.0 kcal/mole differs from this by so great an amount as to permit this structure for ozone to be eliminated. A similar discrepancy between the value −103.8 calculated for the struc-

and the observed enthalpy of formation 0.16 kcal/mole

of the molecule O_4 eliminates this structure. Evidence is now available from spectroscopic, electron diffraction, and x-ray diffraction studies

[27] Pauling, *loc. cit.* (20).

also showing that these single-bonded structures are not correct for O_3 and O_4.

The use of bond energies in the discussion of molecules containing multiple bonds and of molecules that cannot be represented satisfactorily by one valence-bond structure will be presented in Chapters 6 and 8.

Values of the single-bond energies of elements (C—C, N—N, etc.) are represented graphically in Figure 3-6. The sequences Li, Na, K, Rb, Cs, and C, Si, Ge, Sn, show a reasonable decrease in bond-energy value with increase in atomic number. We might well expect that for similar bonds, as in these sequences of congeners, the bond energy would be greatest for the smallest atom and would decrease with in-

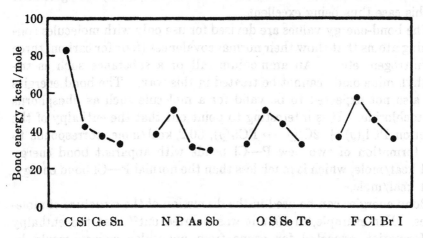

FIG 3-6 —The values of single-bond energies of elements.

crease in atomic size. The values for N, O, and F show a large deviation by comparison with the values for their congeners. An explanation of this peculiarity, proposed by Pitzer, will be discussed in the following chapter (Sec. 4-10).

3-6. THE ELECTRONEGATIVITY SCALE OF THE ELEMENTS

The Formulation of the Electronegativity Scale.—In Section 3-4 it was pointed out that the values of the difference between the energy $D(A—B)$ of the bond between two atoms A and B and the energy expected for a normal covalent bond, assumed to be the arithmetic mean or the geometric mean of the bond energies $D(A—A)$ and $D(B—B)$, increase as the two atoms A and B become more and more unlike with respect to the qualitative property that the chemist calls *electronegativity*, the power of an atom in a molecule to attract electrons to itself. Thus both Δ, the deviation from the arithmetic mean, and Δ', the deviation from the geometric mean, increase rapidly in the sequence HI,

HBr, HCl, HF, in which the halogen changes from iodine, which is recognized by its general chemical properties to be only a little more electronegative than hydrogen, to fluorine, the most electronegative of all the elements.

The property of the electronegativity of an atom in a molecule is different from the electrode potential of the element, which depends on the difference in free energy of the element in its standard state and in ionic solution, and it is different from the ionization potential of the atom, and from its electron affinity; although it is related to these properties in a general way.

It has been found possible to formulate an electronegativity scale of the elements by the analysis of the values of Δ or Δ' given by the single-bond energies. In Table 3-6 values of Δ' are listed for the bonds between nonmetallic atoms whose energies are given in Table 3-4. These are obtained in the same way as those in Table 3-3. It is seen on inspection that the values of Δ' do not satisfy an additivity relation; they cannot be represented as differences of terms characteristic of the two atoms in the bond. However, the square roots of the Δ' values do satisfy approximately a relation of this sort. In the table values of $0.18\sqrt{\Delta'}$ are given. These are the square roots of the Δ' values expressed in units equal to 30 kcal/mole, that is, $\sqrt{\Delta'/30}$. In the original formulation of the electronegativity scale[28] the electron volt, 23 kcal/mole, was used as the unit of energy, and the arithmetic mean was used. This procedure, which was followed also in the first and second editions of this book, leads to a convenient range of electronegativity values. The same values are obtained with the postulate of the geometric mean and the unit 30 kcal/mole.

The electronegativity values selected for the elements occurring in Table 3-6 are given in Table 3-7. Their differences are determined by the values of $0.18\sqrt{\Delta'}$ in Table 3-6. An additive constant has been so chosen as to give the first-row elements C to F the values 2.5 to 4.0.

In constructing Table 3-7 all of the available data were considered and the x value (x = electronegativity) that led to the best general agreement for each atom was selected. These values are given only to one decimal place on the scale; it is my opinion that this is the limit of their reliability.

The original table contained the electronegativity values given below immediately after the symbols of the corresponding elements:

H	0.00	2.05	2.1	Br	0.75	2.80	2.8
P	.10	2.15	2.1	Cl	.94	2.99	3.0
I	.40	2.45	2.5	N	.95	3.00	3.0
S	.43	2.48	2.5	O	1.40	3.45	3.5
C	.55	2.60	2.5	F	2.00	4.05	4.0

[28] Pauling, *loc. cit.* (20).

TABLE 3-6.—EXTRA IONIC ENERGY OF BONDS AND ELECTRONEGATIVITY
DIFFERENCES OF THE BONDED ATOMS

Bond	Δ'	$0.18\sqrt{\Delta'}$	$x_A - x_B$	Bond	Δ'	$0.18\sqrt{\Delta'}$	$x_A - x_B$
C—H	5.8	0.4	0.4	Si—S	7.8	0.6	0.7
Si—H	4.0	.4	.3	Si—F	90.0	1.7	2.2
N—H	30.1	1.0	.9	Si—Cl	36.2	1.1	1.2
P—H	3.3	0.3	.0	Si—Br	25.0	0.9	1.0
As—H	0.8	.2	.1	Si—I	11.8	.6	0.7
O—H	41.8	1.2	1.4	Ge—Cl	50.8	1.3	1.2
S—H	8.3	0.5	0.4	N—F	27.0	0.9	1.0
Se—H	−1.6	—	.3	N—Cl	0.5	.1	0.0
Te—H	−1.9	—	.0	P—Cl	24.5	.9	.9
H—F	72.9	1.5	1.9	P—Br	16.7	.7	.7
H—Cl	25.4	0.9	0.9	P—I	8.3	.5	.4
H—Br	18.2	0.8	.7	As—F	77.0	1.6	2.0
H—I	10.1	.6	.4	As—Cl	25.8	0.9	1.0
C—Si	10.0	.6	.7	As—Br	18.0	.8	0.8
C—N	13.2	.7	.5	As—I	7.5	.5	.5
C—O	31.5	1.0	1.0	O—F	9.3	.5	.5
C—S	−2.4	—	0.0	O—Cl	4.6	.4	.5
C—F	50.2	1.3	1.5	S—Cl	5.3	.4	.5
C—Cl	9.1	0.5	0.5	S—Br	2.2	.3	.3
C—Br	4.0	.4	.3	Cl—F	14.5	.7	1.0
C—I	2.6	.3	.0	Br—Cl	0.6	.1	0.2
Si—O	50.7	1.3	1.7	I—Cl	4.5	.4	.5
				I—Br	1.7	.3	.3

Following these there are given the values obtained by adding 2.05 to them—representing only a change in origin of the scale, from $x_H = 0.00$ to $x_H = 2.05$. It is seen that the new values are within 0.05 of the old except for carbon, which has been decreased by 0.10.

TABLE 3-7.—ELECTRONEGATIVITY VALUES FOR SOME ELEMENTS

H			
2.1			
C	N	O	F
2.5	3.0	3.5	4.0
Si	P	S	Cl
1.8	2.1	2.5	3.0
Ge	As	Se	Br
1.8	2.0	2.4	2.8
			I
			2.5

The differences in electronegativity of atoms for the bonds in Table 3-6 are given in the columns headed $x_A - x_B$. If the extra ionic energy $\Delta'(A\text{---}B)$ were given accurately by the equation

$$\Delta'(A\text{---}B) = 30(x_A - x_B)^2, \qquad (3\text{-}10)$$

and the bond energy (in kcal/mole) by the equation

$$D(A\text{---}B) = \{D(A\text{---}A) \cdot D(B\text{---}B)\}^{1/2} + 30(x_A - x_B)^2, \qquad (3\text{-}11)$$

the values in the two columns headed $0.18\sqrt{\Delta'}$ and $x_A - x_B$ would be equal. It is seen that this is approximately true, the average deviation between the two being 0.1 for 42 pairs.

Only three of the 45 bonds give negative values of Δ': Se—H, Te—H, and C—S. These negative values may be attributed in part to real deviations from our postulates, which are expected to have only approximate validity, and perhaps in part to small errors in the values of the bond energy.

The relation of the electronegativity values of Table 3-7 to the periodic system is the expected one. Fluorine and oxygen are by far the most electronegative of the atoms, with fluorine much more electronegative than oxygen. It is interesting that nitrogen and chlorine have the same electronegativity, as have also carbon, sulfur, and iodine. The contours of equal electronegativity cut diagonally across the periodic table, from the upper left- to the lower right-hand region.

3-7. HEATS OF FORMATION OF COMPOUNDS IN THEIR STANDARD STATES; THE COMPLETE ELECTRONEGATIVITY SCALE

The method just described for formulating the electronegativity scale cannot be used for the remaining elements in general because of lack of knowledge of enthalpies of formation of their compounds as gases and of the values of single-bond energies for the elements themselves. The following extension of the method can, however, be used.

Except for nitrogen and oxygen, which are discussed below, the elements in their standard states do not differ very much in energy from states involving normal single covalent bonds between the atoms. It is known that the standard states of bromine, iodine, sulfur, carbon (diamond), and many other nonmetallic elements are those in which the atoms are attached to adjacent atoms by single bonds. Moreover, the standard states of the metals too are probably not much different from states involving single bonds; there is a close resemblance in properties of the metallic bond and the covalent bond (Chap. 11).

Many elements in their standard states are, however, liquids or crystals, rather than gases, and many compounds in which we are interested are liquids or crystals. The energy of a liquid or a crystal may be described as involving not only the bond energies but also the energy of the van der Waals interaction of adjacent nonbonded atoms. As an approximation we may assume that the energy of the van der Waals stabilization of a substance in its standard state is approximately equal to the van der Waals stabilization of the elements from which the substance is formed, in their standard states, and that accordingly the enthalpy of formation referred to standard states is approximately equal to the enthalpy of formation of the gaseous compound from gaseous elements. Moreover, except with first-row atoms, the formation of stable double bonds and triple bonds is unusual, and we may assume with reasonable confidence that a substance of unknown bond type does not contain multiple bonds of sufficiently greater energy than the corresponding number of single bonds to introduce very great error in the electronegativity calculations.

In the absence of knowledge of the bond energies of the elements the method of the geometric mean cannot be applied. However, for most bonds there is not much difference between the geometric mean and the arithmetic mean—only in substances such as the alkali hydrides, where the bond energies H—H and M—M are very different, do these two mean values differ greatly. We shall accordingly use the arithmetic mean, and assume that the bond energy $D(A—B)$ is given by the equation

$$D(A—B) = \tfrac{1}{2}\{D(A—A) + D(B—B)\} + 23(x_A - x_B)^2 \quad (3\text{-}12)$$

The contribution of this bond to the heat of formation of the substance is accordingly equal to $23(x_A - x_B)^2$; and, except for the corrections for nitrogen and oxygen that we shall now discuss, the heat of formation would be obtained by summing this expression over all the bonds in the molecule.

The standard state $N_2(g)$ for nitrogen is far more stable than it would be if the molecule involved single N—N bonds. From the bond-energy value 38.4 kcal/mole for N—N and the value $2N \rightarrow N_2 + 226.0$ kcal /mole, we see that this extra stability of the standard state amounts to 110.8 kcal/mole for N_2, or 55.4 kcal/mole per nitrogen atom. Similarly, the values 33.2 kcal/mole for O—O and $2O \rightarrow O_2 + 118.3$ kcal /mole lead to an extra stability of 52.0 kcal/mole for O_2 in its standard state, or 26.0 per oxygen atom. These correction terms for nitrogen and oxygen are due to the fact that N_2 contains a triple bond that is much more stable than three single bonds, and O_2 a bond of special character (Chap. 10) that is more stable than two single bonds. Accordingly the enthalpy of formation of a substance in its standard state can be calculated approximately by use of the expression

$$Q = 23 \sum (x_A - x_B)^2 - 55.4 n_N - 26.0 n_O \quad (3\text{-}13)$$

in which n_N is the number of nitrogen atoms in the molecule and n_O the number of oxygen atoms, and the indicated summation is to be carried over all the bonds in the molecule; the value of Q is given in kcal/mole. The equation does not apply to substances containing double or triple bonds.

It is the unusual stability of multiple bonds for oxygen and nitrogen, stabilizing their normal states, that often leads to negative values of the enthalpy of formation of substances. The enthalpy of formation of a molecule containing an atom of nitrogen held by single bonds to other atoms with the same electronegativity should be about -55.4 kcal/mole; the compound would accordingly be very unstable relative to the elements. Nitrogen trichloride is such a compound; in the molecule of this substance the bonds are normal covalent bonds, similar to N—N and Cl—Cl single bonds; it is not the weakness of the N—Cl bonds, but rather the extraordinary strength of the triple bond in N_2 that makes nitrogen trichloride unstable. Its measured enthalpy of formation, in solution in carbon tetrachloride, is -54.7 kcal/mole, in close agreement with the expected value. In nitrogen trifluoride the ionic resonance energy of the N—F bonds is great enough to overcome this handicap and to give the molecule NF_3 a positive enthalpy of formation (27.2 kcal/mole). For OF_2 and Cl_2O (with $x_A - x_B = 0.5$) the ionic character is not enough to counteract the term -26.0 kcal/mole for the oxygen atom; these substances have negative enthalpies of formation, whereas the enthalpies of formation of other normal oxides are positive.

By the use of Equation 3-13 the difference in electronegativity of two elements can be calculated from the enthalpy of formation of the compounds formed by them, and in this way, through study of the compounds of the element with elements with electronegativity values given in Table 3-8, the electronegativity of the element can be evalu-

TABLE 3-8.—THE COMPLETE ELECTRONEGATIVITY SCALE[a]

Li 1.0	Be 1.5	B 2.0											C 2.5	N 3.0	O 3.5	F 4.0
Na 0.9	Mg 1.2	Al 1.5											Si 1.8	P 2.1	S 2.5	Cl 3.0
K 0.8	Ca 1.0	Sc 1.3	Ti 1.5	V 1.6	Cr 1.6	Mn 1.5	Fe 1.8	Co 1.8	Ni 1.8	Cu 1.9	Zn 1.6	Ga 1.6	Ge 1.8	As 2.0	Se 2.4	Br 2.8
Rb 0.8	Sr 1.0	Y 1.2	Zr 1.4	Nb 1.6	Mo 1.8	Tc 1.9	Ru 2.2	Rh 2.2	Pd 2.2	Ag 1.9	Cd 1.7	In 1.7	Sn 1.8	Sb 1.9	Te 2.1	I 2.5
Cs 0.7	Ba 0.9	La-Lu 1.1-1.2	Hf 1.3	Ta 1.5	W 1.7	Re 1.9	Os 2.2	Ir 2.2	Pt 2.2	Au 2.4	Hg 1.9	Tl 1.8	Pb 1.8	Bi 1.9	Po 2.0	At 2.2
Fr 0.7	Ra 0.9	Ac 1.1	Th 1.3	Pa 1.5	U 1.7	Np-No 1.3										

[a] The values given in the table refer to the common oxidation states of the elements. For some elements variation of the electronegativity with oxidation number is observed; for example, Fe^{II} 1.8, Fe^{III} 1.9; Cu^{I} 1.9, Cu^{II} 2.0; Sn^{II} 1.8, Sn^{IV} 1.9. For other elements see W. Gordy and W. J. O. Thomas, *J. Chem. Phys.* **24**, 439 (1956).

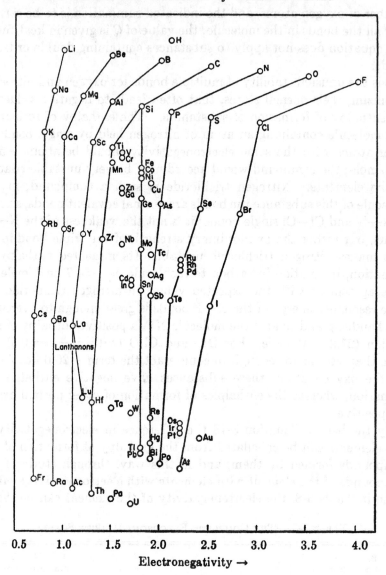

Fɪɢ. 3-7.—Electronegativity values of the elements.

ated. For example, the enthalpies of formation of $BeCl_2$, $BeBr_2$, BeI_2, and BeS from the elements in their standard states are 122.3, 88.4, 50.6, and 55.9 kcal/mole, respectively. These values lead to 1.56, 1.33, 1.03, and 1.06 for the electronegativity differences, and hence to 1.44, 1.47, 1.47, and 1.44 for the electronegativity of beryllium. The value 1.5 has been accepted for the element. The values given in Table 3-8 (except those in Table 3-7) have been obtained in this way.[29]

[29] Values agreeing closely with those in Table 3-8 have been reported also by

The way in which the electronegativities are related to the periodic table is shown in Figure 3-7. In the first short period, Li to F, the electronegativity values differ by a constant amount, 0.5. In the following periods the differences are seen to be smaller for the metals than for the nonmetals, and the values change less for a given column of the periodic table in the metallic region than in the nonmetallic region.

The electronegativity scale brings a certain amount of systematization into the field of inorganic thermochemistry, in which on first survey little order can be detected. It is possible to calculate rough values expected for enthalpies of formation of compounds by the use of the electronegativity values of Table 3-8, which vary in a regular way from element to element in the periodic system, the enthalpies of reaction of the elements to form compounds being attributed in the main to the extra resonance energy that results from the partial ionic character of the bonds between unlike atoms and that increases as the atoms become more and more unlike. This order is brought out of apparent lack of order in the thermochemical values largely through the corrections for the two elements nitrogen and oxygen, which are alone among the elements in having their standard states much different in stability from single-bonded states.

3-8. RELATION TO OTHER PROPERTIES

The property of electronegativity that we have been discussing represents the attraction of a neutral atom in a stable molecule for electrons. The first ionization energy of an atom, the energy of the reaction $X^+ + e^- \rightarrow X$, may be considered as the average of the electron attraction of the atom and the positive ion, and the electron affinity, the energy of the reaction $X + e^- \rightarrow X^-$, may be thought of similarly as the average of the electron attraction of the atom and the negative ion. It was pointed out by Mulliken[30] that the average of the first ionization energy and the electron affinity of an atom should be a measure of the electron attraction of the neutral atom and hence of its electronegativity. For multivalent atoms the significance of these energy quantities

---

M. Haissinsky, *J. phys. radium* **7**, 7 (1946); . HA. Skinner, *Trans. Faraday Soc.* **41**, 645 (1945); W. Gordy, *J. Chem. Phys.* **14**, 305 (1946); W. Gordy, *Phys. Rev.* **69**, 604 (1946); K. S. Pitzer, *J.A.C.S.* **70**, 2140 (1948); M. L. Huggins, *ibid.* **75**, 4123 (1953); Gordy and Thomas, *loc. cit* (T3-8). Huggins' paper contains a detailed discussion of bond-energy values in relation to electronegativity differences for the nonmetallic elements. In a second paper, *J.A.C.S.* **75**, 4126 (1953), he gives a detailed discussion of the relation between bond energy and interatomic distance. A survey of electronegativity values suggested before 1956 is contained in the paper by Gordy and Thomas.

[30] R. S. Mulliken, *J. Chem. Phys.* **2**, 782 (1934); **3**, 573 (1935).

is complicated by the nature of the states of the atoms and ions, and corrections must be made that need not be discussed here. For univalent atoms (hydrogen, the halogens, the alkali metals) the treatment is straightforward. The values of the energy quantities concerned are given in Table 3-9, it being assumed that the electron affinity of the alkali metals is zero.[31] It is seen that the values of x are closely proportional to those of the sum of the two energy quantities except for hydrogen, which, with its unique electronic structure, might be expected to misbehave. This comparison and others were used in fixing the origin for the electronegativity scale.

TABLE 3-9.—COMPARISON OF ELECTRONEGATIVITY WITH AVERAGE OF IONIZATION ENERGY AND ELECTRON AFFINITY[a]

	Ionization energy	Electron affinity	Sum/125	x
F	403.3	83.5	3.90	4.0
Cl	300.3	87.3	3.10	3.0
Br	274.6	82.0	2.86	2.8
I	242.2	75.7	2.54	2.5
H	315.0	17.8	2.66	2.1
Li	125.8	0	1.01	1.0
Na	120.0	0	.96	0.9
K	101.6	0	.81	.8
Rb	97.8	0	.78	.8
Cs	91.3	0	.73	.7

[a] All values are for $-\Delta H°$ at 25°C.

Another property that might be expected to be closely related to the electronegativity is the work function of metals—the amount of energy required to remove an electron from the metal, as given by the limiting frequency of light in the photoelectric effect or by the energy term in the Boltzmann factor of the theoretical expression for thermionic emission. It has been pointed out by Gordy and Thomas[32] that there is a reasonably good linear correlation between the work function and the electronegativity of the element, the work function, ψ, (in electron volts) being given by the equation

$$\psi = 2.27x + 0.34 \qquad (3\text{-}14)$$

The most obvious correlation of the electronegativity scale with the general chemical properties of the elements bears on their division into metals and nonmetals. It is seen that the value $x = 2$ represents ap-

[31] References for the values of the electron affinity are given in Chap. 13.
[32] Gordy and Thomas, *loc. cit.* (T3-8).

proximately the point of separation, the metals being elements with smaller and the nonmetals those with larger electronegativity than 2.

Nuclear magnetic resonance studies have shown that the amount of energy required to change the magnetic moment of a nucleus from one orientation to another depends to some extent upon the nature of the bonds formed by the atom; the change is described as due to diamagnetic shielding of the nuclear magnetic moment by the electrons that are close to the nucleus. Gutowsky and collaborators[33] have correlated the nuclear magnetic shielding of the fluorine nucleus in fluorine compounds with the electronegativity of the atom to which the fluorine atom is bonded. The relation between electronegativity and diamagnetic shielding of the proton in hydrogen compounds has also been discussed by Gutowsky and collaborators.[34] A similar effect of shielding of the magnetic moment of the proton in substituted ethanes has been shown by Shoolery[35] to be related to the electronegativity of the atoms of the groups attached to the carbon atoms. Nuclear spin coupling constants in relation to partial ionic character of bonds and hybrid character of bond orbitals has been discussed by Karplus and Grant.[35a]

Another effect that has been correlated with electronegativity is the interaction of the electric quadrupole moment of a nucleus and the electric quadrupole field produced by the electrons in the neighborhood of the nucleus. It was pointed out by Townes and Dailey[36] and by Gordy[37] that the quadrupole interaction energy is dependent upon the partial ionic character of the bonds, as determined by the difference in electronegativity of the bonded atoms, and a detailed discussion of the correlation has been given by Dailey and Townes.[38] These considerations will be discussed briefly in the following section, in which we take up the question of the relation between partial ionic character of a bond and the electronegativity difference of the bonded atoms.

3-9. THE ELECTRONEGATIVITY OF ATOMS AND THE PARTIAL IONIC CHARACTER OF BONDS

It would be convenient in discussing bonds to be able to make quantitative statements about their nature—to say that certain bonds are essentially covalent, with only 5 percent or 10 percent of ionic char-

[33] H. S. Gutowsky and C. J. Hoffman, *J. Chem. Phys.* **19**, 1259 (1951); H. S. Gutowsky, D. W. McCall, B. R. McGarvey, and L. H. Meyer, *ibid.* **19**, 1328; A. Saika and C. P. Slichter, *ibid.* **22**, 26 (1954).

[34] H. S Gutowsky, *J. Chem. Phys.* **19**, 1266 (1951); L. H. Meyer, A. Saika, and H. S. Gutowsky, *J.A.C.S.* **75**, 4567 (1953).

[35] J. N. Shoolery, *J. Chem. Phys.* **21**, 1899 (1953).

[35a] M. Karplus and D. M. Grant, *Proc. Nat. Ac. Sci. U. S.* **45**, 1269 (1959).

[36] C. H. Townes and B. P. Dailey, *Phys. Rev.* **78**, 346A (1950).

[37] W. Gordy, *J. Chem. Phys.* **19**, 792 (1951).

[38] B. P. Dailey and C. H. Townes, *J. Chem. Phys.* **23**, 118 (1955).

acter, that others are about equally ionic and covalent, and that others are essentially ionic. It is difficult to formulate a reliable relation between the partial ionic character of a bond and the difference in electronegativity of the two atoms between which the bond is formed, or the extra ionic resonance energy of the bond. The difficulty arises primarily from the fact that the description of a bond as a hybrid between a normal covalent bond and an extreme ionic bond is only a rough approximation. We cannot hope to formulate an expression for the partial ionic character of bonds that will be accurate.

In the first edition of this book the following equation was proposed for the amount of ionic character of the single bond between atoms A and B, with electronegativities x_A and x_B:

$$\text{Amount of ionic character} = 1 - e^{-1/4(x_A - x_B)} \qquad (3\text{-}15)$$

This curve corresponds to the amounts 4, 11, 19, and 60 percent of ionic character for HI, HBr, HCl, and HF, respectively. The values for the first three of these hydrogen halides are closely equal to those indicated by the electric dipole moments of the molecules, as given in Table 3-1. At the time when the equation was formulated the value of the dipole moment of HF was not known, and an estimate of 60 percent was made for the partial ionic character in this molecule. As shown in Table 3-1, the dipole moment of HF corresponds to only 45 percent partial ionic character.

Equation 3-15 leads to the amounts of ionic character for various values of the electronegativity difference given in Table 3-10. The function is shown as the curve in Figure 3-8, together with the experimental values of the ratio of the observed electric dipole moment to the product of the electronic charge and internuclear distance for a number of diatomic molecules composed of univalent elements. Points are

TABLE 3-10.—RELATION BETWEEN ELECTRONEGATIVITY DIFFERENCE AND AMOUNT OF PARTIAL IONIC CHARACTER OF SINGLE BONDS

$x_A - x_B$	Amount of ionic character	$x_A - x_B$	Amount of ionic character
0.2	1 percent	1.8	55 percent
.4	4	2.0	63
.6	9	2.2	70
.8	15	2.4	76
1.0	22	2.6	82
1.2	30	2.8	86
1.4	39	3.0	89
1.6	47	3.2	92

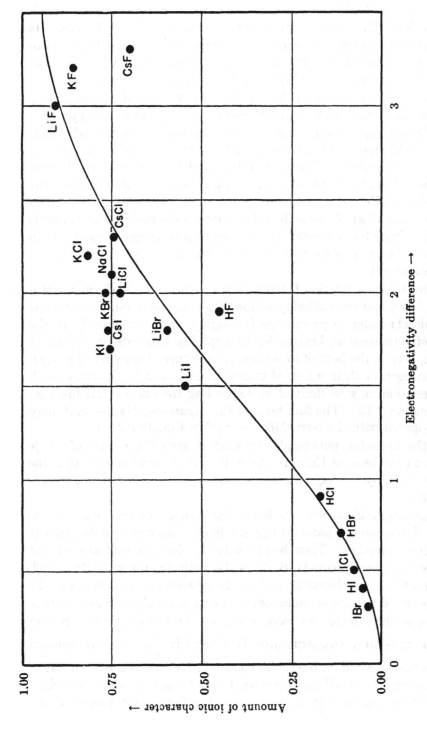

Fig. 3-8.—Curve relating the amount of ionic character of a bond to the electronegativity difference of the two atoms. Experimental points, based upon observed values of the electric dipole moment of diatomic molecules, are shown for 18 bonds.

shown for the hydrogen halogenides, halogen halogenides, and alkali halogenides. The experimental values for the electric dipole moments for the alkali halogenides are those determined by methods of microwave spectroscopy or molecular-beam spectroscopy. It is seen that the curve agrees only roughly with the experimental points.

Efforts have also been made to determine the partial ionic character of bonds from the experimental values of the interaction energy of the electric quadrupole moment of an atomic nucleus and the electric quadrupole field produced in the neighborhood of the nucleus by the electrons in the molecule. The interpretation of the observed quadrupole coupling constants is not straightforward, however, because both the partial ionic character of the bond and the hybridization of the bond orbitals (see Chap. 4) must be taken into consideration. The amounts of ionic character suggested by the quadrupole coupling constants in diatomic molecules are somewhat larger than those given by the electric dipole moments.[39]

There is also a relation between the amount of ionic character of a single bond and the enthalpy of formation of the bond. The amount of ionic character in percentage is roughly equal numerically to the heat of formation in kcal/mole. In applying this rule one must, of course, correct the heat of formation for the special stability of oxygen and nitrogen in their standard states, as expressed in Equation 3-13. This relation may be derived by expanding the exponential function in Equation 3-15. The first term in the expansion, $\frac{1}{4}(x_A - x_B)^2$, may be compared with the term $23(x_A - x_B)^2$ of Equation 3-13.

In the following paragraphs we shall discuss the nature of single bonds on the basis of Equation 3-15; it must be remembered that the values given by this equation, and shown in Table 3-10, are only approximate.

According to Equation 3-15 bonds between atoms with electronegativity difference 1.7 have 50 percent ionic character and 50 percent covalent character. Thus bonds between fluorine and any of the metals or of the elements H, B, P, As, Te, with electronegativity near 2, are largely ionic in character, and bonds between oxygen and any of the metals are 50 percent or more ionic. For a molecule such as HF, containing a single bond, we have discussed the bond type in terms of resonance between two structures, H^+F^- and $H:\overset{..}{\underset{..}{F}}:$. A more complex discussion is needed for molecules containing several bonds. In water, for example, the O—H bond is predicted to have 39 percent ionic character, corresponding to $x_O - x_H = 1.4$. The water molecule can

---

[39] For a discussion of this question see Dailey and Townes, *loc. cit.* (38). A summary is also given by Townes and Schawlow, *op. cit.* (7).

therefore be described as resonating among four electronic structures, one completely covalent, two with one bond to hydrogen ionic and one bond covalent, and one with both bonds ionic. If the bonds were independent of one another, the structures would make contributions of 37 percent, 24 percent, 24 percent, and 15 percent, respectively (note that the sum of 24 percent and 15 percent, corresponding to the two structures for which one of the bonds is ionic, is 39 percent, the expected amount of ionic character for the bond). It is probable, however, that the electrostatic interactions in the completely ionic structure, involving a doubly charged oxygen ion, cut down its contribution somewhat, and that the contributions of the completely covalent structure and each of the half-and-half structures are increased correspondingly. If the doubly ionic structure makes no contribution and the ratio of the others remains the same, their contributions become 44 percent, 28 percent, and 28 percent, respectively.

We may calculate the electric dipole moment expected for the water molecule with inclusion of the doubly ionic structure and with neglect of it. By using the O—H interatomic distance 0.965 Å and H—O—H bond angle 104.5°, we find that the calculated values of the dipole moment are 2.21 D with inclusion of the doubly ionic structure and 1.59 D with its neglect. The experimental value, 1.86 D, lies between these values; it corresponds to inclusion of the doubly ionic structure to such an extent as to make the partial ionic character of each bond equal to 33 percent rather than 39 percent.

It must be remembered that calculations such as the foregoing have only approximate significance and that there is no need to try to make them with precision.

A similar description involving resonance among many electronic structures is to be used for other molecules containing more than one bond of intermediate type. Thus for the ammonium ion, $[NH_4]^+$, we consider 16 structures: one completely covalent structure, four structures with one bond ionic, six with two bonds ionic, four with three bonds ionic, and one completely ionic. In the following discussion we shall in general not mention explicitly the resonance of the molecule among these structures; but it is to be borne in mind that the amounts of ionic character of bonds in molecules are to be interpreted in this way. In the following chapters of the book single bonds will as a rule be represented by the symbol A—B; this symbol represents covalent-ionic resonance, and the structural formula

$$\left[\begin{array}{c} H \\ | \\ H—N—H \\ | \\ H \end{array} \right]^+$$

for the ammonium ion comprises within itself all the 16 structures mentioned above, with contributions such as to give each N—H bond a suitable amount of ionic character. From the difference in electronegativity of nitrogen and hydrogen, 0.9, we estimate about 18 percent partial ionic character for each bond, corresponding to a positive charge of about 0.18 on each hydrogen atom and about 0.28 on the nitrogen atom. We accordingly conclude that the unit positive charge of the ammonium ion is divided about equally among all five atoms.

We may now summarize the conclusions that can be drawn about the partial ionic character of the single bonds formed by the elements.

The Alkali Metals.—Bonds of the alkali metals with all nonmetals are essentially ionic (with more than 50 percent ionic character—electronegativity difference greater than 1.7) except for Li—I, Li—C, and Li—S, with about 43 percent ionic character.

The Alkaline-Earth Metals.—Magnesium, calcium, strontium, and barium form essentially ionic bonds with the more nonmetallic elements. Beryllium bonds have the following amounts of ionic character: Be—F, 79 percent; Be—O, 63 percent; Be—Cl, 44 percent; Be—Br, 35 percent; Be—I, 22 percent.

The Third-Group Elements.—The B—F bond has about 63 percent ionic character, B—O 44 percent, B—Cl 22 percent, and so forth. Boron forms normal covalent bonds with hydrogen. The aluminum bonds are similar to those of beryllium in ionic character.

The Fourth-Group Elements.—The C—F bond, with 44 percent ionic character, is the most ionic of the bonds of carbon with nonmetallic elements. The Si—F bond has 70 percent ionic character, and Si—Cl 30 percent. The Si—O bond is of especial interest because of its importance in the silicates. It is seen to have 50 percent ionic character, the value of $x_0 - x_{Si}$ being 1.7.

The Remaining Nonmetallic Elements.—The bonds formed by fluorine with all of the metals are essentially ionic in character, and those with the intermediate elements (H, B, P, etc.) have a little more than 50 percent ionic character. The C—F, S—F, and I—F bonds are expected to have 44 percent ionic character. In CF_4, SF_6, IF_5, and IF_7 the amounts of ionic character of the bonds are probably somewhat less than this value because of the transfer of positive charge to the central atom, which increases its x value and decreases the ionic character of the bonds.

The bonds of oxygen with all metals are largely ionic.

Since the nonmetallic elements in each row of the periodic table are separated by intervals of 0.5, the bonds formed by a nonmetallic atom with immediate neighbors in the same row have 6 percent ionic character and those with its neighbors once removed 22 percent.

3-10. THE ENTHALPY CHANGE IN ORGANIC REARRANGEMENTS AND THE ELECTRONEGATIVITY SCALE

By the use of the electronegativity scale and Equation 3-12 it is possible to predict rough values for the enthalpy change of reactions in which only single bonds are broken and formed. It is thus possible to use the electronegativity scale in a simple way in discussing the exothermic or endothermic character of some organic reactions, especially organic rearrangements, as illustrated by the examples given in the following paragraphs.[40]

In the preceding sections of this chapter we have discussed simple reactions such as

$$H_2 + Cl_2 \rightarrow 2HCl$$

This reaction is exothermic to the extent of 44.12 kcal/mole. The enthalpy of reaction can be discussed in a rough way with the use of Equation 3-12 and the differences in electronegativity of the atoms H and Cl. The enthalpy of formation of HCl is predicted by the equation to be $23 (3.0 - 2.1)^2$ kcal/mole, which is 19 kcal/mole, approximately equal to the observed value, 22 kcal/mole.

We may represent this reaction by a simple diagram. The symbols of the elements involved are placed in a horizontal line, at loci representing their electronegativity values. Above the symbols arcs are drawn to indicate the single bonds in the reactants, and below the symbols arcs are drawn to represent the single bonds in the product:

We see from this diagram that for the reaction of hydrogen and chlorine to form hydrogen chloride the two spans of zero length (for the two bonds H—H and Cl—Cl) are converted into two longer spans (for the two bonds H—Cl). (The word span is used to represent the difference in electronegativity of the two atoms.) Application of Equation 3-12 shows that such a reaction is predicted to be exothermic.

In general the electronegativity bond diagrams indicate exothermicity in the direction represented by the longest span. The reason for this is that the extra ionic resonance energy of the bond is proportional to the square of the span, and the square of the longest span is usually great enough to be determinative. This principle can be applied in the discussion of molecular rearrangement involving only single bonds and

[40] L. Pauling, *Biochemistry of Nitrogen: A Collection of Papers on Biochemistry of Nitrogen and Related Subjects Dedicated to Artturi Ilmari Virtanen,* Suomalainen Tiedeakatemia, Helsinki, 1955, pp. 428–432.

having such a nature that resonance energy (Chap. 7) is not significantly changed.

As a first example we may take the nitrosamine rearrangement, as
given by the equation

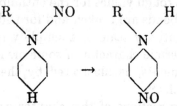

The diagram for this rearrangement is H⌒C⌒N, and the predicted enthalpy of rearrangement is 21 kcal/mole.   The same diagram
and the same predicted enthalpy of rearrangement apply to the rearrangement of nitroamines to nitranilines, such as

In this rearrangement there is some change in resonance energy, but
it is predicted to be small compared with the enthalpy change due to
change in the nature of the single bond.

Another example is the Hoffmann rearrangement of alkylamino-
benzenes to *p*-alkylanilines:

The diagram for this rearrangement is H⌒C⌣N, and the predicted enthalpy of rearrangement is 9 kcal/mole.

The rearrangement of phenylallylether according to the reaction

is represented by the diagram H⌒C⌣O, and the predicted enthalpy of rearrangement is 18 kcal/mole.

As a last example we may mention the rearrangement of an arylhydroxylamine, such as N-phenylhydroxylamine, which when heated with sulfuric acid is converted into *p*-aminophenol:

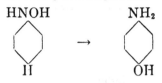

The diagram for this rearrangement is H   C   N   O, and the

predicted enthalpy of rearrangement is 32 kcal/mole.

It must be kept in mind that the equilibrium constant for a reaction is determined by the free-energy change accompanying the reaction, and not just by the enthalpy change. However, for similar reactions the entropy change may often be considered to be essentially the same, permitting a comparison to be made in terms of the enthalpy change.

3-11. THE CORRELATION OF COLOR AND BOND CHARACTER

Often the color of compound is different from that of the ions into which it might be dissociated; thus lead iodide is yellow, although both plumbous ion and iodide ion are colorless. In 1918 Bichowsky,[41] in a paper dealing with "valence colors" of atoms, suggested that this change in color is the result of the sharing of electrons between bonded atoms, and this idea has been extended by Pitzer and Hildebrand,[42] who proposed the postulate that the extent of deviation of the color of a compound from that of the ions into which it might dissociate may be taken as a measure of the deviation of the bonds from pure ionic bonds.

The color of a substance is determined by its absorption spectrum. The colors associated with light of different wavelengths in the visible region of the spectrum are given in Table 3-11, together with their complementary colors. All colorless ions have absorption bands in the ultraviolet; if through a perturbing influence (the formation of a bond with increasingly great covalent character) a single absorption band of an ion were increased in wavelength so as to pass through the visible spectrum, the color of the ion by transmitted light would go through the sequence lemon yellow, yellow, orange, red, purple, and so on. If Pitzer and Hildebrand's postulate is valid this sequence of colors may thus be used as a measure of the amounts of covalent character of compounds of atoms with colorless ions.

(The color of a substance may of course be due to several absorption bands; in particular, the color green results from absorption in both the red and the blue spectral regions.)

[41] F. R. Bichowsky, *J.A.C.S.* **40**, 500 (1918).
[42] K. S. Pitzer and J. H. Hildebrand, *J.A.C.S.* **63**, 2472 (1941).

TABLE 3-11.—SPECTRAL COLORS AND COMPLEMENTARY COLORS

Wavelength	Spectral colors	Their complementary colors
3900 Å		
4000		
4100	Violet	Lemon yellow
4200		
4300	Indigo	Yellow
4400		
4500		
4600		
4700		
4800	Blue	Orange
4900		
5000	Blue-green	Red
5100		
5200		
5300	Green	Purple
5400		
5500		
5600	Lemon yellow	Violet
5700		
5800	Yellow	Indigo
5900		
6000		
6100	Orange	Blue
6200		
6300		
6400		
6500		
6600		
6700		
6800	Red	Blue-green
6900		
7000		
7100		
7200		
7300		
7400		
7500		

Table 3-12, which is similar to a table presented by Pitzer and Hildebrand, gives information about compounds of sulfur and halogens with atoms that are known to form colorless ions or to form with fluorine analogous compounds that are colorless. The numbers in the table are the enthalpies of formation per M—X bond in kcal/mole, these being also, as mentioned earlier in this chapter, approximately equal to the percentages of ionic character of the bonds. There is seen

TABLE 3-12.—COLOR OF SUBSTANCES IN RELATION TO COVALENT CHARACTER OF
BONDS AS SHOWN BY ENTHALPY OF FORMATION (IN KCAL/MOLE)
(Compounds are colorless if color is not given)

Electronegativity→		3.0	2.8	2.5	2.5
↓		Cl	Br	I	S
0.9	Na^I	98	86	69	45
1.2	Mg^{II}	77	62	43	42
1.5	Al^{III}	55	42	25	20
1.6	Zn^{II}	50	39	25	24
1.7	Cd^{II}	47	38	24	17
					Yellow
1.8	Sn^{II}	41	31	19	6
				Yellow	Brown
1.8	Pb^{II}	43	33	20	11
				Yellow	Black
1.9	Ag^I	30	24	15	4
			Light yellow	Yellow	Black
1.9	Sb^{III}	30	21	8	7
			Yellow	Red	Orange, black
2.0	As^{III}	27	16	5	6
			Yellow	Red	Red, yellow
2.2	Pt^{IV}	16	10	5	7
		Red	Brown	Brown	Black

to be a close correlation between the amount of ionic character of the bonds and the color of the substances; with few exceptions the colorless substances are those with more than 20 percent ionic character; and, moreover, the color deepens with decreasing ionic character, passing through yellow (20 to 10 percent) and orange to red and black.

Thus far we have been discussing the color of electronegative atoms, with absorption bands that shift from the ultraviolet into the visible for increasing amounts of covalent-bond character. For electropositive atoms, for which the process of covalent-bond formation involves gaining electrons from a donor rather than losing them, the opposite effect is observed: with increase in covalent-bond character the absorption bands shift toward the violet. The colorless cupric ion, with an absorption band in the infrared, becomes blue on hydration to $[Cu(H_2O)]_4^{++}$; compare the colors of anhydrous $CuSO_4$ (colorless) and $CuSO_4 \cdot 5H_2O$ or a cupric ion solution (blue). The color deepens further on formation of the still more covalent complex $[Cu(NH_3)_4]^{++}$ (deep blue). Similarly, the yellow nickelous ion (with a band in the violet as well as the far red) becomes green on hydration, blue on formation of the complex $[Ni(NH_3)_4(H_2O)_2]^{++}$, and violet as this changes to $[Ni(NH_3)_6]^{++}$ in 15 N ammonium hydroxide solution.

CHAPTER 4

The Directed Covalent Bond;

Bond Strengths and Bond Angles[1]

MUCH progress has been made in the development of a detailed understanding of the nature of covalent bonds through the consideration of the atomic orbitals (bond orbitals) that can be used as the basis of the quantum-mechanical treatment of the bonds.

4-1. THE NATURE AND BOND-FORMING POWER OF ATOMIC ORBITALS

The energy of a covalent bond is largely the energy of resonance of two electrons between two atoms (Sec. 1-5). Examination of the form of the resonance integral shows that the resonance energy increases in magnitude with increase in the *overlapping* of the two atomic orbitals involved in the formation of the bond, the word "overlapping" signifying the extent to which the regions in space in which the two orbital wave functions have large values coincide. (Since the square of an orbital wave function is the probability distribution function for the electron, the overlapping is essentially a measure of the amount of interpenetration of the bond-electron distributions of the two atoms.) Consequently it is expected that *of two orbitals in an atom the one that can overlap more with an orbital of another atom will form the stronger bond with that atom, and, moreover, the bond formed by a given orbital will tend to lie in that direction in which the orbital is concentrated.*

The orbitals of an atom differ from one another in their dependence

---

[1] The argument of this and the following chapter is taken for the most part from my paper "The Nature of the Chemical Bond: Application of Results Obtained from the Quantum Mechanics and from a Theory of Paramagnetic Susceptibility to the Structure of Molecules," *J.A.C.S.* **53**, 1367 (1931), and the preliminary communication, *Proc. Nat. Acad. Sci. U S.* **14**, 359 (1928).

on the distance r of the electron from the nucleus and on the polar angles θ and ϕ, that is, on their angular distribution. The dependence on r has been discussed for the hydrogen atom in Section 1-4. It is this dependence that in the main determines the *stability* of the atomic orbital, and the primary significance of the orbital for bond formation

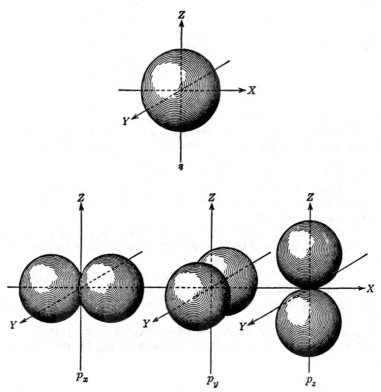

FIG. 4-1.—Representations of the relative magnitudes of sp orbitals in dependence on angle.

can be discussed in terms of stability. Stable bonds are formed only with use of stable atomic orbitals—the $1s$ orbital for hydrogen, the $2s$ and $2p$ orbitals for the first-row atoms, and so on.

The different stable orbitals of an atom which can be used for bond formation do not differ very much from one another in their dependence on r, but they may show a great difference in their angular distribution. This can be seen from Figure 4-1, representing the angular distribution of an s orbital and the three p orbitals.[2] The s orbital is spherically symmetrical, and hence can form a bond in one direction as well as in

[2] This figure gives a general idea of the distribution of an electron occupying these orbitals so far as orientation is concerned, but it does not show the dependence on r.

TABLE 4-1.—OBSERVED VALUES OF BOND ANGLES IN HYDRIDES

Substance	Method[a]	Bond angle	Experimental value	Reference
H_2O	I, M	HOH	104.45° ±0.10°	1
H_3N	I	HNH	107.3° ±0.2°	2
H_2S	Sp, M	HSH	92.2° ±0.1°	3
H_3P	M	HPH	93.3° ±0.2°	4
H_2Se	M	HSeH	91.0° ±1°	5
H_3As	M	HAsH	91.8° ±0.3°	6
H_2Te	I	HTeH	89.5° ±1°	7
H_3Sb	M	HSbH	91.3° ±0.3°	8

[a] I = infrared spectroscopy, M = microwave spectroscopy.

1. R. Mecke and W. Baumann, *Physik. Z.* **33**, 833 (1932); B. T. Darling and D. M. Dennison, *Phys. Rev.* **57**, 128 (1940); D. W. Posener and M. W. P. Strandberg, *ibid.* **95**, 374 (1954). The angle is the same in D_2O: W. S. Benedict, N. Gailor, and E. K. Plyler, *J. Chem. Phys.* **24**, 1139 (1956).

2. G. Herzberg, *Infrared and Raman Spectra*, Van Nostrand Co., New York, 1945; the microwave value is 107.3° ±0.2°: M. T. Weiss and M. W. P. Strandberg, *Phys. Rev.* **83**, 567 (1951).

3. B. L. Crawford, Jr., and P. C. Cross, *J. Chem. Phys.* **5**, 371 (1937); C. A. Burrus, Jr., and W. Gordy, *Phys. Rev.* **92**, 274 (1953); H. C. Allen, Jr., and E. K. Plyler, *J. Chem. Phys.*, **25**, 1132 (1956).

4. C. C. Loomis and M. W. P. Strandberg, *Phys. Rev.* **81**, 798 (1951).

5. A. W. Jache, P. W. Moser, and W. Gordy, *J. Chem. Phys.*, **25**, 209 (1956).

6. Ref. 4; also G. S. Blevins, A. W. Jache, and W. Gordy, *Phys. Rev.* **97**, 684 (1955).

7. K. Rossman and J. W. Straley, *J. Chem. Phys.* **24**, 1276 (1956).

8. Ref. 4; also A. W. Jache, G. S. Blevins, and W. Gordy, *Phys. Rev.* **97**, 680 (1955).

any other, whereas the three p orbitals are directed along the three Cartesian axes and will tend to form bonds in these directions.[3]   Moreover, the p orbitals are concentrated in these directions and have magnitudes $\sqrt{3}$ times as great as that of an s orbital, so far as the angular dependence is concerned.   Since the radial part of an s orbital and that of the p orbitals of the same shell do not differ much, the p orbitals can overlap the orbital of another atom more effectively than can the s orbital of the same shell; *p bonds are stronger than s bonds.*   It has been found on quantitative study of a simple problem of this type[4] that the energy of a bond is about proportional to the product of the magnitudes of bond orbitals of the two atoms (in their angular dependence); an s-p bond would have bond energy about $\sqrt{3}$ times that of an s-s

[3] The orientation of the axes of reference for the orbitals of an atom is of course arbitrary; we should say only that the bond directions for the three p orbitals are at right angles to one another.

[4] L. Pauling and J. Sherman, *J.A.C.S.* **59**, 1450 (1937).

bond, and a *p-p* bond would be stronger than an *s-s* bond by a factor of about 3. It is convenient to call the magnitude of a bond orbital in its angular dependence the *strength* of the bond orbital, with value 1 for an *s* orbital and 1.732 for a *p* orbital.

The conclusion that *p bonds tend to be at right angles to one another*[5] is verified to some extent by experiment (Table 4-1). In water, with

the structure
$$\overset{\text{H}}{\underset{\cdot\cdot}{:\text{O}:}}\text{H},$$
the bond angle is 104.5°. We expect the bonds

to be *p* bonds rather than *s* bonds for the following reason: A 2*s* electron of oxygen is more stable than a 2*p* electron by about 200 kcal /mole; and if the *s* orbital were used in bond formation (being then occupied effectively by only one electron) rather than for an unshared pair the molecule would be made unstable to this extent. The difference of 14.5° between the observed value of the bond angle and the expected value of 90° is probably to be attributed in the main to the partial ionic character of the O—H bonds, estimated in the preceding chapter to be 39 percent. This would give a resultant positive charge to the hydrogen atoms, which would repel one another and thus cause an increase in the bond angle. This effect is discussed in the more detailed treatment of bond angles that is given in Section 4-3. The large value for ammonia, 107°, may be attributed to the same cause.

In hydrogen sulfide, phosphine, and the hydrides of their heavier congeners, in which the bonds are nearly normal covalent bonds, the bond angles are observed to be close to 90° (Table 4-1). The values given in the table apply to the deuterium compounds as well as to the protium compounds.

With larger atoms attached to the central atom the bond angles lie between 94° and 111° (Table 4-2). The increase above 90° may be attributed to steric repulsion of these larger atoms (Sec. 4-3).

4-2. HYBRID BOND ORBITALS; THE TETRAHEDRAL CARBON ATOM

From the foregoing discussion it might be inferred that the quadrivalent carbon atom would form three bonds at right angles to one another and a fourth weaker bond (using the *s* orbital) in some arbitrary direction. This is, of course, not so; and, instead, it is found on quantum-mechanical study of the problem that *the four bonds of carbon are equivalent and are directed toward the corners of a regular tetrahedron,*[6] as had been inferred from the facts of organic chemistry.

[5] This conclusion was first given by J. C. Slater, *Phys. Rev.* **37**, 481 (1931).

[6] Pauling, *loc. cit.* (1); Slater, *loc. cit.* (5); J. H. Van Vleck, *J. Chem. Phys.* **1**, 177, 219 (1933); **2**, 20 (1934); R. S. Mulliken, *ibid.* 492; H. H. Voge, *ibid.* **4**, 581 (1936); **16**, 984 (1948); etc.

TABLE 4-2.—OBSERVED VALUES OF BOND ANGLES

Substance[a]	Method[b]	Bond angle	Experimental value
OF_2	E	FOF	$103.2° \pm 1°$
Cl_2O	E	ClOCl	$110.8° \pm 1°$
$(CH_3)_2O$	E	COC	$111° \pm 3°$
$(CH_3)_3N$	E	CNC	$108° \pm 4°$
$(CH_3)_2NCl$	E	CNCl	$107° \pm 2°$
CH_3NCl_2	E	CNCl	$109° \pm 2°$
		ClNCl	$108° \pm 2°$
S_8	X	SSS	$107.6° \pm 1°$
S_8	E	SSS	$105° \pm 2°$
SCl_2	E	ClSCl	$102° \pm 3°$
$(CH_3)_2S$	E	CSC	$105° \pm 3°$
P (black)	X	PPP	$99° \pm 1°$
			$102° \pm 1°$
$P(CH_3)_3$	M	CPC	$99.1° \pm 0.2°$
PF_3	E	FPF	$104° \pm 4°$
PCl_2F	E	ClPCl	$102° \pm 3°$
PCl_3	M	ClPCl	$100.0° \pm 0.3°$
PBr_3	E	BrPBr	$101.5° \pm 1.5°$
PI_3	E	IPI	$102° \pm 2°$
Se	X	SeSeSe	$104° \pm 2°$
Se_8	X	SeSeSe	$105° \pm 1°$
As	X	AsAsAs	$97° \pm 2°$
$As(CH_3)_3$	E	CAsAs	$96° \pm 5°$
AsF_3	M	FAsF	$102° \pm 2°$
$AsCl_3$	M	ClAsCl	$98.4° \pm 0.5°$
$AsBr_3$	E	BrAsBr	$100.5° \pm 1.5°$
AsI_3	E	IAsI	$101° \pm 1.5°$
Te	X	TeTeTe	$104° \pm 2°$
$TeBr_2$	E	BrTeBr	$98° \pm 3°$
Sb	X	SbSbSb	$96° \pm 2°$
$SbCl_3$	M	ClSbCl	$99.5° \pm 1.5°$
$SbBr_3$	E	BrSbBr	$97° \pm 2°$
SbI_3	E	ISbI	$99° \pm 1°$
Bi	X	BiBiBi	$94° \pm 2°$
$BiCl_3$	E	ClBiCl	$100° \pm 6°$
$BiBr_3$	E	BrBiBr	$100° \pm 4°$

[a] For references see Sutton, *Interatomic Distances*. The value for $P(CH_3)_3$ is from D. R. Lide, Jr., and D. E. Mann, *J. Chem. Phys.* 28, 572 (1958).

[b] The letters E, M, and X designate electron diffraction of gas molecules, microwave spectroscopy of gas molecules, and x-ray diffraction of crystals, respectively.

A rigorous quantum-mechanical treatment of directed valence bonds has not been given, for the reason that the Schrödinger wave equation has not been rigorously solved for any complicated molecule. Several approximate treatments have, however, been carried out, leading in a reasonable way to results such as those described below. Of these treatments we shall describe only the simplest one, which is, indeed, the most powerful one, in that it leads directly to the largest number of satisfactory results.

The simple theory that we use is based on the reasonable postulate given at the beginning of this chapter about the dependence of the bond-forming power (the strength) of a bond orbital on its angular distribution. From it, by the use of general quantum-mechanical principles, there is derived the whole body of results about directed valence, including not only the tetrahedral arrangement of the four single bonds of the carbon atom but also octahedral and square configurations of bonds (as well as other configurations), together with rules for the occurrence of these configurations, the strengths of the bonds, and the relation of configuration to magnetic properties. In this way a single reasonable postulate is made the basis of a large number of the rules of stereochemistry and is found to lead to several new stereochemical results.

There are four orbitals in the valence shell of the carbon atom. We have described these as the $2s$ and the three $2p$ orbitals, with bond strengths 1 and 1.732, respectively. These are, however, not the orbitals used directly in bond formation by the atom. (They are especially suited to the description of the free carbon atom; if quantum theory had been developed by the chemist rather than the spectroscopist it is probable that the tetrahedral orbitals described below would play the fundamental role in the theory, in place of the s and p orbitals.) Now in general a wave function for a system can be constructed by adding together other functions, the wave function for the normal state being the one that minimizes the energy of the system. The energy of a system of a carbon atom and four attached atoms is minimized by making the bond energies as large as possible. It is found that a bond orbital formed by linear combination of s and p orbitals, taken with a certain ratio of numerical coefficients, has a bond strength greater than that for an s or p orbital alone, the strength of the best s-p hybrid bond orbital being as great as 2. The angular dependence of this orbital is shown in Figure 4-2. It is seen that the orbital is greatly concentrated in the bond direction (its axis of rotational symmetry), and it can be understood that this orbital would overlap greatly with the orbital of another atom and would form a very strong bond. We expect this hybridization to take place in order that the bond energy may be a maximum.

A surprising result of the calculations, of great chemical significance, is the following: when it is sought to make the energy of a second bond as large as possible, by forming another hybrid orbital of maximum bond-forming power, it is found that this second best bond orbital is equivalent to the first, with strength 2, and that *its bond direction makes the tetrahedral angle 109°28' with that of the first.* Moreover, a third and a fourth equivalent orbital can be constructed, the four being directed toward the corners of a regular tetrahedron; but then no more orbitals are left in the valence shell. It is convenient to call these four best *s—p* bond orbitals *tetrahedral orbitals.*

The postulate of the tetrahedral carbon atom in classical stereochemistry requires that the atom have a configuration that is tetrahedral but is not necessarily that of a regular tetrahedron; so long as

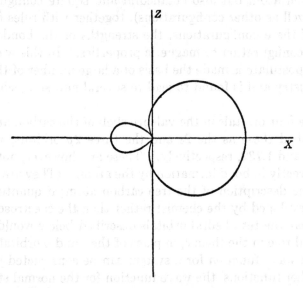

FIG. 4-2.—The angular dependence of a tetrahedral orbital with bond direction along the *x* axis.

the four bonds have a general tetrahedral orientation the phenomenon of optical activity is accounted for. There is no need for the R_1—C—R_2 bond angle in $CR_1R_2R_3R_4$ to be near 109°28'; it might be 150° or more. The result of the bond orbital treatment described above requires, however, that the carbon bond angles be close to the tetrahedral angle, since change from this value is associated with loss in bond strength of the carbon orbitals and hence with decrease in stability of the system. It is of great interest that in hundreds of molecules containing a carbon atom attached by four single bonds to atoms of different kinds experimental values of bond angles have been determined that almost without exception lie within 2° of the value 109°28' corresponding to tetrahedral orbitals. A few of these values are given in Table 4-3.

TABLE 4-3.—OBSERVED VALUES OF BOND ANGLES
FOR QUADRICOVALENT ATOMS[a]

Substance[b]	Method[c]	Bond angle[d]	Experimental value
CH_3Cl	M	HCH	110.5° ± 0.5°
CH_2Cl_2	M	HCH	112.0° ± 0.3°
		ClCCl	111.8° ± 0.3°
$CHCl_3$	M	ClCCl	110.4° ± 1°
CH_3Br	M	HCH	111.2° ± 0.5°
$CHBr_3$	M	BrCBr	110.8° ± 0.3°
CH_3I	M	HCH	111.4° ± 0.1°
CHI_3	E	ICI	113.0° ± 1°
CH_2F_2	M	FCF	108.3° ± 0.1°
		HCH	111.9° ± 0.4°
CHF_3	M	FCF	108.8° ± 0.8°
	E	FCF	108.5° ± 0.5°
CH_2ClF	M	ClCF	110.0° ± 0.1°
		HCCl	109.1° ± 0.2°
		HCH	111.9° ± 0.5°
$CHClF_2$	E	FCF	110.5° ± 1°
		ClCF	110.5° ± 1°
$CClF_3$	E	FCF	108.6° ± 0.4°
CCl_3F	E	ClCCl	111.5° ± 1°
$CBrCl_3$	E	ClCCl	111.2° ± 1°
CH_3OH	M	HCH	109.3° ± 0.8°
CH_3SH	M	HCH	110.3° ± 0.2°
CH_3NH_2	M	HCH	109.5° ± 1°
CH_3CHF_2	M	FCF	109.2° ± 0.1°
		CCF	109.4° ± 0.1°
		HCC	109.8° ± 0.2°
C_2H_6	M, E	HCH	109.3° ± 0.5°
C_2Cl_6	E	ClCCl	109.3° ± 0.5°
F_3CCCCF_3	E	FCF	107.5° ± 1°
Glycine	X[e]	CCN	111.8° ± 0.3°
SiH_3F	M	HSiH	109.3° ± 0.3°
$SiHF_3$	M	FSiF	108.2° ± 0.5°
SiH_3Cl	M + I	HSiH	110.2° ± 0.3°
SiH_2Cl_2	E	ClSiCl	110° ± 1°
$SiHCl_3$	M	ClSiCl	109.4° ± 0.3°
SiH_3Br	M	HSiH	111.3° ± 1°
$SiHBr_3$	E	BrSiBr	110.5° ± 1.5°
SiH_3I	I	HSiH	109.9° ± 0.4°
$SiClF_3$	E	FSiF	108.5° ± 1°
Si_2Cl_6	E	ClSiCl	109.5° ± 1°
CH_3SiHF_2	M[f]	FSiF	106.7° ± 0.5°
		HSiC	116.2° ± 1°
		CSiF	109.8° ± 0.5°
GeH_3Cl	M	HGeH	110.9° ± 1.5°
$GeHCl_3$	M	ClGeCl	108.3° ± 0.2°
$GeClF_3$	M	FGeF	107.7° ± 1.5°
CH_3GeH_3	M[g]	HCH	108.2° ± 0.5°
		HGeH	108.6° ± 0.5°

[a] References for the values in this table, except as noted below, may be found in Sutton, *Interatomic Distances.*

[b] All carbon bond angles reported by Sutton as having standard deviations less than 1° are given in this table.

[c] The letters E, M, I, and X designate electron diffraction, microwave spectroscopy, infrared spectroscopy, and x-ray diffraction, respectively.

[d] Other bond angles about the quadricovalent atom can be evaluated by the approximation that the average value of the six bond angles for small deviations from the tetrahedral value 109.47° is equal to this value.

[e] R. E. Marsh, *Acta Cryst.* 11, 654 (1958). Many more values of bond angles have been reported from x-ray studies of crystals.

[f] J. D. Swalen and B. P. Stoicheff, *J. Chem. Phys.* 28, 671 (1958).

[g] V. W. Laurie, *Bull. Am. Phys. Soc.* 3, 213 (1958).

An interesting feature of these values is the surprisingly large value of the HCH angle. With this exception, the bond angles clearly reflect the differences in the van der Waals radii of the ligands. For example, the XCX values (X = halogen) for the trihalogenomethanes HCF_3, $HCCl_3$, $HCBr_3$, and HCI_3 are 108.8°, 110.4°, 110.8°, and 113.0°, respectively; the angle increases with increase in size of the halogen atom, as would be expected because of the van der Waals repulsion, which is greater for two halogen atoms than for a halogen atom and a hydrogen atom. However, although the hydrogen atom is smaller than any halogen atom, the HCH angle is found in general to be larger than the HCX angles; for the six methyl and methylene halogenides in the table the average value of the HCH angle is 111.5° and that of the HCX angles is 108.4°. This difference can be accounted for as the result of the difference in size of the atoms, when the difference in C—X and C—H bond lengths is also taken into consideration.

For quadricovalent silicon, germanium, and tin (and also for atoms such as nitrogen in a substituted ammonium ion) the same tetrahedral orientation of bonds is expected, since $3s$-$3p$, $4s$-$4p$, and $5s$-$5p$ hybridization is the same as that for the $2s$-$2p$ system. Observed values of bond angles in unsymmetrical compounds of these substances are also included in Table 4-3.

There are many symmetrically substituted compounds (CH_4, $C(CH_3)_4$, CCl_4, $Si(CH_3)_4$, $Ge(CH_3)_4$, $Sn(CH_3)_4$, etc.) in which the bond angles are known to be tetrahedral; these are not included in the table, since they provide no serious test of the theory.

A still more surprising result about the significance of the concept of the carbon atom as a regular tetrahedron is provided by the methylethylenes. The picture of the carbon-carbon double bond as involving the sharing of an edge by two regular tetrahedra leads to the tetrahedral value 125°16' for the single-bond:double-bond angle. The electron-diffraction value for this angle in both isobutene and tetramethylethylene is 124°20'±1°, and the microwave value for phosgene, $Cl_2C\!=\!O$, is 124.3°.

Derivation of Results about Tetrahedral Orbitals.—The results about tetrahedral bond orbitals described above are derived in the following way. We assume that the radial parts of the wave functions ψ_s and ψ_{p_x}, ψ_{p_y}, ψ_{p_z} are so closely similar that their differences can be neglected. The angular parts are

$$\left.\begin{aligned} s &= 1 \\ p_x &= \sqrt{3}\,\sin\theta\cos\phi \\ p_y &= \sqrt{3}\,\sin\theta\sin\phi \\ p_z &= \sqrt{3}\,\cos\theta \end{aligned}\right\} \tag{4-1}$$

θ and ϕ being the angles used in spherical polar coordinates. These functions are *normalized to* 4π, the integral

$$\int_0^{2\pi} \int_0^{\pi} f^2 \sin\theta d\theta d\phi$$

of the square of the function taken over the surface of a sphere having the value 4π. The functions are *mutually orthogonal*, the integral of the product of any two of them (sp_z, say) over the surface of a sphere being zero.

Now we ask whether a new function

$$f = as + bp_x + cp_y + dp_z \qquad (4\text{-}2)$$

normalized to 4π (this requiring that $a^2 + b^2 + c^2 + d^2 = 1$) can be formed which has a larger bond strength than 1.732, and, if so, what function of this type has the maximum bond strength. The direction of the bond is immaterial; let us choose the z axis. It is easily shown that p_x and p_y do not increase the strength of a bond in this direction, but decrease it, so they are ignored, the function thus assuming the form

$$f_1 = as + \sqrt{1 - a^2}\,p_z \qquad (4\text{-}3)$$

in which d is replaced by $\sqrt{1 - a^2}$ for normalization. The value of this in the bond direction $\theta = 0$ is, on substituting the expressions for s and p_z,

$$f_1(\theta = 0) = a + \sqrt{3(1 - a^2)}$$

This is made a maximum by differentiating with respect to a, equating to zero, and solving, the value $a = \frac{1}{2}$ being obtained. Hence the best bond orbital in the z direction is

$$f_1 = \frac{1}{2}s + \frac{\sqrt{3}}{2}p_z = \frac{1}{2} + \frac{3}{2}\cos\theta. \qquad (4\text{-}4)$$

This orbital has the form shown in Figure 4-2. Its strength is seen to be 2 by placing $\theta = 0$, $\cos\theta = 1$.

We now consider the function

$$f_2 = as + bp_x + dp_z$$

which is orthogonal to f_1, satisfying the requirement

$$\int_0^{2\pi} \int_0^{\pi} f_1 f_2 \sin\theta d\theta d\phi = 0,$$

and which has the maximum value possible in some direction. (This direction will lie in the xz plane; i.e., $\phi = 0$, since p_y has been left out.) It is found on solving the problem that the function is

$$f_2 = \frac{1}{2}s + \frac{\sqrt{2}}{\sqrt{3}}p_x - \frac{1}{2\sqrt{3}}p_z \qquad (4\text{-}5)$$

This function is seen on examination to be identical with f_1 except that it is rotated through $109°28'$ from f_1. In the same way two more functions can be constructed, each identical with f_1 except for orientation.

An equivalent set of tetrahedral bond orbitals, differing from these only in orientation, is

$$t_{111} = \tfrac{1}{2}(s + p_x + p_y + p_z)$$
$$t_{1\bar{1}\bar{1}} = \tfrac{1}{2}(s + p_x - p_y - p_z)$$
$$t_{\bar{1}1\bar{1}} = \tfrac{1}{2}(s - p_x + p_y - p_z)$$
$$t_{\bar{1}\bar{1}1} = \tfrac{1}{2}(s - p_x - p_y + p_z)$$

The strength of an s-p hybrid orbital increases with increase in the amount of p involved from 1 (pure s) to a maximum value of 2 (tetrahedral orbital) and then decreases to 1.732 (pure p), in the way shown by the dashed curves in Figure 4-3, in which the square of the bond strength (that is, the product of the strengths of equivalent orbitals of two atoms forming a bond) is shown as a function of the nature of the orbitals. That the strength of the orbital is a measure of its bond-forming power is shown by the approximation of these curves to the full curves, which represent the calculated energy of a one-electron bond as a function of the nature of the bond orbitals.[7]

Quantum-mechanical Description of the Quadrivalent Carbon Atom. —The description that is given above of the quadrivalent carbon atom as forming four sp^3 bonds is somewhat idealized. In a later section (Sec. 4-5) it is pointed out that the bond orbitals have some d and f character. Moreover, the four valence electrons are not closely described by the electron configuration sp^3, even aside from the contribution of configurations involving d and f orbitals.

The most stable atomic energy levels of the carbon atom are shown in Figure 4-4. The three lowest levels, with Russell-Saunders symbols ^3P, ^1D, and ^1S, correspond to the configuration $2s^22p^2$. This configura-

---

[7] Pauling and Sherman, *loc. cit.* (4). Calculations of the bonding power of $2s$-$2p$ hybrid orbitals have been reported also by R. S. Mulliken, *J. Chem. Phys.* **19**, 900, 912 (1951). C. A. Coulson and G. R. Lester, *Trans. Faraday Soc.* **51**, 1605 (1955), have made calculations for excited states of the hydrogen molecule-ion that agree reasonably well with the postulate that the strength S of an orbital determines its bond-forming power.

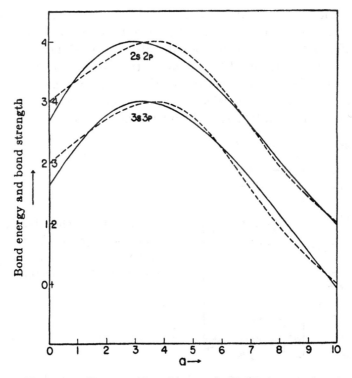

Fɪɢ. 4-3.—Square of bond strength (dashed curves) and calculated bond-energy values (full curves) for hybrid sp orbitals varying from pure p orbitals ($a = 0$, left) to pure s orbitals ($a = 10$, right). The upper pair of curves are for L orbitals ($2s$ and $2p$), and the lower, with shifted vertical scale, for M orbitals ($3s$ and $3p$).

tion has two unpaired electrons and can be the basis of the bivalent state of the carbon atom. The quadrivalent carbon atom requires contribution of the configuration $2s2p^3$. The six atomic levels based upon this configuration are shown in the figure. Their promotion energy (energy relative to the lowest state) ranges from 100 kcal/mole to 345 kcal/mole, with an average of 208 kcal/mole. The difference in energy of a $2p$ electron and a $2s$ electron of carbon is given by Slater,[8] in his compilation of values of one-electron energies, as 199 kcal/mole; this is the promotion energy of the configuration $2s2p^3$.

A detailed quantum-mechanical treatment of methane has been carried out by Voge.[9] He took into consideration the configurations $2s^22p^2$, $2s2p^3$, and $2p^4$ and minimized the energy of the molecule in

[8] J. C. Slater, *Phys. Rev.* **98**, 1093 (1955).

[9] Voge, *loc. cit.* (6).

order to find the best wave function based on these carbon-atom configurations. He found that his best wave function was one to which
the configuration $2s2p^3$ contributed only about 60 percent, the remainder being contributed by the other two configurations. The energy of the valence state relative to that of the normal state of the
isolated carbon atom was calculated to be about 100 kcal/mole, that
for the pure quadrivalent state being 162 kcal/mole.

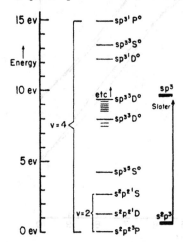

Fig. 4-4.—The low-lying
atomic-energy levels of the
carbon atom, as given by
spectroscopic investigations.

The contributions of $2s^22p^2$ and $2p^4$ might be described as represent-

ing resonance with bicovalent structures such as $\left|\ :\!C\right.$ H H / H H . It is my

opinion, however, that it is not worth while in general to introduce such
a complication into the discussion of the structure of simple molecules.

4-3. THE EFFECT OF AN UNSHARED PAIR ON HYBRIDIZATION

Since tetrahedral orbitals form stronger bonds than other *s-p* orbitals
it might be thought that hybridization to tetrahedral orbitals would
always occur in bond formation. The tendency to use the best bond
orbitals is, however, resisted in the case of atoms with an unshared pair
(or more than one) by the tendency to keep the unshared pair in the *s*
orbital, which is more stable than the *p* orbitals. In OF_2, for example,
the use of tetrahedral orbitals in bond formation would require that
half of the *s* orbital (which is divided equally among the four tetrahedral orbitals) be used for shared pairs and only half for unshared
pairs. Since a shared pair counts as only one electron for each atom,
this would involve the loss of one-quarter of the extra stability due to a
pair of *s* electrons, and the atom will strive to prevent this. On the

other hand, the bonds will strive to be as strong as possible. In consequence a compromise will be reached such as to minimize the energy for
the molecule as a whole; the bond orbitals will have a small amount of
s character and be intermediate between p bonds and tetrahedral
bonds, with the unshared pair utilizing most of the s orbital. The
bond strengths of these bonds will be between 1.732 and 2, and the
angles between the bond directions will be increased somewhat from
the p-bond value 90° toward the tetrahedral value 109°28'.

A simple quantitative treatment of the amount of hybridization of
bond orbitals of this kind can easily be carried out. Let us represent
the bond orbital as

$$\psi = \alpha s + \sqrt{1 - \alpha^2} p_z \qquad (4\text{-}6)$$

in which for convenience the bond is taken to extend along the z axis.
The coefficients of s and p_z have values that normalize ψ (sum of the
squares equals unity). The orbital has the amount α^2 of s character
and $(1 - \alpha^2)$ of p character.

The strength S of the bond orbital is given by

$$S = \alpha + \sqrt{1 - \alpha^2}\sqrt{3}. \qquad (4\text{-}7)$$

The bond energy is assumed to be proportional to S^2. We may use b
for the coefficient of proportionality.

There is, however, also another energy term that must be taken into
consideration—the energy of the unshared pair of electrons. If the
bond orbital is a p orbital the unshared pair occupies the s orbital, and
only a single electron (one of the two involved in forming the bond)
occupies the p orbital, whereas if the bond orbital is an s orbital there
are two electrons occupying the p orbital and only one the s orbital.
Hence the energy of the atom itself depends upon the degree of hybridization of the bond orbital; it increases by the amount $\alpha^2(E_p - E_s)$,
where $E_p - E_s$, the difference in energy of a p electron and an s electron
in the atom, is called the *s-p promotion energy*. Since this energy term
is opposite in sign to the principal bond-energy term bS, the effective
bond energy is

$$B = b(\alpha + \sqrt{1 - \alpha^2}\sqrt{3})^2 - \alpha^2(E_p - E_s). \qquad (4\text{-}8)$$

The hybridization parameter α is to be chosen so as to make the
energy of the molecule a minimum, that is, to make the bond energy B
have its maximum value. Hence we differentiate B with respect to α
and equate to zero, to obtain the equation

$$2b(\alpha + \sqrt{1 - \alpha^2}\sqrt{3})\left(1 - \frac{\sqrt{3}\alpha}{\sqrt{1 - \alpha^2}}\right) - 2\alpha(E_p - E_s) = 0 \qquad (4\text{-}9)$$

The bond-energy coefficient b can be eliminated between Equations 4-8 and 4-9, to give the following equation, in which terms in powers of α higher than the square have been neglected:

$$B = (E_p - E_s)(\sqrt{3}\alpha + 3\alpha^2) \qquad (4\text{-}10)$$

This equation is accurate to 1 percent for values of α not greater than 0.25.

The sp promotion energy can be obtained from spectroscopic values of atomic energy levels. For the valence shell of all atoms its value is approximately 180 kcal/mole.

It is found by substitution in Equation 4-10 that the bond energies 110.6 kcal/mole for O—H and 93.4 kcal/mole for N—H correspond to $\alpha = 0.247$ and 0.219, respectively. The value of α^2 gives the amount of s character in the bond orbital; hence the bond orbitals in water and ammonia are estimated by this calculation to have about 5 percent or 6 percent of s character.

We may predict the corresponding bond angles by finding the directions in which two mutually orthogonal bond orbitals have their maximum strength. For convenience let us take one of the orbitals along the z axis and the other in the xz plane. They then have the general form

$$\psi_1 = \alpha s + \beta_1 p_z$$
$$\psi_2 = \alpha s + \beta_2 p_x + \gamma_2 p_z.$$

Each function is to be normalized; hence $\alpha^2 + \beta_1^2 = 1$ and $\alpha^2 + \beta_2^2 + \gamma_2^2 = 1$. The condition that the two functions be orthogonal is $\alpha^2 + \beta_1\beta_2 = 0$. Hence we find the values $\beta_1 = \sqrt{1 - \alpha^2}$, $\beta_2 = -\alpha^2/\sqrt{1 - \alpha^2}$, and $\gamma_2 = \sqrt{1 - 2\alpha^2}/\sqrt{1 - \alpha^2}$. It is found on examining ψ_2 that its maximum is in the direction with β_2 and γ_2 proportional to direction cosines relative to the z and x axes, and hence that the cosine of the bond angle is $-\alpha^2/(1 - \alpha^2)$; the bond angle itself is then approximately $90° + 57° \, \alpha^2$, for small values of α^2.

Calculated and observed values of the bond angles are compared in Table 4-4; the agreement is only rough. The lack of better agreement may be due in part to the simplified nature of the treatment, in particular, to the neglect of the d and f character of the bond orbitals that will be discussed in Section 4-5.

The largest deviations are shown by water and ammonia, the molecules with bonds that have a considerable amount of ionic character. A possible explanation of part of the deviation is that the repulsion of the charges of the hydrogen atoms causes the angle to increase.[10]

---

[10] For a discussion of these bond angles based on the electron distribution of the central atom see J. W. Linnett and A. J. Poë, *Trans. Faraday Soc.* **47**, 1033 (1951).

TABLE 4-4.—COMPARISON OF OBSERVED AND CALCU-
LATED VALUES OF BOND ANGLES

Molecule	Value of α	Bond angle	
		Calculated	Observed
H_2O	0.247	93.5°	104.45°
H_2S	.194	92.1°	92.2°
H_2Se	.164	91.5°	91.0°
H_2Te	.146	91.2°	89.5°
NH_3	.219	92.7°	107.3°
PH_3	.185	91.9°	93.3°
AsH_3	.148	91.2°	91.8°

Some support of this suggestion is given by the fact that smaller values of the oxygen bond angle are found in some other molecules: 101.5° ± 0.5° in H_2O_2 (neutron diffraction of the crystal[11]), 103° in OF_2 (electron diffraction[12]).

Contribution of Unshared Electron Pairs to the Electric Dipole Moments of Molecules.—In the preceding chapter we discussed the dipole moments of molecules, in relation to the partial ionic character of bonds, without considering the possible contribution of unshared electron pairs. A simple treatment based on hybrid orbitals provides some justification of this procedure.

In the foregoing treatment of the water molecule, which we shall use as an example, each of the two bond orbitals of the oxygen atom has been calculated to have 6 percent s character and 94 percent p character. Each of the two unshared-pair orbitals then has 44 percent s character and 56 percent p character. The maxima for the unshared-pair orbitals lie in directions making an angle of 142° with one another and such that their resultant is opposed to that for the two bond orbitals, which have their maxima at 93.5° with one another. The component for the four unshared-pair electrons is determined by the direction cosine −0.34, and that of the two bonding electrons of the oxygen atom by the direction cosine 0.68; hence the contribution of the four unshared-pair electrons to the dipole moment is just balanced by that of the two bonding electrons.[13]

[11] W. R. Busing and H. A. Levy, *Am. Cryst. Ass'n Meeting*, Milwaukee, June 1958.

[12] J. A. Ibers and V. Schomaker, *J. Phys. Chem.* **57**, 699 (1953).

[13] In this discussion the orbital moments (along the orbital axes) have been assumed to be equal for bond orbitals and unshared-pair orbitals. Calculation of the moment (average value of cosine of angle with bond axis) for the sp hybrid orbitals leads to a value only half as great for the bond orbital as for the unshared-pair orbital. However, equality is found when the d-character (4 percent) and

A similar treatment of other molecules, such as ammonia, leads also to the conclusion that the electric moments of the unshared electron pairs and the bonding electrons largely neutralize one another.

4-4. ORBITALS FOR INCOMPLETE *s-p* SHELLS

In boron trimethyl, $B(CH_3)_3$, only three of the four orbitals of the valence shell are used. If the best bond orbitals were utilized the C—B—C bond angles would be near 109°28′. However, the molecule obtains added stability by using the *s* orbital as completely as possible, and this tends to cause the *s* orbital to be divided among three bond orbitals, which are found by the simple theoretical treatment to be coplanar and to make angles of 120° with one another.[14] Whether this effect would take place completely or whether the bonds would resist this weakening to some extent cannot be predicted; it is found experimentally[15] that the boron trimethyl molecule is planar, with 120° angles, indicating that the bond orbitals are those obtained by dividing the *s* orbital among three.

The formation of a fourth bond by boron would strengthen the bonds (all bond orbitals becoming tetrahedral) and stabilize the molecule; we can thus understand the ability of boron trimethyl to add ammonia to

give the compound

$$\begin{array}{ccc} H_3C & & H \\ & \diagdown \quad \diagup & \\ & B{-}N & \\ & \diagup \quad \diagdown & \\ H_3C & & H \end{array}$$

The boron halogenides, whose structures are discussed in Section 9-5, similarly add molecules containing unshared pairs to give products such as

$$\begin{array}{c} Cl \\ \diagdown \\ Cl{-}B{-}N{\equiv}C{-}CH_3, \\ \diagup \\ Cl \end{array}$$

formed from boron trichloride and methyl cyanide. The electric dipole moment of ammonia-borane, H_3BNH_3, is 4.9 D (measured in dioxane solution);[16] this value is a reasonable one for the structure

---

f character (2 percent) of the bond orbital are taken into consideration (Sec. 4-5).

[14] The three bond orbitals (taken in the xy plane) are

$$\frac{1}{\sqrt{3}}\,s + \frac{\sqrt{2}}{\sqrt{3}}\,p_x, \quad \frac{1}{\sqrt{3}}\,s - \frac{1}{\sqrt{6}}\,p_x + \frac{1}{\sqrt{2}}\,p_y, \quad \text{and} \quad \frac{1}{\sqrt{3}}\,s - \frac{1}{\sqrt{6}}\,p_x - \frac{1}{\sqrt{2}}\,p_y.$$

Their bond strength is 1.991, only slightly less than that of tetrahedral orbitals.

[15] H. A. Lévy and L. O. Brockway, *J.A.C.S.* **59**, 2085 (1937).

[16] J. R. Weaver, S. G. Shore, and R. W. Parry, *J. Chem. Phys.* **29**, 1 (1958).

$$\begin{array}{ccc} H & & H \\ \diagdown & & \diagup \\ H{-}\bar{B}{-}\overset{+}{N}{-}H. \\ \diagup & & \diagdown \\ H & & H \end{array}$$

In a molecule such as mercury dimethyl the two bond orbitals would be expected to use all the s orbital between them.[17] The simple treatment shows that the two bonds would then be opposed. This is substantiated by the observation of a linear configuration for gas mole-

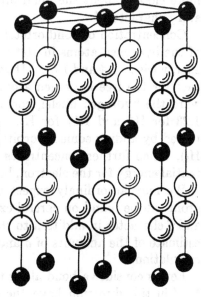

Fig. 4-5.—The arrangement of atoms of gold (small circles), carbon, and nitrogen in the hexagonal crystal AuCN.

cules of $HgCl_2$, $HgBr_2$, HgI_2, CH_3HgCl, CH_3HgBr, and $Hg(CH_3)_2$, and also of a linear configuration for the molecules Hg_2Cl_2, Hg_2Br_2, and Hg_2I_2 (with electronic structures $:\overset{..}{X}{-}Hg{-}Hg{-}\overset{..}{X}:$) in crystals. A similar configuration is expected for bicovalent complexes of univalent copper, silver, and gold. It has been verified for the $[AuCl_2]^-$ ion[18] and the $[Ag(CN)_2]^-$ ion,[19] and also for AuCN crystal[20] (Fig. 4-5). In

[17] The two corresponding bond orbitals (with bond directions taken along the x axis) are

$$\frac{1}{\sqrt{2}}(s + p_x) \quad \text{and} \quad \frac{1}{\sqrt{2}}(s - p_x);$$

their bond strength is 1.932.

[18] N. Elliott and L. Pauling, *J.A.C.S.* **60**, 1846 (1938).

[19] J. L. Hoard, *Z. Krist.* **84**, 231 (1932).

[20] G. S. Zhdanov and E. A. Shugam, *Zhur. Fiz. Khim.* **19**, 519 (1945); *Acta Physicochim. U.R.S.S.* **20**, 253 (1945).

the AuCN crystal there are covalent bonds between both the nitrogen atom and the carbon atom of the cyanide group and the gold atom, so as to form very long —Au—C≡N—Au—C≡N—Au—··· molecules, which are packed in a hexagonal array.

4-5. CONCENTRATION OF BOND ORBITALS

The description of the bond orbitals of the carbon atom as sp^3 tetrahedral hybrid orbitals is satisfactory in many respects, but it can be improved. One improvement is that of concentrating the orbitals more closely about the bond direction by the introduction of some d and f character.[21]

Concentration of bond orbitals was recognized[22] in the bond orbitals of the hydrogen atoms in H_2^+ and H_2 as evaluated by minimizing the energy (Secs. 1-4 and 1-5). It was found that the best $1s$ orbital is not that for the free hydrogen atom; instead, the orbitals are shrunk toward the hydrogen nuclei, corresponding to effective nuclear charges 1.23 for H_2^+ and 1.17 for H_2. Moreover, much improvement is obtained by adding some $2p$ orbital (2 percent for H_2^+ and 1 percent for H_2), which further concentrates the orbital into the region of low potential energy for the electron, between the two nuclei.

We may accordingly expect that the bond orbitals of the carbon atom will be found on careful examination to be hybrids with some d and f character in addition to their principal sp^3 character. A rough estimate of the amounts of d and f character can be made by a simple calculation.

Let us consider a bond along the z axis. The only orbitals that extend in this direction have the following angular wave functions; all others have nodes along the z axis:

$$s = 1$$
$$p_z = \sqrt{3}\cos\theta$$
$$d_z = \sqrt{5/4}(3\cos^2\theta - 1)$$
$$f_z = \sqrt{7/4}(5\cos^3\theta - 3\cos\theta)$$
$$g_z = (3/8)(35\cos^4\theta - 30\cos^2\theta + 3)$$
$$\text{etc.}$$

(These functions are normalized to 4π.)

If the radial parts of the terms of a hybrid bond orbital of the form

$$\psi = \alpha s + \beta p_z + \gamma d_z + \delta f_z \qquad (4\text{-}11)$$

[21] L. Pauling, *Proc. Nat. Acad. Sci. U. S.* **44**, 211 (1958).
[22] L. Pauling, *J.A.C.S.* **53**, 1367 (1931).

were the same, the bond-forming power would be represented by the bond-strength function

$$S = \alpha + \sqrt{3}\beta + \sqrt{5}\gamma + \sqrt{7}\delta \qquad (4\text{-}12)$$

In fact, the radial parts are somewhat different, but we may use this function in carrying out our rough calculation.

The orbital 4-11 has γ^2 as its amount of d character and δ^2 as its amount of f character. Let P_d be the promotion energy of the d orbital (the energy required to promote an electron from the sp^3 orbital to the d orbital) and P_f be that of the f orbital. Then the bond energy must be corrected by the effective promotion energy $\gamma^2 P_d + \delta^2 P_f$. We may take the energy of the bond itself as proportional to S (equal to bS; the value $b = 36$ kcal/mole leads to the correct C—C single-bond energy); the effective bond energy is then given by the equation

$$\text{Effective bond energy} = bS - \gamma^2 P_d - \delta^2 P_f \qquad (4\text{-}13)$$

Values of the promotion energy can be estimated in the following way. In H_2^+ and H_2 the $2p$ orbital that minimizes the energy, when combined with the $1s$ orbital, has an effective nuclear charge such as to make $\bar{r}$, the average value of the distance of the electron from the nucleus, about 40 percent greater than for the normal hydrogen atom. This $2p$ orbital, with effective nuclear charge $Z' = 2.4$, can be described as a hybrid of the true $2p$, $3p$, $4p$, $\cdots$ orbitals of the hydrogen atom, including the p orbitals of the energy continuum (energy greater than that needed for ionization). We assume that the d, f, and g orbitals that contribute to the bond orbitals of the carbon atom have the form $3d$, $4f$, and $5g$, but with effective nuclear charges such as to make the value of $\bar{r}$ equal to $\frac{4}{3}$ that for the sp^3 orbitals. A simple calculation then leads to promotion energies $P_d = 0.67\ I$, $P_f = 1.37\ I$, and $P_g = 2.21\ I$, where I is the ionization energy of sp^3 electrons of the carbon atom. With I given the value 260 kcal/mole, the promotion energies are found to be 174, 356, and 575 kcal/mole, respectively.

When the effective bond energy (Equation 4-13) is minimized relative to γ and δ, their values are found to be 0.20 and 0.14, respectively. (In this calculation α is kept constant at 0.50; the value of β, satisfying the normalization equation $\alpha^2 + \beta^2 + \gamma^2 + \delta^2 = 1$, is 0.83.) Hence the best bond orbital is given by this calculation as having about 4 percent d character and 2 percent f character.[23]

The bond orbital found in this way, $\psi = 0.50\ s + 0.83 p_z + 0.20\ d_z + 0.14\ f_z$, is shown by the full curve in Figure 4-6. It is seen that,

[23] An earlier calculation (Pauling, *loc. cit.* [21]), made with a larger estimated value of P_d, gave 2 percent d character. The g character can be neglected (calculated 0.8 percent).

even though the amount of d and f character is small (total 6 percent), the orbital is significantly more concentrated in the bond direction than the sp^3 orbital (dashed curve). Its strength, 2.76, is 38 percent greater than that of an sp^3 orbital.

A bond formed by two such orbitals, with energy proportional to S^2, has bond energy nearly double (1.90 times) that of a bond formed by two sp^3 orbitals. The d and f character may have significant effect on

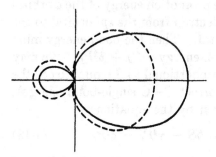

Fig. 4-6.—Tetrahedral bond orbital with 4 percent d character and 2 percent f.

many properties of bonds, such as bond angles. Especially significant is the restriction of rotation about single bonds, which is discussed in Section 4-7.

4-6. ELECTRON DISTRIBUTION IN COMPLETED SHELLS

Atoms and ions with noble-gas electron configurations have usually been described as having spherical symmetry. For some considerations this description is satisfactory; for others, however, it is advantageous to consider the atoms or ions to have a shape other than spherical—the helium atom can be described as deformed to a prolate ellipsoid of revolution, and the neon atom and other noble-gas atoms as deformed to a shape with cubic symmetry.

If the structure of the helium atom were exactly described by the symbol $1s^2$ and that of neon by $1s^2 2s^2 2p^6$ these atoms would have spherically symmetrical electron distributions.[24] However, the mutual repulsion of the two electrons in the atom causes them to avoid one another; the wave function for the atom corresponds to a larger probability for the two electrons to be on opposite sides of the nucleus than on the same side (for the same values of the distances of the two electrons from the nucleus, there is greater probability that the angle described at the nucleus by the vectors to the electrons is greater than 90° than that it is less than 90°). This effect, which is called correla-

---

[24] It was proved by A. Unsöld, *Ann. Physik* **82**, 355 (1927), that the sum of the squares of the wave functions of a subshell (such as the three $2p$ orbitals) is independent of θ and ϕ, and hence is spherically symmetric.

tion of the electrons, stabilizes the helium atom[25] by 26 kcal/mole. Correlation is achieved by the assumption of some p character (and smaller amounts of $d, f, \cdots$ character) by the electrons.

The electron distribution of the helium atom in field-free space is, of course, spherically symmetric. The atom has, however, a large polarizability in a quadrupole electric field, which we may ascribe to the partial orientation of the prolate ellipsoid.

A consequence of the Pauli exclusion principle is that electrons with parallel spins tend to avoid one another; it is said that about an electron there is a Fermi hole which other electrons tend to avoid. In consequence, four electrons with parallel spins occupying the four sp^3 orbitals of an atom tend to assume relative positions corresponding to the corners of a tetrahedron about the nucleus.[26] Hence the carbon atom in the state $1s^2 2s 2p^3\ ^5S$ may be described as tetrahedral. The effect of correlation is to increase the tetrahedral character by the assumption by the orbitals of some $d, f, \cdots$ character, which concentrates them about the tetrahedral directions.

It has been assumed[26,27] that the neon atom and other atoms with an $s^2 p^6$ outer shell also may be described as tetrahedral. However, the four sp^3 electrons with positive spin are independent of the four with negative spin, and the two corresponding tetrahedra are free to assume arbitrary relative orientation.[28] Correlation requires that the most stable relative orientation be the inverse one; hence the neon atom and the other $s^2 p^6$ atoms can be described as cubic. Their polarizability in a cubic multipole electric field is large and that in a tetrahedral field small.[29]

Cuthbert and Linnett[30] have suggested that the stability of the cubic closest packed arrangement of atoms in crystals of neon, argon, krypton, and xenon (helium crystallizing instead in hexagonal closest packing)[31] is explicable by the tetrahedral electron distribution of the atoms

[25] H. Shull and P.-O. Löwdin, *J. Chem. Phys.* **25**, 1035 (1956). The correlation energy is nearly the same in H_2, H^-, Li^+, etc., as in He. See also P. G. Dickens and J. W. Linnett, *Quart. Rev. Chem. Soc.* **11**, 291 (1957).

[26] K. Artmann, *Z. Naturforsch.* **1**, 426 (1946); H. K. Zimmerman, Jr., and P Van Rysselberghe, *J. Chem. Phys.* **17**, 598 (1949); J. E. Lennard-Jones, *ibid* **20**, 1024 (1952); J. Lennard-Jones and J. A. Pople, *Discussions Faraday Soc.* **1951**, 9.

[27] Linnett and Poë, *loc. cit.* (10).

[28] Lennard-Jones, *loc. cit.* (26).

[29] The phase when two electrons (largely $2s$ in character) are close to the nucleus may be described as octahedral; the other six outer electrons (largely $2p$) would tend to be near the corners of an octahedron. Both the cubic and the octahedral aspects of the atom contribute to its cubic polarizability.

[30] J. Cuthbert and J. W. Linnett, *Trans. Faraday Soc.* **54**, 617 (1958).

[31] Helium 3 crystallizes also with the cubic body-centered arrangement: A. F. Schuch, E. R. Grilly, and R. L. Mills, *Phys. Rev.* **110**, 775 (1958).

with four unshared electron pairs in the outer shell.   The large cubic polarizability, discussed above, may provide the explanation.

In methyl fluoride two electrons with opposed spins are concentrated along the C—F bond.   The fluorine atom is, in consequence of correlation, presumably not cylindrically symmetrical about the bond direction, but somewhat hexafoliate.   In water and dimethyl ether the two unshared electron pairs of the oxygen atom, despite the effect of correlation, are directed toward two corners of the tetrahedron that has its other two corners determined by the two bonds.

4-7. RESTRICTED ROTATION ABOUT SINGLE BONDS

The single-bond orbitals discussed above are cylindrically symmetrical about the bond direction, and hence the energy of a molecule, insofar as it is determined by the bond orbital, should be independent of the orientation of the two parts of a molecule about the axis of a single bond.   Other interactions also between the two parts of a molecule such as ethane might be expected not to depend much on this orientation, so that the molecule might show essentially free rotation about the single bond.   This is in agreement with chemical experience—no case of isomerism corresponding to restriction of rotation about a pure single bond has been reported.

It has been found, however, that the forces restricting rotation about single bonds, though not large enough to permit the isolation of isomers, are large enough to be of significance in structural chemistry, as, for example, in the calculation of entropy values from structural information.   It was shown by Kemp and Pitzer[32] that the value of the entropy of ethane indicates very strongly that as the two methyl groups are rotated about the single carbon-carbon bond relative to one another the potential energy of the molecule changes by about 3 kcal /mole, the potential function having three maxima and three minima in a complete rotation, corresponding to the trigonal symmetry of the methyl groups.   Further evidence supporting this restriction of rotation was reported by Kistiakowsky, Lacher, and Stitt.[33]   Values close to 3 kcal/mole have been reported also for several other alkanes.

Attack on the problem of development of a theory of the potential barriers was begun by Eyring,[34] who made approximate quantum-mechanical calculations of the interactions of the hydrogen atoms of the two methyl groups.   Various suggestions and calculations about the importance of van der Waals repulsion between attached groups,

[32] J. D. Kemp and K. S. Pitzer, *J.A.C.S.* **59**, 276 (1937).

[33] G. B. Kistiakowsky, J. R. Lacher, and F. Stitt, *J. Chem. Phys.* **7**, 289 (1939).

[34] H. Eyring, *J.A.C.S.* **54**, 3191 (1932).

electrostratic interactions of the charge distributions in the bonds between the two carbon atoms and the attached groups, and intrinsic lack of cylindrical symmetry in the axial chemical bond itself have been summarized by Wilson,[35] who tested these hypotheses by comparison with the values of the potential barriers that have been determined experimentally.

It seems likely that the potential barrier in ethane and similar molecules results from the exchange interactions (repulsions) of electrons involved in the other bonds (adjacent bonds) formed by each of the two atoms.[36] It is found on evaluation of the interaction energy of a methyl group with another group that the energy is independent of the orientation of the methyl group about its bond to the other group if the bond orbitals are sp hybrids. If, however, they have some d and f character, as described in the preceding section, a potential energy function with three maxima and three minima is produced. The height of the potential hump is proportional to the amounts of d and f character and to the number of adjacent bonds (with concentrated bond orbitals) formed by the two atoms that are connected by the single bond under consideration. For the C—C single bond this effect is estimated to produce a potential energy maximum of about 3 kcal/mole, as observed in ethane and similar molecules.

Many precise values of the height of the potential energy maximum have been obtained by the methods of microwave spectroscopy, especially through the efforts of E. B. Wilson, Jr., and his collaborators. The values 3.30 for CH_3CH_2F and 3.18 for CH_3CHF_2[37] agree roughly with that for ethane.

It is predicted by the theory that the minima of the potential energy function correspond to the staggered configuration (repulsion of the adjacent bonds) rather than to the eclipsed configuration. The staggered configuration was found for many hydrocarbon chains in crystals by x-ray diffraction studies. In addition, it was pointed out by V. Schomaker[38] that the variation of values of enthalpy of hydrogenation of unsaturated cyclic hydrocarbons can be interpreted in a convincing manner on the assumption that the staggered orientation about single bonds is the stable one. Microwave studies have verified the staggered

[35] E. B. Wilson, Jr., *Proc. Nat. Acad. Sci. U. S.* **43**, 816 (1957).

[36] Pauling, *loc. cit.* (21). This theory is similar to the earlier proposals of G. B. Kistiakowsky, J. R. Lacher, and W. W. Ransom, *J. Chem. Phys.* **6**, 900 (1938), and K. S. Pitzer, *Quantum Chemistry*, Prentice-Hall, New York, 1953, p. 168.

[37] D. R. Herschbach, *J. Chem. Phys.* **25**, 358 (1956).

[38] See L. Pauling, *The Nature of the Chemical Bond*, 2nd ed., Cornell University Press, 1940, p. 91.

configuration for CH_3CH_2Cl,[39] CH_3CF_3,[40] CH_3SiH_3,[41] CH_3SiH_2F,[42] $(CH_3)_2O$,[43] and several other molecules.

Moreover, the height of the barrier changes from molecule to molecule in the expected way. The barrier interaction involves the same integral over the radial parts of the wave functions as the axial bond itself, and it would accordingly be expected that for molecules in which the bond orbitals have similar hybrid character the barrier height would be a constant fraction of the bond energy. In particular, for different substituted ethanes essentially the same barrier would be found, provided that the substituent groups are not large enough to cause steric effects, which would increase the barrier height. Approximate constancy is observed for ethane and substituted ethanes, as mentioned above.

The energy of a carbon-silicon bond and that of a carbon-germanium bond are about three quarters as great as that of a carbon-carbon bond, and it would therefore be expected that the height of the barrier in molecules containing these bonds would be about 2.3 kcal/mole; observed values are somewhat smaller: 1.70 for CH_3SiH_3, 1.66 for $(CH_3)_2SiH_2$, 1.56 for CH_3SiH_2F, 1.32 for CH_3SiHF_2, and 1.2 for CH_3GeH_3.

Only bond orbitals, and not orbitals for unshared pairs (except a small amount due to correlation), have d and f character, and accordingly an OH group is expected to interact with a methyl group one-third as strongly as a methyl group itself would, and an NH_2 group two-thirds as strongly. For CH_3OH and CH_3NH_2 the expected barrier heights are hence about 1.0 and 2.0 kcal/mole, respectively; the observed values are 1.07 for methanol[44] and 1.90 for methylamine.[45]

Propylene epoxide, CH_3CHOCH_2, contains a methyl group adjacent to a three-membered ring. Two of the bond orbitals of the ring carbon atom to which the methyl carbon atom is bonded can be described as bent toward each other, in such a way as to decrease somewhat the interaction energy restricting the rotation of the methyl group. The ob-

[39] R. S. Wagner and B. P. Dailey, *J. Chem. Phys.* **23**, 1355 (1955); **26**, 1588 (1957).

[40] W. F. Edgell, G. B. Muller, and J. W. Amy, *J.A.C.S.* **79**, 2391 (1957).

[41] R. W. Kilb and L. Pierce, *J. Chem. Phys.* **27**, 108 (1957); D. Kivelson, *ibid.* **22**, 1733 (1954).

[42] L. Pierce, quoted by E. B. Wilson, Jr., *Proc. Nat. Acad. Sci. U. S.* **43**, 816 (1957).

[43] Electron-diffraction investigation, K. Kimura and M. Kubo, *Nature*, **183**, 533 (1959).

[44] E. V. Ivash and D. M. Dennison, *J. Chem. Phys.* **21**, 1804 (1953).

[45] D. R. Lide, Jr., *J. Chem. Phys.* **22**, 1613 (1954); K. Shimoda, T. Nishikawa, and T. Itoh, *J. Phys. Soc. Japan* **9**, 974 (1954).

served value[46] for the height of the barrier, 2.56 kcal/mole, is somewhat less than that for the substituted ethanes, as expected from this consideration.

For nitromethane, CH_3NO_2, and methyl difluoroborane, CH_3BF_2, symmetry requires that the potential barrier have six maxima. The experimental values for the barrier height are very small, 0.006 and 0.014 kcal/mole, respectively.[47,48] These small barriers result from interactions of higher order than those in molecules such as ethane.

In ethanes with large substituent atoms, such as chlorine or bromine, the steric effects of van der Waals repulsions may cause the potential energy maxima to be higher than they are in ethane itself and the fluoroethanes. The value 2.91 kcal/mole found for methyl chloroform by infrared spectroscopy[49] is not larger than that for the fluoroethanes, and the conclusion might be drawn that the steric effects between chlorine atoms and hydrogen atoms on adjacent carbon atoms are small. On the other hand, the values 3.560 ± 0.012 kcal/mole for CH_3CH_2Cl and 3.567 ± 0.030 kcal/mole for CH_3CH_2Br have been reported from microwave-spectroscopic studies,[50] and it is likely that these values are more reliable than the infrared value, and that there is some steric repulsion between the halogen atom and hydrogen atoms on the adjacent carbon atom, which increases the barrier height. Moreover, electron-diffraction studies of 1,2-dichloroethane,[51] 1,2-dibromoethane,[52] 1,2-chlorobromoethane,[52] and 2,3-dibromobutane[53] have shown that for all of these molecules the stable orientation is that in which the two halogen atoms are on opposite sides of the carbon-carbon axis. For 1,1,2-trichloroethane[54] the stable orientation is similar, the 2-chlorine atom being nearly opposite one of the 1-chlorine atoms. For 1,1,2,2-tetrachloroethane[55] two staggered forms are present. A study of dipole moments has been reported[56] to show that these forms differ in enthalpy by 0.0 ± 0.2 kcal/mole, whereas for 1,1,2-trichloroethane the cis configuration (with the 2-chlorine atom in the staggered configuration be-

[46] J. D. Swalen and D. R. Herschbach, *J. Chem. Phys.* **27**, 100 (1957).

[47] E. Tannenbaum, R. D. Johnson, R. J. Myers, and W. D. Gwinn, *J. Chem. Phys.* **22**, 949 (1954); E. Tannenbaum, R. J. Myers, and W. D. Gwinn, *ibid.* **25**, 42 (1956).

[48] R. E. Naylor, Jr., and E. B. Wilson, Jr., *J. Chem. Phys.* **26**, 1057 (1957).

[49] K. S. Pitzer and J. L. Hollenberg, *J.A.C.S.* **75**, 2219 (1953).

[50] D. R. Lide, Jr., *J. Chem. Phys.* **30**, 37 (1959).

[51] J. Y. Beach and K. J. Palmer, *J. Chem. Phys.* **6**, 639 (1938).

[52] J. Y. Beach and A. Turkevich, *J.A.C.S.* **61**, 303 (1939).

[53] D. P. Stevenson and V. Schomaker, *J.A.C.S.* **61**, 3173 (1939).

[54] A. Turkevich and J. Y. Beach, *J.A.C.S.* **61**, 3127 (1939).

[55] V. Schomaker and D. P. Stevenson, *J. Chem. Phys.* **8**, 637 (1940).

[56] J. R. Thomas and W. D. Gwinn, *J.A.C.S.* **71**, 2785 (1949).

tween the two 1-chlorine atoms) is at least 4 kcal/mole less stable than the skew configuration. Values of rotational isomerization energies between 1.5 and 2.0 kcal/mole have been reported[57] for chloroethanes in solution. The height of the potential energy maximum in hexachloroethane[58] has been found to be at least 7 kcal/mole; the difference between this value and that for ethane (3 kcal/mole) is due to the steric repulsion between chlorine atoms. Calculated barriers due to steric repulsions have been given by Mason and Kreevoy.[59]

Restricted Rotation about Single Bonds between Atoms with Unshared Electron Pairs.—In a molecule such as H_2O_2 the repulsion of the unshared electron pairs of the two oxygen atoms may largely determine the orientation of the groups about the bond axis. If the two unshared pairs of each atom occupy the two oppositely directed orbitals formed by hybridization of an *s* orbital and a *p* orbital, their repulsion would cause the plane of these orbitals for one oxygen atom to be perpendicular to that for the other oxygen atom, and the dihedral angle formed by the bonds H—O—O and O—O—H would then be 90°, as was pointed out by Penney and Sutherland.[60] With each of the unshared-pair orbitals assumed to have 44 percent *s* character and 56 percent *p* character, and their maxima making an angle of 142°, as described in Section 4-3, a value of the dihedral angle differing by a few degrees from 90° might be expected. The experimental value 89° $\pm$ 2° has been reported[61] from a neutron-diffraction investigation of crystals of hydrogen peroxide in which the positions of the hydrogen atoms were located. Hydrogen bonds are present in the crystal, and the value of the dihedral angle determined by the hydrogen-bonded oxygen atoms is 93.8°.

Values for the dihedral angle between 100° and 106° have been found for many molecules containing S—S, Se—Se, and Te—Te bonds. Rhombic sulfur[62] and sulfur vapor[63] contain molecules S_8 that are staggered eight-membered rings with bond angle 105° and dihedral angle 102°. Two crystalline forms of selenium[64] contain similar molecules, with bond angle 106° and dihedral angle 101°. Another form of sele-

[57] J. Powling and H. J. Bernstein, *J.A.C.S.* **73**, 1815 (1951).

[58] D. A. Swick, I. L. Karle, and J. Karle, *J. Chem. Phys.* **22**, 1242 (1954).

[59] E. A. Mason and M. M. Kreevoy, *J.A.C.S.* **77**, 5808 (1955).

[60] W. G. Penney and G. B. B. M. Sutherland, *J. Chem. Phys.* **2**, 492 (1934)

[61] Busing and Levy, *loc. cit.* (11).

[62] B. E. Warren and J. T. Burwell, *J. Chem. Phys.* **3**, 6 (1935).

[63] C. S. Lu and J. Donohue, *J.A.C.S.* **66**, 818 (1944).

[64] R. D. Burbank, *Acta Cryst.* **4**, 140 (1951); **5**, 236 (1952); R. E. Marsh, L. Pauling, and J. D. McCullough, *ibid.* **6**, 71 (1953).

nium consists of infinite helical molecules, with three atoms per turn; the bond angle is 105° and the dihedral angle 102°. For the similar crystal tellurium the bond angle is 102° and the dihedral angle 100°. Other reported values of the dihedral angle are 106° for S—S—S—C in dimethyl trisulfide[65] and 82° for S—S—S—C in diiododiethyltrisulfide.[66]

Rhombohedral sulfur contains S_6 molecules, and some S_6 in equilibrium with S_8 and S_2 is found in sulfur vapor. The dihedral angle in S_6 is 71° (assuming bond angle 104°, slightly strained from S_8). The difference in enthalpy[67] of S_6 and S_8 is 1.10 kcal per S—S bond. If the energy as a function of dihedral angle δ is assumed to have the simple form $A \cos \delta + B \cos^2 \delta$, this difference in enthalpy and the value $\delta = 102°$ for the minimum lead to $A = 1.64$ kcal/mole and $B = 3.95$ kcal/mole. The heights of the potential energy barrier are 5.6 kcal/mole at $\delta = 0°$ (cis configuration) and 2.5 kcal/mole at $\delta = 180°$ (trans).

Fibrous sulfur can be made by chilling viscous sulfur from about 350°C and then stretching. The x-ray pattern indicates[68] that it consists of helical chains with $3\frac{1}{2}$ atoms per turn, bond angle 106°, and dihedral angle 80° (Fig. 4-7). Treatment with solvent leads to a more stable fibrous form[69] with $3\frac{1}{3}$ atoms per turn, bond angle 106°, and dihedral angle 85°. The values of the dihedral angle correspond to strain energy about 0.5 kcal/mole per bond, which is compatible with the stability relative to S_8 (no strain) and S_6 (strain energy 1.10 kcal/mole per bond). Rhombohedral sulfur (S_6) spontaneously changes on standing to fibrous sulfur, which then changes to rhombic sulfur (S_8).

It seems likely that the stable value of the dihedral angle for group VI elements is in general about 102°, and that the barrier (at 180°) is about 2 or 3 kcal/mole. The value 0.32 kcal/mole reported for H_2O_2 by microwave spectroscopy[70] seems small in comparison with the values discussed above, although some decrease, perhaps 50 percent, might be expected because of the partial ionic character of the O—H bonds.

It is hard to make a prediction about hydrazine, although the staggered configuration with the unshared electron pairs in the trans position might be expected from the foregoing considerations. The configuration has not yet been determined for the gas molecule; the ar-

[65] J. Donohue and V. Schomaker, *J. Chem. Phys.* **16**, 92 (1948).

[66] J. Donohue, *J.A.C.S.* **72**, 2701 (1950).

[67] Bureau of Standards Tables.

[68] J. J. Trillat and H. Forestier, *Bull. soc. chim. France* **51**, 248 (1932); K. H. Meyer and Y. Go, *Helv. Chim. Acta* **17**, 1081 (1934); M. L. Huggins, *J. Chem. Phys.* **13**, 37 (1945); L. Pauling, *Proc. Nat. Acad. Sci. U. S.* **35**, 495 (1949).

[69] J. A. Prins, J. Schenk, and P. A. M. Hospel, *Physica* **22**, 770 (1956); J. Schenk, *ibid.* **23**, 325 (1957).

[70] J. T. Massey and D. R. Bianco, *J. Chem. Phys.* **22**, 442 (1954).

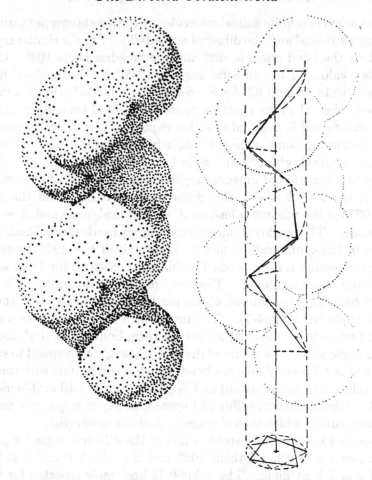

Fig. 4-7.—The helix with seven atoms in two turns, proposed as the repeating unit in the chains of fibrous sulfur.

rangement of nitrogen atoms in the crystal has been described as favoring an eclipsed configuration.[71]   For hydroxylamine the eclipsed configuration, with the unshared electron pair of the amino group over the O—H bond, is probably the stable one.   There is some experimental evidence supporting it.[72]

4-8. ORBITALS AND BOND ANGLES FOR MULTIPLE BONDS

For a double bond, as in ethylene, two orbitals of each of the two bonded atoms are required.   There are two alternative ways in which

[71] R. L. Collin and W. N. Lipscomb, *Acta Cryst.* **4,** 10 (1951).

[72] P. A. Giguère and I. D. Liu, *Can. J. Chem.* **30,** 948 (1952); E. A. Meyers and W. N. Lipscomb, *Acta Cryst.* **8,** 583 (1955).

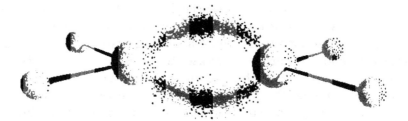

Fig. 4-8.—A representation of the ethylene molecule with
the double bond shown as two bent single bonds.

these orbitals have usually been described. In the first[73] the two orbitals for each atom have been assumed to be essentially tetrahedral
orbitals, extending toward two corners of a tetrahedron defining an
edge shared with another tetrahedron representing the other atom with
which the double bond is formed. This leads to a description of the
double bond, as involving two bent single bonds, that is closely similar
to the one used by organic chemists for many decades (Fig. 4-8).
Baeyer,[74] for example, explained the instability of the carbon-carbon

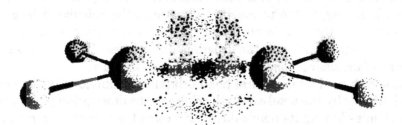

Fig. 4-9.—A representation of the ethylene molecule with
the double bond shown as a sigma bond and a pi bond.

double bond relative to two single bonds as resulting from the strain
energy of bending the two bonds that constitute the double bond.

The other description of the double bond[75] is in terms of a σ bond,
formed by a σ orbital for each atom directed toward the other atom,
and a π bond, formed by a π orbital for each atom, as shown in Figure
4-9.

When the quantum mechanical equations are examined it is found
that the two descriptions of the double bond are identical in the molecular-orbital treatment based on s-p hybrids.[76] They are not identical

[73] L. Pauling, *J.A.C.S.* **53**, 1367 (1931); Slater, *loc. cit.* (5).

[74] A. Baeyer, *Ber.* **18**, 2269 (1885).

[75] E. Hückel, *Z. Physik* **60**, 423 (1930); W. G. Penney, *Proc. Roy. Soc. London*
A144, 166 (1934); **A146**, 223 (1934).

[76] G. G. Hall and J. Lennard-Jones, *Proc. Roy. Soc. London* **A205**, 357 (1951).

in the valence-bond treatment, especially when the bond orbitals are concentrated about their bond directions by the assumption of d and f character, as described in Section 4-5. In the case of the bent-bond orbitals the d and f character increases the electron density in two regions, one on each side of the internuclear axis (near the tetrahedron corners determining the shared edge). In the case of the σ-π description the d and f character increases the electron density in three regions, one on the internuclear axis and one on either side of it. The greater separation of the electrons for the bent-bond structure with concentrated bond orbitals than for the σ-π structure may stabilize the bent-bond structure enough to make it the better approximation to use in discussing multiple bonds in general.[77] In addition, it has the advantage of being the more closely related to single bonds, whose properties are well known.

The bent-bond structure accounts in a simple way for some of the properties of the double bond and the triple bond, such as the bond lengths. The carbon-carbon single-bond, double-bond, and triple-bond lengths are 1.54, 1.33, and 1.20 Å, respectively. If the multiple bonds are represented by arcs with constant curvature, with length 1.54 Å and beginning in the tetrahedral directions, the calculated lengths are 1.32 Å for the double bond and 1.18 Å for the triple bond, in approximate agreement with the experimental values. There is no similar way of discussing bond lengths for the σ-π structure.

Inasmuch as the tetrahedral bond orbitals are individually the best bond orbitals by the bond-strength criterion, the description of the carbon atom as having its bonds directed toward the corners of a regular tetrahedron would be expected to apply to a carbon atom forming two single bonds and one double bond as well as to one forming four single bonds. Hence for the bent-bond structure the value 125.27° is predicted between a single bond and a double bond of a carbon atom. For the σ-π structure a smaller value might be predicted; for example, Coulson[78] has described the orbitals for the σ bond and the two single bonds as the trigonal orbitals described in Section 4-4, which give the value 120°. Values close to 125.27° have been found by experiment for many molecules; some of the most reliable of the experimental values are given in Table 4-5. Values equal to $125 \pm 3°$ have been reported for C—C=C in other unsaturated molecules, for O—C=O in many other carboxylic acids, and for N—C=O in many other amides and peptides.[79]

[77] L. Pauling, Kekulé Address, London, Sept. 15, 1958.

[78] C. A. Coulson, *Valence*, Clarendon Press, Oxford, 1952, p. 195.

[79] For a summary, see R. B. Corey, *Fortschr. Chem. org. Naturstoffe* **8**, 310 (1951); L. Pauling and R. B. Corey, *ibid.* **11**, 180 (1954).

TABLE 4-5.—EXPERIMENTAL VALUES OF SINGLE-BOND:
DOUBLE-BOND ANGLE FOR QUADRICOVALENT ATOMS

Compound	Method[a]	Angle	Value	Reference
Propylene	M	C—C=C	124.75° ±0.3°	1
$F_2C=CH_2$	M	F—C=C	125.2° ±0.2°	2
CCl_2O	M	Cl—C=O	124.3° ±0.1°	3
$Cl_2C=CH_2$	M	Cl—C=O	123.2° ±0.5°	4
CH_3CHO	M	C—C=O	123.9° ±0.3°	5
Propynal, HCCCHO	M	C—C=O	123.8° ±0.2°	5a
HCOOH	M	O—C=O	124.5° ±0.5°	6
Glycine	X	O—C=O	125.5° ±0.3°	7
Alanine	X	O—C=O	125.6° ±0.5°	8
Oxamide	X	N—C=O	125.7° ±0.3°	9
Formamide	M	N—C=O	123.58° ±0.35°	10
α-Glycylglycine	X	N—C=O	124.2° ±1.0°	11
N,N′-Diglycylcystine	X	N—C=O	125 3° ±1.0°	12
OHCNHNHCHO	X	N—C=O	124.9° ±0.4°	13
Dithio-oxamide	X	N—C=S	124.8° ±0.5°	14
CH_3NO_2	M	O—N=O	127° ±4°	15
$O_2NNHC_2H_4NHNO_2$	X	O—N=O	125° ±3°	16
p-Dinitrobenzene	X	O—N=O	124° ±3°	17

[a] M denotes microwave spectroscopy of gas molecules and X denotes x-ray diffraction of crystals.

1. D. R. Lide, Jr., and D. E. Mann, *J. Chem. Phys.* **27,** 868 (1957).
2. W. F. Edgell, P. A. Kinsey, and J. W. Amy, *J.A.C.S.* **79,** 2691 (1957).
3. G. W. Robinson, *J. Chem. Phys.* **21,** 1741 (1953).
4. S. Sekino and T. Nishikawa, *J. Phys. Soc. Japan* **12,** 43 (1957).
5. R. W. Kilb, C. C. Lin, and E. B. Wilson, Jr., *J. Chem. Phys.* **26,** 1695 (1957).
5a. C. C. Costain and J. R. Morton, *J. Chem. Phys.* **31,** 389 (1959).
6. R. Trambarulo and P. M. Moser, *J. Chem. Phys.* **22,** 1622 (1954); R. G. Lerner, J. P. Friend, and B. P. Dailey, *ibid.* **23,** 210 (1955).
7. R. E. Marsh, *Acta Cryst.* **11,** 654 (1958).
8. J. Donohue, *J.A.C.S.* **72,** 949 (1950).
9. E. M. Ayerst and J. R. C. Duke, *Acta Cryst.* **7,** 588 (1954).
10. R. J. Kurland, *J. Chem. Phys.* **23,** 2202 (1955).
11. E. W. Hughes, A. B. Biswas, and J. N. Wilson, unpublished research, Calif. Inst. Tech
12. H. L. Yakel, Jr., and E. W. Hughes, *Acta Cryst.* **7,** 291 (1954).
13. Y. Tomiie, C. H. Koo, and I. Nitta, *Acta Cryst.* **11,** 774 (1958).
14. B. Long, P. Markey, and P. J. Wheatley, *Acta Cryst.* **7,** 140 (1954).
15. Tannenbaum, Johnson, Myers, and Gwinn, *loc. cit.* (47).
16. F. J. Llewellyn and F. E. Whitmore, *J. Chem. Soc.* **1948,** 1316.
17. F. J. Llewellyn, *J. Chem. Soc.* **1947,** 884.

Ethylene is an exception; the angle C=C—H is reported as 122.0° by electron diffraction[80] and 121.3° by infrared spectroscopy;[81] it is thus closer to the value corresponding to trigonal quantization than to that corresponding to tetrahedral quantization. The low value, corresponding to a high value of the angle HCH (116°, 117.4°), may be the result of large van der Waals repulsion of the hydrogen atoms (Sec. 7-12). A similar behavior is shown by formaldehyde, for which the H—C=O angle[82] is 119.2° ± 1.0°. This very low value for formaldehyde may be associated with the large amount of ionic character (totaling about 44 percent) of the carbon-oxygen bond.

It has been pointed out[83] that the angle —N̈= in pyrimidines is about 11° less than the angle —C=. Also, in *s*-triazine, $C_3N_3H_3$, which is a planar molecule with trigonal symmetry, the two angles have the values[84] 113.2° ± 0.4° and 126.8° ± 0.4°, respectively, with difference −13.6°, and in *s*-tetrazine, $C_2N_4H_2$, they have the values[85] 115.9° ± 0.7° and 127.4° ± 0.7°, respectively, with difference −11.5°. (In all of these planar six-membered rings the bond angles have average value 120°, and hence it is the difference that is significant.) We conclude that the —N= angle is about 12° less than the —C=angle (125.27°), and hence that its normal value is 113°.

Other experimental values for the angle —N̈= are given in Table 4-6. It may be seen that they are in general compatible with the value 113° described above as the normal value.

The explanation of the deviation of the value from 125.27° (the tetrahedral value) is that the bond orbitals of the tricovalent nitrogen atom are not tetrahedral orbitals, but are orbitals with only about 5 percent of *s* character, plus small amounts of *d* and *f* character (Secs. 4-3 and 4-5). For these orbitals a value for the single-bond:double-bond angle intermediate between the values 90° for pure *p* orbitals and 125.27° for tetrahedral orbitals would be expected.

Another sort of evidence supporting the bent-bond structure of the double bond is provided by the information about restricted rotation. The bent-bond structure for propylene leads to the expectation that the potential function hindering the rotation of the methyl group would be nearly the same as in ethane, but with the barrier a little smaller than in ethane because two of the bonds on the adjacent carbon atom (the bent bonds) are distorted; the stable orientation would be the stag-

[80] L. S. Bartell and R. A. Bonham, *J. Chem. Phys.* **27**, 1414 (1957).

[81] H. C. Allen, Jr., and E. K. Plyler, *J.A.C.S.* **80**, 2673 (1958).

[82] Calculated from the spectroscopic value of the moment of inertia about the symmetry axis with the assumption that the C—H distance is 1.08 ± 0.01 Å.

[83] L. Pauling and R. B. Corey, *Arch. Biochem. Biophys.* **65**, 164 (1956).

[84] P. J. Wheatley, *Acta Cryst.* **8**, 224 (1955).

[85] F. Pertinotti, G. Giacomello, and A. M. Liquori, *Acta Cryst.* **9**, 510 (1956).

gered one—that is, staggered with respect to the bond to hydrogen (and the two bent bonds), and hence eclipsed with respect to the axis of the double bond. The σ-π structure predicts a much lower barrier, because the σ bond would cancel the bond to the hydrogen atom and the π electron, extending to both sides of the plane, would have nearly the same interaction for the staggered as for the eclipsed orientation.

TABLE 4-6.—EXPERIMENTAL VALUES FOR THE SINGLE-BOND: DOUBLE-BOND
ANGLE OF THE TRICOVALENT NITROGEN ATOM

Substance	Method[a]	Angle	Value	Reference
NOF	M	F—N=O	$110° \pm 5°$	1
NClO	M	Cl—N=O	$113° \pm 2°$	2
NBrO	E	Br—N=O	$117° \pm 3°$	3
cis-NO(OH)	I	O—N=O	$114° \pm 2°$	4
trans-NO(OH)	I	O—N=O	$118° \pm 2°$	4
NO_2^-	X	O—N=O	$115.4° \pm 1.7°$	5
N_2F_2	E	F—N=N	$115° \pm 5°$	6
$(CH_3)_2N_2$	E	C—N=N	$110° \pm 10°$	7
Cyanuric triazide, C_3N_{12}	X	C—N=C	$113° \pm 5°$	8

[a] M, microwave spectroscopy; E, electron diffraction; I, infrared spectroscopy; X, x-ray diffraction of crystals.

1. D. W. Magnuson, *J. Chem. Phys.* **19**, 1071 (1951).
2. J. D. Rogers, W. J. Pietenpol, and D. Williams, *Phys. Rev.* **83**, 431 (1951).
3. J. A. A. Ketelaar and K. J. Palmer, *J.A.C.S.* **59**, 2629 (1937). These authors also reported $116° \pm 2°$ for Cl—N=O in NClO.
4. L. H. Jones, R. M. Badger, and G. E. Moore, *J. Chem. Phys.* **19**, 1599 (1951).
5. G. B. Carpenter, *Acta Cryst.* **8**, 852 (1955).
6. S. H. Bauer, *J.A.C.S.* **69**, 3104 (1947).
7. H. Boersch, *Sitzber. Akad. Wiss. Wien* **144**, 1 (1935).
8. I. E. Knaggs, *Proc. Roy. Soc. London* **A150**, 576 (1935); E. W. Hughes, *J. Chem. Phys.* **3**, 1, 650 (1935).

In fact, for propylene[86] the height of the barrier is 1.98 kcal/mole and for 1-methyl-2-fluoroethylene[87] it is 2.15 kcal/mole, and the configuration as predicted for the bent-bond structure has been verified for propylene[88] and for acetylcyanide.[89] In aldehydes and related substances the height of the barrier is somewhat less than in propylene, as would be expected because of the partial ionic character of the C=O

[86] Lide and Mann, *loc. cit.* (T4-5).
[87] S. Siegel, *J. Chem. Phys.* **27**, 989 (1957).
[88] D. R. Herschbach and L. C. Krisher, *J. Chem. Phys.* **28**, 728 (1958).
[89] L. C. Krisher and E. B. Wilson, Jr., *Am. Chem. Soc. Meeting*, Boston, April 1959.

bond: 1.15 kcal/mole for acetaldehyde,[90] 1.08 for CH_3COF,[91] 1.35 for CH_3COCl,[92] and 1.27 for CH_3COCN.[93] (These values suggest about 40 percent partial ionic character.) The expected configuration has been verified for all four substances.

The bent-bond model of the triple bond gives this bond a threefold symmetry axis and leads to the prediction that the two methyl groups in dimethylacetylene should be somewhat restricted in their mutual rotation, with the eclipsed configuration stable. Restriction of rotation about the single bonds in conjugated systems is also expected, and the nature of the stable configurations can be predicted from the theory described above. These systems are discussed in Chapters 6 and 8.

4-9. PARTIAL IONIC CHARACTER OF MULTIPLE BONDS

The expectation from the bent-bond description of multiple bonds is that each of the bent single bonds of the multiple bond would have the same amount of partial ionic character as an unstrained single bond between the two atoms. This expectation is borne out reasonably well by experiment. For example, for formaldehyde, H_2CO, the H—C bond is expected from the electronegativity difference of the atoms to have 4 percent ionic character and each of the C—O bonds to have 22 percent. The electric dipole moment is then calculated (H—C bond length 1.09 Å, C=O 1.23 Å, angles 120°) to be 2.81 D, somewhat larger than the experimental value, 2.27 D. If the reasonable assumption is made that for each C—O single bond the ionic-covalent ratio is 22/78 but that the structure with both bonds ionic makes no contribution, the calculated value of the moment is 2.34 D, in good agreement with experiment. Similar agreement is found also for other molecules; it is necessary, however, to be alert for the possibility of significant contribution by resonance structures, as discussed in later chapters of this book.

4-10. THE EFFECT OF UNSHARED PAIRS ON BOND ENERGIES AND BOND LENGTHS[94]

The bond energies of the bonds N—N, O—O, and F—F show a striking abnormality; their values are much smaller than would be expected from extrapolation of the corresponding values for their congeners, whereas the values for Li—Li and C—C (and also for Be—Be and

[90] K. T. Hecht and D. M. Dennison, *J. Chem. Phys.* **26**, 31 (1957).
[91] Kilb, Lin, and Wilson, *loc. cit.* (T4-5).
[92] J. D. Swalen, *J. Chem. Phys.* **24**, 1072 (1956).
[93] Krisher and Wilson, *loc. cit.* (89).
[94] K. S. Pitzer, *J.A.C.S.* **70**, 2140 (1948); R. S. Mulliken, *ibid.* **72**, 4493 (1950); **77**, 884 (1955).

B—B, not given in Chapter 3) are about as expected (Sec. 3-5). The bond lengths of these bonds are also larger than would be expected in comparison with those of their congeners (Chap. 7). These abnormalities are to be attributed to the strong repulsion of the unshared electron pairs of the bonded atoms (and of the electrons involved in the other bonds in the case of N—N and O—O).

The quantum-mechanical theory of valence leads to the result that the interaction energy of two unshared electron pairs on adjacent

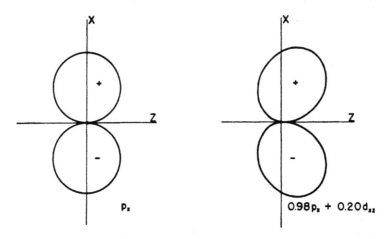

Fɪɢ. 4-10.—The effect of hybridization of a π orbital. At the left is shown the angular dependence of orbital strength for a pure $p\pi$ orbital, p_x (x axis vertical). At the right is shown a π orbital with 4.5 percent d character. It is seen that the d character increases overlap of the orbital with a similar orbital to the right (bonding overlap) and decreases overlap with a similar orbital to the left (nonbonding overlap).

atoms is -2 times the bond resonance energy of the bond that would be formed by a shared pair occupying the same orbitals; it is hence a destabilizing interaction. (The factor for a shared pair on one atom and a bonding electron on the other is -1, and for a bonding electron [to other atoms] on each atom $-\frac{1}{2}$.) The amount of destabilization is, of course, decreased by hybridization that decreases the overlap of the orbitals occupied by unshared pairs, as shown in Figure 4-10, whereas for bond orbitals hybridization increases the overlap.

For F—F, for example, the bond energy 100 kcal/mole would be expected from extrapolation of the sequence I—I, Br—Br, Cl—Cl, with Sn—Sn, Ge—Ge, Si—Si, C—C as a guide. This value is 63 kcal/mole greater than the actual value, 36.6 kcal/mole. It therefore corresponds to a value of the energy of repulsion of the unshared electron pairs of the fluorine atoms that is reasonable in comparison with the bond energy.

Similar effects are not shown by the heavier atoms. It is likely that the orbitals occupied by the unshared pairs of these atoms have a larger amount of d and f character, such that there is much less overlap than for the elements of the first row. For chlorine, for example, the $3d$ orbitals can be hybridized with $3s$ and $3p$ with a much smaller promotion energy than is required for the d character of the $2s$ and $2p$ orbitals of fluorine, which must go beyond its valence shell (to $3d$) for d character.

CHAPTER 5

Complex Bond Orbitals; The Magnetic Criterion for Bond Type

THE hybrid bond orbitals discussed in the preceding chapter have been described as having only a small amount of d and f character. The bonds in many molecules and complex ions, especially those involving atoms of the transition elements, can be discussed in a simple way in terms of hybrid orbitals with a large amount of d character (and, in a few cases, f character). These bonds and a magnetic criterion for bond type are discussed in the following sections.

5-1. BONDS INVOLVING d ORBITALS

The first-row atoms can form no more than four stable bond orbitals. For the second-row atoms the s and p orbitals of the M shell are much more stable than the d orbitals, and in general contribute preponderantly to the bond orbitals, but the promotion energy for the d orbitals (which also are in the M shell) is small enough to permit these orbitals to take a larger part in bond formation than for the first row atoms.

The existence of compounds such as PF_5, PF_3Cl_2, PCl_5, $[PF_6]^-$, and SF_6 suggests that one or two $3d$ orbitals are here being used together with the $3s$ orbital and the three $3p$ orbitals (all hybridized to bond orbitals) for bond formation. It seems probable, however, that for fluo-

rides the completely covalent structures such as $F-P$ ⟨with F, F above and F, F below⟩ are of little significance, and that the molecules instead resonate mainly among

145

structures such as

$$\text{F—P}^+\begin{array}{c}:\!\!\overset{\cdot\cdot}{\underset{\cdot\cdot}{\text{F}}}\!:^{-}\quad\text{F}\\ \diagdown\diagup\\ \diagup\diagdown\\ \text{F}\quad\text{F}\end{array}$$

, and so on, involving at most four cova-

lent bonds.   (The four covalent bonds resonate among the five positions, making all bonds in the molecule nearly equivalent in bond type.)
A theoretical treatment of PCl_5 has been carried out that has indicated
that the normal state of the molecule involves not only these structures
but also, in considerable amount, the quinquecovalent structure with
five sp^3d orbitals for the phosphorus atom (Sec. 5-9).

Heavier atoms such as tin form complexes $[MX_6]^{--}$ with chlorine,
bromine, and even iodine; it is likely that some use is made of the d orbitals of the valence shell of the central atom in these complexes.

It is, however, the d orbitals of the shell with total quantum number
one less than that of the valence shell that are of great significance for
bond formation.   In the transition elements the inner d orbitals have
about the same energy as the s and p orbitals of the valence shell; and
if they are not completely occupied by unshared electron pairs they
play a very important part in bond formation.   For the hexamminoco-
baltic ion, for example, structures such as

$$\left[\begin{array}{ccc} \text{H}_3\text{N} & & \text{NH}_3\\ & \diagdown\ \diagup & \\ \text{H}_3\text{N}\!-\!&\!\!\text{Co}\!\!&\!-\!\text{NH}_3\\ & \diagup\ \diagdown & \\ \text{H}_3\text{N} & & \text{NH}_3 \end{array}\right]^{+++}$$

are written.   It is seen on counting electrons that the cobalt atom
(with atomic number 27) holds 24 unshared electrons in addition to the
six pairs shared with nitrogen.   The number of available orbitals is
such that six of the stable orbitals (of the krypton shell) can be used for
bond formation with enough remaining for the unshared pairs.   This
is seen from the following diagram:

$1s$	$2s$	$2p$	$3s$	$3p$	$3d$	$4s$	$4p$

The 24 unshared electrons occupy the orbitals $1s$, $2s$, three $2p$, $3s$, three
$3p$, and three of the $3d$ orbitals, leaving two $3d$ orbitals, the $4s$ orbital,
and the three $4p$ orbitals available for use as bond orbitals.

For the atoms of the first transition group (the iron group) there is
little difference in energy of the $3d$ orbitals and the $4s$ and $4p$ orbitals
(see Fig. 2–19), so that the question as to how these orbitals can be com-

bined to form good bond orbitals becomes an interesting one. Similarly, the orbitals 4*d*, 5*s*, and 5*p* have about the same energy for atoms of the palladium group, and 5*d* 6*s*, and 6*p* for atoms of the platinum group. The following discussion of *dsp* hybridization applies to all three transition groups.

5-2. OCTAHEDRAL BOND ORBITALS

It is found on analysis of the problem that when only two *d* orbitals are available for combination with the *s* and *p* orbitals six equivalent bond orbitals of strength 2.923 (nearly as great as the maximum 3 for the best *spd* hybrid) can be formed, and that these six orbitals have their bond directions toward the corners of a regular octahedron. We accordingly conclude that complexes such as $[Co(NH_3)_6]^{+++}$, $[PdCl_6]^{--}$, and $[PtCl_6]^{--}$ should be octahedral in configuration. This conclusion is of course identical with the postulate made by Werner to account for isomerism in complexes with different substituent groups,[1] and verified also by the x-ray examination of $Co(NH_3)_6I_3$, $(NH_4)_2PdCl_6$, $(NH_4)_2$ $PtCl_6$, and other crystals (see Fig. 5-1).

A polar graph of an octahedral bond orbital is shown in Figure 5-2, from which its great concentration in the bond direction, leading to large overlapping and the formation of a very strong bond, can be seen.

---

[1] There is only one form of a monosubstituted octahedral complex MA_5B A disubstituted complex MA_4B_2 can exist in two isomeric forms, cis and trans:

Cis form.   Trans form.

Two forms can be shown by a trisubstituted complex MA_3B_3:

Optically active stereoisomers can be obtained of a complex such as $M(C_2O_4)_3$, containing oxalate groups which occupy two adjacent octahedral corners:

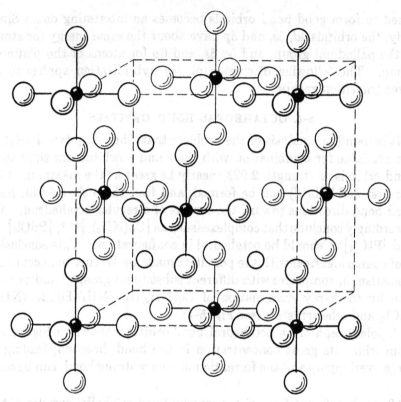

FIG. 5-1.—The structure of the cubic crystal K₂PtCl₆. Octahedral complexes PtCl₆ have their centers at the corners and the centers of the faces of the cubic unit of structure. The potassium ions are at positions $\frac{1}{4}$ $\frac{1}{4}$ $\frac{1}{4}$, etc.; that is, they are at the centers of the eight small cubes with edges one-half as great as those of the large cube. Only four of the eight potassium ions in the cube are represented in the drawing. The chlorine atoms have coordinates $u00$, $0u0$, $00u$, etc., in which u is a parameter determining the Pt—Cl bond length. Its value can be determined by analysis of the x-ray photographs of the crystal. The value of the parameter for most substances of this class is approximately 0.25. For this value of the parameter the chlorine atoms and potassium ions occupy the positions corresponding to cubic closest packing of spheres (Sec. 11-5).

It is interesting to note, as was pointed out to me some years ago by J. L. Hoard, that these considerations lead to an explanation of the difference in stability of cobalt(II) and cobalt(III) as compared with iron (II) and iron (III) in covalent octahedral complexes. The formation of covalent complexes does not change the equilibrium between bipositive and tripositive iron very much, as is seen from the values of the oxida-

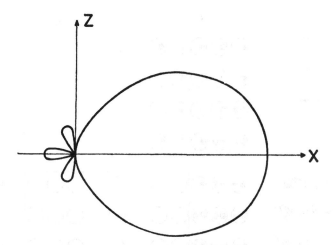

Fig. 5-2.—The angular dependence of an octahedral d^2sp^3 bond orbital with bond direction along the x axis.

tion-reduction potentials,[2] whereas a great change is produced in the equilibrium between bivalent and trivalent cobalt:

$$
\begin{array}{lcl}
 & & \text{Potential} \\
\text{Fe}^{++} = \text{Fe}^{+++} + e^-, & & -0.77 \text{ v.} \\
{[\text{Fe(CN)}_6]}^{----} = {[\text{Fe(CN)}_6]}^{---} + e^-, & & -0.36 \text{ v.} \\
\text{Co}^{++} = \text{Co}^{+++} + e^-, & & -1.84 \text{ v.} \\
{[\text{Co(CN)}_6]}^{----} = {[\text{Co(CN)}_6]}^{---} + e^-, & & +0.83 \text{ v.}
\end{array}
$$

$$-0.41 \text{ v.}$$

$$-2.67 \text{ v.}$$

The effect is so pronounced that covalent compounds of cobalt(II) can decompose water with liberation of hydrogen, whereas the cobalt(III) ion decomposes water with liberation of oxygen, being one of the most powerful oxidizing agents know. The explanation is contained in Figure 5-3. In the ions Co^{++}, Co^{+++}, Fe^{++}, and Fe^{+++} there is room for all unshared electrons in the $3d$ orbitals and inner orbitals. When octahedral bonds are formed in the covalent complexes, with use of two of the $3d$ orbitals, only three $3d$ orbitals are left for occupancy by unshared electrons. These are enough for bipositive and tripositive iron and for tripositive cobalt, but they can hold only six of the seven outer unshared electrons of bipositive cobalt. The seventh electron must

[2] W. M. Latimer, *The Oxidation States of the Elements and Their Potentials in Aqueous Solutions*, Prentice-Hall, New York, 1952.

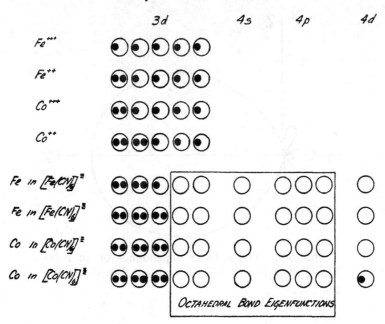

Fig. 5-3.—Occupancy of orbitals by electrons in hypoligating and hyperligating octahedral complexes of bipositive and tripositive iron and cobalt.

accordingly occupy an outer unstable orbital, causing the complex to be unstable.

The five d orbitals, in their angular dependence, are

$$
\begin{aligned}
d_{z^2} &= \sqrt{5/4}\,(3\cos^2\theta - 1) \\
d_{yz} &= \sqrt{15}\,\sin\theta\cos\theta\cos\phi \\
d_{zx} &= \sqrt{15}\,\sin\theta\cos\theta\sin\phi \\
d_{xy} &= \sqrt{15/4}\,\sin^2\theta\sin 2\phi \\
d_{x^2+y^2} &= \sqrt{15/4}\,\sin^2\theta\cos 2\phi
\end{aligned}
\qquad (5\text{-}1)
$$

The set of six equivalent octahedral orbitals formed from two d orbitals, the s orbital, and the three p orbitals is

$$
\psi_1 = \frac{1}{\sqrt{6}}\,s + \frac{1}{\sqrt{2}}\,p_z + \frac{1}{\sqrt{3}}\,d_{z^2}
$$

$$
\psi_2 = \frac{1}{\sqrt{6}}\,s - \frac{1}{\sqrt{2}}\,p_z + \frac{1}{\sqrt{3}}\,d_{z^2}
\qquad (5\text{-}2)
$$

$$\psi_3 = \frac{1}{\sqrt{6}}\, s + \frac{1}{\sqrt{2}}\, p_z - \frac{1}{\sqrt{12}}\, d_{z^2} + \frac{1}{2}\, d_{x^2+y^2}$$

$$\psi_4 = \frac{1}{\sqrt{6}}\, s - \frac{1}{\sqrt{2}}\, p_z - \frac{1}{\sqrt{12}}\, d_{z^2} + \frac{1}{2}\, d_{x^2+y^2}$$

$$\psi_5 = \frac{1}{\sqrt{6}}\, s + \frac{1}{\sqrt{2}}\, p_{xy} - \frac{1}{\sqrt{12}}\, d_{z^2} - \frac{1}{2}\, d_{x^2+y^2}$$

$$\psi_6 = \frac{1}{\sqrt{6}}\, s - \frac{1}{\sqrt{2}}\, p_y - \frac{1}{\sqrt{12}}\, d_{z^2} - \frac{1}{2}\, d_{x^2+y^2}$$

(5-2)
(continued)

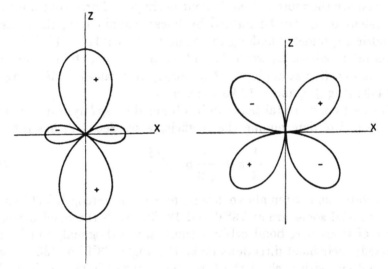

FIG. 5-4.—The angular dependence of the d_{z^2} orbital.   FIG. 5-5.—The angular dependence of the d_{zx} orbital.

The orbital d_{z^2} has the angular dependence shown in Figure 5-4. It is cylindrically symmetrical about the z axis and consists of two positive lobes extending in the directions $+z$ and $-z$ and a negative belt about the xy plane. The nodal zones are at 54°44' and 125°16' with the z direction. The strength of the orbital is 2.236, which is $\sqrt{5}$.

In his thorough discussion of *spd* hybrid orbitals Hultgren[3] proved several interesting theorems. One of them is that the best bond orbital that can be formed by hybridization of orbitals constituting one or

[3] R. Hultgren, *Phys. Rev.* **40**, 891 (1932). A general discussion of bond orbitals in relation to symmetry has been given by G. E. Kimball, *J. Chem. Phys.* **8**, 188 (1940). For a survey of chemical bonds involving d orbitals see D. P. Craig, A. Maccoll, R. S. Nyholm, L. E. Orgel, and L. E. Sutton, *J. Chem. Soc.* **1954**, 332.

more complete subshells has strength equal to the square root of the number of orbitals: $\sqrt{1}$ for s, $\sqrt{3}$ for p, $\sqrt{5}$ for d, $\sqrt{4}$ for sp^3, $\sqrt{9}$ for sp^3d^5, and so forth. He also proved that equivalent orthogonal best bond orbitals can be formed provided that the direction of the maximum of each coincides with a node of the others.

The four other d orbitals described in Equation 5-1 differ in shape from d_{z^2}. The four are equivalent except for spatial orientation. The angular dependence of one of them (d_{xz}) is shown in Figure 5-5. It has four equivalent lobes, with extrema in the directions $+x$ and $-x$ (positive lobes) and $+y$ and $-y$ (negative lobes). The strength (value in these directions) is 1.936. The five d orbitals (unlike the three p orbitals) are therefore not equivalent in shape. Three (but not more) equivalent to d_{z^2} can be formed by linear combination, their axes of cylindrical symmetry making the nodal angles 54°44' or 125°16'. Orbitals intermediate between d_{z^2} and $d_{x^2+y^2}$ can be obtained by linear combination (for example, $\frac{2}{3}\sqrt{2}d_{z^2}+\frac{1}{3}d_{x^2+y^2}$ has values 2.108 along $\pm z$, -0.409 along $\pm x$, and -1.699 along $\pm y$).

The best bond orbital that can be obtained by spd hybridization (for the method of determining the coefficients see Sec. 4-2) has the form

$$\frac{1}{3}s + \frac{1}{\sqrt{3}}p_z + \frac{\sqrt{5}}{3}d_z. \qquad (5\text{-}3)$$

This orbital, as written above, has its maximum (strength 3.000) along z. Its nodal zones are at 73°9' and 133°37' with the bond direction. Three of these best bond orbitals (mutually orthogonal) can be constructed; their bond directions make the angles 73°9' or 133°37' with one another, with each of the three bond angles having independent choice between these two values (except that three bonds at 133°37' cannot be formed).

The five d orbitals described in Equation 5-1 are related in a simple way to the six octahedral directions $\pm x$, $\pm y$, and $\pm z$, as may be seen by converting the angular functions to functions of x/r, y/r, and z/r. It is seen that d_{xy}, d_{yz}, and d_{zx} vanish in these six directions, and hence their incorporation in bond orbitals in these directions would decrease the strength of the orbitals. It is the other two d orbitals, d_{z^2} and $d_{x^2+y^2}$, that can be used effectively in forming single bonds in the octahedral directions.

The orbitals d_{xy}, d_{yz}, and d_{zx} can, however, be used in case that the central atom of the complex forms multiple bonds with the ligands. Some of the octahedral complexes of the transition metals contain bonds with a large amount of double-bond character. These complexes will be discussed in Chapter 9.

The magnetic moment of octahedral complexes can often be used to distinguish between those in which there are d^2sp^3 octahedral bonds and those with a different electronic structure, in the way that will be discussed in Section 5-5. An alternative method of treatment of these complexes is mentioned in Section 5-8.

5-3. SQUARE BOND ORBITALS

In a covalent complex of bivalent nickel such as the nickel cyanide ion $[Ni(CN)_4]^{--}$ the 26 inner electrons of the nickel atom can be placed in pairs in the $1s$, $2s$, three $2p$, $3s$, three $3p$, and four of the $3d$ orbitals. This leaves available for use in bond formation the fifth $3d$ orbital as well as the $4s$ and three $4p$ orbitals. It is found on hybridizing these orbitals that four strong bond orbitals directed to the corners of a square can be formed.[4] The four orbitals (written with the bonds directed along $+x$, $-x$, $+y$, and $-y$) are

$$\left. \begin{aligned}
\psi_1 &= \frac{1}{2}s + \frac{1}{\sqrt{2}}p_x + \frac{1}{2}d_{xy} \\
\psi_2 &= \frac{1}{2}s - \frac{1}{\sqrt{2}}p_x + \frac{1}{2}d_{xy} \\
\psi_3 &= \frac{1}{2}s + \frac{1}{\sqrt{2}}p_y - \frac{1}{2}d_{xy} \\
\psi_4 &= \frac{1}{2}s - \frac{1}{\sqrt{2}}p_y - \frac{1}{2}d_{xy}
\end{aligned} \right\} \qquad (5\text{-}4)$$

They have the bond strength 2.694, much greater than that of sp^3 tetrahedral orbitals (2.000). These four square orbitals are formed with use of only two of the $4p$ orbitals; the other p orbital might accordingly also be used by the nickel atom to form another (rather weak) bond.

From this argument these nickel complexes are expected to have a square planar configuration, rather than the tetrahedral configuration usually assumed for four groups about a central atom. In 1931, when the argument was first presented,[5] this configuration had not been recognized for any complexes of nickel. The foregoing discussion is also applicable to the coordination complexes of palladium(II) and platinum(II), with suitable change in the total quantum numbers of the atomic orbitals. For these complexes the square configuration had been deduced many years ago by Werner from the observed existence of isomers and had been later verified by Dickinson[6] by the x-ray in-

---

[4] The best spd orbitals directed toward the corners of a square are $d^{1.4}s^{.9}p^2$ hybrids, with strength 2.943 (H. Kuhn, *J. Chem. Phys.*, **16**, 727 [1948]).

[5] L. Pauling, *J.A.C.S.* **53**, 1367 (1931).

[6] R. G. Dickinson, *J.A.C.S.* **44**, 2404 (1922). The square configuration was then found for the tetramminopalladous cation in $[Pd(NH_3)_4]Cl_2 \cdot H_2O$ by B. N Dickinson, *Z. Krist.* **88**, 281 (1934).

vestigation of crystals of the chloropalladites and chloroplatinites (Fig. 5-6).

Evidence for the square configuration for $[Ni(CN)_4]^{--}$ and other complexes of quadriligated nickel(II) was provided in 1931 only by the magnetic properties of salts containing this ion and by the observed

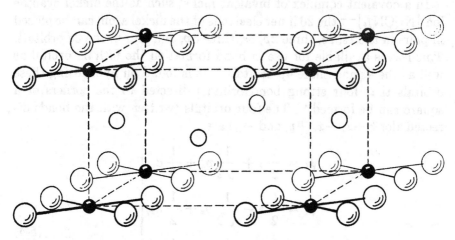

Fɪɢ. 5-6.—The structure of the tetragonal crystals K_2PdCl_4 and K_2PtCl_4. The small circles represent palladium or platinum atoms, those of intermediate size (unshaded) potassium ions, and the largest chlorine atoms. The four chlorine atoms about each palladium or platinum atom lie at the corners of a square.

isomorphism of $K_2Ni(CN)_4 \cdot H_2O$ and $K_2Pd(CN)_4 \cdot H_2O$. In the last few years many investigations have been carried out that show the presence of this configuration in nickel complexes. The first of these, made by Sugden,[7] was the synthesis of cis and trans forms of the nickel compound of benzylmethylglyoxime, with the following configurations:

$$
\begin{array}{cc}
\text{Cis form.} & \text{Trans form.}
\end{array}
$$

[7] S. Sugden, *J. Chem. Soc.* 1932, 246.

Similar pairs of isomers of compounds of nickel as well as of palladium and of platinum with other groups have also been obtained.[8] The x-ray study of crystalline potassium nickel dithio-oxalate[9] has shown it to be isomorphous with the palladium and platinum compounds and has provided a detailed verification of the planar structure

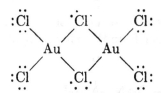

for the complex ion. Isomorphism has been shown for many other sets of substances,[10] such as $BaM(CN)_4 \cdot 4H_2O$, with M = Ni, Pd, and Pt, and $Na_2M(CN)_4 \cdot 3H_2O$, with M = Ni and Pd, and a complete x-ray study of the structure of the latter crystals, showing the presence of planar $[M(CN)_4]^{--}$ ions, has been made.[11] The planar configuration for this ion has been verified[12] also in the crystals $Sr[Ni(CN)_4] \cdot 5H_2O$ and $Ni(CN)_2 \cdot NH_3 \cdot C_6H_6$.

A survey of the magnetic evidence for the dsp^2 quadricovalent state of nickel(II) is given in Section 5-6.

The square planar structure has been verified[13] for the $[AuBr_4]^-$ ion in $KAuBr_4 \cdot H_2O$, for the $[AuCl_4]^-$ ion in the compounds $Cs_2AgAuCl_6$ and $Cs_2Au_2Cl_6$ described in a following paragraph, and for the molecule[14] $(CH_3)_3PAuBr_3$. The gold(III) chloride dimer has been shown[15] to have the planar structure represented by the formula

$$
\begin{array}{ccccc}
:\!\ddot{C}l & & \cdot\dot{C}l\cdot & & \ddot{C}l: \\
& \diagdown \quad \diagup & & \diagdown \quad \diagup & \\
& Au & & Au & \\
& \diagup \quad \diagdown & & \diagdown & \\
:\!\ddot{C}l & & \cdot\dot{C}l\cdot & & \ddot{C}l: \\
\end{array}
$$

[8] K. A. Jensen, *Z. anorg. Chem.*, **221**, 6 (1934).

[9] E. G. Cox, W. Wardlaw, and K. C. Webster, *J. Chem. Soc.* **1935**, 1475; N. Elliott, dissertation, Calif. Inst. Tech., 1938. See also E. G. Cox, F. W. Pinkard, W. Wardlaw, and K. C. Webster, *J. Chem. Soc.* **1935**, 459; and Cox, Wardlaw, and Webster, *loc. cit.*, for x-ray work on related crystals.

[10] H. Brasseur, A. de Rassenfosse, and J. Pierard, *Z. Krist.* **88**, 210 (1934), and later papers.

[11] H. Brasseur and A. de Rassenfosse, *Mem. soc. roy. sci. Liège* **4**, 397, 447 (1941).

[12] H. Lambot, *Bull. soc. roy. sci. Liège* **12**, 439, 522 (1943); J. H. Rayner and H. M. Powell, *J. Chem. Soc.* **1952**, 319.

[13] E. G. Cox and K. C. Webster, *J. Chem. Soc.* **1936**, 1635.

[14] M. F. Perutz and O. Weisz, *J. Chem. Soc.* **1946**, 438.

[15] E. S. Clark, *U. Cal. Radiation Lab. Reports* **1955**, 3190.

The structure of the dimer of diethylmonobromogold, $(C_2H_5)_4Au_2Br_2$, is similar,[16] with the two bromine atoms in the bridging positions and

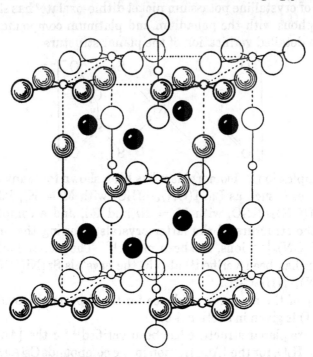

Fig. 5-7.—The structure of the tetragonal crystals $Cs_2AgAuCl_6$ and $Cs_2AuAuCl_6$. Large full circles represent cesium atoms, large open circles chlorine atoms, and small circles gold or silver atoms.

the ethyl groups in the end positions. Di-*n*-propylgold cyanide, $[Au(C_3H_7)_2CN]_4$, has been found[17] to have the planar structure

$$
\begin{array}{ccc}
\text{R} & & \text{R} \\
| & & | \\
\text{R—Au—C}\equiv\text{N—Au—R} \\
| & & | \\
\text{N} & & \text{C} \\
\| & & \| \\
\text{C} & & \text{N} \\
| & & | \\
\text{R—Au—N}\equiv\text{C—Au—R} \\
| & & | \\
\text{R} & & \text{R}
\end{array}
\qquad (\text{R is —C}_3\text{H}_7)
$$

The tetragonal crystals[18] $Cs_2AgCl_2AuCl_4$ and $Cs_2AuCl_2AuCl_4$, with

[16] A. Burawoy, C. S. Gibson, G. C. Hampson, and H. M. Powell, *J Chem. Soc.* **1937**, 1690.

[17] R. F. Phillips and H. M. Powell, *Proc. Roy. Soc. London* **A173**, 147 (1939).

[18] N. Elliott and L. Pauling, *J.A.C.S.* **60**, 1846 (1938).

the structure shown in Figure 5-7, contain square complexes $[AuCl_4]^-$ of tripositive gold, as well as linear complexes $[AgCl_2]^-$ or $[AuCl_2]^-$ of unipositive silver or gold.

An interesting case of infinite polymerization is provided by palla-

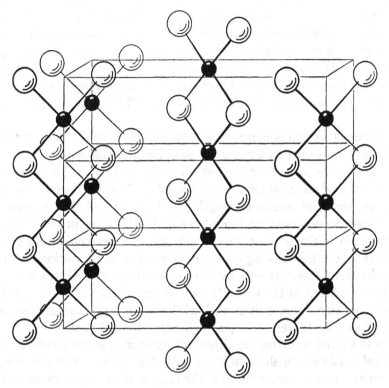

Fig. 5-8.—The structure of the $PdCl_2$ crystal. Small circles represent palladium atoms and large circles chlorine atoms.

dous chloride. In crystals of this substance[19] (Fig. 5-8) there are planar strings of indefinite length with the configuration

$$\cdots Pd \diagdown_{Cl}^{Cl} Pd \diagdown_{Cl}^{Cl} Pd \diagdown_{Cl}^{Cl} Pd \cdots.$$

These strings contain rectangular $PdCl_4$ groups that share edges in such a way as to lead to the composition $PdCl_2$. The $PdCl_4$ groups are only slightly distorted from the square configuration, the Cl—Pd—Cl angles being 93° and 87°.

It may be mentioned that quadricovalent complexes of quadriposi-

[19] A. F. Wells, *Z. Krist.* **100**, 189 (1938).

tive platinum and palladium might be expected to have the tetrahedral rather than the square configuration, since two d orbitals are available for bond orbital formation, instead of only one as in the bipositive complexes. Molecules of the substances[20] tetramethylplatinum, $Pt(CH_3)_4$, and hexamethyldiplatinum, $Pt_2(CH_3)_6$, might be expected to be similar in structure to neopentane and hexamethylethane, respectively, with a platinum-platinum bond in the second. However, an investigation of the structure of crystalline tetramethylplatinum and trimethylplatinum chloride has shown that tetramers $Pt_4(CH_3)_{16}$ and $Pt_4(CH_3)_{12}Cl_4$ exist in the crystals, with each platinum atom forming six octahedral bonds. The structure of these substances will be discussed in Chapter 10.

The configuration of complexes of copper(II) presents an interesting problem. It is seen on analysis of the problem that copper(II) would be expected to form four covalent bonds of the dsp^2 coplanar type rather than tetrahedral sp^3 covalent bonds. The dsp^2 bonds are much stronger than sp^3 bonds (strength 2.694 instead of 2). Copper(II) has, however, one electron more than nickel(II), and in the usual assignment of unshared electrons to orbitals this electron would occupy the fifth $3d$ orbital, making it unavailable for bond formation. On the other hand, no loss of energy by the copper atom occurs if the unshared electron is placed in the third $4p$ orbital and the $3d$ orbital is used for bond formation, inasmuch as each of the five orbitals under discussion (one $3d$, one $4s$, three $4p$) is occupied either by a shared pair or by the single unshared electron on either formulation (single electron in $3d$ with sp^3 bonds single electron in $4p$ with dsp^2 bonds), and the interaction energy of a shared pair with the copper atom is the same as that of a single unshared electron, if the bonds are normal covalent bonds. (There is some loss of energy if they have some ionic character, with the copper atom positive.) The greater strength of dsp^2 bonds than of sp^3 bonds is accordingly the determining factor; a complex of copper(II) involving four covalent bonds will have the square rather than the tetrahedral configuration.

The planar configuration of quadricovalent copper(II) was discovered by Cox and Webster[21] in the compounds of copper with β-diketones (copper disalicylaldoxime, copper acetylacetonate, copper benzoylacetonate, the copper salt of dipropionylmethane) and by Tunell, Posnjak, and Ksanda[22] in the mineral tenorite, CuO. In crystalline cupric chloride dihydrate (Fig. 5-9) there are molecules with the planar configuration[23]

[20] H. Gilman and M. Lichtenwalter, *J.A.C.S.* **60**, 3085 (1938).
[21] E. G. Cox and K. C. Webster, *J. Chem. Soc.* **1935**, 731.
[22] G. Tunell, E. Posnjak, and C. J. Ksanda, *Z. Krist.* **90**, 120 (1935).
[23] D. Harker, *Z. Krist.* **93**, 136 (1936).

$$\begin{array}{c} H_2O \\ | \\ Cl\!-\!Cu\!-\!Cl, \\ | \\ H_2O \end{array}$$

and the same group occurs[24] in crystalline $K_2CuCl_4 \cdot 2H_2O$.

A careful determination of the structure of $CuCl_2 \cdot 2H_2O$ has been made[25] by neutron diffraction, which locates the hydrogen atoms.

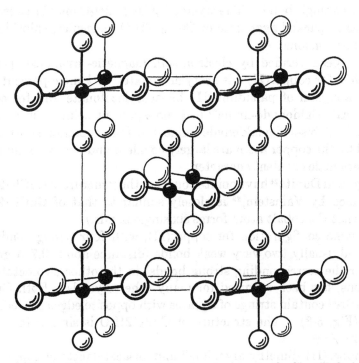

FIG. 5-9.—The crystal structure of $CuCl_2 \cdot 2H_2O$. Small circles represent copper atoms, circles of intermediate size represent oxygen atoms of water molecules, and large circles represent chlorine atoms.

These atoms were found to lie in the same plane as the other atoms of the molecule. The dimensions $O\!-\!H = 0.95$ Å and angle $H\!-\!O\!-\!H = 108°$ are essentially the same as those of the free water molecule. This is the result that would be expected not for a single bond between the oxygen atom and the copper atom, which would make the $Cu\!-\!O\!-\!H$ angles about 105°, but rather for a double bond or for single-bond:double-bond resonance.

[24] S. B. Hendricks and R. G. Dickinson, *J.A.C.S.* **49**, 2149 (1927); L. Chrobak, *Z. Krist.* **88**, 35 (1934).

[25] S. W. Peterson and H. A. Levy, *J. Chem. Phys.* **26**, 220 (1957).

In $K_2CuCl_4 \cdot 2H_2O$ each copper atom has as nearest neighbors two chlorine atoms at the distance 2.32 Å and two oxygen atoms (of the water molecules) at 1.97 Å, with the planar configuration represented in the text. There are also two neighboring chlorine atoms above and below the plane of this group at the distance 2.95 Å; this distance is much greater than is expected for a covalent bond. The discussion of Section 7-10 indicates that these two interactions correspond to a bond number about 0.1; that is, they represent bonds about 10 percent as strong as a single bond. The crystal $K_2CuCl_4 \cdot 2H_2O$ may be considered as a closely packed aggregate of $CuCl_2 \cdot 2H_2O$ molecules, chloride ions, and potassium ions.

It has been verified by electron-spin magnetic-resonance spectroscopy that for copper(II) bis-acetylacetonate[26] in ½ mole-percent solution in a crystal of palladium(II) bis-acetylacetonate and for copper (II) bis-salicylaldehyde-imine[27] in ½ mole-percent solution in a crystal of nickel(II) bis-salicylaldehyde-imine the four coplanar short bonds formed by the copper atom are largely covalent and the two long bonds have very little covalent character.

(Orgel and Dunitz[28] have pointed out that the structure of $CoCl_2 \cdot 2H_2O$, determined by Vainstein,[29] is closely similar to that of $CuCl_2 \cdot 2H_2O$, except that the cobalt atom forms six single bonds.)

The same configuration for copper(II), with four strong bonds in a plane and usually two very weak bonds (distance about 0.7 Å greater than for the corresponding strong bonds) in the other two octahedral directions, has been reported for many other crystals. Both $CuCl_2$[30] and $CuBr_2$[31] contain strings of squares with opposite edges shared, as in $PdCl_2$ (Fig. 5-8). The structure of $CuF_2 \cdot 2H_2O$ is similar to that of $CuCl_2 \cdot 2H_2O$.[32]

In copper(II) dimethylglyoxime[33] and bisacetylacetone copper(II)[34] the copper atom forms four dsp^2 bonds with adjacent nitrogen or oxygen atoms and no weak bonds; the other two octahedral directions point toward carbon and hydrogen atoms.

An exceptional complex is $[CuCl_4]^{--}$ in the crystal Cs_2CuCl_4. The

[26] A. H. Maki and B. R. McGarvey, *J. Chem. Phys.* 29, 31 (1958).

[27] A. H. Maki and B. R. McGarvey, *J. Chem. Phys.* 29, 35 (1958).

[28] L. E. Orgel and J. D. Dunitz, *Nature* 179, 462 (1957).

[29] B. I. Vainstein, *Doklady Akad. Nauk S.S.S.R.* 68, 301 (1949).

[30] A. F. Wells, *J. Chem. Soc.* 1947, 1670.

[31] L. Helmholz, *J.A.C.S.* 69, 886 (1947).

[32] S. Geller and W. L. Bond, *Am. Cryst. Ass'n Meeting*, Milwaukee, June 1958.

[33] S. Bezzi, E. Bua, and G. Schiavinato, *Gazz. chim. ital.* 81, 856 (1951).

[34] Cox and Webster, *loc. cit.* (21). E. A. Shugam, *Doklady Akad. Nauk S.S.S.R.* 1951, 853; H. Koyama, Y. Saito, and H. Kuroya, *J. Inst. Polytech. Osaka City Univ.* 4, 43 (1953).

reported structure[35] is intermediate between a square plane and a regular tetrahedron; the four chlorine atoms alternate at heights 0.76 Å above and below the median plane, with height zero corresponding to the planar configuration and 1.80 Å to the regular tetrahedron (the Cu—Cl bond length is 2.22 Å). A theoretical study of this complex has been reported.[36] It is possible that the distortion from the planar configuration is caused by the interactions with the cesium ions in the crystal. The planar configuration of $CuCl_4$ groups (each sharing two chlorine atoms with adjacent groups) has been found[37] for a closely related substance, $CsCuCl_3$.

Quadricovalent complexes of silver(II) should have the same planar configuration as those of copper(II). This has been verified[38] for the silver(II) salt of picolinic acid, which is isomorphous with the copper(II) salt and which shows moreover the high birefringence expected for a parallel arrangement of planar molecules with the structure

No compounds of gold(II) are known.

5-4. THE MAGNETIC CRITERION FOR BOND TYPE

The usefulness of the magnetic susceptibility of substances in giving information about their electronic structure was emphasized by G. N. Lewis in his early work on valence. In 1925 Welo and Baudisch[39] discussed the magnetic properties of complex ions and suggested a simple rule, which was used later by Sidgwick[40] and other investigators: that the magnetic moment of a complex (as found by measuring its magnetic susceptibility—see App. X) is equal to that of the atom with the

[35] L. Helmholz and R. F. Kruh, *J.A.C.S.* **74**, 1176 (1952).
[36] G. Felsenfeld, *Proc. Roy. Soc. London* **A236**, 506 (1956).
[37] A. F. Wells, *J. Chem. Soc.* **1947**, 1662.
[38] E. G. Cox, W. Wardlaw, and K. C. Webster, *J. Chem. Soc.* **1936**, 775.
[39] L. A. Welo and O. Baudisch, *Nature* **116**, 606 (1925).
[40] N. V. Sidgwick, *The Electronic Theory of Valency*, Clarendon Press, Oxford, 1927.

same number of electrons as the central atom of the complex, including two for each covalent bond that the central atom forms. For example, the ferrocyanide ion, $[Fe(CN)_6]^{----}$, has zero magnetic moment. The ion Fe^{++} has 24 electrons, to which we add 12 for the six covalent bonds to the cyanide ions, to obtain the total 36; this is the electron number for krypton, which is diamagnetic ($\mu = 0$).

This simple rule is satisfactory for many substances, but there are also many exceptions. For example, the complex $[Ni(CN)_4]^{--}$ is diamagnetic, although the rule would make it paramagnetic (resembling Se, $Z = 34$, with magnetic moment about 2.8 magnetons).

It has been found[41] that the magnetic moments of complexes can be discussed in a generally satisfactory way by assigning the atomic electrons (the electrons that are not involved in bond formation) to the stable orbitals that are not used as bond orbitals. The assignment is made in the way corresponding to maximum stability, as given by Hund's rules for atoms (App. IV); in particular, electrons are introduced into equivalent orbitals in such a way as to give the maximum number of unpaired electron spins compatible with the Pauli exclusion principle. Observed values of the magnetic moment can often be used in selecting one from among several alternative electronic structures for a complex. Application of this magnetic criterion to octahedral and square complexes is made in the following sections.

5-5. THE MAGNETIC MOMENTS OF OCTAHEDRAL COMPLEXES

There are three kinds of electronic structures that may be expected for the octahedral complexes MX_6 of the iron-group transition elements (and also for those of the palladium and platinum groups).

The first kind is that in which no $3d$ orbitals are involved in bond formation; the bonds may be formed with use of the $4s$ orbital and the three $4p$ orbitals (four sp^3 bonds resonating among the six positions), or with use of these four orbitals and two $4d$ orbitals. For this structure all five $3d$ orbitals of M are available for occupancy by the atomic electrons, and the expected magnetic moment is close to that for the monatomic ion M^{+z}. In earlier discussions[42] this kind of structure was described as essentially ionic; this description may, however, be misleading, and we shall here refer to complexes with this kind of structure as *hypoligated* complexes (the ligands are bonded less strongly than in complexes with the other structures).

In structures of the second kind, which occur only rarely, one of the $3d$ orbitals is used in bond formation, leaving four for occupancy by atomic electrons. It was mentioned in earlier editions of this book

[41] Pauling, *loc. cit.* (5).
[42] Pauling. *loc. cit.* (5); also earlier editions of this book.

that the observed magnetic moment of ferrihemoglobin hydroxide[43] indicates that the octahedrally ligated iron(III) atoms in this molecule have a structure of this kind (with three unpaired electrons). Two other substances that probably represent the second kind of octahedral ligation are known;[44] they are iron(II) phthalocyanine and iron(III) protoporphyrin chloride.

Octahedral structures of the third kind are those in which d^2sp^3 bonds are formed with use of two of the $3d$ orbitals, leaving only three for occupancy by atomic electrons. Complexes with these structures were formerly described as essentially covalent; here we shall describe them as *hyperligated* complexes (complexes with strong bonds).

The way in which the magnetic criterion can be used to distinguish between hypoligated and hyperligated octahedral complexes can be illustrated for iron(II). The Fe^{++} ion has six electrons outside the argon shell. For hypoligated complexes five $3d$ orbitals are available, and the stable disposition of the six electrons among the five orbitals leaves four unpaired, with one orbital occupied by a pair; the corresponding magnetic moment due to the spins of four electrons is 4.90 magnetons. The hydrated iron(II) ion, $[Fe(OH_2)_6]^{++}$, is observed to have $\mu = 5.25$, and hence this ion is a hypoligated octahedral complex. On the other hand, the hyperligated octahedral complexes of iron(II) must place the six electrons in three orbitals, and hence must have $\mu = 0$, as is observed for $[Fe(CN)_6]^{----}$.

The magnetic moments predicted for the normal states of the monatomic ions Fe^{++}, Co^{++}, and so on are due in part to spin and in part to orbital motion. Their values may be calculated for the predicted stable Russell-Saunders state (Chap. 2 and App. IV) as $g\sqrt{J(J+1)}$, where J is the total angular momentum quantum number and g is the Landé g-factor appropriate to the Russell-Saunders state (App. IV). For example, the normal state of Fe^{++} is $^5D_4$, for which $g = 1.500$ and $\mu = 6.70$. In complexes, however, the orbital magnetic moment of the complex is in large part quenched, and the moment approaches the value due to the spin alone, which is $\sqrt{n(n+2)}$, in which n is the number of electrons with unpaired spins. For $n = 4$ the spin moment is 4.90, as mentioned above. The experimental value for the hexahydrated iron(II) ion in solution and in several crystals is 5.25, showing that the orbital moment is largely quenched.

The value of the spin moment for iron-group ions rises to a maximum of 5.92, corresponding to five unpaired electrons, and then decreases, as shown in Table 5-1.

The observed values for the iron-group ions in aqueous solution are

[43] C. D. Coryell, F. Stitt, and L. Pauling, *J.A.C.S.* **59**, 633 (1937).
[44] J. S. Griffith, *Discussions Faraday Soc.* **26**, 81 (1959).

TABLE 5-1.—MAGNETIC MOMENTS OF IRON-GROUP
IONS IN AQUEOUS SOLUTION

Ion	Number of 3d electrons	Number of unpaired electrons	Calculated spin moment[a]	Observed moment[a]
K+, Ca++, Sc+++, Ti⁴⁺	0	0	0.00	0.00
Ti+++, V⁴⁺	1	1	1.73	1.78
V+++	2	2	2.83	2.80
V++, Cr+++, Mn⁴⁺	3	3	3.88	3.7–4.0
Cr++, Mn+++	4	4	4.90	4.8–5.0
Mn++, Fe+++	5	5	5.92	5.9
Fe++	6	4	4.90	5.2
Co++	7	3	3.88	5.0
Ni++	8	2	2.83	3.2
Cu++	9	1	1.73	1.9
Cu+, Zn++	10	0	0.00	0.00

[a] In Bohr magnetons.

seen from the table to agree reasonably well with the theoretical
values. The deviations observed can be explained as resulting from
contributions of the orbital moments of the electrons.

In many crystalline salts of these elements values of μ are observed
that are close to those for the aqueous ions; some of these are given
in Table 5-2. For central atoms with more than three 3d electrons

TABLE 5-2.—MAGNETIC MOMENTS OF IRON-GROUP IONS
IN SOLID COMPOUNDS

Substance	Calculated spin moment[a]	Observed moment[a]	Substance	Calculated spin moment[a]	Observed moment[a]
CrCl₃	3.88	3.81	CoCl₂	3.88	5.04
Cr₂O₃·7H₂O		3.85	CoSO₄·7H₂O		5.06
CrSO₄·6H₂O	4.90	4.82	(NH₄)₂Co(SO₄)₂·6H₂O		5.00
MnCl₂	5.92	5.75	Co(N₂H₄)₂SO₄·H₂O		4.31
MnSO₄		5.87	Co(N₂H₄)₂(CH₂COO)₂		4.56
MnSO₄·4H₂O		5.87	Co(N₂H₄)₂Cl₂		4.93
Fe₂(SO₄)₃		5.86			
NH₄Fe(SO₄)₂		5.86	NiCl₂	2.83	3.3
(NH₄)₃FeF₆		5.88	NiSO₄		3.42
(NH₄)₃FeF₆·H₂O		5.91	Ni(N₂H₄)₂SO₄		3.20
FeCl₃		5.84	Ni(N₂H₄)₂(NO₂)₂		2.80
			Ni(NH₃)₆SO₄		2.63
FeCl₂	4.90	5.23			
FeCl₂·4H₂O		5.25	CuCl₂	1.73	2.02
FeSO₄		5.26	CuSO₄		2.01
FeSO₄·7H₂O		5.25	Cu(NH₃)₄(NO₃)₂		1.82
(NH₄)₂Fe(SO₄)₂·6H₂O		5.25	Cu(NH₃)₄SO₄·H₂O		1.81
Fe(N₂H₄)₂Cl₂		4.87			

[a] In Bohr magnetons.

this agreement substantiates the assignment of the octahedral complexes to the hypoligated class.

The observation that an iron(II) complex contains four unshared electrons does not require that the bonds in the complex be of the extreme ionic type. As many as four rather weak covalent bonds could be formed with use of the $4s$ and $4p$ orbitals without disturbing the $3d$ shell, and a corresponding amount of covalent character of the bonds would not change the magnetic moment of the complex. Similarly the octahedral d^2sp^3 bonds could have some ionic character without relinquishing their hold on the two $3d$ orbitals. At some point in the change in bond type from the ionic extreme to the octahedral covalent extreme the discontinuity in the nature of the normal state will occur, and the argument given above permits us to describe the octahedral complexes with four unpaired electrons as hypoligated and those with no unpaired electrons as hyperligated.[45]

The decision between hypoligation and hyperligation is determined by competition between two factors. The factor favoring hypoligation is the resonance interaction that stabilizes atomic states with a large number of unpaired electrons (large multiplicity, as given by Hund's first rule, Sec. 2-7). The factor favoring hyperligation is the bond energy, as determined by the bond-forming power of the ligands and the strength of the bond orbitals of the central atom.

In Table 5-3 there are given observed values of magnetic moments of some compounds containing octahedral complexes, not only of the iron-group elements but also of the palladium-group and platinum-group elements, to which the discussion is also applicable. It is seen that the octahedral complexes of iron with fluorine and with water are hypoligated, whereas those with the cyanide, nitrite, and dipyridyl groups are hyperligated.[46] All of the complexes of cobalt(III) that

---

[45] It has been shown by J. H. Van Vleck, *J. Chem. Phys.* **3**, 807 (1935), that extremely strong ionic forces might lead to pairing of the $3d$ electrons. This phenomenon does not occur, however, in the complex of iron with the most electronegative of all atoms, fluorine, and so it is not to be expected to occur in any complex.

[46] In some derivatives of ferrohemoglobin and ferrihemoglobin the iron atoms (bivalent or trivalent) are shown to be surrounded octahedrally (probably by four nitrogen atoms of the porphyrin complex, one nitrogen atom of the globin, and the attached group) by the observed values of the magnetic moments, which correspond to d^2sp^3 octahedral bonds (oxyferrohemoglobin, $\mu = 0.0$; carbonmonoxyferrohemoglobin, $\mu = 0.0$; ferrihemoglobin cyanide, $\mu = 2.5$; ferrihemoglobin hydrosulfide, $\mu = 2.3$). In other derivatives the bonds are essentially ionic (ferrohemoglobin, $\mu = 5.4$; ferrihemoglobin, $\mu = 5.8$; ferrihemoglobin fluoride, $\mu = 5.9$). The value $\mu = 4.5$ for ferrihemoglobin hydroxide suggests a structure with three unpaired electrons, not known to be represented among the simpler iron complexes. The derivatives of the prosthetic group of hemo-

TABLE 5-3.—OBSERVED MAGNETIC MOMENTS OF OCTAHEDRAL COMPLEXES
OF TRANSITION ELEMENTS[a]

Hyperligated complexes	μ^b calculated	μ^b observed	Hypoligated complexes	μ^b calculated	μ^b observed
$K_4[Cr^{II}(CN)_6]$	2.83	3.3			
$K_3[Mn^{III}(CN)_6]$		3.0			
$K_4[Mn^{II}(CN)_6]$	1.73	2.0	$Mn^{II}(NH_3)_6Br_2$	5.92	5.9
$K_3[Fe^{III}(CN)_6]$		2.33	$(NH_4)_3[Fe^{III}F_6]$	5.92	5.9
$K_4[Fe^{II}(CN)_6]$	0.00	0.00	$(NH_4)_2[Fe^{III}F_5 \cdot H_2O]$		5.9
$Na_3[Fe^{II}(CN)_5 \cdot NH_3]$		.00	$[Fe^{II}(H_2O)_6](NH_4SO_4)_2$	4.90	5.3
$[Fe^{II}(dipyridyl)_3]SO_4$		.00			
$K_3[Co^{III}(CN)_6]$		.00			
$[Co^{III}(NH_3)_3F_3]$		.00[c]	$K_3[Co^{III}F_6]$	4.90	5.3[c]
$[Co^{III}(NH_3)_6]Cl_3$		.00	$[CoF_3(OH_2)_3] \cdot 1/2H_2O$		4.47[d]
$[Co^{III}(NH_3)_5Cl]Cl_2$		.00			
$[Co^{III}(NH_3)_4Cl_2]Cl_2$		.00			
$[Co^{III}(NH_3)_2(NO_2)_3]$		.00			
$[Co^{III}(NH_3)_6 \cdot H_2O]_2(C_2O_4)_3$		.00			
$[Co^{III}(NH_3)_5CO_3]NO_3 \cdot 3/2H_2O$		.00			
$K_2Ca[Co^{II}(NO_2)_6]$	1.73	1.9	$[Co^{II}(NH_3)_6]Cl$	3.88	4.96
$K_2[Pd^{IV}Cl_6]$	0.00	0.00			
$[Pd^{IV}Cl_4(NH_3)_2]$		.00			
$Na_2[Ir^{III}Cl_3(NO_2)_3]$		.00			
$[Ir^{III}(NH_3)_5NO_2]Cl_2$		.00			
$[Ir^{III}(NH_3)_4(NO_2)_2]Cl$		.00			
$[Ir^{III}(NH_3)_3(NO_2)_3]$		.00			
$K_2[Pt^{IV}Cl_6]$		.00			
$[Pt^{IV}(NH_3)_2]Cl_4$		.00			
$[Pt^{IV}(NH_3)_5Cl]Cl_3$		.00			
$[Pt^{IV}(NH_3)_4Cl_2]Cl_2$		.00			
$[Pt^{IV}(NH_3)_3Cl_3]Cl$		.00			
$[Pt^{IV}(NH_3)_2Cl_4]$		.00			

[a] The values quoted are taken in the main from W. Biltz, *Z. anorg. Chem.* 170, 161 (1928); D. M. Bose, *Z. Physik* 65, 677 (1930); and the *International Critical Tables.* Values for many other complexes are tabulated by P. W. Selwood, *Magnetochemistry*, Interscience Publishers, New York, 1956.

[b] In Bohr magnetons.

[c] Private communication from Prof. G. H. Cartledge of the University of Buffalo.

[d] H. C. Clark, B. Cox, and A. G. Sharpe, *J. Chem. Soc.* 1957, 4132.

have been investigated are hyperligated except that with fluorine, $[CoF_6]^{---}$, which is hypoligated. It is interesting that in the sequence $[Co(NH_3)_6]^{+++}$, $[Co(NH_3)_3F_3]$, $[CoF_6]^{---}$ the transition from hyperligation to hypoligation occurs between the second and third complex.

Bipositive cobalt forms hypoligating bonds with water and hyperligating bonds with nitrite groups.[47]

globin (hemin, ferroheme, hemochromogens) are in part ionic and in part covalent in structure (L. Pauling and C. D. Coryell, *Proc. Nat. Acad. Sci. U. S.* 22, 159 [1936]; 22, 210 [1936]; Coryell, Stitt, and Pauling, *loc. cit.* [43]).

[47] A detailed discussion of the magnetic moments and structure of complexes of cobalt(II) has been made by B. N. Figgis and R. S. Nyholm, *J. Chem. Soc.* 1959, 338.

The magnetic method cannot be applied to tripositive chromium, the structures of the two extreme types having the same number of unpaired electrons and entering into resonance with each other. The chemical properties of the chromium complexes indicate that chromium, like the other iron-group elements, forms hyperligating bonds

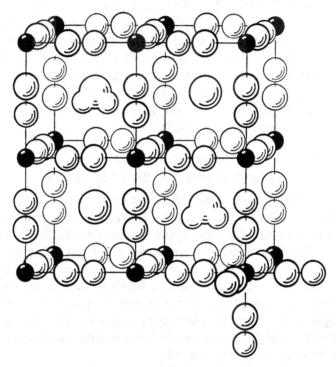

Fig. 5-10.—A drawing representing the front half of the cubic unit of structure of the crystal prussian blue, $KFeFe(CN)_6 \cdot H_2O$. The structure can be described by reference to the eight small cubes that constitute the cubic unit of structure. Alternate corners of the small cubes are occupied by iron(II) and iron(III) atoms. The cyanide groups lie along the edges of the small cubes; each cyanide group forms a single bond with two iron atoms, defining the edge of the cube. Water molecules and potassium ions alternate at the centers of the small cubes. The structure can be described as a three-dimensional latticework of iron atoms and cyanide groups defining cubical cells that contain the water molecules and potassium ions.

with groups such as cyanide and hypoligating bonds with water and ammonia.[48] The complexes of the iron-group elements are discussed further in Section 9-7.

[48] That chromium forms d^2sp^3 hyperligating bonds with oxalate in the ion $[Cr(C_2O_4)_2]^{+++}$ has been inferred by C. H. Johnson (*Trans. Faraday Soc.* **28**, 845) [1932]) from the following facts: The chromium trioxalate complex and the cobaltic trioxalate complex can be resolved into optical isomers, whereas

All the octahedral complexes of the elements of the palladium and platinum groups that have been investigated are diamagnetic, showing the strong tendency of these elements to form hyperligating bonds.

The magnetic properties of prussian blue and similar substances are of unusual interest. X-ray investigations[49] have shown that substances such as $KFeFe(CN)_6 \cdot H_2O$ form cubic crystals in which iron atoms lie at the points of a simple cubic lattice, each being connected with its six neighbors by CN groups extending along the cube edges (Fig. 5-10). The potassium ions and water molecules lie in the cubes outlined in this way. The magnetic susceptibility shows that half of the iron atoms, presumably those bonded to the carbon atoms of the six adjacent cyanide groups, form hyperligating bonds, whereas the other iron atoms form hypoligating bonds.[50]

5-6. THE MAGNETIC MOMENTS OF TETRAHEDRAL AND SQUARE COORDINATED COMPLEXES

The bipositive nickel atom forming four covalent dsp^2 bonds has only four $3d$ orbitals available for the eight unshared $3d$ electrons, which must thus form four pairs, the square complex NiX_4 being diamagnetic. Bipositive nickel in a complex involving only the $4s$ and $4p$ orbitals (electrostatic bonds or weak covalent bonds) distributes the eight $3d$ electrons among the five $3d$ orbitals in such a way as to leave two electrons unpaired, the complex having a magnetic moment of 2.83 Bohr magnetons. From this argument it is seen that the assignment of nickel complexes to the tetrahedral and square coplanar classes can be made by magnetic measurements.

The crystals $K_2Ni(CN)_4$ and $K_2Ni(CN)_4 \cdot H_2O$, shown by isomorphism to contain the planar complex $[Ni(CN)_4]^{--}$, are diamagnetic. Many other nickel complexes, some of which have been shown to be planar by the methods mentioned in Section 5-3, have been found to satisfy the magnetic criterion. These include the nickel glyoximes,[51]

---

resolution has not been effected for the trioxalates of trivalent manganese, iron, and aluminum. Observed magnetic moments ($K_3Mn(C_2O_4)_3 \cdot 3H_2O$, $\mu = 4.88$; $K_3Fe(C_2O_4)_3 \cdot 3H_2O$, $\mu = 5.75$; $K_3Co(C_2O_4)_3 \cdot 3\frac{1}{2}H_2O$, $\mu = 0.00$) show the manganese and iron complexes to be hypoligated and the cobalt complex to be hyperligated.

[49] J. F. Keggin and F. D. Miles, *Nature* 137, 577 (1936); N. Elliott (unpublished work at the California Institute of Technology) obtained similar results for $KMFe(CN)_6 \cdot H_2O$, with M = Mn, Co, and Ni.

[50] The magnetic susceptibility does not show whether the covalently bonded iron is tripositive or bipositive.

[51] Sugden, *loc. cit.* (7); H. J. Cavell and S. Sugden, *J. Chem Soc.* 1935, 621; L. Cambi and L. Szegö, *Ber.* 64, 2591 (1931).

potassium nickel dithio-oxalate,[52] nickel diacetyldioxime,[53] nickel ethyl-xanthogenate[54] $[Ni(C_2H_5O \cdot CS_2)_2]$, and nickel ethyldithiocarbamate,[54] $[Ni(C_2H_5 \cdot NH \cdot CS_2)_2]$.

On the other hand, the compounds $[Ni(NH_3)_4]SO_3$, $[Ni(N_2H_4)_2](NO_2)_2$, $[Ni(C_2H_4(NH_2)_2)_2](SCN)_2 \cdot H_2O$, and $[Ni(C_5H_7O_2)_2]$ (nickel acetyl-acetone) are paramagnetic, with values of μ between 2.6 and 3.2. In these complexes the four atoms attached to nickel are presumably arranged tetrahedrally; this has not yet been shown, however, by x-ray examination or by the synthesis of isomers. The tetrahedral configuration has been verified by x-ray diffraction[55] for the ion $[NiCl_4]^{--}$.

The values found for the molal paramagnetic susceptibility of hydrated nickel cyanides of composition between $Ni(CN)_2 \cdot 2H_2O$ and $Ni(CN_2 \cdot 4H_2O$ are about one-half those for ionic nickel compounds, indicating that these substances contain square covalent complexes $[Ni(CN)_4]^{--}$ and tetrahedral ionic complexes $[Ni(H_2O)_4]^{++}$ or $[Ni(H_2O)_6]^{++}$ in equal numbers.[56] Anhydrous nickel cyanide is also paramagnetic, with molal susceptibility about 10 percent as great as for compounds of ionic nickel, the value found depending somewhat on the method of preparing the sample. This indicates that about 90 percent of the nickel atoms form square covalent bonds with carbon or nitrogen atoms of the cyanide groups, and the remaining 10 percent of the nickel atoms form hypoligating bonds.

The factors that determine whether the diamagnetic square or the paramagnetic tetrahedral configuration will be assumed by a nickel complex cannot be stated precisely. Groups containing sulfur atoms, which have a strong tendency to form covalent bonds, form square complexes; for nitrogen and oxygen the decision seems to depend on the presence and disposition of double bonds in the group.

The complexes of palladium(II) and platinum(II) are all diamagnetic. Diamagnetism has been verified[57] for $PdCl_2 \cdot 2H_2O$, $PdCl_2 \cdot 2NH_3$, K_2PdCl_4, $K_2Pd(CN)_4$, K_2PdI_4, $K_2Pd(SCN)_4$, $K_2Pd(NO_2)_4$, palladium dimethylglyoxime, and even palladous nitrate in solution (probably containing the ion $[Pd(H_2O)_4]^{++}$). The crystalline substances $PdCl_2$, PdI_2, $Pd(CN)_2$, $Pd(SCN)_2$, and $Pd(NO_3)_2$ are also diamagnetic. Their atomic arrangements are unknown, except that of $PdCl_2$, described in Section 5-3, but it is probable that they all involve square-coordinated palladium. With the cyanide, for example, this could oc-

[52] Elliott, *loc. cit.* (49).

[53] W. Klemm, H. Jacobi, and W. Tilk, *Z. anorg. Chem.* 201, 1 (1931).

[54] Cambi and Szegö, *loc. cit.* (51).

[55] P. Pauling, Ph.D. dissertation, Univ. London, 1960.

[56] L. Cambi, A. Cagnasso, and E. Tremolada, *Gazz. chim. ital.* 64, 758 (1934).

[57] R. B. Janes, *J.A.C.S.* 57, 471 (1935).

cur by continued polymerization to give sheets with the structure

Continued polymerization is observed in crystals[58] of PdO and PtS (cooperite), which contain planar rectangular PtO_4 or PtS_4 groups, with shared O or S, as shown in Figure 5-11. Braggite, (Pt, Pd, Ni)S, and PdS, however, seem to have a related but more complex structure,[59] involving slight distortions from the cooperite configuration. The values reported for the Pd—S bond distances are 2.26, 2.29, 2.34, and 2.43 Å.

Compounds of platinum similar to those of palladium listed above, including also $Pt(NH_3)_4SO_4$, $K_2Pt(C_2O_4)_2 \cdot 2H_2O$, $PtCl_2 \cdot 2CO$, and $PtCl_2 \cdot CO$, are diamagnetic. The last substance is probably a tetramer.

The magnetic method cannot be used to distinguish easily between the square and the tetrahedral configurations for complexes of bipositive copper or silver, since for each configuration one unpaired electron is expected. A small difference between the moments for the two configurations may arise in the following way. The moment of cupric ion in solution, 1.95, is larger than the spin moment of one electron, 1.73, because of a small contribution of the orbital moment. This contribution should be smaller for square complexes than for tetrahedral complexes because of the greater quenching effect of the more unsymmetrical field of the attached groups. There is some indication that this occurs; for $CuSO_4 \cdot 5H_2O$ and $Cu(NO_3)_2 \cdot 6H_2O$ values of μ of 1.95 to 2.02 are reported, whereas for $CuCl_2 \cdot 2H_2O$, $K_2CuCl_4 \cdot 2H_2O$, and $Cu(NH_3)_4(NO_3)_2$ the observed values lie between 1.73 and 1.87. Anisotropy of the paramagnetic susceptibility of $CuSO_4 \cdot 5H_2O$ has

[58] L. Pauling and M. L. Huggins, *Z. Krist.* **87**, 205 (1934); F. A. Bannister and M. H. Hey, *Mineral. Mag.* **23**, 188 (1932).

[59] T. F. Gaskell, *Z. Krist.* **96**, 203 (1937); F. A. Bannister, *ibid.* 201.

been found;[60] the effective magnetic moment of the $[Cu(H_2O)_4]^{++}$ complex is 2.12 Bohr magnetons with the magnetic field normal to the plane of the complex and 1.80 with the field in the plane.

Nickel tetracarbonyl has a tetrahedral configuration; this does not lead to paramagnetism, however, because the neutral nickel atom has two electrons more than bipositive nickel, and the $3d$ orbitals are completely occupied by pairs. $Ni(CO)_4$, like other metal carbonyls and

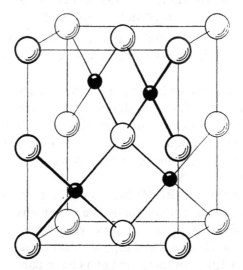

Fig. 5-11.—The structure of the tetragonal crystal PdO. Small circles represent palladium atoms and large circles oxygen atoms. The PdO_4 groups are planar.

related compounds that have been studied magnetically (including $Co(CO)_3NO$, $Fe(CO)_2(NO)_2$, $Fe(CO)_5$, $Fe_2(CO)_9$, $Fe_3(CO)_{12}$, $Cr(CO)_6$, and $Mo(CO)_6$), is diamagnetic.

The color of a complex is closely related to its bond type and coordination type. Lifschitz and his collaborators[61] have prepared many complexes of nickel with stilbenediamine (1,2-diphenylethylenediamine) and with monophenylethylenediamine, two molecules of diamine being combined with one nickel atom in each compound. Some of these substances are yellow in color and some are blue. All the yellow substances are diamagnetic, showing that in these each nickel atom forms square dsp^2 bonds with the four nitrogen atoms of the two attached diamine groups. All the blue substances, on the other hand, are paramagnetic, with susceptibilities corresponding to values close

[60] K. S. Krishnan and A. Mookherji, *Phys. Rev.* **54**, 841 (1938).

[61] I. Lifschitz, J. G. Bos, and K. M. Dijkema, *Z. anorg. Chem.* **242**, 97 (1939).

to 3.0 Bohr magnetons for the nickel atom. This shows that the bonds are hypoligating, the configuration of atoms about each nickel atom probably being octahedral. The probability of octahedral coordination is supported by the fact that the substances $Ni(NH_3)_6X_2$, with X = Cl, Br, I, and NO_3, in which there is octahedral coordination about nickel,[62] are violet in color.

5-7. THE ELECTRONEUTRALITY PRINCIPLE AND THE STABILITY OF OCTAHEDRAL COMPLEXES

Many factors affect the stability of complexes. One important factor, the multiple-bond character of the M—X bonds, will be discussed in Section 9-7.

Another important factor is the partial ionic character of the bonds. In general it may be said that stable complexes are those with structures such that each atom has only a small electric charge, approximating zero (that is, in the range -1 to $+1$). The electroneutrality principle,[63] of which the foregoing statement is a special case, will be discussed further in Section 8-2.

Let us consider the cobalt(III) hexammoniate ion, $[Co(NH_3)_6]^{+++}$. If the Co—N bonds were ionic bonds the entire charge $3+$ would be located on the cobalt atom; and if they were extreme covalent bonds the cobalt atom would have the charge $3-$ and each nitrogen atom the charge $1+$ (Fig. 5-12). In fact, the bonds have partial ionic character such as to make the atoms nearly neutral. If it is assumed, as illustrated in Figure 5-13, that the Co—N bonds have 50 percent and the N—H bonds 17 percent ionic character, the cobalt and nitrogen atoms become neutral and each hydrogen atom has the charge $+\frac{1}{6}$. This distribution of the charge of the complex ion, over the surface of the nearly spherical group, corresponds to electrostatic stability; an electrically charged solid metal sphere has its charge entirely on its surface.

The assumed amount of ionic character of the N—H bond is that corresponding to its electronegativity difference, but that for the Co—N bond is larger (50 percent, whereas the electronegativity difference corresponds to 30 percent).

We may understand from an extension of the foregoing discussion why stable cationic complexes have a peripheral set of hydrogen atoms, as in the hydrates and ammoniates, and the stable anionic complexes have a peripheral set of electronegative atoms, as in $[Co(NO_2)_6]^{---}$, $[Fe(CN)_6]^{----}$, and $[Co(C_2O_4)_3]^{---}$.

The electroneutrality principle provides an explanation of the stability of hydrated ions of the iron-group transition elements with oxida-

[62] R. W. G. Wyckoff, *J.A.C.S.* **44**, 1239, 1260 (1922).
[63] L. Pauling, *J. Chem. Soc.* **1948**, 1461.

tion number $+2$ or $+3$.   The electronegativity values for the elements Ti to Ni lie in the range 1.5 to 1.8, corresponding to 52 to 63 percent of partial ionic character of the bonds to oxygen atoms, and hence to the

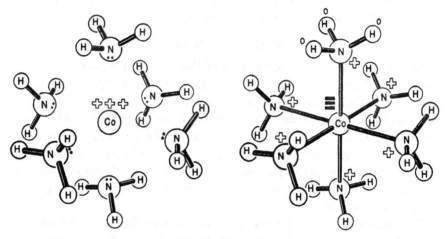

Fig. 5-12.—The representation of alternative extreme types of electronic structures for the octahedral complex ion $[Co(NH_3)_6]^{+++}$.   On the left is a representation of the structure with extreme electrostatic bonds.   The cobalt atom is represented as having a positive electric charge, $3+$.   At the right is represented the structure in which normal covalent bonds are between the cobalt atom and the surrounding nitrogen atoms, as well as between the nitrogen atom and its three attached hydrogen atoms.   This structure places the charge $3-$ on the cobalt atom and $1+$ on each nitrogen atom.

Fig. 5-13.—The distribution of charge in the complex ion $[Co(NH_3)_6]^{+++}$, with the cobalt-nitrogen bonds represented as having 50 percent ionic character and the nitrogen-hydrogen bonds as having 16.7 percent ionic character. This electronic structure leaves the cobalt atom and the nitrogen atoms with zero electric charge.   The total charge of the complex, $3+$, is distributed over the eighteen hydrogen atoms.

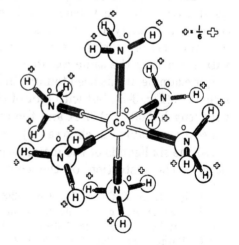

transfer of 2.22 to 2.88 units of negative charge to the metal atom in the hexahydrated complex.   This charge transfer would make the metal atom nearly neutral if its oxidation number were $+2$ or $+3$.

A discussion of other properties of hexahydrated ions and other complexes of the iron-group metals, including paramagnetic resonance and spin-orbit coupling constants evaluated from absorption spectra, has led to the conclusion that in both bipositive and tripositive metal-ion complexes the metal atom is close to electric neutrality.[64]

5-8. LIGAND FIELD THEORY

An interesting and useful method of theoretical treatment of certain properties of complexes and crystals, called the *ligand field theory*, has been applied with considerable success to octahedral complexes, especially in the discussion of their absorption spectra involving electronic transitions.[65] The theory consists in the approximate solution of the Schrödinger wave equation for one electron in the electric field of an atom plus a perturbing electric field, due to the ligands, with the symmetry of the complex or of the position in the crystal of the atom under consideration.

The general theory was worked out in detail by Bethe in a paper that is the starting point used by nearly every investigator.[66] Application of the theory was soon made to the magnetic and optical properties of complexes by Penney and Schlapp, Van Vleck, and others.[67] The theory is usually applied in such a way as to permit the evaluation from experimental quantities of parameters representing the strength of the ligand field and the magnitudes of the interactions of the *d* electrons in the complexes; there is difficulty, however, in the interpretation of these values in terms of structural features.

In some respects the ligand field theory is closely related, at least qualitatively, to the valence-bond theory described in the preceding sections, and many arguments about the structure of the normal state of a complex or crystal can be carried out in either of the two ways, with essentially the same results.[68]

For example, it has been found[69] that CrF_2 crystallizes with the rutile structure (Fig. 3-2), but with four of the Cr—F bonds (lying in a plane) with length 2.00 ± 0.02 Å and the other two with length 2.43 Å (and hence presumably much weaker), whereas in other crystals (MgF_2, TiO_2) the six ligands of the metal atom are at essentially the same distance. The distortion of the coordination polyhedron can be ex-

[64] T. M. Dunn, *J. Chem. Soc.* 1959, 623.

[65] See the several communications on "Ions of the Transition Elements," *Discussions Faraday Soc.* 26, 7-192 (1959).

[66] H. Bethe, *Ann. Physik* 3, 143 (1929).

[67] W. G. Penney and R. Schlapp, *Phys. Rev.* 41, 194 (1932); Van Vleck, *loc. cit.* (45); J. S. Griffith, *Trans. Faraday Soc.* 54, 1109 (1958).

[68] N. S. Gill, R. S. Nyholm, and P. Pauling, *Nature* 182, 168 (1958).

[69] K. H. Jack and R. Maitland, *J. Chem. Soc.* 1957, 232.

plained in a straightforward way by the ligand field theory, or alternatively by the consideration of bond orbitals. The substance has paramagnetic susceptibility corresponding to four unpaired $3d$ electrons per chromium atom. These electrons utilize four of the five $3d$ orbitals. Hence it would be expected (Sec. 5-3) that the atom would make use of the remaining d orbital to form dsp^2 square bonds (with, of course, some partial ionic character).

CrF_3 forms a cubic crystal containing regular octahedral CrF_6 groups, each fluorine atom forming a joint corner of two octahedra; the Cr—F bonds all have length 1.90 Å. The regularity of these octahedra is expected; the three unpaired $3d$ electrons use only three of the $3d$ orbitals, leaving two available for formation of d^2sp^3 octahedral bonds.

In an environment with regular octahedral symmetry the five d orbitals can be divided into two sets. Two orbitals, d_{z^2} and $d_{x^2+y^2}$, interact in an equivalent manner along the x, y, and z axes, and the other three, d_{xy}, d_{yz}, and d_{yz}, interact in a different way with the field. The latter three, which avoid the octahedral ligands, represent a triply degenerate state for a nonbonding electron that is more stable than the doubly degenerate state represented by the first two.

If only one of the orbitals d_{xy}, d_{yz}, d_{xz} is occupied, the structure no longer has regular octahedral symmetry. If three are occupied, as in CrF_3, the regular symmetry is retained. The fourth $3d$ electron in CrF_2 can be described as occupying the d_{z^2} orbital, and repelling the two fluorine atoms along $+z$ and $-z$.

5-9. OTHER CONFIGURATIONS INVOLVING d ORBITALS

In molybdenite, MoS_2, the molybdenum(IV) atom, with only one unshared pair of $4d$ electrons, has four $4d$ orbitals available for bond formation. The configuration of the six sulfur atoms about each molybdenum atom in this crystal[70] is not octahedral, but is that of a trigonal prism with unit axial ratio, as shown in Figure 5-14. The S—Mo—S bond angles have values 82° and 136°, which are not far from those for the strongest dsp bonds (73°09′ and 133°37′); six equivalent orbitals of strength 2.983 with the trigonal-prismatic orientation of bond directions can be constructed.[71]

This configuration occurs also in tungstenite, WS_2, but it has not

---

[70] R. G. Dickinson and L. Pauling, *J.A.C.S.* **45**, 1466 (1923).

[71] Hultgren. *loc. cit.* (3); Kuhn, *loc. cit.* (4); G. H. Duffey, *J. Chem. Phys.* **17**, 1328 (1949). Hultgren mentions that the observed diamagnetism of molybdenite may be explained by the fact that there is only one orbital with large d character orthogonal to the six trigonal-prism bond orbitals; the other two orbitals have smaller d character (more s and p), and are hence less stable for occupancy by nonbonding electrons.

been recognized in any other compounds of molybdenum or tungsten.

Both quadripositive and quinquepositive molybdenum and tungsten form complexes with eight cyanide groups. In these complexes a molybdenum atom has available five $4d$ orbitals, one $5s$ orbital, and

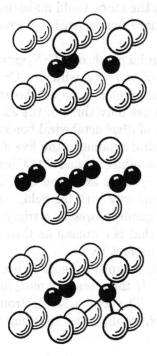

FIG. 5-14.—The structure of the hexagonal crystal molybdenite, MoS_2, showing the arrangement of sulfur atoms (large circles) at the corners of a trigonal prism about each molybdenum atom (small circles).

three $5p$ orbitals, which in combination give rise to nine hybrid orbitals. One of these is occupied by the electron pair or odd electron (for Mo(IV) or Mo(V), respectively), leaving eight orbitals for bond formation. The configuration of the complex $[Mo(CN)_8]^{----}$, as determined experimentally by Hoard and Nordsieck in their x-ray study[72] of crystalline potassium molybdocyanide dihydrate, $K_4Mo(CN)_8$ ·$2H_2O$, is shown in Figure 5-15. It is of interest that the coordination polyhedron is neither the square antiprism, which would be favored by steric interactions of the cyanide groups, nor the cube, which comes to mind because it, like the tetrahedron and octahedron, is a regular polyhedron. Four bonds lie at 34° with the vertical symmetry axis of the complex, and the other four at 73°. The values for the set of eight bond orbitals with maximum sum of strengths (4 with strength 2.995, 4 with 2.968) are 34°33' and 72°47', respectively.[73]

[72] J. L. Hoard and H. H. Nordsieck, *J.A.C.S.* **61**, 2853 (1939).

[73] G. Racah, *J. Chem. Phys.* **11**, 214 (1943). Racah finds that the best eight equivalent *dsp* orbitals have strength 2.9886; they are directed toward the corners of a tetragonal antiprism, with angle 60°54' with the tetragonal axis.

The substances $K_3Re(CN)_8$ and $K_2Re(CN)_8$, recently reported,[74] probably contain Re(V) and Re(VI) with ligancy 8 and the same configuration as the molybdenum octacyanide complexes.

The configuration of the eight oxygen atoms bonded to the thorium atom in thorium(IV) acetylacetonate, $Th(C_5H_7O_2)_4$, has been found[75] to be that of the tetragonal antiprism (Fig. 5-16). Each bidentate ligand connects adjacent corners of one of the squares of the polyhedron.

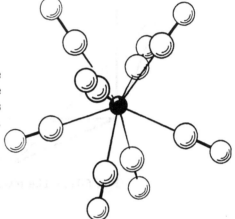

FIG. 5-15.—The structure of the complex ion $[Mo(CN)_8]^{----}$. The carbon atoms of the cyanide groups are bonded to the molybdenum atom.

Five covalent bonds can be formed by the phosphorus atom in the molecules PF_5, PF_3Cl_2, and PCl_5 with use of one $3d$ orbital. All three of these molecules have been shown by the electron-diffraction method[76] to have the configuration of a trigonal bipyramid of halogen atoms with the phosphorus atom at its center (Fig. 5-17). In PF_3Cl_2 the two chlorine atoms are at the apices of the two pyramids and the three fluorine atoms at the corners of their common base. The trigonal bipyramidal configuration for PF_5 has been verified by nuclear magnetic resonance;[77] the same technique has shown that IF_5 and BrF_5 have the square pyramidal configuration described below.

An approximate quantum mechanical treatment[78] of PCl_5 has led to

[74] R. Colton, R. D. Peacock, and G. Wilkinson, *Nature* **182**, 393 (1958).

[75] D. Grdenić and B. Matković, *Nature* **182**, 465 (1958).

[76] L. O. Brockway and J. Y. Beach, *J.A.C.S.* **60**, 1836 (1938); M. Rouault, *Compt. rend.* **207**, 620 (1938); V. Schomaker, unpublished investigation. Brockway and Beach reported the interatomic distances P—F = 1.59 ± 0.03, and P—Cl = 2.05 ± 0.03 Å in PF_3Cl_2 and P—F = 1.57 ± 0.02 Å in PF_5. The two chlorine atoms at the apices of the pyramids in PCl_5 are 2.11 Å from the phosphorus atom, the other three being at 2.04 Å.

[77] H. S. Gutowsky and C. J. Hoffman, *J. Chem. Phys* **19**, 1259 (1951).

[78] L. Pauling and J. I. Fernandez Alonso, *Proc. Nat. Acad. Sci. U. S.*, in press.

Complex Bond Orbitals

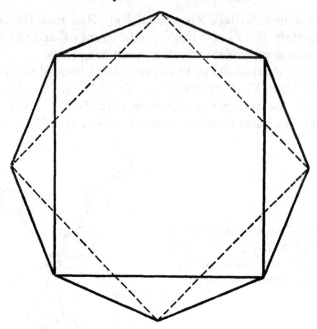

FIG. 5-16.—The tetragonal antiprism.

the conclusion that the molecule can be described as involving resonance of structure A with the five structures B and C and the six structures D, each contributing about 8 percent:

A(1 structure) B(2) C(3) D(6)

(The Cl—Cl bond shown in D is a long bond that does not contribute
much to stabilizing the molecule.)   Only for structure A is a large
amount of d character involved in the bond orbitals.

The same configuration has been reported also for molybdenum
pentachloride,[79] $MoCl_5$, antimony pentachloride,[80] $SbCl_5$, and the tri-
methylstibine dihalides,[81] $(CH_3)_3SbX_2$.   Molybdenum pentachloride

[79] Electron-diffraction study of gas molecules, R. V. G. Ewens and M. W.
Lister, *Trans. Faraday Soc.* **34**, 1358 (1938).

[80] Gas molecules (electron diffraction), M. Rouault, *Ann. phys.* **14**, 78 (1940);
crystal (x-ray diffraction), S. M. Ohlberg, *J.A.C.S.* **81**, 811 (1959).

[81] A. F. Wells, *Z. Krist.* **99**, 367 (1938).

occurs as the dimer, Mo_2Cl_{10}, in the crystal.[82]   Each molybdenum atom has six chlorine atoms ligated about it at the corners of a nearly regular octahedron.   The octahedra of the two molybdenum atoms of the dimer have a common edge, defined by two chlorine atoms bonded to both of the molybdenum atoms. Nb_2Cl_{10} has the same structure.

It is interesting that $AsCl_5$ has never been synthesized.   There are several compounds of elements of the first long period that have never been made, although the corresponding compounds of their congeners

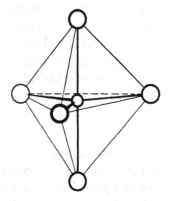

Fig. 5-17.—The structure of the molecule PCl_5, showing the arrangement of the five chlorine atoms at the corners of a trigonal bipyramid about the phosphorus atom.

in adjacent periods are known:[83] other examples are VCl_5, CrF_6, SeO_3, and $HBrO_4$.

The molecule IF_7 has the configuration of a pentagonal bipyramid; five fluorine atoms are arranged in a belt in the equatorial plane about the iodine atom, and the other two are in the axial positions.[84]   The I—F bond lengths are about 1.85 Å.   Bond orbitals for this configuration have been reported by Duffey.[85]   The equatorial sp^3d^3 orbitals have strength 2.976 and the axial orbitals have strength 2.920.

The configuration of the pentagonal bipyramid has been found[86] for the $[UF_7]^{---}$ complex ion in the crystal K_3UF_7.   In the A-modification of the rare-earth sesquioxides[87] and in the ions $[ZrF_7]^{---}$ and $[NbOF_6]^{---}$ each metal atom is surrounded by seven oxygen or fluorine atoms with the configuration of an octahedron distorted by spreading one face and introducing the seventh atom at its center,[88] whereas

[82] D. E. Sands and A. Zalkin, *Acta Cryst.*, in press (1959).

[83] W. E. Dasent, *J. Chem. Educ.* 34, 535 (1957).

[84] S. H. Bauer and F. A. Keidel, reported by Sutton, *Interatomic Distances;* R. C. Lord, M. A. Lynch, W. C. Schumb, and E. F. Slowinski, *J.A.C.S.* 72, 522 (1950); R. D. Burbank and F. N. Bensey, Jr., *J. Chem. Phys.* 27, 981 (1957).

[85] G. H. Duffey, *J. Chem. Phys.* 18, 943 (1950); also R. L. Scott, *ibid.*, 1420.

[86] W. H. Zachariasen, *Acta Cryst.* 7, 792 (1954).

[87] L. Pauling, *Z. Krist.* 69, 415 (1929).

[88] G. C. Hampson and L. Pauling, *J.A.C.S.* 60, 2702 (1938); M. B. Williams and J. L. Hoard, *ibid.* 64, 1139 (1942).

in the ions $[NbF_7]^{--}$ and $[TaF_7]^{--}$ the configuration of the seven fluorine atoms can be described as obtained by introducing a seventh atom at the center of one of the square faces of a trigonal prism.[89]

The configuration of the tetragonal antiprism has been found[90] for the $[TaF_8]^{---}$ ion in the crystal Na_3TaF_8 (Fig. 5-16).

It is to be borne in mind that the stoichiometric formula of a compound may not indicate uniquely the ligancy of its complexes. Thus Hoard and Martin[91] have shown that the crystal K_3HNbOF_7 does not contain $[NbF_7]^{--}$ groups (which do exist in K_2NbF_7) or $[NbOF_7]^{----}$ groups, but rather octahedral $[NbOF_5]^{--}$ groups and hydrogen bifluoride ions $[HF_2]^-$, the formula being preferably written $K_3HF_2NbOF_5$. Similarly[92] the crystal $(NH_4)_3SiF_7$ contains no complexes $[SiF_7]^{---}$, but instead octahedral complexes $[SiF_6]^{--}$ and fluoride ions F^-.

5-10. CONFIGURATIONS FOR ATOMS WITH UNSHARED ELECTRON PAIRS

A considerable amount of information about the relative orientations of bonds formed by an atom that also possesses one or more unshared electron pairs has been gathered.

In a few cases an unshared electron pair seems to have no effect on bond directions. This is observed[93] for Se(IV) in the octahedral complex ion $[SeBr_6]^{--}$ and for Sb(III) in $[SbBr_6]^{---}$.

Usually, however, an unshared pair seems to occupy one of the corners of a coordination polyhedron and to replace the shared pair of a bond. The molecules NH_3, PCl_3, etc. have pyramidal configurations that might be described as involving bonds directed toward three corners of a tetrahedron with the fourth corner occupied by the unshared pair, and a similar description can be given for H_2O, $(CH_3)_2S$, and related molecules.

The extension of this postulate to atoms with five bonds and one unshared pair suggests that the bonds should be directed toward the five corners of a square pyramid, which with the unshared pair would form an octahedron. BrF_5 has been shown[94] to have this configuration. The bromine atom lies about 0.15 Å below the base of the pyra-

[89] J. L. Hoard, *J.A.C.S.* **61**, 1252 (1939).

[90] J. L. Hoard, W. J. Martin, M. E. Smith, and J. F. Whitney, *J.A.C.S.* **76**, 3820 (1954).

[91] J. L. Hoard and W. J. Martin, *J.A.C.S.* **63**, 11 (1941).

[92] J. L. Hoard and M. B. Williams, *J.A.C.S.* **64**, 633 (1942).

[93] J. L. Hoard and B. N. Dickinson, *Z. Krist.* **84**, 436 (1933); N. Elliott, *J. Chem. Phys.* **2**, 298 (1934).

[94] R. D. Burbank and F. N. Bensey, Jr., *J. Chem. Phys.* **21**, 602 (1953); **27**, 983 (1957).

mid, so that the F—Br—F angles (from the apical fluorine atom to the basal atoms) are about 86°. The unshared electron pair thus occupies a larger volume about the bromine atom than the shared pairs. This distortion from the regular octahedral configuration is the expected consequence of the large amount of s character of the orbital occupied by the unshared pair and the d and f character of the bond orbitals.

In the crystal KICl₄ the four chlorine atoms of the $[ICl_4]^-$ complex are located at the corners of a square about the iodine atom;[95] we may consider that the octahedron about iodine is completed by the two unshared electron pairs of the iodine atom, one above and one below the ICl₄ plane.

Our postulate suggests that the configuration of molecules such as TeCl₄ is similar to that of PCl₅ and related molecules; that is, that the four bonds and one unshared pair occupy the five corners of a trigonal bipyramid. The unshared pair would probably occupy one of the equatorial positions rather than one of the apical positions. Such a configuration has, indeed, been found for tellurium tetrachloride,[96] with chlorine atoms at the two apices and at two of the three equatorial positions of a trigonal bipyramid, the unshared pair occupying the third. In crystalline SeBr₂(C₆H₅)₂ the configuration is the same,[97] with the bromine atoms in the apical positions. This configuration has also been found[98] for Te(CH₃)₂Cl₂. As in other molecules with unshared pairs, the unshared pair occupies a larger volume about the central atom than the shared pairs: the values of the bond angles are Cl—Te —Cl = 172.3° ± 0.3°, Cl—Te—C = 87.5° ± 1.0°, and C—Te—C = 98.2° ± 1.1°.

The x-ray investigation[99] of the crystal KIO₂F₂ has shown that the $[IO_2F_2]^-$ ion can also be described as having the trigonal bipyramidal configuration, with the two oxygen atoms and the unshared electron pair in the equatorial positions and the two fluorine atoms at the apices.

The bromine trifluoride molecule can be similarly described from a careful microwave study[100] as a trigonal bipyramid with fluorine atoms at the two apical positions and one equatorial position (and unshared pairs at the other two equatorial positions), the four atoms thus being

[95] R. C. L. Mooney, *Z. Krist.* **98**, 377 (1938).

[96] D. P. Stevenson and V. Schomaker, *J.A.C.S.* **62**, 1267 (1940).

[97] J. D. McCullough and G. Hamburger, *J.A.C.S.* **63**, 803 (1941).

[98] G. D. Christofferson, R. A. Sparks, and J. D. McCullough, *Acta Cryst.* **11**, 782 (1958).

[99] L. Helmholz and M. T. Rogers, *J.A.C.S.* **62**, 1537 (1940).

[100] D. W. Magnuson, *J. Chem. Phys.* **27**, 223 (1957). An agreeing crystal-structure study has also been reported: Burbank and Bensey, *J. Chem. Phys.* **27**, 983 (1957).

coplanar. The two small F—Br—F bond angles have the value 86°12.6′; the Br—F bond lengths are 1.721 Å (to one F atom) and 1.810 Å (to the other two). A similar structure has been found[101] for ClF_3; the structural parameters are F—Cl—F = 87.5°, Cl—F = 1.598 ± 0.005 Å (one bond) and 1.698 ± 0.005 Å (two bonds).

It has been pointed out[102] that the electrical conductivity of liquid BrF_3 strongly indicates the presence of the ions $[BrF_2]^+$ (tetrahedral, two unshared electron pairs at two tetrahedron corners) and $[BrF_4]^-$ (octahedral, two unshared pairs at two corners). The cation is present[103] in BrF_2SbF_6 and $(BrF_2)_2SnF_6$, which are acids in solution in BrF_3, and the anion is present[104] in $KBrF_4$, $AgBrF_4$, and $Ba(BrF_4)_2$, which are bases in this solution.

We conclude from the foregoing examples that in general the configurations of molecules with unshared electron pairs are similar to those of molecules with only shared pairs in the valence shell, except that the unshared pairs occupy a larger volume than the shared pairs, thus causing a decrease in the values of the bond angles. This effect has been discussed in detail in Section 4-3 for the simple case of *sp* hybridization.

[101] Microwave: D. F. Smith, *J. Chem. Phys.* **21**, 609 (1953); crystal structure: R. D. Burbank and F. N. Bensey, Jr., *ibid.* 602.

[102] A. A. Banks, H. J. Emeléus, and A. A. Woolf, *J. Chem. Soc.* 1949, 2861.

[103] A. A. Woolf and H. J. Emeléus, *J. Chem. Soc.* 1949, 2865. Also BrF_2AuF_4: A. G. Sharpe, *ibid.* 2901.

[104] A. G. Sharpe and H. J. Emeléus, *J. Chem. Soc.* 1948, 2135.

CHAPTER 6

The Resonance of Molecules among
Several Valence-Bond Structures

ONE of the most interesting and useful applications of the theory of resonance is in the discussion of the structure of molecules for which no one valence-bond structure is satisfactory. An introduction to this discussion is presented in the following sections. The chapter ends with a reply to some critical comments that have been made about the theory.

6-1. RESONANCE IN NITROUS OXIDE AND BENZENE

For many molecules it is possible to formulate valence-bond structures that are so reasonable and that account so satisfactorily for the properties of the substances that they are accepted by everyone without hesitation. The structures given on the next page may be shown for illustration. The physical and chemical properties of substances and the configurations of molecules associated with structures of this type are well understood, and this understanding forms the basis for a large part of chemical reasoning.

It is sometimes found, however, that an unambiguous assignment of a single valence-bond structure to a molecule cannot be made: two or more structures that are about equally stable may suggest themselves, no one of which accounts in a completely satisfactory way for the properties of the substance. Under these circumstances some new structural concepts and symbols might be introduced; we might, for

example, use the symbol ⬡ for benzene, without attempting to

interpret this symbol in terms of single and double bonds. With the development of the idea of quantum-mechanical resonance a more il-

183

Oxygen fluoride

$$
\begin{array}{c}
:\ddot{F}: \\
| \quad\ddot{} \\
:O\!-\!\ddot{F}: \\
\ddot{}\quad\ddot{}
\end{array}
$$

Trimethylamine

$$
\begin{array}{c}
H \\
| \\
H\!-\!\overset{|}{C}\!-\!H \\
\qquad\quad H \\
\qquad\quad | \\
:N\!-\!\!-\!\!-\!\overset{|}{C}\!-\!H \\
\qquad\quad | \\
\qquad\quad H \\
H\!-\!\overset{|}{C}\!-\!H \\
| \\
H
\end{array}
$$

Ethane

$$
\begin{array}{c}
H \quad H \\
| \quad\ | \\
H\!-\!\overset{|}{C}\!-\!\overset{|}{C}\!-\!H \\
| \quad\ | \\
H \quad H
\end{array}
$$

Ethylene

$$
\begin{array}{c}
H \qquad\qquad H \\
\diagdown \qquad\quad \diagup \\
C\!=\!C \\
\diagup \qquad\quad \diagdown \\
H \qquad\qquad H
\end{array}
$$

luminating and useful solution to this difficulty has been found: *the actual normal state of the molecule is not represented by any one of the alternative reasonable structures, but can be represented by a combination of them,* their individual contributions being determined by their nature and stability. The molecule is then described as *resonating among the several valence-bond structures.*[1]

[1] The idea of the quantum-mechanical resonance of molecules among several valence-bond structures was developed in 1931: see J. C. Slater, *Phys. Rev.* **37**, 481 (1931); E. Hückel, *Z. Physik* **70**, 204 (1931); **72**, 310 (1931); **76**, 628 (1932); **83**, 632 (1933); L. Pauling, *J.A.C.S.* **53**, 1367, 3225 (1931); **54**, 988, 3570 (1932); *Proc. Nat. Acad. Sci. U. S.* **18**, 293 (1932); L. Pauling and G. W. Wheland, *J. Chem. Phys.* **1**, 362 (1933); etc. During the twentieth century rapid progress has been made in the development of chemical theories that bear some relationship to the theory of resonance. A small resemblance to it is shown by Thiele's theory of partial valence (J. Thiele, *Ann. Chem.* **306**, 87 [1899]), and it is much more closely approximated by Arndt's theory of intermediate stages (F. Arndt, E. Scholz, and F. Nachtwey, *Ber.* **57**, 1903 [1924]; F. Arndt, *ibid.* **63**, 2963 [1930]) and the theory of the mesomeric state developed by English and American organic chemists (T. M. Lowry, *J. Chem. Soc.* **123**, 822, 1866 [1923]; H. J. Lucas and A. Y. Jameson, *J.A.C.S.* **46**, 2475 [1924]; R. Robinson *et al., J. Chem. Soc.* **1926**, 401; C. K. Ingold and E. H. Ingold, *ibid.* 1310; etc.).

The resonance of molecules among various electronic structures has been discussed in detail in Chapter 3 for the case that the resonating structures differ in regard to bond type (ionic and covalent). The resonance under discussion in this chapter is not greatly different; it involves structures that differ in the distribution of bonds rather than in their type.

The nitrous oxide molecule may be used as an example. This molecule is linear, with the oxygen atom at one end. It contains 16 valence electrons; and it is seen that these can be assigned to the stable L orbitals of the atoms in any of the following reasonable ways:

$$A \qquad \overset{-}{:}\!\overset{..}{N}\!\!=\!\!\overset{+}{N}\!\!=\!\!O:$$

$$B \qquad \overset{-}{:}N\!\!=\!\!\overset{+}{N}\!\!=\!\!\overset{..}{O}:$$

$$C \qquad :N\!\!\equiv\!\!\overset{+}{N}\!\!-\!\!\overset{..}{\underset{..}{O}}\overset{-}{:}$$

Each of these three structures involves four covalent bonds (counting a double bond as two and a triple bond as three), and a separation of charge to adjacent atoms. (Structures A and B differ in that in A the double bond between N and N is formed with use of p_z orbitals and that between N and O with p_y orbitals, and in B they are reversed; see Sec. 4-8.) Other structures that might be written are recognized at once as being much less stable than these, such as

$$D \qquad \overset{--}{:}\!\overset{..}{N}\!\!-\!\!\overset{+}{N}\!\!\equiv\!\!\overset{+}{O}:$$

on which instability is conferred by the arrangement of electric charges, and

$$E \qquad \overset{-}{:}\!\overset{..}{N}\!\!=\!\!\overset{..}{N}\!\!-\!\!\overset{..}{\overset{..}{O}}\overset{+}{:}$$

$$F \qquad :\!\overset{..}{N}\!\!-\!\!\overset{..}{N}\!\!=\!\!\overset{..}{O}:$$

with instability arising from the smaller number of covalent bonds.

A decision cannot be made between structures A, B, and C, which are, indeed, so closely similar in nature that there can be no large energy difference between them. Moreover, they satisfy the other conditions for resonance: they involve the same number of unpaired electrons (zero), and they correspond to about the same equilibrium configuration of the nuclei (linear, for a central tetrahedral atom forming either two double bonds or a single bond and a triple bond). We accordingly expect the normal state of the molecule to correspond to *resonance among structures A, B, and C*, with small contributions by the other less stable structures, which can be neglected in our discussion.   The molecule is more stable than it would be if either A, B, or C alone represented its normal state by an amount of energy equal to the resonance energy among the three structures.   Its interatomic distances and force constants are not those corresponding to any one structure alone, but to resonance among them (Chap. 7).   Its electric dipole moment is not large, but is close to zero, the opposed moments of the structures canceling each other; the experimental value, from the microwave spectrum,[2] is $0.166 \pm 0.002\ D$, with the direction not known.

The value of the electric dipole moment provides an illustration of the significance of resonance as compared with tautomerism.   If nitrous oxide gas were a tautomeric mixture of molecules of types A, B, and C, its dielectric constant would be large, since the molecules of each type would have large dipole moments and would make a large contribution to the dielectric constant of the gas.   The frequency of resonance among the structures is, however, very large, of the order of magnitude of electronic frequencies in general, and the nuclei do not have time to orient themselves in an applied electric field in order to contribute to the dielectric constant of the medium before the electrons of the molecule have run through their phases, which results in a very small average electric moment.

The discussion in Section 1-3 about the element of arbitrariness in the concept of resonance may be recalled at this point with reference to the nitrous oxide molecule and the other molecules that are described in this chapter as resonating among several valence-bond structures. It is not necessary that the structures A, B, and C be used as the basis of discussion of the nitrous oxide molecule.   We might say instead that the molecule cannot be satisfactorily represented by any single valence-bond structure, and abandon the effort to correlate its structure and properties with those of other molecules.   By using valence-bond structures as the basis for discussion, however, with the aid of the concept of resonance, we are able to account for the properties of the molecule in terms of those of other molecules in a straightforward and simple way.   It is for this practical reason that we find it convenient to speak of the resonance of molecules among several electronic structures.

It is to be emphasized again that in writing three valence-bond structures for the nitrous oxide molecule and saying that it resonates among them we are making an effort to extend the valence-bond picture to molecules to which it is not applicable in its original form, and that we are not required to do this but choose to do it in the hope of obtaining a satisfactory description of these unusual molecules, permitting us to correlate and "understand" the results of experiments on their chemi-

[2] R. G. Shulman, B. P. Dailey, and C. H. Townes, *Phys. Rev.* **78**, 145 (1950).

cal and physical properties and to make predictions in the same way as for molecules to which a single valence-bond structure can be assigned. Nitrous oxide does not consist of a mixture of tautomeric molecules, some with one and some with another of the structures written above; instead, all the molecules have the same electronic structure, this being of such a nature that it cannot be satisfactorily represented by any one valence-bond diagram but can be reasonably well represented by three. The properties of the molecule are essentially those expected for an average of the three valence-bond structures, except for the stabilizing effect of the resonance energy.

To represent the molecule we may use the symbol $\{:\overset{\cdot\cdot}{\overset{-}{N}}{=}\overset{+}{N}{=}O:,$ $:\overset{-}{N}{=}\overset{+}{N}{=}\overset{\cdot\cdot}{O}:, :N{\equiv}\overset{+}{N}{-}\overset{\cdot\cdot-}{O}:\}$, the resonating structures being enclosed in brackets. I do not believe that it is wise to attempt to simplify the symbol further—to write, for example, $N{\equiv}N{=}O$, even though, as we shall see later, the N—N bond and the N—O bond have properties approaching those of a triple bond and a double bond, respectively (Chap. 8). If the formula $N{\equiv}N{=}O$ were to be used, it would be confused with formulas for nonresonating molecules. This formula suggests that the nitrogen atom can form five covalent bonds, which is not true. Moreover, the formula carries with it no stereochemical implications—we do not know the relative orientation to expect for a double bond and a triple bond—whereas the resonating formula shows at once that the molecule is linear.[3]

Benzene provides an interesting and important illustration of a resonating molecule. The two Kekulé structures for benzene, *A* and *B*, are the most stable valence-bond structures that can be written for the known hexagonal

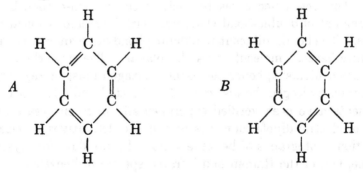

---

[3] English structural chemists make use of formulas such as $:N{\equiv}N{-}\overset{\cdot\cdot}{O}:;$ the arrows indicate changes in positions of electron pairs, corresponding to resonance with the structure $:\overset{\cdot\cdot}{N}{=}N{=}\overset{\cdot\cdot}{O}:$. F. Arndt and B. Eistert, *Ber.* **71, 237** (1938), have suggested the use of the double arrow ↔ to indicate resonance, writing $:\overset{\cdot\cdot}{N}{=}N{=}\overset{\cdot\cdot}{O}:{\leftrightarrow}:N{\equiv}N{-}\overset{\cdot\cdot}{O}:$ for normal nitrous oxide.

planar configuration.   Other structures, such as the Dewar structures

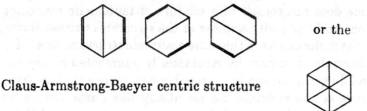

or the

Claus-Armstrong-Baeyer centric structure

correspond to diminished stability because of their weak bonds between distant atoms, and they need not be considered in a simple discussion. The Kekulé structures individually do not provide a satisfactory representation of the benzene molecule; the high degree of unsaturation that by comparison with hexene or cyclohexene would be expected for a molecule containing three double bonds is not shown by benzene, which instead is surprisingly stable and unreactive.   It is resonance that gives to benzene its aromatic properties.   The two Kekulé structures are equivalent, and have the same energy; they accordingly enter into complete resonance.   The molecule is stabilized in this way by the resonance energy of about 37 kcal/mole (Sec. 6-3).   The unsaturation of a compound containing a double bond is due, from the thermodynamic point of view, to the instability of a double bond relative to two single bonds, amounting to about 19 kcal/mole per double bond,[4] or 57 kcal/mole for three double bonds.   The resonance energy removes the greater part of this instability and gives to the molecule a degree of saturation approaching that of the paraffins.

The stereochemical properties of the benzene molecule can be predicted from the concept of resonance between the two Kekulé structures.   This resonance gives to each of the carbon-carbon bonds a large amount of double-bond character, with its stereochemical implications.   The bonds adjacent to a double bond are planar; accordingly the entire benzene molecule must be planar.   The six carbon-carbon bonds are equivalent; hence the carbon hexagon must be regular, and the carbon-hydrogen bonds must be directed radially.   All these statements have been verified experimentally in recent years by the study of electric dipole moments of benzene derivatives, of electron-diffraction photographs of benzene vapor, of x-ray data for crystalline benzene, and of the Raman and infrared spectra of benzene.

6-2. RESONANCE ENERGY

The assignment of a resonating structure to a molecule can sometimes be made on the basis of theoretical arguments, as in the two cases

[4] The values 147 and 83.1 kcal/mole for the energy of the $C{=}C$ and $C{-}C$ bonds, respectively, are given in Secs. 6-2 and 3-5.

discussed above. In general such an assignment should be supported by experimental evidence, such as that provided by chemical properties, resonance energies, interatomic distances, force constants of bonds, bond angles, electric dipole moments, and so forth. If the reasonable valence-bond structures are not equivalent, an estimate of the magnitudes of the contributions of different structures to the normal state of a molecule may be made from this information.

Of these methods of studying resonance in molecules the most fruitful at present are the determination and interpretation of values of

TABLE 6-1.—VALUES OF BOND ENERGIES FOR MULTIPLE BONDS

Bond	Bond energy	Compounds
C=C	147 kcal/mole	
N=N	100	Azoisopropane[a]
O=O	96	$^1\Delta$ state of O_2
C=N	147	n-Butylisobutylideneamine[a]
C=O	164	Formaldehyde
	171	Other aldehydes
	174	Ketones
C=S	114	
C≡C	194	
N≡N	226	N_2
C≡N	207	Hydrogen cyanide
	213	Other cyanides

[a] From heats of combustion reported by G. E. Coates and L. E. Sutton, *J. Chem. Soc.* 1948, 1187. The value 95 ±5 kcal/mole is given by the estimate 26 ±5 kcal/mole for the enthalpy change of the reaction $N_2H_4(g) \rightarrow N_2H_2(g) + H_2(g)$ given by the mass spectrographic work of S. N. Foner and R. L. Hudson, *J. Chem. Phys.*, 28, 719 (1958).

interatomic distances, discussed in the following chapter, and the calculation of values of resonance energies from thermochemical data. It is to the latter that we now turn our attention.

Values of Bond Energies for Multiple Bonds.—In Section 3-5 there is given a table of values of bond energies for single bonds. In the construction of this table care was taken to make use of data for only those molecules to each of which an unambiguous assignment of a valence-bond formula could be made. This consideration of bond energies is extended in Table 6-1, which contains values for some multiple bonds, obtained by methods similar to those described in Section 3-5.

By the addition of the suitable quantities from Tables 3-4 and 6-1 an approximate value can be predicted for the heat of formation of a gas molecule from the elements in the state of monatomic gases, provided that the molecule in its normal state is well represented by a single electronic structure of the valence-bond type. For example, the heat of formation of acetylene from the elements in their standard states is -53.9 kcal/mole, and that from atoms is 393.7 kcal/mole. The sum of bond energies $2C$—$H + C$≡C is 393 kcal/mole. The error of 1 kcal/mole is indicative of the degree of reliability of the bond energy values, which are averages of those found by consideration of the thermochemical data for many substances.

Ionic Resonance Energy and Partial Ionic Character of Multiple Bonds.—The energy values given in Table 6-1 for bonds between unlike atoms include the extra partial ionic resonance energy of the multiple bonds. In Section 4-9 it was pointed out that the observed value of the electric dipole moment of acetone indicates that each of the two bent bonds constituting the carbon-oxygen double bond has about 22 percent ionic character, as given for carbon-oxygen bonds by the electronegativity scale. Hence the molecule should be represented by the resonance structure $\{(H_3C)_2C::\overset{..}{O}:, (H_3C)_2C^+:\overset{..}{\underset{..}{O}}:^-\}$, with the second structural formula representing two structures, corresponding to the ionic aspect of one or the other half of the double bond.

The values of the resonance energy corresponding to the partial ionic character of carbon—oxygen bonds are given by the following calculation:

C—O	84		C=O	164 to 174
$\frac{1}{2}$(C—C + O—O)	58		$\frac{1}{2}$(C=C + O=O)	122
Δ	26		Δ	42 to 52

In this calculation the value 96 kcal/mole has been used for the oxygen-oxygen double bond. This is the enthalpy of dissociation of the $^1\Delta$ state of the oxygen molecule, which is 22.4 kcal/mole less stable than the normal state, the structure of which is discussed in Chapter 10. It is seen that the value of Δ, the resonance energy due to partial ionic character, is twice as great for the carbon-oxygen double bond as for the carbon-oxygen single bond. This result substantiates our earlier conclusion that each half of a double bond has the amount of ionic character indicated by the difference in electronegativity of the two atoms forming the bond.

Because of the large amount of ionic resonance for multiple bonds between atoms with large difference in electronegativity, it is to be expected that these bonds will be affected more strongly (as to amount of ionic character) by neighboring bonds than are single bonds of small ionic character; and it is observed that the values calculated for the energies of multiple bonds vary for different compounds. This has been taken into consideration to some extent by the tabulation of more than one value for some bonds; it may lead, however, to a greater error in energy calculations for molecules containing multiple bonds than for those containing only single bonds.

The Nitrogen-Nitrogen Triple Bond.—There is an interesting regularity in the values of the bond energies of Table 6-1 that throws some light on the problem of the cause of the striking thermodynamic stability of molecular nitrogen. The differences in energy of the symmetrical double bonds C=C, N=N, and O=O (the $^1\Delta$ state of O_2) and the corresponding single bonds (Table 3-4) are nearly the same (65, 62, and 63 kcal/mole, respectively), indicating a close similarity in the nature of the bonds. We might expect a similar regularity to hold for the energy difference of the triple bonds and corresponding double bonds. In fact, the difference for C≡C and C=C is 47 kcal/mole, but the difference between N≡N and N=N is not also 47, but 126 kcal /mole. The molecule N_2 is hence seen to be about 79 kcal/mole more stable than would be expected from the consideration of the energies of related molecules.

This argument is represented in the following diagram:

C—C	83	*45*	38	N—N
	64		*62*	
C=C	147	*47*	100	N=N
	47		*(47)*	
C≡C	194	*(47)*	*(147)*	N≡N

The differences of bond energies, given in italics, seem reasonable: a double bond has energy 62 to 64 kcal/mole more than the corresponding single bond, and a triple bond has energy 47 kcal/mole more than the corresponding double bond. Similarly, a carbon-carbon bond has energy 44 to 47 kcal/mole more than the corresponding nitrogen-nitrogen bond. But these regularities are illusory; they have been obtained by taking 147 kcal/mole as the energy of the N≡N bond, instead of the correct value 226 kcal/mole.

We conclude that there is an abnormality in the structure of the nitrogen molecule such as to increase the N≡N bond energy from 147 to 226 kcal/mole. This abnormality is not shown by the N=N and N—N bonds. Its nature is not known. It is responsible for the great stability of the nitrogen molecule that causes many nitrogen compounds to be explosive and causes elementary nitrogen to be a major constituent of the atmosphere.

The abnormality is found also in the nitrosyl group (Sec. 10-3).

Empirical Values of Resonance Energies.—The tables of bond energies permit the calculation of values of the heats of formation of molecules to which a single valence-bond structure can be assigned that agree with the experimental values to within a few kcal/mole. On carrying out a similar calculation for a resonating molecule on the basis of any single valence-bond structure that can be formulated, it is found that *in every case the actual energy of formation of the molecule is greater than the calculated value;* that is, the molecule is actually more stable than it would be if it had the valence-bond structure assumed for it in making the bond-energy calculation. This result is required by the fundamental quantum-mechanical principle on which the concept of resonance is based (Sec. 1-3); it provides, however, a pleasing confirmation of the arguments used in the construction of the table of bond energies.

The difference between the observed heat of formation and that calculated for a single valence-bond structure for a molecule with use of the table of bond energies is an empirical value of the resonance energy of the molecule relative to the assumed valence-bond structure.

It is desirable that the structure used as the basis for the resonance-energy calculation be the most stable (or one of the most stable) of those among which resonance occurs. It is not always convenient for this choice to be made, for the following reason. The tabulated bond energies are designed for use only between atoms with zero formal charge; no simple method of calculating the heats of formation of molecules containing charged atoms has been devised, because of the difficulties introduced by the Coulomb energy terms for the separated charges. For this reason there is no empirical value available for the resonance energy of the nitrous oxide molecule; the stable structures involve atoms with formal charges.

It must be remembered that one of the conditions for resonance of molecules among several electronic structures is that the configuration of the molecule (the arrangement of the nuclei) remain constant during the electronic resonance; it is the composite electronic structure that provides a single potential function determining the equilibrium configuration and modes of oscillation for the molecule. It is not possible for an amide to resonate between the structures

$$R-C\overset{\displaystyle \ddot{O}:}{\underset{\displaystyle \ddot{N}H_2}{\Big\langle}} \quad \text{and} \quad R-C\overset{\displaystyle :\ddot{O}H}{\underset{\displaystyle \ddot{N}H}{\Big\langle}}$$

We use the structures

$$R-C\overset{\displaystyle \ddot{O}:}{\underset{\displaystyle \ddot{N}H_2}{\Big\langle}} \quad \text{and} \quad R-C\overset{\displaystyle :\ddot{O}:^-}{\underset{\displaystyle \overset{+}{N}H_2}{\Big\langle}}$$

to describe the molecule.

The heat of formation of the benzene gas molecule from separated atoms is found from the heat of combustion (789.2 kcal/mole) and the heats of formation of the products of combustion, water and carbon dioxide, to have the value 1323 kcal/mole.   The sum of the bond energies 6C—H + 3C—C + 3C=C gives the value 1286 kcal/mole for the heat of formation of a hypothetical molecule with the Kekulé struc-

ture ⬡ or ⬡ , involving noninteracting double bonds.   The

difference between these, 37 kcal/mole, is the resonance energy of the molecule.

In calculating resonance energies it is for simplicity and convenience only that the thermochemical data are converted into energies of formation of molecules from separated atoms and compared with sums of bond energies; the same results can be obtained by dealing directly with heats of formation from elementary substances in their standard states or with heats of combustion or of hydrogenation reactions or other reactions, the resonating substance being compared with suitable nonresonating substances.   This may be illustrated by the calculation of the resonance energy of benzene from data obtained in the important series of direct measurements of heats of hydrogenation carried out by Kistiakowsky and his collaborators.[5]   The value expected for the heat of hydrogenation of a hypothetical molecule with a Kekulé structure involving noninteracting double bonds is 85.77 kcal/mole, three times the heat of hydrogenation of cyclohexene:

$$C_6H_{10} + H_2 \rightarrow C_6H_{12} + 28.59 \text{ kcal/mole.}$$

The value observed for the heat of hydrogenation of benzene is much less than this:

$$C_6H_6 + 3H_2 \rightarrow C_6H_{12} + 49.80 \text{ kcal/mole.}$$

The difference, 35.97 kcal/mole, is the resonance energy for benzene, which stabilizes the molecule relative to the individual Kekulé structures.   The agreement with the value found above, 37 kcal/mole, provides sound substantiation of the magnitude of the benzene resonance energy.[6]

[5] G. B. Kistiakowsky, J. R. Ruhoff, H. A. Smith, and W. E. Vaughan, *J.A.C.S.* **57**, 876 (1935); **58**, 137, 146 (1936); etc.

[6] A surprisingly large variability in bond-energy values is shown to exist by the range of values (26.6 to 30.1 kcal/mole) found for the heats of hydrogenation of different olefins. The double-bond energy value given in Table 6-1 corresponds to an average olefin, with heat of hydrogenation 29.8 kcal/mole (calculated with use of the C—C, C—H, and H—H bond-energy values). The comparison of benzene with cyclohexene is obviously reasonable.

As a second illustration let us consider the carbon monoxide molecule. For years there was discussion as to which of the structures :C⚌Ö: and :C≡O: is the better one. We say that resonance occurs between them; or, splitting C⚌O into its constituents, among the four structures (*a*) :C̈:Ö:, (*b*) :C::Ö:, (*c*) :C::O:, and (*d*) :C:::O:. From the discussion of the carbon-oxygen double bond in ketones the conclusion can be drawn that the structures *a*, *b*, and *c* make about equal contributions, *a* being nearly as stable as *b* and *c* despite its smaller number of covalent bonds because the great electronegativity of oxygen stabilizes a structure containing negatively charged oxygen. The fourth structure *d* would also be expected to be significant because of its stabilization through the formation of a triple covalent bond, which counteracts the instability resulting from the unfavorable distribution of charge. The observed very small value of the electric dipole moment provides evidence that the contribution made by structure *d* is about the same as that made by structure *a*. Structures *b* and *c* would have no large dipole moment, whereas those of *a* and *d* are very large, about equal in magnitude, and opposed in direction; and only if *a* and *d* contribute about equally would the moment of the molecule be small.

We may well inquire how it is possible for four structures as different in character as *a*, *b*, *c*, and *d* to contribute about equally to the normal state of the molecule. The answer, indicated above, is this: the four structures have about the same energy, as the result of two opposing effects, those of the number of covalent bonds and of the separation of charge. In the sequence *a*, *b* (and *c*), *d* the number of covalent bonds changes from one to three; this would tend to make *a* the least stable and *d* the most stable. However, the charge distribution for *a*, with the more electronegative atom negative, is the favorable one, and this stabilizes the structure, making it nearly equal in energy to *b*; whereas the charge distribution for *d* is unfavorable, the more electronegative atom having a positive charge, which counteracts the extra stability of the triple covalent bond and brings structure *d* also into approximate energy equality with *b*.

The resonance energy of carbon monoxide relative to the structure :C⚌Ö: (which itself corresponds to resonance between $^+$:C:Ö:$^-$, :C::Ö:, and :C::O:) can be found by comparing its heat of formation from atoms, 257 kcal/mole, with the ketone value of the double-bond energy, 174 kcal/mole.[7] The very large difference, 83 kcal/mole,

---

[7] The bond-energy values are not designed for use with a molecule as unconventional as carbon monoxide, containing bivalent carbon. It seems probable, however, that the error involved in this application is not great.

represents the energy of resonance, relative to the structure $:C{=}\ddot{O}:$. It is the very large resonance energy in carbon monoxide that stabilizes the substance despite its lack of saturation of the carbon valences.

The empirical resonance-energy values[8] given in Table 6-2 are discussed in the following sections of this chapter and in later chapters.

TABLE 6-2.—EMPIRICAL RESONANCE-ENERGY VALUES

Substance	Resonance energy (in kcal/mole)	Reference structure
Benzene, C_6H_6	37	
Naphthalene, $C_{10}H_8$	75	
Anthracene, $C_{14}H_{10}$	105	
Phenanthrene, $C_{14}H_{10}$	110	
Biphenyl, $C_{12}H_{10}$	5[a]	
Dihydronaphthalene, $C_{10}H_{10}$	3[a]	
Cyclopentadiene, C_5H_6	4	
1,3,5-Triphenylbenzene, $C_{24}H_{18}$	20[a]	

[a] Extra resonance energy, not including that within the benzene ring.

[8] L. Pauling and J. Sherman, *J. Chem. Phys.* 1, 606 (1933).

TABLE 6-2.—(continued)

Substance	Resonance energy (in kcal/mole)	Reference structure
Styrene, C_8H_8	5[a]	⬡—CH=CH₂
Stilbene, $C_{14}H_{12}$	7[a]	⬡—CH=CH—⬡
Phenylacetylene, C_6H_5CCH	10[a]	⬡—C≡CH
Azulene, $C_{10}H_8$	46	
Cyclooctatetraene, C_8H_8	5	
Pyridine, C_5H_5N	43	
Quinoline, C_9H_7N	69	
Pyrrole, C_4H_5N	31	
Indole, C_8H_7N	54	
1,4-Diphenylbutadiene-1,3	11[a]	
Carbazole, $C_{12}H_9N$	91	

[a] Extra resonance energy, not including that within the benzene ring

TABLE 6-2.—(continued)

Substance	Resonance energy (in kcal/mole)	Reference structure
Furan, C_4H_4O	23	
Thiophene, C_4H_4S	31	
Tropolone, C_7H_5OOH	36	
Acids, $RCOOH$	28	
Esters, $RCOOR'$	24	
Amides, $RCONH_2$	21	
Urea, $CO(NH_2)_2$	37	
Dialkylcarbonates, R_2CO_3	42	
Phenol, C_6H_5OH	7°	

TABLE 6-2.—(continued)

Substance	Resonance energy (in kcal/mole)	Reference structure
Aniline, $C_6H_5NH_2$	6[a]	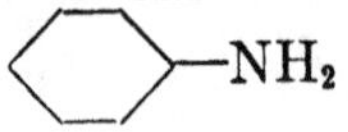
Benzaldehyde, C_6H_5CHO	4[a]	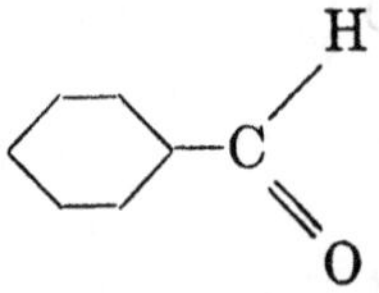
Phenyl cyanide, C_6H_5CN	5[a]	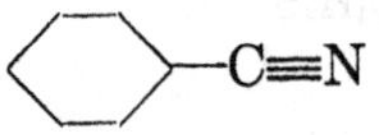
Benzoic acid, C_6H_5COOH	4[b]	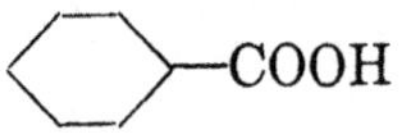
Acetophenone, $C_6H_5COCH_3$	7[a]	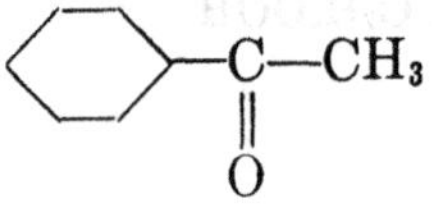
Benzophenone, $C_6H_5COC_6H_5$	10[a]	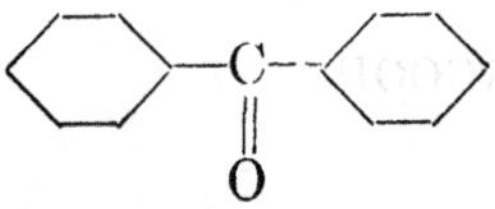
Carbon monoxide, CO	83	$C{=}O$
Carbon dioxide, CO_2	36	$O{=}C{=}O$
Carbon oxysulfide, SCO	20	$S{=}C{=}O$
Carbon disulfide, CS_2	11	$S{=}C{=}S$
Alkyl cyanates, RNCO	7	$R{-}N{=}C{=}O$

[a] Extra resonance energy, not including that within the benzene ring.
[b] Extra resonance energy, not including that within the benzene ring or the carboxyl group.

6-3. THE STRUCTURE OF AROMATIC MOLECULES

In the foregoing discussion of the structure of benzene the stability and characteristic aromatic properties of the substance have been attributed to resonance of the molecule between the two Kekulé structures. A similar treatment, which provides a similar explanation of their outstanding properties, can be given the condensed polynuclear aromatic hydrocarbons.

For naphthalene the conventional valence-bond structure is the Erlenmeyer structure:

I

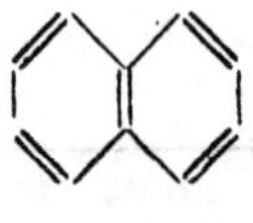

There are two other structures that differ from this only in a redistribution of the bonds:

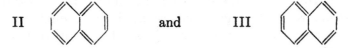

These three structures, the most stable valence-bond structures that can be formulated for naphthalene, are seen to have about the same energy and to correspond to about the same molecular configuration. It is to be expected then that they will be combined to represent the normal state of the naphthalene molecule, to which they should contribute about equally. Resonance among these three stable structures should stabilize the molecule to a greater extent than does the Kekulé resonance in benzene, involving two equivalent structures; it is seen from Table 6-2 that the resonance energy of naphthalene, 75 kcal/mole, is indeed much greater than that of benzene.

For anthracene four stable valence-bond structures can be formulated,

and for phenanthrene five,

The observed values of the resonance energy are 105 kcal/mole for anthracene and 110 kcal/mole for phenanthrene. These are reasonable in comparison with those of benzene and naphthalene, and also with each other, the angular ring system having a larger number of stable resonating structures and a larger resonance energy than the linear ring system.

The higher condensed ring systems can be similarly represented as resonating among many valence-bond structures. The resonance energy increases in rough proportion to the number of hexagonal rings in the system. In addition, it is somewhat greater for the branched and angular ring systems than for the corresponding linear ones, the former resonating among more stable valence-bond structures than the latter (as in the case of phenanthrene and anthracene).

The configurations of the molecules are those expected for the resonating structures. Through resonance each bond acquires some double-bond character, which causes the adjacent bonds to strive to be coplanar. The molecules are thus brought into completely planar configurations, with 120° bond angles. This has been verified for naphthalene and anthracene and many larger aromatic hydrocarbons by careful x-ray studies.

The general chemical properties of the substances are also accounted for. The stabilization of the molecules by resonance gives them aromatic character in the same way as for benzene.

A simple consideration of the resonating structures leads to an explanation for observed differences in behavior of different carbon-carbon bonds in these molecules. In benzene we may say that each bond has $\frac{1}{2}$ double-bond character, since it occurs as a single bond in one Kekulé structure and as a double bond in the other. This does not mean that the bond behaves half the time as a double bond, but rather that it is a bond of a new type, very different from a double bond, and with properties intermediate between those of a double bond and a single bond. (The properties are not the average of those for the two bond types; consideration must also be given the stabilizing effect of the resonance energy.)

In naphthalene, anthracene, and phenanthrene the amounts of double-bond character, found by averaging the stable resonating structures, are shown in the diagrams at right. In naphthalene the 1,2 bonds have $\frac{2}{3}$ double-bond character and the 2,3 bonds $\frac{1}{3}$ double-bond character. These numbers cannot be given a simple quantitative interpretation in terms of chemical reactivity; they do demand, however, that qualitative relations be satisfied. The 1,2 bonds in naphthalene must be closer to ordinary double bonds in their properties than are the benzene bonds, which in turn are much more like double bonds than are the 2,3 bonds in naphthalene, the last, indeed, having practically no double-bond properties. These statements are in agreement with general chemical experience. A hydroxyl group on carbon atom 2 of the system

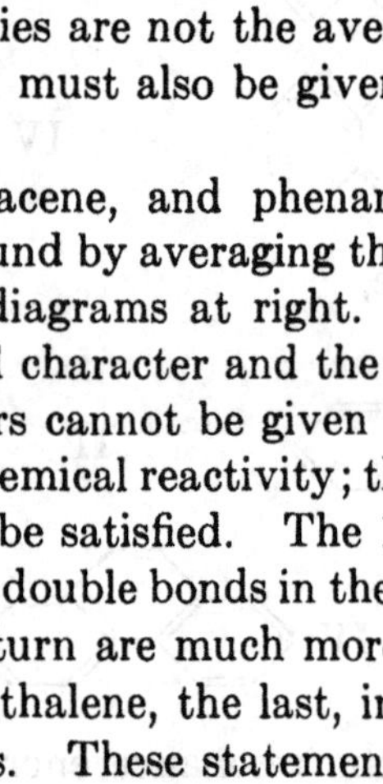

will induce substitution on carbon atom 3 on attack by certain reagents (bromine, diazomethane) rather than on carbon atom 1, the double

Naphthalene

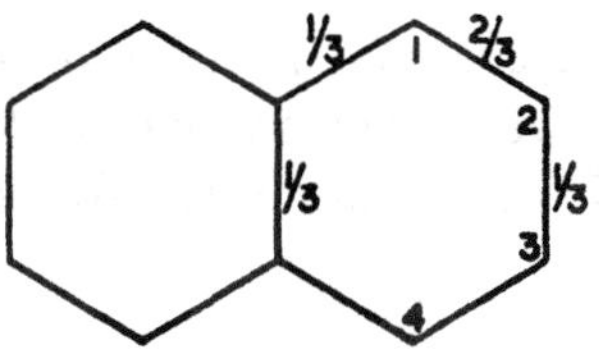

Anthracene

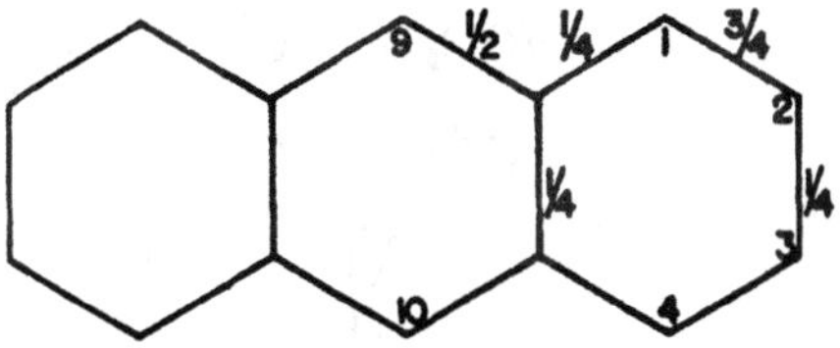

Phenanthrene

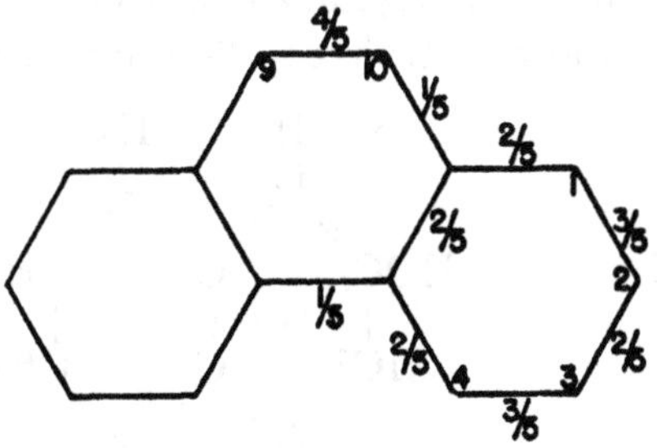

bond serving as the path for the directing influence. This phenomenon can be used to test the extent to which different carbon-carbon bonds have the properties of a double bond. It has been found[9] that with hydroxyl in position 2 of naphthalene reaction occurs readily in position 1, whereas even when position 1 is blocked with methyl reaction does not occur at position 3. This shows strong double-bond properties for the 1,2 bond and very weak ones for the 2,3 bond, as expected. Moreover, it has also been found[10] that the 1,2 bonds in anthracene have stronger double-bond properties than the 1,2 bond in naphthalene, and that the double-bond properties of the 9,10 bond in phenanthrene are stronger still, in agreement with the amounts of double-bond character. For this reason phenanthrene, despite its greater thermodynamic stability, consequent to its greater resonance energy, is more reactive than anthracene.

An interesting related phenomenon involving the benzene ring, discovered by Mills and Nixon,[11] can be discussed similarly.[12] By attaching saturated hydrocarbon rings of different sizes to two *ortho* positions

[9] L. F. Fieser and W. C. Lothrop, *J.A.C.S.* **57**, 1459 (1935), and earlier references quoted by them.

[10] L. F. Fieser and W. C. Lothrop, *J.A.C.S.* **58**, 749 (1936).

[11] W. H. Mills and I. G. Nixon, *J. Chem. Soc.* **1930**, 2510.

[12] L. E. Sutton and L. Pauling, *Trans. Faraday Soc.* **31**, 939 (1935).

of the benzene molecule it is possible to make the ring react as though the double bonds were fixed in the positions corresponding to one or the other of the Kekulé structures.   Mills and Nixon found 5-hydroxy-hydrindene (I)

on reaction with the phenyldiazonium ion to undergo substitution in the 6 position, and *ar*-tetrahydro-β-naphthol (II)

to undergo substitution in the 1 position.   These results were originally interpreted as showing complete fixation of one or the other of the Kekulé structures, resulting from the influence of the five-membered side ring (with 108° angles) in bringing a single bond into position, with its normal tetrahedral angle of 109°28′, and thus minimizing the strain energy, and from the opposite influence of the six-membered ring, which favors large angles.   We see, however, that the stabilization of one Kekulé structure over the other need not be complete in order for one of the bonds adjacent to the hydroxyl-substituted carbon atom to assume enough additional double-bond character to dominate the reaction.   The effect of the side rings in stabilizing one Kekulé structure relative to the other probably causes it to contribute a few percent more than the other to the normal state of the molecule, and this slight superiority gives one bond much stronger double-bond properties than the other for the orientation of substituents.[13]

[13] Sutton and Pauling, *loc. cit.* (12).   For further discussion see N. V. Sidgwick and H. D. Springall, *Chem. & Ind.* (London) 55, 476 (1936); *J. Chem. Soc.* 1936, 1532; L. F. Fieser and W. C. Lothrop, *J.A.C.S.* 58, 2050 (1936); W. Baker, *Ann. Repts. Chem. Soc.* 33, 281 (1936); *J. Chem. Soc.* 1937, 476; W. C. Lothrop, *J.A.C.S.* 62, 132 (1940); R. T. Arnold and H. E. Zaugg, *ibid.* 63, 1317 (1941).

The Quantitative Treatment of Resonance in Aromatic Molecules.—
It has been found possible to carry out the quantitative discussion of
resonance in aromatic molecules by simplifying the problem in the
following way: Of the four valence orbitals of the carbon atom shown
in Figure 4-1 before hybridization, three lie in the plane of the ring
(s, p_x, and p_y, the plane of the ring being taken as the xy plane). These
can be combined to give three bond orbitals that are coplanar and
make 120° angles with one another,[14] and are thus adapted to the
formation by the carbon atom of single covalent bonds to the two adjacent carbon atoms in the ring and to the attached hydrogen atom.
It is assumed that this single-bond framework of the molecule,

H
H   C   H
C   C
C   C
H   C   H
H

remains unchanged; for each atom the fourth orbital and its electron
then remain to be considered.

The fourth orbital is the p_z orbital shown in Figure 4-1. It possesses
lobes above and below the plane of the ring. Let us assume that each
of the six p_z orbitals is occupied by one electron (this involving neglect
of ionic structures). The problem is to calculate the interaction energy
of the six electrons in the six orbitals.

If there were only two orbitals and two electrons, as in the hydrogen
molecule, the interaction energy would be just the resonance energy
associated with the interchange of the two electrons between the two
orbitals. This is the situation in ethylene; the two p_z electrons here
convert the single bond into a double bond. Let us designate this p_z
resonance energy by the symbol α.[15]

The resonance energy for the two Kekulé structures in benzene can
be calculated in terms of α by neglecting all interactions except those

[14] These orbitals are given in a footnote of Sec. 4-4 for the s orbital divided
equally among the three bonds. In benzene it is probable that the strong C—C
bonds, with interatomic distances smaller than the single-bond value, use more
of the s orbital than does the H—C bond.

[15] In this discussion, contrary to the usual custom, α has been used to represent
the magnitude of the resonance energy of two p_z electrons, taken with positive
sign.

between adjacent atoms in the ring.[16]   The way in which this calculation is carried out is described in Appendix V.   The resonance energy
for the two Kekulé structures is found to have the value 0.9 α, this
being the extra stability of the ring relative to one of the Kekulé structures.

However, it is found on examination of the problem that consideration must also be given, in addition to the Kekulé structures A and B,
to the three structures C, D, and E of the Dewar type:

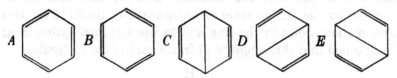

The three structures C, D, and E are less stable than the Kekulé
structures and make much smaller contributions to the normal state
of the benzene molecule.   They increase the resonance energy from
0.9α to 1.11α.   By equating this to the empirical resonance energy of
benzene, 37 kcal/mole, α is found to have the value 33 kcal/mole.

A similar treatment of naphthalene[17] leads to the value 2.04α, which
on equation to the empirical resonance energy 75 kcal/mole fixes α at
37 kcal/mole, in approximate agreement with the result for benzene.
Calculations for anthracene and phenanthrene[18] lead to 2.95α and
3.02α, respectively, for the resonance energy, giving $\alpha = 36$ and 35
kcal/mole on comparison with the empirical values.

A second method of treatment, called the *molecular-orbital* treatment
to differentiate it from the valence-bond treatment described above,
has also been applied to the problem.[19]   With it the six electrons are
not combined in pairs to form bonds, but move independently from
atom to atom.   The calculated resonance energies are expressed in
terms of an energy quantity β, their values being 2.00β for benzene and
3.68β for naphthalene.   These lead, on comparison with the empirical
values, to $\beta = 20$ kcal/mole for both substances, the ratio for the two
being given satisfactorily by this treatment as well as by the valence-
bond treatment.   For anthracene and phenanthrene the theory gives
as values of the resonance energy 5.32β and 5.45β, corresponding again
to $\beta = 20$ kcal/mole (within 0.5 kcal/mole) in each case.

There is, moreover, a reasonable relation between α and β.   The first
quantity is the energy of interchange of two p_z electrons, analogous to
that of the hydrogen molecule, and the second is the energy of resonance of one electron between two p_z orbitals, analogous to that of the

[16] E. Hückel, *loc. cit.* (1); Pauling and Wheland, *loc. cit.* (1).
[17] Pauling and Wheland, *loc. cit.* (1); J. Sherman, *J. Chem. Phys.* 2, 488 (1934).
[18] M. B. Oakley and G. E. Kimball, *J. Chem. Phys.* 17, 706 (1949).
[19] Hückel, *loc. cit.* (1).

hydrogen molecule-ion. The ratio of bond energies in H_2^+ and H_2 is 0.59, and that of β and α is 0.57; the agreement of these two ratios is excellent.

The valence-bond treatment described above involves neglect of the partial ionic character of the bonds in the benzene molecule, and the molecular-orbital treatment overemphasizes it.[20]

The agreement of the two treatments with each other and with the empirical resonance-energy values makes it probable that the point of view presented above regarding the structure of aromatic molecules will not need extensive revision in the future, although it may be subjected to further refinement.

The Orientation of Substituents in Aromatic Molecules.—When a substituent is introduced into an aromatic molecule it may enter into certain of the available positions more readily than into others. This phenomenon has been extensively studied, and empirical rules have been formulated that describe the experimental results fairly satisfactorily.

In a monosubstituted benzene C_6H_5R the groups $R = CH_3$, F, Cl, Br, I, OH, NH_2 are ortho-para directing, and the groups $R = COOH$, CHO, NO_2, SO_3H, $[N(CH_3)_3]^+$ are meta directing for the electrophilic reagents causing substitution.[21] Most ortho-para-directing groups activate the molecule so that substitution takes place more readily than in benzene itself, and most meta-directing groups have a deactivating effect. In naphthalene substitution occurs at the α position, in furan, thiophene, and pyrrole at the α position, and in pyridine at the β position, all of these molecules except pyridine being more active than benzene.

During the last 15 years a qualitative theory has been developed[22] that accounts satisfactorily for the phenomenon in its major features, and a quantitative treatment based on quantum mechanics has been

[20] A comparison of the two methods of quantitative discussion of aromatic molecules has been published by G. W. Wheland, *J. Chem. Phys.* 2, 474 (1934); see also G. W. Wheland, *Resonance in Organic Chemistry*, John Wiley and Sons, New York, 1955.

[21] R. Robinson (*Outline of an Electrochemical [Electronic] Theory of the Course of Organic Reactions*, Institute of Chemistry of Great Britain and Ireland, London 1932), following Lapworth's suggestion, has classified reagents causing substitution as cationoid or anionoid, the former resembling reactive cations and the latter reactive anions in their behavior. Cationoid (electrophilic) reagents include acids, reactive cations such as diazonium cations, alkyl halides, quaternary ammonium compounds, etc. Anionoid reagents include reactive anions ($[NH_2]^-$, $[OH]^-$, $[CN]^-$, $[OR]^-$, etc.), molecules containing unshared electron pairs (nitrogen atom of ammonia and amines), etc.

[22] Many workers, including Fry, Stieglitz, Lapworth, Lewis, Lucas, Lowry, Robinson, and Ingold, have contributed to the theory. For a review, see C. K. Ingold, *Chem. Revs.* 15, 225 (1934).

carried out,[23] with a degree of success which provides strong support for the theory.

The theory is based on consideration of the distribution of electric charge in the molecule in which substitution is taking place. In a benzene molecule the six carbon atoms are equivalent, and the charge distribution is accordingly such as not to make one carbon atom different from another. In the molecule C_6H_5R, with R attached to carbon atom 1, the electron distribution will in general be affected by the group R in such a way as to change the charges on the ortho (2 and 6), meta (3 and 5), and para (4) carbon atoms. Moreover, the electron distribution may also be changed somewhat on the approach of the substituting group R' to one of the carbon atoms ("polarization" of the molecule by the group R'); in benzene the polarization of one carbon atom by the group would be the same as for another, but in a substituted benzene the polarization would in general vary from atom to atom, and so might cause a difference in the behavior of different positions. The fundamental postulate of the theory of orientation of substituents is the following: *In an aromatic molecule undergoing substitution by an electrophilic group R' the rate of substitution of R' for hydrogen on the nth carbon atom increases with increase in the negative charge on the nth carbon atom when the group R' approaches it.*

Substitution by an electrophilic reagent is thus assumed to take place preferentially at that carbon atom on which the negative charge is the largest. This assumption is a reasonable one, in view of the electron-seeking character of these reagents.

There are two principal ways in which the charge distribution can be affected by the group R, for each of which it has been assumed, and has been verified by quantum-mechanical calculations,[24] that the ortho and para carbon atoms are about equally affected, the meta carbon atoms being affected to a much smaller extent.

The first effect of the group R, called the *inductive effect*, results whenever the electronegativity of the group is larger than or smaller than that of hydrogen. In the former case electrons are attracted to the group and to the attached carbon atom 1; it can be seen from the following argument that they are removed to a larger extent from the ortho and para carbon atoms than from the meta carbon atoms. The electronegative group attracts electrons from carbon atom 1, and this in turn attracts electrons from the other atoms of the ring. This effect is then continued around the ring, carried in part by the single bonds in the plane of the ring and in part by the six aromatic (p_z) electrons. The contribution of the latter is of such a nature as to affect the ortho and

[23] G. W. Wheland and L. Pauling, *J.A.C.S.* **57**, 2086 (1935).

[24] Wheland and Pauling, *loc. cit.* (23). This was first shown, for the inductive effect alone, by E. Hückel, *Z. Physik* **72**, 310 (1931).

para carbon atoms preferentially. The transfer of negative charge from the other atoms of the ring to carbon atom 1 by action of the six aromatic electrons can be described as resulting from resonance with ionic structures. There are only three stable ionic structures of this type,

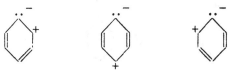

and they lead to the removal of electrons equally from the two ortho atoms and the para atom. Consequently the rate of substitution at the ortho and para positions will be greatly decreased and that at the meta positions somewhat decreased; the group R will be meta directing, with deactivation. An example of such a group is $[N(CH_3)_3]^+$, in trimethylphenylammonium ion; the nitrogen atom is more electronegative than hydrogen, and its electronegativity is further intensified in this case by its positive charge. The same effect is seen in pyridine; the nitrogen atom attracts electrons mainly from the α and γ carbon atoms, and consequently pyridine substitutes in the β positions, and is less active than benzene. Toluene shows the opposite effect. Electric moment measurements show that the methyl group loses electrons to the ring;[25] these go mainly to the ortho and para carbon atoms, which are thus activated; in consequence, toluene substitutes in these positions, and the substitution occurs with greater ease than in benzene.

We might expect that F, Cl, Br, I, OH, and NH_2 would be meta directing, inasmuch as these groups all are more strongly electronegative than hydrogen. Actually they are all ortho-para directing. The inductive effect is in these cases overcome by another effect, called the *resonance effect* (or sometimes the *mesomeric* or *electromeric effect*).

Let us consider a molecule C_6H_5X in which the group X possesses an unshared pair of electrons on the atom adjacent to the benzene ring. The stable structures among which resonance analogous to that in benzene occurs are the Kekulé structures, A and B (smaller contributions are also made by other structures, which will be ignored here for the sake of simplicity).

---

[25] This result is surprising, since in the electronegativity scale carbon is more electronegative than hydrogen; this is a resonance effect, called hyperconjugation, which is discussed in Sec. 8-9.

In addition to these, however, there are three structures *F*, *G*, and *H*
that can be written for these benzene derivatives but not for benzene
itself.   These structures

$$F \qquad G \qquad H$$

are not so stable as *A* and *B*, because, although they contain the same
number of double bonds, they involve an unstable separation of charge.
They make a significant, although small, contribution to the normal
state of the molecule.   The extra resonance energy resulting from their
contribution is about 6 kcal/mole (Table 6-2, phenol and aniline).
As a result of this conjugation of the unshared pair of the group X
with the benzene ring there is built up a negative charge on each of the
ortho carbon atoms and the para carbon atom, this effect being superimposed on the inductive effect of the group.   For the groups listed
above the resonance effect is stronger than the inductive effect, making
the groups ortho-para directing.[26]
   In benzaldehyde and many other similar molecules, on the other
hand, the resonance effect directs toward the meta positions, this resulting whenever the substituted group R contains an electronegative
atom and a double or triple bond conjugated with the benzene ring
(R = COOH, CHO, NO_2, $COCH_3$, SO_3H, CN, etc.).   The structures
leading to this effect, *F'*, *G'*, and *H'*, are of the types

$$F' \qquad G' \qquad H' \qquad ,$$

[26] Resonance of this type is often indicated by the use of arrows; the method is
obvious from the examples

etc.

which decrease the electron density on the aromatic nucleus, especially at the ortho and para positions, and thus lead to reaction at the meta positions, but at a slower rate than for benzene itself. The extra resonance energy due to these structures is about 5 kcal/mole.

The foregoing discussion for monosubstituted benzene can be summarized as follows: When resonance does not occur, substitution is usually determined by the inductive effect, an electron-attracting group being meta directing and an electron-repelling group ortho-para directing. The resonance effect, which when present is usually more powerful than the inductive effect, is meta directing when the group contains an electronegative atom and a double bond conjugated with the benzene ring, and ortho-para directing when the group contains an unshared electron pair on the atom adjacent to the benzene ring.

In a few cases (naphthalene, for example) it is necessary to consider also the polarization of the molecule by the attacking group; as yet no general qualitative rules have been formulated for this effect, although some quantitative calculations have been carried out. The effect can be treated qualitatively by consideration of the number of stable ionic structures placing an unshared pair on the carbon atom being attacked. For the α position of naphthalene there are seven:

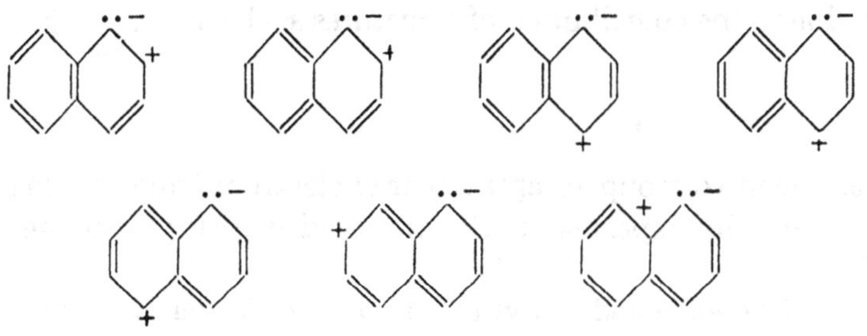

and for the β position only six:

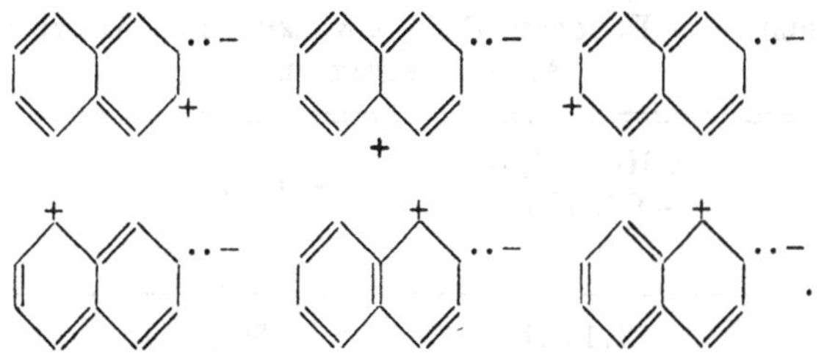

Accordingly polarization by the attacking group will be greater for the α position, and substitution will tend to occur there.

The Effect of Resonance on the Electric Dipole Moments of Molecules.—It was pointed out by Sutton[27] in 1931 that resonance of the

[27] L. E. Sutton, *Proc. Roy. Soc. London* **A133**, 668 (1931); *Trans. Faraday Soc.* **30**, 789 (1934).

types discussed in the preceding section for molecules such as chloro-benzene and nitrobenzene would cause the electric dipole moments of these molecules to have values differing from those of the corresponding alkyl derivatives, and that a test of the resonance theory of orientation of substituents could be carried out by analysis of dipole-moment data.

The dipole-moment vectors ($+ \to -$) in both R—Cl and R—NO$_2$ (with R an alkyl group—preferably sec-propyl or tert-butyl for comparison with phenyl compounds) are directed along the R—Cl or R—N axes. In chlorobenzene resonance to structures such as

is seen to lead to a decrease in moment, the dipole vector for these less important structures being opposed in direction to that of the primary structure; this is verified by experiment, the change in moment from the alkyl chlorides to phenyl chloride being $-$ 0.58 D. Similar changes are shown by the bromides and iodides (Table 6-3).

For the meta-directing nitro group an increase in moment is expected, due to the contribution of structures such as

in which the nitro group accepts a pair of electrons from the ring. This too is verified by experiment, the observed increase in moment being 0.64 D.

As would be expected, vinyl and naphthyl derivatives are observed to have nearly the same dipole moments as the phenyl derivatives

TABLE 6-3.—ELECTRIC DIPOLE MOMENTS OF ALKYL AND
ARYL DERIVATIVES

Group	—CH(CH$_3$)$_2$ or —C(CH$_3$)$_3$ μ	—C$_6$H$_5$ μ	Difference
—Cl	2.14 D	1.56 D	$-$ 0.58 D
—Br	2.15	1.54	$-$ 0.61
—I	2.13	1.38	$-$ 0.75
—NO$_2$	3.29	3.93	$+$ 0.64
—CHO	2.46	2.75	$+$ 0.29
—NO	2.51	3.14	$+$ 0.63

(1.66 D for vinyl chloride, 1.59 D for naphthyl chloride), verifying the close similarity in the powers of conjugation of these three groups.

The discussion of dipole-moment values in relation to resonance is continued in Chapter 8.

6-4. THE STRUCTURE AND STABILITY OF THE HYDROCARBON FREE RADICALS

Ever since the discovery by Gomberg in 1900 of the dissociation of hexaphenylethane into triphenylmethyl radicals the search for a theoretical explanation of the phenomenon has been carried on. The modern theory of the stability of the aromatic free radicals attributes it in the main to the resonance of the free valence among many atoms.[28]

The hexaalkylethanes, which do not dissociate appreciably, have the valence-bond structure

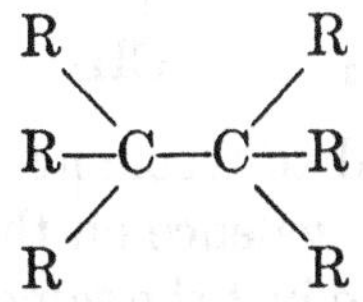

$$
\begin{array}{ccc}
R & & R \\
\backslash & & \diagup \\
R\!-\!C\!\!-\!\!C\!-\!R \\
\diagup & & \backslash \\
R & & R
\end{array}
$$

and the corresponding free radicals the structure

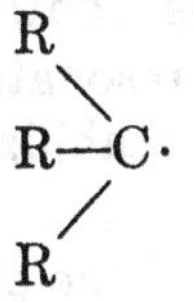

$$
\begin{array}{c}
R \\
\backslash \\
R\!-\!C\cdot \\
\diagup \\
R
\end{array}
$$

the odd electron (free valence) being located on the methyl carbon atom. The introduction of an aryl group, however, provides additional structures for the radical; it is principally the energy of resonance among these that stabilizes the free radical and increases the degree of dissociation of the substituted ethane.

For simplicity, let us consider the molecule $C_6H_5CH_2\!-\!CH_2C_6H_5$, 1,2-diphenylethane, and let us restrict the discussion of resonance to the structures with the greatest stability (those with the maximum number of double bonds). For the undissociated molecule there is resonance among the four Kekulé structures

[28] Pauling and Wheland, *loc. cit.* (1); *J. Chem. Phys.* **2**, 482 (1934); E. Hückel, *Z. Physik* **83**, 632 (1933). The quantitative theory was foreshadowed by a somewhat similar qualitative discussion by C. K. Ingold, *Ann. Repts. Chem. Soc.* **25**, 152 (1928); H. Burton and C. K. Ingold, *Proc. Leeds Phil. Lit. Soc.* **1**, 421 (1929).

whereas each of the free radicals can resonate among the five structures

If the radical were restricted to resonance between the Kekulé structures A and B, with the free valence on the methyl carbon, resonance would stabilize the radicals to just the same extent as the undissociated molecules, which would then have only the same tendency to dissociate as a hexaalkylethane. But actually the five structures A, B, C, D, and E (each with three double bonds) contribute about equally to the structure of the radical, which thus resonates among five structures instead of two and is correspondingly stabilized by the additional resonance energy.

The extra resonance energy of the phenylmethyl radical is not large enough to lead to appreciable dissociation of 1,2-diphenylethane. However, an experimental value of the resonance energy has been obtained by Szwarc[29] by analysis of the measured rates of pyrolysis of toluene relative to those for methane. The values obtained by Szwarc for the enthalpy of breaking of the carbon-hydrogen bond are 77.5 kcal/mole for toluene (to form $C_6H_5CH_2\cdot$ and $\cdot H$) and 102 kcal/mole for methane. The difference, 24.5 kcal/mole, is the resonance energy with structures C, D, and E. In the same way resonance-energy values about 26 kcal/mole are found for the propylene radical ($CH_2{=}CH{-}CH_2\cdot$, $\cdot CH_2{-}CH{=}CH_2$) and similar radicals.[30]

In the triphenylmethyl radical the odd electron can resonate among nine positions (the ortho and para positions of the three phenyl groups) in addition to that on the methyl carbon atom. It is found on evaluat-

[29] M. Szwarc, *J. Chem. Phys.* **16**, 128 (1948).

[30] Calculated and experimental values of resonance energy of radicals are given by A. Brickstock and J. A. Pople, *Trans. Faraday Soc.* **50**, 901 (1954).

ing the the extra resonance energy for this radical by the two methods described in Section 6-3 that it is of the order of magnitude of one-half the carbon-carbon single-bond energy; the enhanced stability of two radicals is accordingly great enough to overcome in large part the energy of the bond, and the substance shows a large degree of dissociation.

It has been pointed out by Adrian[31] that there is steric hindrance in the triphenylmethyl radical, the phenyl groups being twisted about the bond to the central atom through about 32°. This twist decreases the calculated resonance energy from 35 kcal/mole (for planarity) to 21 kcal/mole. The steric repulsion energy of the two halves of hexaphenylethane is estimated to be about 36 kcal/mole, and the enthalpy of dissociation is about 16.5 kcal/mole.

Whereas for the phenyl group three structures with the free valence in the group can be written (C, D, and E), there are seven structures for the α-naphthyl group and six for the β-naphthyl group, these being analogous to those shown in the last paragraph of the preceding section. This suggests that the α-naphthyl group should be the more effective of the two in promoting dissociation. The quantitative treatment leads to the same expectation, which is borne out by the results of experimental studies of the degree of dissociation of hexaarylethanes, the order found being hexaphenylethane < tetraphenyldi-β-naphthylethane < tetraphenyldi-α-naphthylethane. The biphenyl group is about as effective as the β-naphthyl group.

In recent years valuable data regarding the degree of dissociation of hexaarylethanes have been obtained by the magnetic method; the concentration of the triarylmethyl radicals as determined by measuring the magnetic susceptibility of the solution, to which the unpaired spin of the odd electron in the radical makes a paramagnetic contribution. This method, first used by Taylor[32] at the suggestion of G. N. Lewis, has been extensively applied by Müller[33] and Marvel[34] and their collaborators.

It has been found that hexa-p-alkylphenylethanes in solution dissociate to a somewhat greater extent than hexaphenylethane itself, the magnitude of the enhancement of the degree of dissociation by the

[31] F. J. Adrian, *J. Chem. Phys.* **28**, 608 (1958).

[32] N. W. Taylor, *J.A.C.S.* **48**, 854 (1926).

[33] E. Müller, I. Müller-Rodloff, and W. Bunge, *Ann. Chem.* **520**, 235 (1935); E. Müller and I. Müller-Rodloff, *ibid.* **521**, 89 (1935).

[34] M. F. Roy and C. S. Marvel, *J.A.C.S.* **59**, 2622 (1937); C. S. Marvel, E. Ginsberg, and M. B. Mueller, *ibid.* **61**, 77 (1939); C. S. Marvel, M. B. Mueller, and E. Ginsberg, *ibid.* 2008; C. S. Marvel, W. H. Rieger, and M. B. Mueller, *ibid.* 2769; C. S. Marvel, M. B. Mueller, C. M. Himel, and J. F. Kaplan, *ibid.* 2771.

p-alkyl groups being uncertain because of instability of the radicals.[35] An explanation for this effect, involving resonance of the radical to structures such as

$$\underset{\text{Ar}}{\overset{\text{Ar}}{\text{C}}} = \underbrace{} = \underset{\text{R}}{\overset{\text{R}}{\text{C}}} \cdot \text{R},$$

in which a single bond within the alkyl group is broken, was proposed by Wheland.[36] This explanation applies also to the effect of the *t*-butyl group in increasing the degree of dissociation when it is present as a direct substituent in the ethane. Steric repulsion is no doubt also important.

Resonance of this type gives to all the bonds in the triarylmethyl radicals partial double-bond character, and the radicals would strive to assume a completely planar configuration. Consideration of the dimensions of the groups shows, however, that steric effects prevent this; the phenyl groups in triphenylmethyl must be rotated somewhat out of the median plane, as mentioned above. The steric interactions of large substituent groups in the substituted ethanes weaken the carbon-carbon bond somewhat and are responsible in part for the unusual properties of these substances.[37] The very large degree of dissociation reported for hexa-*o*-methylphenylethane[38] (as compared with hexa-*p*-methylphenylethane) is a steric effect.

The substance triptycene,

has been synthesized[39] and shown not to have the activity of aliphatic hydrogen toward potassium exchange, chlorination, and oxidation that

[35] Marvel, Rieger, and Mueller, also Marvel, Mueller, Himel, and Kaplan, *loc. cit.* (34).

[36] G. W. Wheland, *loc. cit.* (20).

[37] H. E. Bent and E. S. Ebers, *J.A.C.S.* **57**, 1242 (1935); Wheland, *loc. cit.* (20).

[38] Marvel, Mueller, Himel, and Kaplan, *loc. cit.* (34).

[39] P. D. Bartlett, M. J. Ryan, and S. G. Cohen, *J.A.C.S.* **64**, 2649 (1942).

characterizes triphenylmethane. This behavior is as predicted from the inability of the benzene rings to approximate the planar configuration with a central carbon atom.

It is interesting that the electron distribution in radicals is such as to place a fraction of an unpaired electron on each carbon atom of the conjugated system.[40] The wave function (App. VI) corresponding to the structural representation $\{CH_2{=}CH{-}CH_2\cdot, \cdot CH_2{-}CH{=}CH_2\}$ is $\psi = (1/\sqrt{6})\{+-+) - (++-) + (+-+) - (-++)\}$, and its square is $\psi^2 = \frac{1}{6}\{4(+-+)^2 + (++-)^2 + (-++)^2\}$. This gives $\frac{2}{3}\uparrow$ as the spin density on carbon atoms 1 and 3, and $\frac{1}{3}\downarrow$ on carbon atom 2, in approximate agreement with the results of magnetic resonance experiments. The failure of the valence-bond structures to indicate clearly the distribution of unpaired electrons is a weakness that might be of importance in the consideration of the chemical reactivity of radicals. Possibly an improved system will be devised by someone.

6-5. THE NATURE OF THE THEORY OF RESONANCE[41]

Although the theory of resonance in chemistry is now about 30 years old, there seem still to be some misunderstandings about its nature. In particular, the theory is criticized on the ground that it is artificial—that the individual valence-bond structures that, according to the theory, contribute to the normal state of a molecule such as benzene are idealizations, and do not have independent existence; and it has been suggested that for this reason the theory should be abandoned. In fact, however, the theory of resonance is no more artificial than the classical structure theory of organic chemistry, and the contributing valence-bond structures in the theory of resonance are not more ideal (imaginary) than the structural elements of classical theory, such as the double bond.

The essential identity in character of the theory of resonance and the classical structure theory of organic chemistry, which has before been referred to only briefly,[42] will be discussed in detail in the following paragraphs.

The theory of resonance has been applied to many problems in chemistry. In addition to its use in the discussion of the normal covalent bond (involving the interchange of two electrons, with opposed spins, between two atoms) and to the structure of molecules for which a single valence-bond structure does not provide a satisfactory description, it has rendered service to chemistry by leading to the discovery of several

[40] H. M. McConnell, *J. Chem. Phys.* 29, 244 (1958).

[41] From L. Pauling, *Perspectives in Organic Chemistry*, ed. by A. R. Todd, Interscience Publishers, New York, 1956, pp. 1–8.

[42] L. Pauling, *Modern Structural Chemistry, The Nobel Prizes*, Stockholm, 1954.

previously unrecognized structural features, including the one-electron bond, the three-electron bond, the partial ionic character of covalent bonds between unlike atoms (resonance between a normal covalent structure and an ionic structure), hybridization of bond orbitals (resonance between bonds formed by s, p, and d orbitals), hyperconjugation (no-bond resonance, first discussed by Wheland[43] in 1934), and fractional bonding in metals. It is interesting that these aspects of the resonance theory have not been subjected to serious criticism; instead, the criticism has been concentrated on the application of the resonance theory to the structure of those molecules for which a single valence-bond structure does not provide a satisfactory description and which, according to the resonance theory, may be described as involving resonance among several structures of the valence-bond type.

The Russian criticism of the theory of resonance seems to be based largely upon the fact that the contributing resonance structures do not have real existence.[44] Essentially the same point is made in Hückel's book *Structural Chemistry of Inorganic Compounds;*[45] in the last paragraph of Volume I, in a note of criticism of the theory of resonance supplied by the translator of the English edition, L. H. Long, the complaint is expressed in the following words:

Enough has been said to indicate a serious need for a reply on the part of the advocates of the resonance theory to the various objections which have been raised in recent years. In the absence of a convincing response, the resonance theory stands in danger of being largely discredited, at least in so far as it has been applied hitherto. At best it provides a picture which can be described no less accurately in other terms. At worst, the picture it gives is highly misleading. It must further never be forgotten that the theory ultimately depends upon the use of limiting structures which, by admission, have no existence in reality.

Let us first consider, as an example, the substance cyclohexene. For many years chemists all over the world have been in complete agreement about the structural formula to be assigned to this substance.

[43] Wheland, *loc. cit.* (20).

[44] D. N. Kursanov, G. Gonikberg, B. M. Dubinin, M. I. Kabachnik, E. D. Kaverzneva, E. N. Prilezhaeva, N. D. Sokolov, and R. K. Freidlina, *Report of the Commission of the Institute of Organic Chemistry of the Academy of Sciences, U.S.S.R., for the Investigation of the Present State of the Theory of Chemical Structure, Uspekhi Khim.* **19,** 529 (1950), English translation by I. S. Bengelsdorf, *J. Chem. Educ.* **29,** 2 (1952); V. N. Tatevskii and M. I. Shakhparanov, "About a Machistic Theory in Chemistry and Its Propagandists," *Voprosi Filosofii* **3,** 176 (1949), English translation by I. S. Bengelsdorf, *J. Chem. Educ.* **29,** 13 (1952); I. M. Hunsberger, "Theoretical Chemistry in Russia," *J. Chem. Educ.* **31,** 504 (1954).

[45] Trans. by L. H. Long, Elsevier, New York, 1950, vol. I.

The molecule of cyclohexene is described as containing a ring of six carbon atoms; five pairs of adjacent carbon atoms in the ring are described as being connected by carbon-carbon single bonds, and one pair by a carbon-carbon double bond. In addition, four of the carbon atoms are connected to two hydrogen atoms each by carbon-hydrogen single bonds, and the other two are connected to one hydrogen atom each. The properties of the substance have been correlated with this structural formula; for example, the unsaturation of the substance is attributed to the presence of a double bond.

Now let us consider benzene. There is no single valence-bond structure that accounts satisfactorily for the properties of benzene. The simple description of benzene that is given by the theory of resonance involves two valence-bond structures, the two Kekulé structures

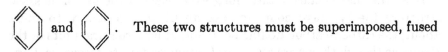 and . These two structures must be superimposed, fused

together, to represent the molecule of benzene, with consideration given also to the stabilizing effect of resonance—that is, the benzene molecule does not have a structure midway between two Kekulé structures, but rather a structure that is changed from the intermediate structure in the way corresponding to energetic stabilization. By comparing the experimental value of the heat of formation of benzene with a value predicted for a single Kekulé structure with use of bond energies, the energy of stabilization through resonance is found to be about 37 kcal/mole. It is this stabilization that makes benzene more resistant to hydrogenation, and hence less unsaturated, than the olefins.

The several structures that are used in the description of a molecule such as benzene by application of the theory of resonance are idealizations, and do not have existence in reality. This fact has been advanced, as in the quotation above, as an argument against the theory of resonance. If the argument were to be accepted, and the theory of resonance were in consequence to be abandoned, it would be necessary also, for the sake of consistency, to abandon the whole structure theory of organic chemistry, because the structural elements that are used in classical structure theory (as in the discussion of cyclohexene given above), the carbon-carbon single bond, the carbon-carbon double bond, the carbon-hydrogen bond, and so on, also are idealizations, having no existence in reality. There is no rigorous way of showing by experiment that two of the carbon atoms in the cyclohexene molecule are connected by a double bond. Indeed, we may say that the cyclohexene molecule is a system that can be shown experimentally to be resolvable into six carbon nuclei, ten hydrogen nuclei, and 46 electrons, and that can be shown to have certain other structural properties, such as values 1.33 Å, 1.54 Å, and so on, for the average distances between nuclei in the molecule in its normal state; but it is not resolvable by any experimental technique into one carbon-carbon double bond, five carbon-carbon single bonds, and ten carbon-hydrogen bonds—these bonds are theoretical constructs, idealizations, with the aid of which the chemist during the past one hundred years has developed a convenient and extremely valuable theory. The theory of resonance constitutes an extension of this classical structure theory of organic chemistry; it is based upon the use of the same idealizations, the bonds between atoms, as classical structure theory, with the important extension that in describing the benzene molecule two arrangements of these bonds must be used, rather than only one.

It is true that chemists, after long experience in the use of classical structure theory, have come to talk about, and probably to think about, the carbon-carbon double bond and other structural units of the theory as though they were real. Reflection leads us to recognize, however, that they are not real, but are theoretical constructs in the same way as the individual Kekulé structures for benzene. It is not possible to isolate a carbon-carbon bond and to subject it to experimental investigation. There is, indeed, no rigorous definition of the carbon-carbon double bond. We cannot accept, as a rigorous definition, the statement that the carbon-carbon double bond is a bond between two carbon atoms that involves four electrons, because there is no experimental method of determining precisely the number of electrons that are involved in the interaction of two carbon atoms in a molecule, and, of course, this interaction has rigorously to be considered as being dependent on the nature of the entire molecule. We might define the double bond as the bond between the two carbon atoms in the ethylene molecule; but this definition would not be useful because in fact the ethylene molecule is different from every other molecule, and there is no other molecule in which two carbon atoms are related in exactly this way. We know, of course, that the two carbon atoms in a molecule that are connected by a double bond in the structural formula of the molecule that would be written by any chemist usually have an average internuclear distance of about 1.33 Å, whereas singly bonded carbon atoms lie about 1.54 Å apart, and triply bonded carbon atoms about 1.20 Å apart; but these distances vary somewhat from molecule to molecule, and no way is known of assigning a range within which the internuclear distance should lie for a "real" carbon-carbon double bond, and outside of which it should lie for a bond of another sort. Despite the fact that the structural units that they use in classical structure theory, such as the carbon-carbon double bond, are idealizations, chemists have for nearly a century striven, with continued success, to develop the theory based upon the use of these structural units, and this theory has become more and more powerful. The incorporation of the theory of resonance into chemical structure theory has been a part of this continuing progress.

There has been especially strong objection to the concept of resonance energy. The resonance energy for benzene, for example, is calculated with the use of assumed bond-energy values, which are added together to give the energy of a hypothetical molecule with a single Kekulé structure. The system of bond energies is not very accurate, and there is for this reason some uncertainty in the values of resonance energy obtained with its use. It may be pointed out, however, that this feature is not restricted to the resonance theory; the bond-energy system has been used also in classical chemistry. In 1920 Fajans[46] discussed the heats of combustion of aliphatic hydrocarbons and other substances (not involving resonance) with the use of a set of bond energies that he had formulated. Some ways in which bond energies can be used in the prediction of properties of substances, including especially those to which classical structure theory applies, have been mentioned by many authors, such as Lucas.[47] The use of bond-energy values in the discussion of molecular rearrangements, especially for nonresonating molecules, has also been pointed out recently.[48]

I feel that the greatest advantage of the theory of resonance, as compared with other ways (such as the molecular-orbital method) of discussing the structure of molecules for which a single valence-bond structure is not enough, is that it makes use of structural elements with which the chemist is familiar. The theory should not be assessed as inadequate because of its occasional unskillful application. It becomes more and more powerful, just as does classical structure theory, as the chemist develops a better and better chemical intuition about it.

The theory of resonance should not be identified with the valence-bond method of making approximate quantum-mechanical calculations of molecular wave functions and properties. The theory of resonance is essentially a chemical theory (an empirical theory, obtained largely by induction from the results of chemical experiments). Classical structure theory was developed purely from chemical facts, without any help from physics. The theory of resonance was also well on

[46] K. Fajans, *Ber.* **53**, 643 (1920); **55**, 2826 (1922); *Z. physik. Chem.* **99**, 395 (1921).

[47] H. J. Lucas, *Organic Chemistry*, American Book Co., New York, 1953 (1st ed. 1935).

[48] L. Pauling, in *Biochemistry of Nitrogen*, Suomalainen Tiedeakatemia, Helsinki, 1955; see Sec. 3-10.

its way toward formulation before quantum mechanics was discovered. Already in 1899 Thiele had developed his theory of partial valence, which must be considered as a first step toward the development of the resonance theory, and by 1924 Lowry, Arndt, and Lucas had made suggestions about the change in structure of molecules during reaction that reflect to some extent the spirit of the theory of resonance. The suggestion made in 1926 by Ingold and Ingold[49] that molecules in their normal state have structures that differ from those corresponding to a single valence-bond structure was made on the basis of chemical considerations, without essential assistance from quantum mechanics. It is true that the idea of resonance energy was then provided by quantum mechanics, that many applications of the theory of resonance (such as the hybridization of bond orbitals) require the penetrating understanding of atomic and molecular structure that has been provided by quantum mechanics, and that, moreover, the approximate quantum-mechanical calculations such as those made by Hückel[50] for aromatic molecules were of great value in indicating how the chemical theory of resonance should be developed; but the theory of resonance in chemistry has gone far beyond the region of application in which any precise quantum-mechanical calculations have been made, and its great extension has been almost entirely empirical, with only the valuable and effective guidance of fundamental quantum-mechanical principles.

The theory of resonance in chemistry is an essentially qualitative theory, which, like the classical structure theory, depends for its successful application largely upon a chemical feeling that is developed through practice. We may believe the theoretical physicist who tells us that all the properties of substances should be calculable by known methods—the solution of the Schrödinger equation. In fact, however, we have seen that during the 35 years since the Schrödinger equation was discovered only a few accurate nonempirical quantum-mechanical calculations of the properties of substances in which the chemist is interested have been made. The chemist must still rely upon experiment for most of his information about the properties of substances. Experience has shown that he can be immensely helped by the use of the simple chemical structure theory. The theory of resonance is a part of the chemical structure theory, which has an essentially empirical (inductive) basis; it is not just a branch of quantum mechanics.

[49] Ingold and Ingold, *loc. cit.* (1)
[50] Hückel, *loc. cit.* (1).

CHAPTER 7

Interatomic Distances and Their Relation

to the Structure of Molecules

and Crystals

7-1. INTERATOMIC DISTANCES IN NORMAL COVALENT
MOLECULES: COVALENT RADII

As a result of the development of the x-ray method of studying the structure of crystals and the band-spectroscopic method and especially the electron-diffraction method of studying gas molecules, a large amount of information about interatomic distances in molecules and crystals has been collected. It has been found that the values of interatomic distances corresponding to covalent bonds can be correlated in a simple way in terms of a set of values of *covalent bond radii* of atoms, as described below.[1]

---

[1] Interatomic-distance values obtained in various ways are reliable to an extent determined by the nature of the method. Spectroscopic values for diatomic molecules are usually accurate to within 0.001 Å; those for polyatomic molecules are somewhat less reliable. Many accurate interatomic distances (to within 0.001 Å) for moderately complex molecules, such as methyl cyanide, have in recent years been determined by the methods of microwave spectroscopy. Electron-diffraction values for gas molecules may be assigned probable errors of from 0.005 to 0.05 Å or more, depending on the care with which the investigation has been carried out and the complexity of the molecule. X-ray values for crystals may be reliable to 0.001 Å, if the interatomic distance is determined directly by the size of the unit cell (as in diamond). In general, however, they depend also on some additional parameters evaluated with use of intensity data; they are then reliable to 0.005 Å only in exceptional cases. The probable errors for x-ray crystal structure values involving several parameters are around 0.005 Å for investigations carried out carefully in recent years, and 0.05 Å or more for others. X-ray diffraction values for gas molecules are reliable only to 0.1 or 0.2 Å. Tables of values of interatomic distances are given by P. W. Allen and L. E. Sutton, *Acta Cryst.* **3**, 46 (1950), G. W. Wheland, *Resonance in*

The values found for the equilibrium distance between two atoms A and B connected by a covalent bond of fixed type (single, double, etc.) in different molecules and crystals are in most cases very nearly the same, so that it becomes possible to assign a constant value to the A—B bond distance for use in any molecule involving this bond.   For example, the carbon—carbon distance in diamond (representing a single covalent bond) is 1.542 Å, and the values found in the seven molecules given in Table 7-1, as well as in many others, lie between 1.53

TABLE 7-1.—EXPERIMENTAL VALUES OF CARBON-CARBON
SINGLE-BOND DISTANCES[a]

Substance	C—C distance
Diamond	1.542 Å
Ethane	1.533
Propane	1.54
n-Butane	1.534
Neopentane	1.54
n-Heptane	1.532
Cyclohexane	1.53
Adamantane, $C_{10}H_{16}$	1.54

[a] These values are good to about ±0.01 Å. The value for ethane is that obtained by K. Hedberg and V. Schomaker, *J.A.C.S.* **73**, 1482 (1951), by combining the results of electron-diffraction and microwave studies. The next five hydrocarbon values are from the electron-diffraction study by L. Pauling and L. O. Brockway, *ibid.* **59**, 1223 (1937), and R. A. Bonham and L. S. Bartell, *J.A.C.S.* **81**, 3491 (1959).   The value for adamantane is due to W. Nowacki and K. Hedberg, *J.A.C.S.* **70**, 1497 (1948), and W. Nowacki, *Helv. Chim. Acta* **28**, 1233 (1945).

and 1.54 Å, being equal to the diamond value to within their probable errors.   This constancy is of interest in view of the varied nature of the molecules.

It will be pointed out later (Chap. 8) that the interaction of a methyl group and a double bond or an aromatic group (hyperconjugation) causes some shortening of the single bond, about 0.03 Å.   There is a larger shortening, about 0.08 Å, for a single bond adjacent to a triple bond.   A large shortening is also observed for a single bond between two double bonds or aromatic nuclei, forming a conjugated system. Some shortening is also observed in small rings (1.524 Å in cyclo-

Organic Chemistry, John Wiley and Sons, New York, 1955, and Sutton, *Interatomic Distances*.   Most of the values for crystals are from *Strukturbericht*, vols. I to VII (1913 to 1939), and *Structure Reports*, vol. 8 and later volumes.   A useful compilation is R. W. G. Wyckoff, *Crystal Structures*, Interscience Publishers, New York, 1948 on

propane); this effect, which may be attributed to the bending of the bonds, has already been discussed (Sec. 4-8).

In cyclobutane the carbon—carbon distance[2] is found to be 1.568 ± 0.020 Å, somewhat larger than the normal value. The presumably correct explanation that has been proposed[2] is that the bonds are stretched because of the repulsion of the atoms separated by the diagonal of the square. Similar distances (1.555 ± 0.010 Å and 1.563 ± 0.010 Å) have been reported[3] in two other molecules containing four-membered rings, bicycloheptane and the polycyclohydrocarbon $C_{13}H_{14}$.

Similar constancy is shown by other covalent bond distances (with certain exceptions that will be discussed later). For the carbon—oxygen single bond, for example, the value 1.43 Å has been reported for methanol,[4] ethanol, ethylene glycol, dimethyl ether, paraldehyde, metaldehyde, and many other molecules; this value is accepted as standard for the C—O bond.

Moreover, covalent bond distances are often related to one another in an additive manner; the bond distance A—B is equal to the arithmetic mean of the distances A—A and B—B. For example, the C—C distance in diamond is 1.542 Å and the Cl—Cl distance in Cl_2 is 1.988 Å. The arithmetic mean of these, 1.765 Å, is identical with the C—Cl distance 1.766 ± 0.003 Å found in carbon tetrachloride to within the the probable error of the experimental value.[5] In consequence, it becomes possible to assign to the elements *covalent radii* such that the sum of two radii is approximately equal to the equilibrium internuclear distance for the two corresponding atoms connected by a single covalent bond.

These covalent radii are for use in molecules in which the atoms form covalent bonds to a number determined by their positions in the periodic table—carbon four, nitrogen three, and so on. It is found empirically that the radii are applicable to covalent bonds with considerable ionic character; for extreme ionic bonds, however, ionic radii are to be used (Chap. 13), and in some molecules, discussed in later sections, the partial ionic character plays an important part in determining the interatomic distances.

[2] J. D. Dunitz and V. Schomaker, *J. Chem. Phys.* **20**, 1703 (1952).

[3] C. Wong, A. Berndt, and V. Schomaker, unpublished research, Cal. Inst. Tech.

[4] The value C—O = 1.427 ± 0.007 Å from a microwave study has been reported by P. Venkateswarlu and W. Gordy, *J. Chem. Phys.* **23**, 1200 (1955). The other values, which are in general reliable to ±0.02 Å, are from older electron-diffraction and x-ray investigations.

[5] L. S. Bartell, L. O. Brockway, and R. H. Schwendeman, *J. Chem. Phys.* **23**, 1854 (1955).

The radii are so chosen that their sums represent average internuclear distances for bonded atoms in molecules and crystals at room temperature. The atoms carry out thermal oscillations, which cause the internuclear distances to vary about their average values. At room temperature these are only slightly different from the values corresponding to the minima in the potential energy functions.

Values of the single-bond covalent radii of the nonmetallic elements are given in Table 7-2. These values, which were originally obtained

TABLE 7-2.—COVALENT RADII FOR ATOMS

	C	N[a]	O[a]	F[a]
Single-bond radius	0.772	0.70	0.66	0.64 Å
Double-bond radius	.667			
Triple-bond radius	.603			
	Si	P	S	Cl
Single-bond radius	1.17	1.10	1.04	0.99
Double-bond radius	1.07	1.00	0.94	.89
Triple-bond radius	1.00	0.93	.87	
	Ge	As	Se	Br
Single-bond radius	1.22	1.21	1.17	1.14
Double-bond radius	1.12	1.11	1.07	1.04
	Sn	Sb	Te	I
Single-bond radius	1.40	1.41	1.37	1.33
Double-bond radius	1.30	1.31	1.27	1.23

[a] See also Table 7-5.

largely from x-ray diffraction studies of crystals, may be tested by comparison with the results of the many recent investigations of gas molecules as well as of crystals.[6]

[6] Shortly after the formulation of a rough set of atomic radii for use in crystals of all types (W. L. Bragg, *Phil. Mag.* 40, 169 [1920]) it was recognized that the effective radius of an atom depends on its structure and environment, and especially on the nature of the bonds that it forms with adjacent atoms. Between 1920 and 1927 a complete set of values of ionic radii, for use in ionic molecules and crystals, was developed by Landé, Wasastjerna, Goldschmidt, and Pauling; this work is discussed in Chap. 13. In 1926 M. L. Huggins (*Phys. Rev.* 28, 1086 [1926]) published a set of atomic radii for use in crystals containing covalent bonds. V. M. Goldschmidt in the same year published values of atomic radii obtained from metals as well as from covalent nonmetallic crystals ("Geochemische Verteilungsgesetze der Elemente," *Skrifter Norske Videnskaps-Akad. Oslo, I, Mat.-Naturv. Kl.*, 1926); he later collected these and additional values into a table of radii for use in metals and intermetallic compounds (*Trans. Faraday Soc.* 25, 253 [1929]); see Chap. 11). A survey of the interatomic-distance values for covalent crystals was then made by L. Pauling and M. L. Huggins (*Z. Krist.* 87, 205 [1934]), leading to the formulation of the tables of tetrahedral radii, octahedral radii, and square radii given and described in Sec. 7-9, and by making small changes in the values of some of the tetrahedral radii in

TABLE 7-3.—SINGLE-BOND DISTANCES AND RADII FOR ELEMENTS

Bond	Substance	Method[a]	One-half of observed distance	Assigned radius
C—C	Diamond	X-ray	0.772 Å	0.772 Å
Si—Si	Si(c)	X-ray	1.17	1.17
Ge—Ge	Ge(c)	X-ray	1.22	1.22
Sn—Sn	Sn(c)	X-ray	1.40	1.40
P—P	$P_4(g)$	ED[b]	1.10	1.10
	P(c, black)	X-ray[c]	1.09, 1.10	
As—As	$As_4(g)$	ED[b]	1.22	1.21
	As(c)	X-ray	1.25	
Sb—Sb	Sb(c)	X-ray	1.43	1.41
S—S	$S_8(g)$	ED[d]	1.04	1.04
	$S_8(c)$	X-ray[e]	1.05, 1.02	
Se—Se	$Se_8(c, \alpha)$	X-ray[f]	1.17	1.17
	$Se_8(c, \beta)$	X-ray[g]	1.17	
	Se(c, gray)	X-ray	1.16	
Te—Te	Te(c)	X-ray	1.38	1.37
F—F	$F_2(g)$	ED[h]	0.718	0.64
	$F_2(g)$	Raman[i]	.709	
Cl—Cl	$Cl_2(g)$	Sp	.994	.99
Br—Br	$Br_2(g)$	Sp	1.140	1.14
I—I	$I_2(g)$	Sp	1.333	1.33

[a] X-ray signifies the x-ray study of crystals, ED the electron-diffraction study of gas molecules, and Sp the spectroscopic study of gas molecules. References are not given for the older x-ray and spectroscopic values; they may be obtained from standard compilations.

[b] L. R. Maxwell, S. B. Hendricks, and V. M. Mosley, *J. Chem. Phys.* **3**, 698 (1935).

[c] R. Hultgren and B. E. Warren, *Phys. Rev.* **47**, 808 (1935); approximately the same value is found also in amorphous red phosphorus, amorphous black phosphorus, and liquid phosphorus: C. D. Thomas and N. S. Gingrich, *J. Chem. Phys.* **6**, 659 (1938).

[d] C.-S. Lu and J. Donohue, *J.A.C.S.* **66**, 818 (1944).

[e] B. E. Warren and J. T. Burwell, *J. Chem. Phys.* **3**, 6 (1935); S. C. Abrahams, *Acta Cryst.* **8**, 661 (1955).

[f] R. D. Burbank, *Acta Cryst.* **4**, 140 (1951).

[g] R. E. Marsh, L. Pauling, and J. D. McCullough, *Acta Cryst.* **6**, 71 (1953).

[h] M. T. Rogers, V. Schomaker, and D. P. Stevenson, *J.A.C.S.* **63**, 2610 (1941).

[i] D. Andrychuk, *Can. J. Phys.* **29**, 151 (1951).

the way indicated by the data available at that time for a few normal-valence molecules, values of single-bond normal covalent radii differing only slightly from those given in Table 7-2 were obtained (see also L. Pauling, *Proc. Nat. Acad. Sci. U. S.* **18**, 293 [1932]). Since then the electron-diffraction and microwave study of gas molecules and further x-ray work on molecular crystals have provided many interatomic-distance values for testing and refining the table of radii.

TABLE 7-4.—THE COVALENT RADIUS OF HYDROGEN

Molecule	Method[a]	Distance M—H	Radius of hydrogen
H_2	Sp	0.74 Å	0.37 Å
HF	Sp	.918	.28
HCl	Sp, M	1.27	.28
HBr	Sp, M	1.42	.28
HI	Sp	1.61	.28
H_2O	Sp	0.96	.30
H_2S	Sp	1.34	.30
H_2Se	Sp	1.47	.30
NH_3	Sp	1.01	.31
PH_3	M	1.42	.32
AsH_3	M	1.52	.31
SbH_3	M	1.71	.30
CH_4	Sp	1.095	.32
C_2H_6	ED	1.095	.32
C_2H_4	ED, Sp	1.087	.31
C_2H_2	Sp	1.065	.29
C_6H_6	Sp	1.084	.31
HCN	Sp	1.066	.29
SiH_4	Sp	1.48	.31
GeH_4	Sp	1.53	.31
SnH_4	Sp	1.70	.30

[a] Here Sp means by infrared or ultraviolet spectroscopy and M by microwave spectroscopy. Similar values, in general somewhat less reliable, have also been obtained for many molecules by x-ray diffraction of crystals, neutron diffraction of crystals, electron diffraction of gas molecules (ED), or analysis of vibrational frequencies.

Some of the values are from the papers mentioned in the footnotes to Table 4-1. Other recent papers are the following:

CH_4: D. R. J. Boyd and H. W. Thompson, *Trans. Faraday Soc.* 49, 1281 (1953); H. M. Kaylor and A. H. Nielsen, *J. Chem. Phys.* 23, 2139 (1955).

C_2H_4: L. S. Bartell and R. A. Bonham, *J. Chem. Phys.* 27, 1414 (1957); W. S. Gallaway and E. F. Barker, *ibid.* 10, 88 (1942); H. C. Allen, Jr., and E. K. Plyler, *J.A.C.S.* 80, 2673 (1958).

C_2H_2: B. D. Saksena, *J. Chem. Phys.* 20, 95 (1952); M. T. Christensen, D. R. Eaton, B. A. Green, and H. W. Thompson, *Proc. Roy. Soc. London* A238, 15 (1956).

C_6H_6: B. P. Stoicheff, *Can. J. Phys.* 32, 339, 635 (1954); G. Herzberg and B. P. Stoicheff, *Nature* 175, 79 (1955).

HCN: A. E. Douglas and D. Sharma, *J. Chem. Phys.* 21, 448 (1953); I. R. Dagg and H. W. Thompson, *Trans. Faraday Soc.* 52, 455 (1956).

SiH_4: S. R. Polo and M. K. Wilson, *J. Chem. Phys.* 22, 1559 (1954).

GeH_4: L. P. Lindeman and M. K. Wilson, *J. Chem. Phys.* 22, 1723 (1954).

SnH_4: G. R. Wilkinson and M. K. Wilson, *J. Chem. Phys.* 25, 784 (1956).

Comparison of the radii with half the interatomic distances in elementary molecules or crystals involving single bonds may be made as a first check on the radii (Table 7-3).   For the fourth-row elements, crystallizing with the diamond structure, and the halogens (other than fluorine) the agreement is perfect, since these were the sources of the values in the table.   The crystals P, As, Sb, Se, and Te also show reasonably good agreement.   The electron-diffraction results for P_4, As_4, and S_8, obtained since the table was formulated, provide a good check of the corresponding radii.[7]

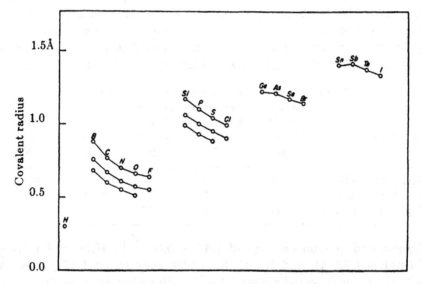

Fɪɢ. 7-1.—Values of covalent radii for the elements.

The radius of the hydrogen atom is more variable than that of other atoms, as can be seen from the experimental values for M—H distances in compounds of hydrogen collected in Table 7-4.   The values are reliable to about 0.01 Å.   The average value is about 0.30 Å.

The dependence of the covalent radii of the elements on atomic number is shown in Figure 7-1.   The relation is a simple one; for

[7] One important assumption made in the original formulation of the table of covalent radii was that the S—S single-bond distance is 2.08 Å, as in the crystals pyrite, FeS_2, and hauerite, MnS_2.   This has been verified subsequently by measurements not only on S_8(c) and S_8(g) (Table 7-3) but also on S_8(l) and S_x (plastic) (S—S = 2.07 Å, 2.08 Å: N. S. Gingrich, *Phys. Rev.* 55, 236 [1939]; *J. Chem. Phys.* 8, 29 [1940]), on H_2S_2 (S—S = 2.05 ± 0.02 Å) and $(CH_3)_2S_2$ (S—S = 2.04 ± 0.03 Å: D. P. Stevenson and J. Y. Beach, *J.A.C.S.* 60, 2872 [1938]), on $(CH_3)_2S_3$ (2.04 ± 0.02 Å: J. Donohue and V. Schomaker, *J. Chem. Phys.* 16, 92 [1948]), and on $(CF_3)_2S_2$ and $(CF_3)_2S_3$ (2.06 ± 0.02 Å: H. J. M. Bowen, *Trans. Faraday Soc.* 50, 444, 452, 463 [1954]).

the first and second rows of the periodic table smooth curves can be drawn through the points, whereas for the other rows there is only a slight discontinuity between the quadrivalent atoms and their neighbors, which may be attributed to the change in the nature of the bond orbitals.[8]

7-2. THE CORRECTION FOR ELECTRONEGATIVITY DIFFERENCE

The first values of covalent radii, as given in Table 7-2, were formulated before experimental values were available for F—F, O—O, and N—N single bonds. An electron-diffraction study of F_2 by Brockway[9] then gave the F—F distance as 1.45 Å (a value substantiated by Rogers, Schomaker, and Stevenson, Table 7-3, who found 1.435 ± 0.010 Å), whereas the accepted radius of fluorine would require 1.28 Å. Similar discrepancies were then reported for O—O, for which

TABLE 7-5.—SCHOMAKER-STEVENSON SINGLE-BOND
RADII OF FIRST-ROW ELEMENTS
(For other elements as in Table 7-2)

	B	C	N	O	F
Single-bond radius	0.81	0.772	0.74	0.74	0.72 Å
Double-bond radius	.71	.667	.62	.62	.60
Triple-bond radius	.64	.603	.55	.55	

Giguère and Schomaker[10] found 1.47 ± 0.02 Å in H_2O_2 (value from old radius 1.32 Å), and N—N, for which they found 1.47 ± 0.02 Å in N_2H_4 (value from old radius 1.40 Å). Since the distances for many bonds between N, O, and F and other atoms are rather well given by the old radii, there are large deviations from additivity in the bond lengths.

It was suggested by Schomaker and Stevenson[11] in 1941 that these deviations result from the partial ionic character of the bonds between unlike atoms. They proposed that the radii for N, O, and F be taken to be those given by the N—N, O—O, and F—F distances (Table 7-5), and that in general the interatomic distance for a bond A—B be taken to be the sum of the radii for the atoms A and B with a correction term -0.09 Å $|x_A - x_B|$, in which $|x_A - x_B|$ is the absolute value of the difference in electronegativity of the two atoms.

[8] Similar agreement is also found for a great many other substances, such as hexamethylene tetramine (C—N = 1.47 Å), etc.; the very extensive table that might be reproduced to show this agreement will be omitted.

[9] L. O. Brockway, *J.A.C.S.* **60**, 1348 (1938).

[10] P. A. Giguère and V. Schomaker, *J.A.C.S.* **65**, 2025 (1943).

[11] V. Schomaker and D. P. Stevenson, *J.A.C.S.* **63**, 37 (1941).

For example, the average value of the Si—C bond length found in the methylsilanes (Table 7-6) is 1.87 Å. The sum of the single-bond radii of carbon and silicon is 1.94 Å, which is 0.07 Å larger. If this value is corrected for the 0.7 difference in electronegativity by the Schomaker-Stevenson term it becomes 1.88 Å, in much better agreement with experiment.

On the other hand, the correction is not needed for the C—Cl bond. The experimental values (1.76 Å for CCl_4, 1.77 Å for $CHCl_3$ and CH_2Cl_2, and 1.781 Å for CH_3Cl) agree reasonably well with the value 1.765 Å corresponding to additivity, and disagree with the value 1.720 Å obtained by applying the correction.

TABLE 7-6.—COMPARISON OF CALCULATED (EQUATION 7-1)
AND OBSERVED BOND LENGTHS

C—N	1.47 Å	C—O	1.43 Å	C—F 1.37 Å	Si—C	1.88 Å
$(CH_3)_3N$	1.47	CH_3OH	1.427	CH_3F 1.385	CH_3SiH_3	1.86
$C_6H_{12}N_4$	1.47	Many others	1.43		$(CH_3)_2SiH_2$	1.86
Many others	1.47				$(CH_3)_3SiH$	1.87
					$(CH_3)_4Si$	1.89
Cl—N	1.73 Å	Cl—O	1.69 Å	Cl—F 1.63 Å	O—F	1.42 Å
NCH_2Cl_2	1.73	Cl_2O	1.69	ClF 1.63	OF_2	1.42

There is at present uncertainty as to how to predict bond lengths in a reliable way. In this book we shall assume that the length of a single bond between two atoms A and B can in most cases be reasonably well calculated by use of the radii r_A and r_B as given for light atoms in Table 7-5 and for heavier atoms in Table 7-2, with the equation

$$D(A-B) = r_A + r_B - c\,|\,x_A - x_B\,| \qquad (7\text{-}1)$$

Here the Schomaker-Stevenson coefficient c is to be given the value 0.08 Å for all bonds involving one first row atom (or two such atoms), the value 0.06 Å for bonds between Si, P, or S and a more electronegative atom (not of the first row), 0.04 Å for bonds between Ge, As, or Se and a more electronegative atom (not of the first row), and 0.02 Å for bonds between Sn, Sb, or Te and a more electronegative atom (not of the first row). The electronegativity correction is not to be made between carbon and the elements of the fifth, sixth, and seventh groups (beyond the first row); it seems likely that another effect (double-bond character, Sec. 9-3) overwhelms it.

A comparison of a few calculated and observed bond lengths is given in Table 7-6. Similar agreement has been found for other bonds.

For some bonds, such as C—N and C—O, the length given by Equation 7-1 is the same as that given by the sum of the radii in

Table 7-2 (with the small values for N, O. and F). For others, how-ever, such as Si—C, there is a great difference.

Bonds between heavier atoms and fluorine and other halogens are in general found to be shorter than calculated by Equation 7-1. The nature of these bonds will be discussed further in Chapter 9.

7-3. DOUBLE-BOND AND TRIPLE-BOND RADII

For the carbon-carbon double-bond distance in ethylene Bartell and Bonham[12] have reported the value 1.334 ± 0.003 Å from an electron-diffraction study and have stated that this value is compatible with the spectroscopic values of the moments of inertia.[13] This value corre-sponds to the double-bond radius 0.667 Å for carbon, as given in Table 7-2.

For the length of the carbon-carbon triple bond a number of reliable values in excellent agreement with one another are available. These include the value 1.204 Å for acetylene,[14] 1.207 Å for methylacetylene,[15] 1.211 Å for chloroacetylene,[16] and 1.207 Å for methylchloroacetylene,[17] from infrared and microwave studies, and closely agreeing values from electron-diffraction studies. In Table 7-2 we have accordingly given the value 0.603 Å for the triple-bond radius of carbon. The spectro-scopic value 1.094 Å for N≡N in N_2 leads to 0.547 Å for the triple-bond radius of nitrogen. The sum of these two radii, 1.150 Å, agrees well with the values for the length of the C≡N bond reported for HCN,[18] 1.153 Å, methyl cyanide,[19] 1.156 Å, and methylcyanoacetyl-ene,[20] 1.157 Å.

Some other values of double-bond and triple-bond radii given in Table 7-2 have been obtained from experimental values of interatomic distances, and some have been estimated. In general the double-bond

---

[12] Bartell and Bonham, *loc. cit.* (T7-4).

[13] Gallaway and Barker, *loc. cit.* (T7-4). H. W. Thompson, *Trans. Faraday Soc.* **35**, 697 (1939), has reported 1.331 ± 0.005 Å, and Allen and Plyler, *loc. cit.* (T7-4), have reported 1.337 ± 0.003 Å, from infrared spectroscopy. The value 1.330 ± 0.005 Å had been obtained earlier by electron diffraction by V. Schomaker and was reported in the second edition of this book.

[14] G. Herzberg and J. W. T. Spinks, *Z. Physik* **91**, 386 (1934); Saksena, also Christensen *et al.*, both *loc. cit.* (T7-4).

[15] L. F. Thomas, E. I. Sherrard, and J. Sheridan, *Trans. Faraday Soc.* **51**, 619 (1955).

[16] A. A. Westenberg, J. H. Goldstein, and E. B. Wilson, Jr., *J. Chem. Phys.* **17**, 1319 (1949).

[17] C. C. Costain, *J. Chem. Phys.* **23**, 2037 (1955).

[18] See footnote to Table 7-4 for references.

[19] Thomas *et al.*, *loc. cit.* (15); M. D. Danford and R. L. Livingston, *J.A.C.S.* **77**, 2944 (1955).

[20] J. Sheridan and L. F. Thomas, *Nature* **174**, 798 (1954).

radii are about 0.105 Å less and the triple-bond radii are about 0.17 Å less than the single-bond radii.

The usefulness of the multiple-bond radii in the discussion of the electronic structure of molecules will be illustrated in later sections.

7-4. INTERATOMIC DISTANCES AND FORCE CONSTANTS OF BONDS

When the available spectroscopic values for simple molecules are examined, it is seen that the force constants of the bonds, which, to-

TABLE 7-7.—CONSTANTS FOR EQUATION 7-2 FOR USE IN
GAS MOLECULES
(D_e in Å, k in megadynes/cm)

Type of bond		Example	a_{ij}	b_{ij}
Atom i in row	Atom j in row			
0	0	H_2	2.32	0.025 Å
0	1	HF	2.32	.335
0	2	HCl	2.32	.585
0	3	HBr	2.32	.65
1	1	CO	1.75	.68
1	2	PN	1.87	.94
1	3	TiO	2.00	1.06
1	4	SnO	2.04	1.18
1	5	PbO	2.04	1.26
2	2	Cl_2	2.04	1.25
2	4	ICl	1.98	1.48
3	3	Br_2	1.98	1.48
4	4	I_2	2.04	1.76

gether with the nuclear masses, determine the frequencies of vibration of the nuclei in the molecules, are not independent of the corresponding interatomic distances, but are closely related to them. Various equations expressing this relationship have been advanced. We select for discussion that of Badger,[21] which has the form

$$k^{-1/3} = a_{ij}(D_e - b_{ij}) \tag{7-2}$$

Here k is the force constant, D_e the equilibrium internuclear distance, and a_{ij} and b_{ij} are constants, with values determined by the nature of the bonded atoms, as given in Table 7-7.

The equation applies to both the normal states and excited states of molecules.

[21] R. M. Badger, *J. Chem. Phys.* **2**, 128 (1934); **3**, 710 (1935).

As an illustration of the usefulness of the equation we may mention the calculation by Eyster[22] in 1938, when no very reliable value of C═C distance in ethylene was available, of the value 1.325 ± 0.005 Å from the spectroscopic value for the carbon-carbon force constant. (The accepted value is now 1.334 Å).

Equation 7-2 can also be applied to the bonds in crystals. A discussion of the compressibilities of elementary metals and metalloids[23] has permitted the evaluation of the constants a_{ij} and b_{ij} as given in Table 7-8.

TABLE 7-8.—CONSTANTS FOR EQUATION 7-2 FOR USE IN
CRYSTALS OF ELEMENTARY METALS AND METALLOIDS

(D, in Å, k in megadynes/cm)

Element	a_{ij}	b_{ij}
Li, Be, C	2.89	1.13 Å
Na, Mg, Al, Si	3.10	1.73
K, Ca, Ti, V, Ge	2.06	1.46
Cr, Fe, Co, Ni, Cu	13.3	2.31
Rb, Sr, Zr, Nb, Mo, Sn	2.32	1.86
Ru, Rh, Pd, Ag	4.12	2.10
Ba, Ta, W	2.03	1.80
Ce, Ir, Pt, Au, Tl	2.96	1.99

7-5. INTERATOMIC DISTANCES AND RESONANCE[24]

The resonance of the benzene molecule between the two Kekulé structures (the small contributions of other structures being neglected) can be considered to give each of the six carbon-carbon bonds 50 percent single-bond character and 50 percent double-bond character. We would expect that the carbon-carbon interatomic distance would lie between the single-bond value 1.544 Å and the double-bond value 1.334 Å—not midway between, but closer to the lower value, both because of the extra stabilization due to the resonance energy (a strong bond having smaller interatomic distance than a weak bond) and because of the greater effectiveness of the double-bond potential function (with its greater curvature in the neighborhood of its minimum, corresponding to its larger force constant) in determining the position of the minimum of the potential function for the resonating molecule.

[22] E. H. Eyster, *J. Chem. Phys.* **6**, 580 (1938).

[23] J. Waser and L. Pauling, *J. Chem. Phys.* **18**, 747 (1950).

[24] Pauling, *loc. cit.* (6), L. Pauling, L. O. Brockway, and J. Y. Beach, *J.A.C.S.* **57**, 2705 (1935); Pauling and Brockway, *loc. cit.* (T7-1).

The observed value for benzene,[25] 1.397 ± 0.001 Å, is only 0.07 Å greater than the double-bond distance.

It is to be expected that the bending of the bent bonds of the two double bonds in a system $\backslash C{-}C /$ away from their tetrahedral direc-

tions would make the forces of repulsion between them somewhat less than in a saturated molecule, and that for this reason the carbon-carbon single-bond distance in a conjugated system would be less than the normal value 1.544 Å. An estimate of the magnitude of this effect can be made by consideration of the C—H distances in saturated and unsaturated molecules. The C—H distance in methane, ethane, and other saturated molecules is about 1.100 Å; in ethylene, benzene, and other molecules in which the carbon atom forms a double bond it is about 1.085 Å (Table 7-4), representing a decrease in the bond length of about 0.015 Å. In allene, $H_2C{=}C{=}CH_2$, the carbon-carbon double-bond length[26] is 1.310 Å, 0.024 Å less than its normal value, and in glycine[27] the carbon-carbon single bond (adjacent to a double bond with oxygen in the carboxylate ion) has length 1.523 Å, which is 0.021 Å less than the normal value. The average of these observed decreases in bond length is 0.020 Å, and we conclude that, as a result of the decreased repulsion of the bent bonds, the effective radius of the carbon atom, for either another double bond or the two single bonds that it can form in addition to the first double bond, is about 0.020 Å less than the normal radius.

We accordingly conclude that a pure single bond between two carbon atoms forming double bonds with other atoms, as in 1,3-butadiene, $H_2C{=}CH{-}CH{=}CH_2$, would be decreased in length by 0.040 Å to 1.504 Å. In fact, the central carbon-carbon distance in the molecule is somewhat shorter still, about 1.46 Å. The further decrease in length may be attributed to partial multiple-bond character of the bond as discussed in later sections of this and the following chapter.

Similarly, the expected length of the central double bond in butatriene, $H_2C{=}C{=}C{=}CH_2$, is 1.294 Å, a value 0.040 Å less than normal

[25] By high-resolution Raman spectroscopy: Stoicheff, *loc. cit.* (T7-4), 339. The values 1.39 and 1.40 ± 0.03 Å had been reported by R. Wierl, *Ann. Physik* **8**, 521 (1931), 1.390 ± 0.005 Å by L. Pauling and L. O. Brockway, *J. Chem. Phys.* **2**, 867 (1934), and 1.393 Å by V. Schomaker and L. Pauling, *J.A.C.S.* **61**, 1769 (1939), all by electron diffraction.

[26] J. Ovenend and H. W. Thompson, *J. Opt. Soc. Am.* **43**, 1065 (1953); Herzberg and Stoicheff, *loc. cit.* (T7-4); O. Bastiansen, unpublished electron-diffraction investigation (1958).

[27] R. E. Marsh, *Acta Cryst.* **11**, 654 (1958).

value, to correct for the bent-bond effect. The observed length of this bond[28] is 1.284 Å, and that of the other two double bonds (with 1.314 Å expected) is 1.309 Å. These bond lengths show a small shortening from conjugation.

In acetylene and hydrogen cyanide the length of the C—H bond is about 0.04 Å less than it is for methane and its saturated derivatives, and we conclude that the correction to the effective radius of a carbon atom forming a triple bond is −0.040 Å, twice as great as the double-

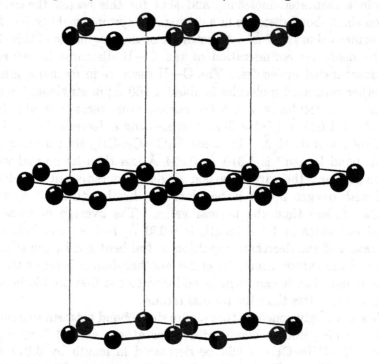

FIG. 7-2.—The arrangement of carbon atoms in the graphite crystal.

bond correction. This is not unreasonable, in that in the triple bond there are three bent bonds, rather than two, and they are bent somewhat more than are the two bonds of the double bond.

An empirical curve relating carbon-carbon interatomic distances with the amounts of single-bond and double-bond character for molecules resonating between structures some of which represent the bond as a single bond and some as a double bond could be used in interpreting observed values of the interatomic distances to obtain information as to the type of the bonds. The pure single-bond distance 1.504 Å (for use in a conjugated system of alternating single and double bonds)

[28] B. P. Stoicheff, *Can. J. Phys.* **35**, 837 (1957); O. Bastiansen, unpublished electron-diffraction investigation (1958).

and the pure double-bond distance 1.334 Å provide the end points of the curve. A third point, at 50 percent double-bond character, is provided by the value 1.397 Å for benzene, and a fourth point by the value 1.420 Å for graphite. The structure of the graphite crystal is shown in Figure 7-2. It consists of hexagonal layers of molecules which are separated by a distance so large (3.40 Å) that there can be no covalent bonds between them; each of the layers is a giant molecule, and the superimposed layer molecules are held together only by weak van der Waals forces. The four valences of each carbon atom are used to form bonds with its three neighbors; the later molecule resonates among many valence-bond structures such as

and in this way each carbon-carbon bond achieves one-third double-bond character.

Through these four points we draw a smooth curve, as shown in Figure 7-3, which we accept as representing the dependence of carbon-carbon interatomic distance on the amount of double-bond character for single-bond:double-bond resonance. The use of the curve in the discussion of the nature of the carbon-carbon bonds in resonating molecules is illustrated in the following chapter.

In view of the reasonable behavior of interatomic-distance values in general, it seems probable that by a suitable translation and change of vertical scale (to give the correct end points) the same function can be used for bonds between other atoms, and also for resonance involving triple bonds. These further uses of the curve are also illustrated in the following chapter.

It is interesting that the curve can be represented with only small deviations from the experimental points by the equation

$$D_n = D_1 - (D_1 - D_2) \frac{1.84(n - 1)}{0.84n + 0.16} \tag{7-3}$$

in which D_n is the value of the interatomic distance for a bond of intermediate type, D_1 that for a single bond, D_2 that for a double bond, and n the bond number. This equation can be derived in the following simple way: Let the potential function for a resonating bond be given

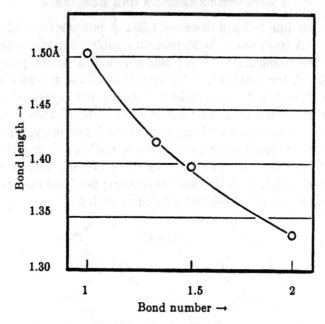

FIG. 7-3.—The relation between interatomic distance and the amount of double-bond character for single-bond:double-bond resonance of carbon-carbon bonds.

as the sum of two parabolic functions, representing single-bond and double-bond potential functions, with coefficients $2 - n$ and $n - 1$, respectively:

$$V(D) = \tfrac{1}{2}(2 - n)k_1(D - D_1)^2 + \tfrac{1}{2}(n - 1)k_2(D - D_2)^2 \quad (7\cdot4)$$

If the derivative of this with respect to D is equated to zero, its equilibrium value (corresponding to the minimum of the potential function) can be found as a function of n and the ratio of the force constants $\cdot k_2/k_1$. The function becomes identical with that of Equation 7-3 when k_2/k_1 is given the value 1.84. This value for the ratio of the force constants is somewhat larger than that given by Badger's rule, 1.58; it is possible that the small increase is needed to compensate for the neglect of resonance energy in the assumed potential function.

In Table 7-9 there are given values for carbon-carbon bond lengths for single-bond:double-bond resonance, calculated by Equation 7-3 with $D = 1.504$ Å (as corrected for the adjacent-bent-bond effect) and $D_2 = 1.334$ Å. These values are for application to conjugated and aromatic systems of single bonds and double bonds, as illustrated in the following section and in later chapters.

Bond Lengths in Aromatic Hydrocarbons.—As an example of the use of the single-bond:double-bond resonance curve of interatomic

TABLE 7-9.—CARBON-CARBON BOND LENGTHS FOR SINGLE-BOND:
DOUBLE-BOND RESONANCE

Bond number	D_n	D_1-D_n	D_n-D_2	Bond number	D_n	D_1-D_n	D_n-D_2
1.00	1.504 Å	0.000 Å	0.170 Å	1.40	1.410 Å	0.094 Å	.076 Å
1.05	1.489	.015	.155	1.45	1.402	.102	.068
1.10	1.475	.029	.141	1.50	1.394	.110	.060
1.15	1.462	.042	.128	1.60	1.380	.124	.046
1.20	1.450	.054	.116	1.70	1.367	.137	.033
1.25	1.439	.065	.105	1.80	1.355	.149	.021
1.30	1.429	.075	.095	1.90	1.344	.160	.010
1.35	1.419	.085	.085	2.00	1.334	.170	.000

distances, we may discuss the observed interatomic distances in the condensed aromatic hydrocarbons.

In general there is reasonably good agreement between the observed carbon-carbon bond distances in these molecules and the values calculated for the amounts of double-bond character given by consideration of the Kekulé structures alone. These calculated values agree closely with values calculated by the molecular-orbital method.[29]

In anthracene there are five nonequivalent kinds of carbon-carbon bonds, with amounts of double-bond character, as calculated by consideration of the Kekulélike structures alone (given equal weight), varying from $\frac{1}{4}$ to $\frac{3}{4}$ (see Chap. 6). In Table 7-10 a comparison is made between the corresponding calculated values of the interatomic dis-

TABLE 7-10.—CALCULATED AND OBSERVED CARBON-CARBON
BOND DISTANCES IN ANTHRACENE

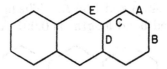

	Double-bond character	D_{calc}	D_{obs}
A	3/4	1.361 Å	1.366 Å
B	1/4	1.439	1.419
C	1/4	1.439	1.433
D	1/4	1.439	1.436
E	1/2	1.394	1.399

[29] W. G. Penney, *Proc. Roy. Soc. London* A158, 306 (1937); H. O. Pritchard and F. H. Sumner, *Trans. Faraday Soc.* 51, 457 (1955).

Interatomic Distances

Table 7-11.—Bond Lengths in 1:14-Benzbisanthrene

Bond	Double-bond character	D_{calc}	D_{obs}	Bond	Double-bond character	D_{calc}	D_{obs}
A	1/30	1.494 Å	1.49 Å	L	1/3	1.423 Å	1.43 Å
B	2/15	1.466	1.47	M	3/10	1.429	1.42
C	3/10	1.429	1.44	N	7/30	1.443	1.44
D	8/15	1.389	1.40	P	2/15	1.466	1.47
E	2/3	1.371	1.36	Q	2/3	1.371	1.37
F	3/10	1.429	1.43	R	1/3	1.423	1.43
G	2/5	1.410	1.40	S	11/30	1.416	1.40
H	7/15	1.399	1.41	T	19/30	1.376	1.37
I	3/10	1.429	1.43	U	2/15	1.466	1.47
J	2/5	1.410	1.42	V	13/15	1.348	1.35
K	7/15	1.399	1.39				

tances, labeled as in the diagram below, and the average observed distances as reported by Cruickshank[30] (these values have probable errors of about 0.005 Å). The agreement between the calculated and the observed values is reasonably good, the mean deviation being 0.008 Å. The same agreement has been reported for the molecular-orbital calculations made by Pritchard and Sumner.

Similar agreement has been found for some of the larger condensed hydrocarbons. As an example of these we may discuss 1:14-benzbisan-threne. In Table 7-11 the amounts of double-bond character, cal-culated by giving equal weight to the 30 Kekulélike structures, are given, together with the corresponding calculated bond distances and the observed values reported by Trotter.[31] The amounts of double-bond character vary from 1/30 to 23/30, and the observed bond lengths vary between 1.49 Å and 1.35 Å. The mean deviation between observed and calculated bond lengths is 0.007 Å.

[30] D. W. J. Cruickshank, *Acta Cryst.* **9**, 915 (1956); **10**, 470 (1957).
[31] J. Trotter, *Acta Cryst.* **11**, 423 (1958).

The 30 Kekulélike structures for this molecule are shown below. The groups of hexagons with circles in their centers represent the corresponding set of Kekulélike structures: two for benzene, three for naphthalene, and four for anthracene. The number under each structural formula is the number of Kekulélike structures corresponding to it.

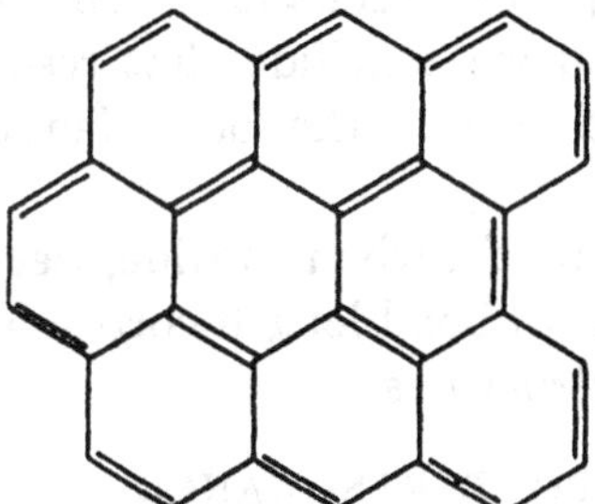

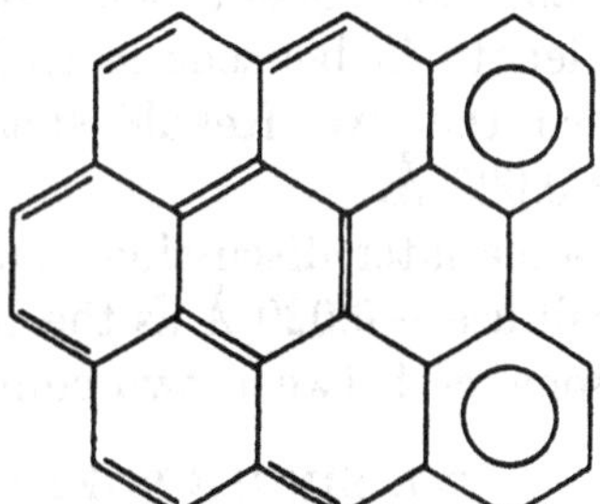

7-6. BOND ORDER AND BOND LENGTH; CHANGE IN BOND LENGTH CAUSED BY RESONANCE BETWEEN TWO EQUIVALENT STRUCTURES

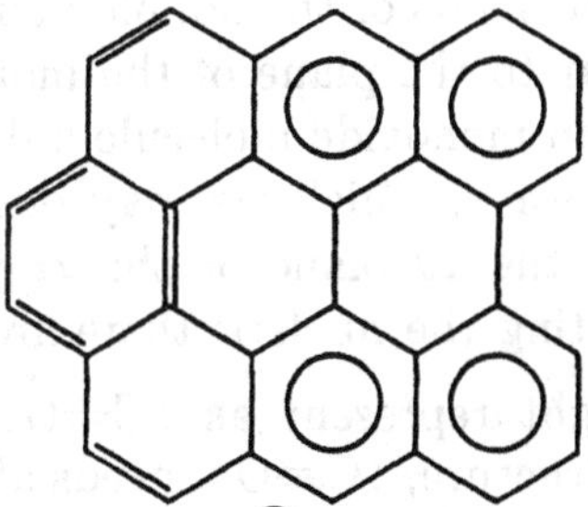

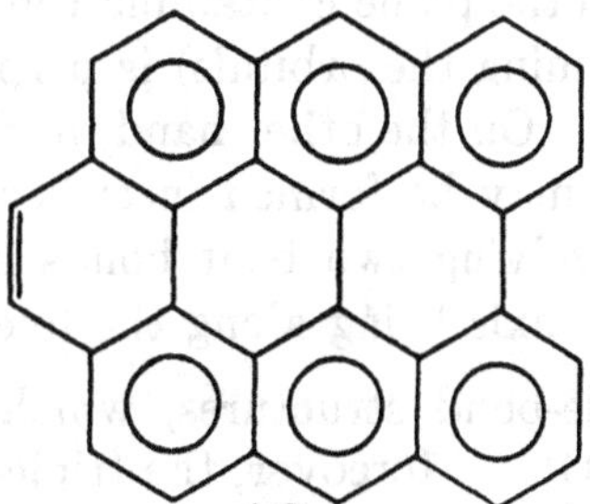

A smooth curve can be passed through the points of a graph representing the lengths of the C—C, C=C, and C≡C bonds. The equation for the curve is

$$D(n') = D_1 - 0.71 \text{ Å} \log n' \tag{7-5}$$

In this equation n' is the *bond order*; it is equal to the bond number n for $n' = 1$, 2, and 3, but has a different interpretation for fractional values. Whereas we have taken the bond number for the carbon-carbon bonds in benzene to be $1\frac{1}{2}$, so that the valence of carbon remains equal to the sum of the bond numbers of the bonds formed by the atom, the bond order is taken to be somewhat larger, reflecting the extra resonance energy of the molecule. The value of n' calculated by molecular orbital theory[32] is $1\frac{2}{3}$, and the observed bond length in benzene inserted in Equation 7-5 leads to $n' = 1.66$, in excellent agreement.

With use of Equation 7-5 we calculate 1.420 Å for the length of a carbon-carbon bond with bond order 1.5. This may be taken as the length that the bonds in benzene would have if there were no stabilization (and consequent bond shortening) by resonance. The actual bond length in benzene is 1.397 Å, and we conclude that resonance between the two Kekulé structures decreases the bond length by about 0.023 Å.

In some later discussion (such as that of carbon dioxide, Sec. 8-1) we shall use -0.020 Å as the correction to bond length corresponding to resonance between two equivalent structures.

7-7. SINGLE-BOND:TRIPLE-BOND RESONANCE

The nuclear configurations of benzene and graphite are such as to permit the formation of unstrained carbon-carbon double bonds in which the plane containing the two bent bonds of the double bond (or containing the orbitals) is perpendicular to the plane of the molecule itself. On the other hand, in the carbon monoxide molecule a double bond may be formed in either of two ways, which we may describe as involving two bent bonds in either the xy plane or the xz plane (the x axis being along the line connecting the nuclei); there are two double-bond structures, which we might represent as $:C\!=\!\overset{..}{O}:$ and $:C\!=\!O:$. Moreover, the triple-bond structure, $:C\!\equiv\!O:$, needs also to be considered.

Detailed discussions of this molecule and of similar molecules will be given in the following chapter. In these discussions we shall have need of an equation relating bond length to bond number for multiple-bond resonance of this sort, which we may call single-bond:triple-bond resonance.[33]

Equation 7-3, for single-bond:double-bond resonance, was derived by consideration of the two potential functions corresponding to the two structures A—B and A=B. In the same way we may derive an equation corresponding to the four structures A—$\overset{..}{B}$, A=$\overset{..}{B}$, A=B, and A≡B (or $\overset{..}{A}$—B, $\overset{..}{A}$=B, A=B, and A≡B), by taking the sum of the four quadratic potential functions $\frac{1}{2}k_i(D - D_i)^2$, with D_i the respective bond lengths and k_i the force constants k_1, k_2, k_2, and k_3. For

[32] C. A. Coulson, *Proc. Roy. Soc. London* A169, 413 (1939).

[33] L. Pauling, *J. Phys. Chem.* 56, 361 (1952). In the discussion given above small changes have been made from that in this article.

$k_1:k_2:k_3$ we use the ratio $1:2:4$ indicated by Badger's rule (Sec. 7-4) and the discussion of Section 7-5. In this way, with the assumption that the weights of the four structures correspond to independent resonance of the bonds in the planes xy and xz, we obtain[34] the following equation:

$$D - D_2 = \frac{(3 - n)^2(D_1 - D_2) - 4(n - 1)^2(D_2 - D_3)}{(n + 1)^2} \qquad (7\text{-}6)$$

Values of $D - D_1$, $D - D_2$, and $D - D_3$ calculated with this equation for $D_1 - D_2 = 0.18$ Å or 0.21 Å and $D_2 - D_3 = 0.09$ Å or 0.12 Å are given in Table 7-12 (these differences are appropriate in most of the applications of the equation). It can be seen that the distance

TABLE 7-12.—BOND LENGTHS FOR SINGLE-BOND:
TRIPLE-BOND RESONANCE

n	$D-D_1{}^a$	$D-D_1{}^b$	n	$D-D_2{}^a$	$D-D_2{}^b$	n	$D-D_3{}^a$	$D-D_3{}^b$
1.00	0.000 Å	0.000 Å	1.35	0.080 Å	0.108 Å	2.4	0.034 Å	0.045 Å
1.05	− 0.018	− 0.024	1.40	.068	.091	2.6	.021	.028
1.10	− 0.035	− 0.047	1.45	.060	.080	2.8	.010	.013
1.15	− 0.050	− 0.067	1.50	.050	.067	3.0	.000	.000
1.20	− 0.063	− 0.084	1.60	.032	.043			
1.25	− 0.076	− 0.101	1.70	.017	.023			
1.30	− 0.088	− 0.117	1.80	.004	.005			
1.35	− 0.099	− 0.132	1.90	− 0.008	− 0.011			
			2.00	− 0.020	− 0.027			
			2.10	− 0.030	− 0.040			
			2.20	− 0.039	− 0.052			
			2.30	− 0.048	− 0.064			
			2.40	− 0.056	− 0.075			
			2.50	− 0.063	− 0.084			

[a] For $D_1-D_2 = 0.18$ Å. $D_2-D_3 = 0.09$ Å.
[b] For $D_1-D_2 = 0.24$ Å, $D_2-D_3 = 0.12$ Å.

for bond number $n = 2$ is 0.02 Å or 0.03 Å less than for a double bond; this decrease is to be attributed to the additional resonance stabilization. The use of Table 7-12 will be illustrated in later chapters.

7-8. THE CONDITIONS FOR EQUIVALENCE OR NONEQUIVALENCE OF BONDS

It often happens that the most reasonable valence-bond structure that can be written for a molecule or crystal is less symmetrical than the

[34] The potential function used is

$$V(D) = \tfrac{1}{2}\alpha_1 k_1(D - D_1)^2 + \tfrac{1}{2}\alpha_2 k_2(D - D_2)^2 + \tfrac{1}{2}\alpha_3 k_3(D - D_3)^2$$

The weights α_1, α_2, α_3 have the relative values $(1 - \alpha)^2:2\alpha(1 - \alpha):\alpha^2$, with $\alpha = (n - 1)/2$, where n is the bond number.

known or predicted arrangement of the atomic nuclei. In this circumstance there is another valence-bond structure equivalent to the first and differing from it only in the distribution of the bonds (or there may be several other equivalent structures). An example is provided by benzene, the two Kekulé structures being the most stable valence-bond structures for this molecule. Another example is sulfur dioxide, for

which the reasonable structures

$$\overset{\cdot\cdot}{S^+} \diagup\diagdown \quad \text{and} \quad \overset{\cdot\cdot}{S^+} \diagup\diagdown \quad \text{may be}$$
$$^-{:}\overset{\cdot\cdot}{O}{:} \quad \overset{\cdot\cdot}{O}{:} \qquad {:}\overset{\cdot\cdot}{O} \quad {:}\overset{\cdot\cdot}{O}{:}^-$$

written; many other molecules of this type are discussed in the following chapters.

Let us consider a molecule A_2B, for which the two equivalent structures A—B—A (I) and A—B—A (II) may be written, each involving two bonds of which one is stronger than the other. We now ask whether the two A atoms in the molecules are to be considered as equivalent or not, and whether or not there are two different A—B bond lengths in the molecule.

This question needs to be discussed in detail. The principles of quantum mechanics require that the normal state of the isolated molecule that is observed over a long period of time have a resonance structure to which the equivalent structures I and II contribute equally. The interpretation of this resonance depends on the magnitude of the resonance energy. If the resonance energy is very large, no experiment can be devised to detect the individual structures I and II. The frequency of resonance is the resonance energy divided by Planck's constant h, and the minimum period of time during which an experiment could be carried out that would distinguish structure I from structure II is h divided by the resonance energy. For resonance between valence-bond structures in benzene and sulfur dioxide the resonance energy is so large that this period is about 10^{-15} seconds, and the bonds are made completely equivalent by resonance. There is only one value for the bond lengths.

If, on the other hand, the resonance integral is very small, it may be convenient to refer to the substance as containing tautomeric or isomeric molecules, with electronic structures that are represented essentially by I or II alone. The relation of resonance to tautomerism and the distinction between tautomerism and isomerism are discussed in the last chapter of this book.

It is convenient for us to draw a rather broad line between resonance (in the sense of resonance of a molecule among alternative valence-bond structures) and tautomerism by the use of the ratio of the resonance frequency to the frequency of nuclear motion. If the resonance frequency is much greater than the frequency of oscillation of the nuclei, the molecule will be represented by the resonating structure { A—B—A, A—B—A } and the two bonds in the molecule will be equivalent, whereas if it is much less the two atoms A will oscillate for some time about equilibrium positions relative to B that are not equivalent, and will then interchange their roles.

The discussion can be made more definite by considering the forces between B and the two attached atoms. Structure I alone corresponds to a potential function that brings the equilibrium position of one atom (A′ for the molecule A′—B—A) closer to B than that for A is, whereas structure II places A closer to B than A′ is. (A and A′ might, indeed, be isotopes, and hence physically distinguishable.) If the resonance energy is small, the molecule will oscillate for a time in a manner corresponding to a small distance A′—B and a larger distance B—A and then oscillate about new equilibrium positions, with B—A smaller than A′—B. This corresponds to tautomerism. If the resonance energy is sufficiently large, however, the potential function for nuclear oscillation will be changed, so that it has for each atom A a single minimum, and the two atoms A and A′ will oscillate in equivalent fashion about equilibrium positions equidistant from B. The two bonds in the molecule are then equivalent. The magnitude of the resonance integral required to achieve this depends on (among other factors) the difference in the equilibrium configurations of the alternative structures. Thus the configurations of the carbon nuclei in benzene corresponding to the two Kekulé structures separately (with C—C = 1.54 Å and C=C = 1.33 Å) would place the carbon atoms only about 0.1 Å from their actual positions (with C—C = 1.39 Å). This distance (0.1 Å) is less than the usual amplitude of nuclear oscillation (about 0.2 Å), so that with each libration of the nuclei the molecule would pass through the configuration appropriate to each of the Kekulé structures. On the other hand, tautomerism may be expected when the stable configurations differ largely from one another, as, for example, with D and L configurations of complex molecules.

Some years ago Braune and Pinnow[35] reported the molecules UF_6, WF_6, and probably MoF_6 to have structures with orthorhombic symmetry, the three pairs of M—F distances having the ratios 1:1.12:1.22. It could not be determined from theoretical considerations that this structure was incorrect; the central atom might have a tendency to form three kinds of bonds in pairs, with the resonance energy among the three corresponding structures so small that the potential function

[35] H. Braune and P. Pinnow, *Z. physik. Chem.* **B35**, 239 (1937).

would not be changed from that for one of the individual structures. This seemed improbable, however, and it was discovered by Schomaker and Glauber[36] that their electron-diffraction study gave an incorrect result because of a failure of the Born theory of electron scattering; the structures are octahedral, with the six bonds equivalent.

The inequality of the apical and equatorial P-Cl distances in phosphorus pentachloride is not cast in doubt by this argument, since these directions are not geometrically equivalent even for equality of the distances.

7-9. TETRAHEDRAL AND OCTAHEDRAL COVALENT RADII

Tetrahedral Radii.—In crystals with the diamond, sphalerite, and wurtzite arrangements (Figs. 7-4, 7-5, and 7-6) each atom is surrounded tetrahedrally by four other atoms. If the atoms are those of fourth-

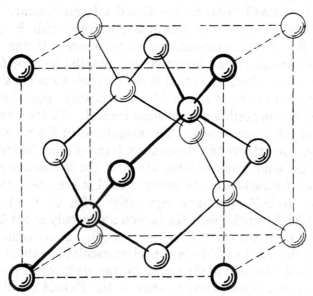

Fig. 7-4.—The arrangement of the carbon atoms in the diamond crystal. Each atom has four near neighbors, which are arranged about it at the corners of a regular tetrahedron.

column elements or of two elements symmetrically arranged relative to the fourth column, the number of valence electrons is right to permit the formation of a tetrahedral covalent bond between each atom and its four neighbors. The diamond arrangement is shown by C, Si, Ge, and

[36] V. Schomaker and R. Glauber, *Nature* 170, 290 (1952); *Phys. Rev.* 89, 667 (1953).

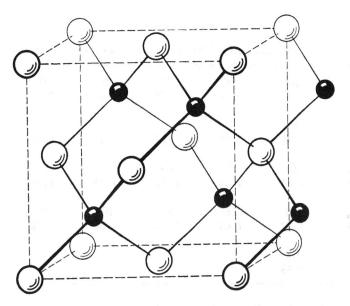

Fig. 7-5.—The arrangement of zinc atoms (small circles) and sulfur atoms (large circles) in sphalerite, the cubic form of zinc sulfide.

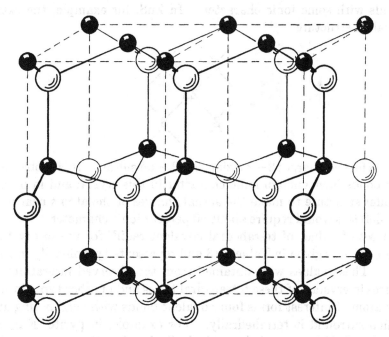

Fig. 7-6.—The arrangement of zinc atoms (small circles) and sulfur atoms (large circles) in wurtzite, the hexagonal form of zinc sulfide. The unit of structure outlined by dashes is the orthohexagonal unit, with edges at right angles to one another. The lengths of the edges of the unit in the basal plane have the ratio $\sqrt{3}/1$, corresponding to the 120° angles of the hexagonal unit.

TABLE 7-13.—TETRAHEDRAL COVALENT RADII

	Be	B	C	N	O	F
	1.06	0.88	0.77	0.70	0.66	0.64
	Mg	Al	Si	P	S	Cl
	1.40	1.26	1.17	1.10	1.04	0.99
Cu	Zn	Ga	Ge	As	Se	Br
1.35	1.31	1.26	1.22	1.18	1.14	1.11
Ag	Cd	In	Sn	Sb	Te	I
1.52	1.48	1.44	1.40	1.36	1.32	1.28
	Hg					
	1.48					

Sn, and the sphalerite or the wurtzite arrangement (or both) by the compounds listed in Table 7-14.

In all of these crystals it is probable that the bonds are covalent bonds with some ionic character. In ZnS, for example, the extreme covalent structure

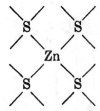

places formal charges $2-$ on zinc and $2+$ on sulfur. It is probable that the bonds have enough ionic character in this crystal and in others of similar structure to make the actual charges of the atoms nearly zero; for ZnS this would require about 50 percent ionic character.

A set of values of tetrahedral covalent radii[37] for use in crystals of these types is given in Table 7-13 and represented graphically in Figure 7-7. These values were obtained from the observed interatomic distances in crystals of these tetrahedral types and of other types in which the atom of interest forms four covalent bonds with neighboring atoms which surround it tetrahedrally. For example, in pyrite, FeS_2, each sulfur atom is surrounded tetrahedrally by three iron atoms and one sulfur atom, with all of which it forms essentially covalent bonds (Fig. 7-8); the substance is a derivative of hydrogen disulfide, H_2S_2. That the Fe—S bonds are essentially covalent is shown by the magnetic cri-

[37] Huggins, also Pauling and Huggins, *loc. cit.* (6).

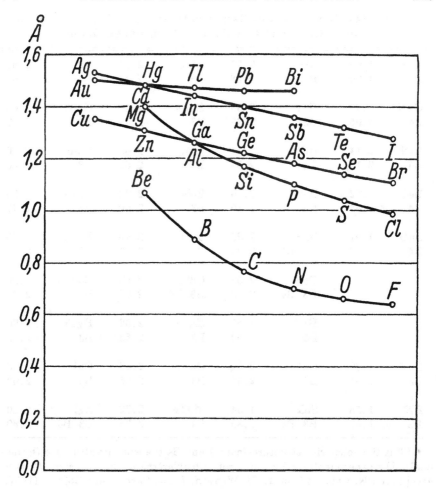

FIG. 7-7.—Values of tetrahedral covalent radii for sequences of atoms.

terion; the substance is only slightly paramagnetic, which corresponds to the formation of octahedral $3d^24s4p^3$ bonds by the ferrous iron ($\mu = 0$) rather than to ionic bonds ($\mu = 4.90$). The sulfur-sulfur distance in this crystal is 2.09 Å, which agrees well with the value from the table, 2.08 Å.

The tetrahedral radius sums are seen to agree closely with the experimentally determined values of the interatomic distances (which were, of course, used in their derivation), the average deviation being 0.01 Å (Table 7-14). For CuF, BeO, AlN, and SiC the observed bond lengths are significantly less than the sum of the radii; the difference is with little doubt due to partial ionic character. The radii have been chosen in such a way as not to need this correction in other cases.

TABLE 7-14.—COMPARISON OF OBSERVED INTERATOMIC DISTANCES IN B3
AND B4 CRYSTALS[a] WITH SUMS OF TETRAHEDRAL RADII

AlN	1.96	AlP	2.36	AlAs	2.44	AlSb	2.62
B4	1.90	B3	2.36	B3	2.44	B3	2.64
GaN	1.96	GaP	2.36	GaAs	2.44	GaSb	2.62
B4	1.95	B3	2.36	B3	2.44	B3	2.63
InN	2.14	InP	2.54	InAs	2.62	InSb	2.80
B4	2.15	B3	2.54	B3	2.62	B3	2.80
BeO	1.72	BeS	2.10	BeSe	2.20	BeTe	2.38
B4	1.65	B3	2.10	B3	2.20	B3	2.41
ZnO	1.97	ZnS	2.35	ZnSe	2.45	ZnTe	2.63
B4	1.97	B3, B4	2.35	B3	2.45	B3	2.63
		CdS	2.52	CdSe	2.62	CdTe	2.80
		B3, B4	2.53	B3, B4	2.63	B3	2.80
		HgS	2.52	HgSe	2.62	HgTe	2.80
		B3	2.52	B3	2.63	B3	2.79
CuF	1.99	CuCl	2.34	CuBr	2.46	CuI	2.63
B3	1.85	B3	2.34	B3	2.46	B3	2.62
BN	1.58	SiC	1.94	MgTe	2.72	AgI	2.80
B3	1.57	B3, B4	1.89	B4	2.76	B3, B4	2.80

[a] B3 is the sphalerite structure (cubic) and B4 the wurtzite structure (hexagonal). The experimental values are from *Strukturbericht* and *Structure Reports*, except for BN, which is from R. H. Wentorf, Jr., *J. Chem. Phys.* **26,** 956 (1957).

The tetrahedral radii for first-row and second-row elements are identical with the normal single-bond covalent radii given in Table 7-2. For the heavier atoms there are small differences, amounting to 0.03 Å for bromine and 0.05 Å for iodine. It is possible that these differences are due to the difference in the nature of the bond orbitals in tetrahedral and normal covalent compounds.

Octahedral Radii.—In pyrite (Fig. 7-8) each iron atom is surrounded by six sulfur atoms, which are at the corners of a nearly regular octahedron, corresponding to the formation by iron of $3d^24s4p^3$ bonds. The iron-sulfur distance is 2.27 Å, from which, by subtraction of the tetrahedral radius of sulfur, 1.04 Å, the value 1.23 Å for the d^2sp^3 octahedral covalent radius of bivalent iron is obtained (Table 7-15).

From similar data for other crystals with the pyrite structure or a closely related structure (of the marcasite or arsenopyrite types), given

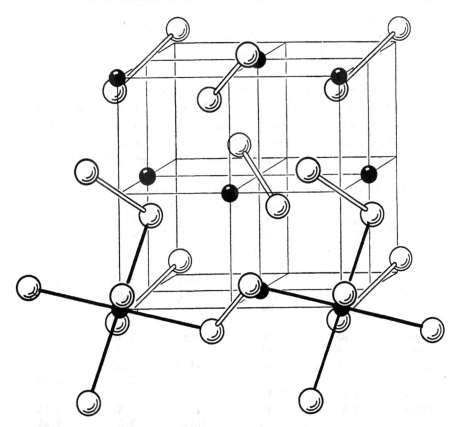

FIG. 7-8.—The arrangement of iron atoms (small circles) and sulfur atoms (large circles) in the cubic crystal pyrite, FeS$_2$. Each iron atom is surrounded octahedrally by six sulfur atoms, and each sulfur atom is surrounded tetrahedrally by one sulfur atom and three iron atoms.

in Table 7-16, values can be obtained for the octahedral radii of other transition-group elements. The elements Fe(II) and Co(III) in the indicated valence states are isoelectronic; it is interesting that there is very little difference in their radii, the decrease in radius with unit in-

TABLE 7-15.—OCTAHEDRAL COVALENT RADII

Fe(II)	1.23 Å	Ru(II)	1.33 Å	Os(II)	1.33 Å
Co(III)	1.22	Rh(III)	1.32	Ir(III)	1.32
Ni(IV)	1.21	Pd(IV)	1.31	Pt(IV)	1.31
Co(II)	1.32			Au(IV)	1.40
Ni(III)	1.30				
Ni(II)	1.39				
Fe(IV)	1.20				

crease in atomic number being only 0.01 Å. The same very small decrease is shown in the two sequences Ru(II), Rh(III), Pd(IV) and Os(II), Ir(III), Pt(IV), which, moreover, have identical values of the radii.

For all of these atoms the number of electrons is such that all of the stable orbitals are occupied by unshared pairs or are used in bond formation. In CoS_2, $CoSe_2$, $NiAsS$, and $AuSb_2$ the atoms Co(II), Ni(III), and Au(IV) contain one more electron than can be fitted into the three $3d$ orbitals ($5d$ for Au) that are left after the d^2sp^3 orbitals are usurped

TABLE 7-16.—INTERATOMIC DISTANCES IN PYRITE-TYPE CRYSTALS

Substance	Distance M—X	Radius of M	Substance	Distance M—X	Radius of M
FeS_2	2.27	1.23	PtP_2	2.38	1.28
			$PtAs_2$	2.49	1.31
$CoAsS$	2.40	1.24	$PtSb_2$	2.67	1.31
	2.26	1.24			
			CoS_2	2.37	1.33
RuS_2	2.35	1.31	$CoSe_2$	2.45	1.31
$RuSe_2$	2.48	1.34			
$RuTe_2$	2.64	1.32	$NiAsS$	2.48	1.30
				2.34	1.30
$PdAs_2$	2.49	1.31			
$PdSb_2$	2.67	1.31	NiS_2	2.42	1.38
			$NiSe_2$	2.53	1.39
OsS_2	2.37	1.33			
$OsSe_2$	2.48	1.34	$AuSb_2$	2.76	1.40
$OsTe_2$	2.65	1.33			
			FeP_2	2.27	1.17
			$FeAs_2$	2.36	1.18
			$FeSb_2$	2.60	1.24

by the bonds. It is not known whether this extra electron is pushed into an outer orbital ($4d$) or whether a compromise is reached, the bonds relinquishing some $3d$ orbital to the electron. The effect of the extra electron is to produce an increase of 0.09 or 0.10 Å in the octahedral covalent radius for each of these atoms. The two extra electrons in Ni(II) produce a total increase of twice as much, 0.18 Å.

A deficiency of electrons, on the other hand, has, as might be expected, little effect on the radius. The value found for Fe(IV) is about 1.20 Å, only slightly less than that for Fe(II). The values[38] 2.36 Å and 2.51 Å for the Os—Cl distance in K_2OsCl_6 and the Os—Br distance in K_2OsBr_6, respectively, lead to the same value, 1.37 Å, for the octahedral

[38] J. D. McCullough, *Z. Krist.* **94**, 143 (1936).

radius of quadrivalent osmium; whereas the value 2.28 Å for the Os—Cl distance[39] in $K_2OsO_2Cl_4$ gives 1.29 Å for the octahedral radius of sexavalent osmium. These lie within 0.04 Å of the radius of bivalent osmium.

The octahedral radii of the table are applicable to complex ions such as $[PtCl_6]^{--}$. The radius sum Pt(IV)—Cl is 2.30 Å, and the several reported experimental values for salts of chloroplatinic acid range from 2.26 Å to 2.35 Å. The radii can also be applied to the sulfides, selenides, and tellurides of quadrivalent palladium and platinum (PdS_2, etc.), which crystallize with the cadmium iodide structure, consisting of layers of MX_6 octahedra so packed together that each X is common to three octahedral complexes. The average deviation between radius sums and reported distances for these substances is about 0.02 Å.

For ferrous iodide, on the other hand, the observed interatomic distance, 2.88 Å, is much greater than the radius sum 2.56 Å. This shows that the bonds in the octahedral complexes (the crystals have

TABLE 7-17.—ADDITIONAL OCTAHEDRAL RADII

| Ti(IV) | 1.36 Å | Sn(IV) | 1.45 Å | Se(IV) | 1.40 Å |
| Zr(IV) | 1.48 | Pb(IV) | 1.50 | Te(IV) | 1.52 |

the cadmium iodide structure) are not d^2sp^3 covalent bonds, and the observed paramagnetism of the substance ($\mu = 5.4$) supports the conclusion that the bonds are essentially ionic. The essentially ionic character of the bonds in all the halides of manganese, iron, cobalt, and nickel is similarly indicated by the magnetic data and the interatomic distances.

From the observed values of interatomic distances in complex ions such as $[SnCl_6]^{--}$, $[PbBr_6]^{--}$, and $[SeBr_6]^{--}$ and from crystals such as TiS_2 with the cadmium iodide structure the octahedral radii given in Table 7-17 have been obtained. These correspond not to d^2sp^3 bonds, involving d orbitals of the shell within the valence shell, but to sp^3d^2 orbitals, use being made of the unstable d orbitals of the valence shell itself.

For Sn(IV) and Pb(IV) these radii are greater than the corresponding tetrahedral sp^3 radii by the factor 1.03.

There is a pair of unshared electrons in Se(IV) (in the ion $[SeBr_6]^{--}$) occupying the 4s orbital, which is therefore not available for use in forming $4s4p^34d^2$ bonds. It has been suggested to me by Dr. J. Y. Beach that the role of the s orbital in bond formation is here being

[39] J. L. Hoard and J. D. Grenko, *Z. Krist.* **87**, 100 (1934).

played by the 5s orbital, the bonds being $4p^3 4d^2 5s$ bonds; the large value of the radius (23 percent greater than the tetrahedral radius of selenium) is to be expected from this point of view. A similar effect is shown by Te(IV).

In a molecule such as $As(CH_3)_3$ the unshared pair of electrons in the valence shell occupies the 4s orbital and in this way plays an important part in determining the configuration of the molecule; the As—C bonds are p bonds (with perhaps a small amount of s character), and make angles of 100° with one another, whereas in $B(CH_3)_3$, with the 2s orbital not occupied by an unshared pair, the sp^2 bonds are coplanar. An s unshared pair in a molecule of this sort is not an "inert pair"[40] so far as stereochemistry is concerned. On the other hand, the 4s pair of selenium in the $[SeBr_6]^{--}$ ion is really an inert pair, since the 5s orbital replaces the 4s orbital for purposes of bond formation, and the configuration (but not the size) of the complex is the same as it would be if the inert pair were not present. A striking case of this behavior is presented by the compounds $(NH_4)_2SbBr_6$ and Rb_2SbBr_6. The observed diamagnetism of the substances[41] shows that $[SbBr_6]^{--}$ ions, which would contain one unpaired electron, are not present moreover, the structure of the crystals is found on x-ray examination to be similar to that of potassium chloroplatinate. The substances accordingly contain the two octahedral complex ions $[SbBr_6]^-$ and $[SbBr_6]^{---}$, the former with $5s5p^3 5d^2$ bonds and the latter with an inert 5s pair of electrons and $5p^3 5d^2 6s$ bonds.

It is probable that the $[Bi(SCN)_6]^{---}$ ion is octahedral, showing a truly inert pair. The configurations of $SeCl_4$, $[AsCl_4]^-$, and similar molecules and complexes have not been determined; it will be interesting to see whether or not the outer unshared pair is stereochemically inert.

Other Covalent Radii.—Bipositive nickel, palladium, and platinum and tripositive gold form four coplanar dsp^2 bonds, directed to the corners of a square, with attached atoms. Examination of the observed values of interatomic distances reveals that *square dsp^2 radii of atoms have the same values as the corresponding octahedral d^2sp^3 radii*, as given in Table 7-15. This is shown by the comparisons on the following page.

Bipositive copper often forms four strong bonds directed toward the corners of a square. The observed interatomic distances correspond to the radius 1.28 Å, about 0.08 Å larger than for square-ligated nickel (II); the increase is to be attributed to the presence of the extra elec-

[40] N. V. Sidgwick, *Ann. Repts. Chem. Soc.* 30, 120 (1933).
[41] N. Elliott, *J. Chem. Phys.* 2, 298 (1934).

Substance	Observed distance	Radius sum
PdO[42]	2.00 Å	1.98 Å
PdS[43]	2.26, 2.29, 2.34, 2.43	2.36
PdCl$_2$[44]	2.31	2.31
K$_2$PdCl$_4$	2.29	2.31
(NH$_4$)$_2$PdCl$_4$	2.35	2.31
PtS	2.32	2.36
K$_2$PtCl$_4$	2.31	2.31
Nickel dimethylglyoxime[45]	1.87, 1.90	1.91

tron, which probably occupies an orbital with some $3d$ character. Examples of observed distances are Cu—Cl = 2.275 Å and Cu—O = 1.925 Å in CuCl$_2\cdot$2H$_2$O[46] and Cu—O = 1.95 Å in bis-acetylacetone copper.[47]

In molybdenite and tungstenite the metal atom is surrounded by six sulfur atoms at the corners of a right trigonal prism with axial ratio unity (Fig. 5-14).[48] From the observed interatomic distances the trigonal-prism radius values of 1.37 Å for Mo(IV) and 1.44 Å for W(IV) are obtained.

The average Mo—C bond distance in K$_4$Mo(CN)$_8\cdot$2H$_2$O is 2.15 Å,[49] which corresponds to the value 1.38 Å for the octacovalent radius of Mo(IV). The close approximation of this value to the trigonal-prism radius indicates that the bond orbitals are nearly the same for the two types of coordination.

In Cu$_2$O and Ag$_2$O crystals (Fig. 7-9) each oxygen atom is surrounded tetrahedrally by four metal atoms, each of which is midway between two oxygen atoms and with which it probably forms two covalent bonds of the sp type.[50] The interatomic distances in these crystals lead to the radius values 1.18 Å for Cu(I) and 1.39 Å for Ag(I), which are 0.17 Å and 0.13 Å less than the corresponding tetrahedral radii. By microwave spectroscopy[51] of the linear molecules H$_3$CHgCl and H$_3$CHgBr the bond lengths C—Hg = 2.061 Å, Hg—Cl = 2.282 Å, C—Hg = 2.074 Å, and Hg—Br = 2.406 Å. These correspond to the values 1.29, 1.29, 1.30, and 1.27 Å for the radius of bicovalent Hg(II),

[42] L. Pauling and M. L. Huggins, *loc. cit.* (6).

[43] T. F. Gaskell, *Z. Krist.* **96**, 203 (1937); F. A. Bannister, *ibid.* 201.

[44] A. F. Wells, *Z. Krist.* **100**, 189 (1938).

[45] L. E. Godycki and R. E. Rundle, *Acta Cryst.* **6**, 487 (1953).

[46] S. W. Peterson and H. A. Levy, *J. Chem. Phys.* **26**, 220 (1957).

[47] H. Koyama, Y. Saito, and H. Kuroya, *J. Inst. Polytech. Osaka City Univ.* **4**, 43 (1953); E. A. Shugam, *Doklady Akad. Nauk S.S.S.R.* 1951, 853; E. G. Cox and K. C. Webster, *J. Chem. Soc.* 1935, 731.

[48] R. G. Dickinson and L. Pauling, *J.A.C.S.* **45**, 1466 (1923).

[49] J. L. Hoard and H. H. Nordsieck, *J.A.C.S.* **61**, 2853 (1939).

[50] The two bond orbitals $\frac{1}{2}(s + p_x)$ and $\frac{1}{2}(s - p_x)$ make complete use of the s orbital. They have opposed bond directions and strength 1.95.

[51] W. Gordy and J. Sheridan, *J. Chem. Phys.* **22**, 92 (1954).

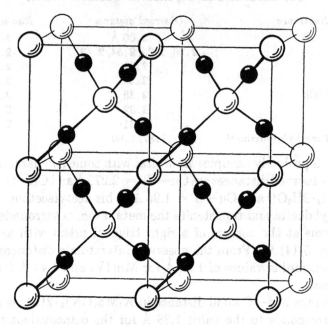

FIG. 7-9.—The arrangement of copper atoms (small circles) and oxygen atoms (large circles) in the cubic crystal cuprite, Cu_2O. Each oxygen atom is surrounded by a tetrahedron of four copper atoms. Each copper atom has two oxygen neighbors, in a linear arrangement with it. Note that the bonds connect the atoms into infinite frameworks, which are interpenetrating but not connected.

with average 1.29 Å, which is 0.19 Å less than its tetrahedral radius.

A discussion of interatomic distances in diatomic hydrides is given in Section 7-11.

The Anomalous Manganese Radius.—At the time of the formulation of the tables of covalent radii it was pointed out[52] that the manganese-sulfur distance in hauerite, MnS_2, with the pyrite structure, is surprisingly large. Its magnitude has since been verified[53] and an x-ray study leading to similarly large interatomic distances has been made of manganese diselenide and manganese ditelluride,[54] which also have the pyrite structure. The values found, 2.59 Å for Mn—S, 2.70 Å for Mn—Se, and 2.90 Å for Mn—Te, correspond to 1.55, 1.56, and 1.58 Å for the octahedral radius of bivalent manganese, whereas by extrapolation of the values in Table 7-15 1.24 Å is obtained for the d^2sp^3 radius of

[52] L. Pauling and M. L. Huggins, *loc. cit.* (6).
[53] F. Offner, *Z. Krist.* **89**, 182 (1934).
[54] N. Elliott, *J.A.C.S.* **59**, 1958 (1937).

univalent manganese, which should not be different from that of bivalent manganese by more than 0.01 or 0.02 Å.

The solution to this difficulty is found by application of the magnetic criterion for bond type. It was shown by Elliott that the available magnetic data for hauerite, interpreted by the Weiss-Curie equation, lead to the value 6.1 Bohr magnetons for the magnetic moment of manganese. This is far from the predicted value for d^2sp^3 bonds, 1.73, but close to that for ionic bonds, 5.92, and it shows that the electronic structure of this crystal, and presumably also of the diselenide and the ditelluride, is entirely different from that of pyrite. The bonds from manganese to the surrounding atoms need not be of the extreme ionic type; resonance with covalent bonds of the $4s4p^34d^2$ type, which have the same magnetic properties, could occur.

The selenium-selenium and tellurium-tellurium distances observed in the manganese compounds provide further evidence for this structural interpretation. They correspond to radii that agree more satisfactorily with the normal-valence radii than with the tetrahedral radii of the nonmetallic atoms, indicating that the atoms are not forming tetrahedral covalent bonds:

	Distance $\dfrac{X-X}{2}$	Tetrahedral radius	Normal covalent radius
Sulfur	1.04 Å	1.04 Å	1.04 Å
Selenium	1.19	1.14	1.17
Tellurium	1.37	1.32	1.37

7-10. INTERATOMIC DISTANCES FOR FRACTIONAL BONDS

Some years ago, in the course of a discussion of interatomic distances in metals, an equation for interatomic distances of fractional bonds, bonds with bond number less than 1, was proposed.[55] This equation is

$$D(n) = D(1) - 0.60 \log n \qquad (7\text{-}7)$$

Here $D(n)$ is the bond length for bond number n (less than 1) and $D(1)$ is the bond length for a single bond of similar type (using similar bond orbitals).

The equation was formulated in the following way. In Section 7-6 it was pointed out that the dependence of bond length on bond order n' over the range $n' = 1, 2, 3$ is $D(n') = D(1) - 0.71 \log n'$. This equation might be used also for values of n' less than 1; for example, for a bond with bond order $\frac{1}{2}$ it makes the bond length 0.21 Å greater than it is for a single bond.

The relation between n' and n, as discussed in Section 7-6, is such that a bond with a given bond number n is expected to be shorter than

[55] L. Pauling, *J.A.C.S.* **69**, 542 (1947).

a bond with the same value of n', because of the stabilizing effect of resonance energy (except for $n' = 1, 2,$ and 3, where n and n' coincide). This shortening is achieved by use of Equation 7-7, which has the co-efficient 0.60 in place of 0.71 in Equation 7-5.

Equation 7-7 receives support from its effective use in the formulation of a set of values of metallic radii from observed interatomic distances in metallic elements and the discussion with these radii of the structure of intermetallic compounds (Chap. 11).

As an example of the use of Equation 7-7, we may select the crystal $CuCl_2 \cdot 2H_2O$, already mentioned in Chapter 5 and in Section 7-9. In this crystal the copper atom forms four strong bonds directed to the corners of a square, two to chlorine atoms, at 2.275 Å, and two to oxygen atoms (of water molecules), at 1.925 Å. These bonds can be accepted as single bonds, with the bond orbitals of copper approximating dsp^2 in character, but probably with somewhat less than 25 percent d character because of competition with the odd electron. In addition, the copper atom has two other chlorine atoms nearby, completing a roughly regular coordination octahedron. These atoms are at the distance 2.95 Å, that is, 0.67 Å greater than the single-bond distance. The corresponding bond number given by Equation 7-7 is 0.07. A similar calculation made with use of the tetrahedral radius 1.35 Å for copper gives 0.10 for the bond number. We conclude that these two bonds are very weak and that the copper atom in this crystal (and also in others, such as $CuCl_2$ and $K_2CuCl_4 \cdot 2H_2O$, in which its environment is similar) is properly described as square quadricovalent.

7-11. VALUES OF SINGLE-BOND METALLIC RADII

In Chapter 11 there will be given a discussion of the observed interatomic distances in crystals of the metallic elements and an account of the derivation from them of a set of values of single-bond metallic radii. These values are shown in Table 7-18. They refer to single covalent bonds for which the bond orbitals have the same hybrid character as in the metals themselves, as discussed in Chapter 11. The relation be-

TABLE 7-18.—SINGLE-BOND METALLIC RADII

Li	Be	B										
1.225	0.889	0.81										
Na	Mg	Al	Si									
1.572	1.364	1.248	1.173									
K	Ca	Sc	Ti	V	Cr	Mn	Fe	Co	Ni	Cu	Zn	Ga	Ge
2.025	1.736	1.439	1.324	1.224	1.176	1.171	1.165	1.162	1.154	1.173	1.249	1.245	1.223
Rb	Sr	Y	Zr	Nb	Mo	Tc	Ru	Rh	Pd	Ag	Cd	In	Sn
2.16	1.914	1.616	1.454	1.342	1.296	1.271	1.246	1.252	1.283	1.339	1.413	1.497	1.399
Cs	Ba	La	Hf	Ta	W	Re	Os	Ir	Pt	Au	Hg	Tl	Pb
2.35	1.981	1.690	1.442	1.343	1.304	1.283	1.260	1.265	1.295	1.336	1.440	1.549	1.538

TABLE 7-19.—EFFECTIVE RADII OF METAL ATOMS IN
DIATOMIC HYDRIDE MOLECULES MH

	Li	Be							
D_e—0.300	1.295	1.043 Å							
Radius	1.225	0.889							

	Na	Mg	Al
	1.587	1.431	1.347
	1.572	1.364	1.248

	K	Ca		Mn	Fe	Co	Ni	Cu	Zn
	1.944	1.702		1.431	1.176	1.243	1.175	1.163	1.295
	2.025	1.736		1.171	1.165	1.157	1.154	1.173	1.249

	Rb	Sr					Ag	Cd	In	Sn
	2.067	1.846					1.317	1.462	1.538	1.485
	2.16	1.914					(1.343)	1.413	1.497	1.399

	Cs	Ba					Au	Hg	Tl	Pb
	2.194	1.932					1.224	1.440	1.570	1.539
	2.33	1.981					1.336	1.440	1.549	1.538

tween these radii and other radii (tetrahedral, octahedral) may be seen by a comparison of their values.

Use will be made of the single-bond metallic radii in some of the following chapters.

Spectroscopic values of interatomic distances are available for many diatomic hydrides. The hydrogen atom cannot form multiple bonds, and accordingly the difference between the interatomic distance for a diatomic hydride and the covalent radius of hydrogen (which we assume to be 0.300 Å) can be taken to be the effective covalent radius of the other atom in the hydride molecule. Values of this difference for the normal states of MH(g) are given in Table 7-19, together with values of the single-bond metallic radius (Table 7-18). It is seen that for many elements the difference between D_e − 0.300 Å and $R(1)$ is small —not over ± 0.020 Å for half of them. The larger differences presumably correspond to changes in the nature of the hybrid bond orbitals, as discussed in Chapter 11.

It is interesting that the difference between the effective radius of manganese in MnH and the single-bond metallic radius, 0.260 Å, is nearly as great as that between the anomalous octahedral manganese radius and the d^2sp^3 octahedral radius (Sec. 7-9). It is likely that in MnH the $3d$ orbitals of the manganese atom are all occupied by electrons and that the bond orbital has very little d character.

7-12. VAN DER WAALS AND NONBONDED RADII OF ATOMS

In a molecule of chlorine, with the electronic structure $: \overset{..}{\underset{..}{Cl}} : \overset{..}{\underset{..}{Cl}} :$, the covalent radius of chlorine may be described as representing roughly the distance from the chlorine nucleus to the pair of electrons that is shared with the other chlorine atom.   In a crystal of the substance the molecules are attracted together by their van der Waals interactions and assume equilibrium positions at which the attractive forces[56] are balanced by the characteristic repulsive forces between atoms, resulting from interpenetration of their electron shells.   Let us call one-half of the equilibrium internuclear distance between two chlorine atoms

FIG. 7-10.—Layers of octahedra corresponding to the composition MX$_2$. There is an atom M in the center of each octahedron, and an atom X at each octahedron corner, shared by three octahedra. The layer is to be considered as extending to infinity.

in such van der Waals contact, corresponding to the relative positions for two molecules, the *van der Waals radius* of chlorine.

The van der Waals radius is expected to be larger than the covalent radius, since it involves the interposition of two electron pairs between the atoms rather than one.   Moreover, the van der Waals radius of chlorine should be about equal to its ionic radius, inasmuch as the bonded atom presents the same face to the outside world in directions away from its bond as the ion, :Cl:⁻, does in all directions.

[56] These are mainly the London dispersion forces, the nature of which we shall not discuss. See F. London, *Z. Physik* **63**, 245 (1930); Also *Introduction to Quantum Mechanics*.

The ionic radius of chlorine has the value 1.81 Å (Chap. 13). The following distances between chlorine atoms of different molecules have been observed in the molecular crystal 1,2,3,4,5,6-hexachlorocyclohexane:[57] 3.60, 3.77, 3.82 Å; these are close to twice the ionic radius. Similar agreement is shown by many other organic crystals and inorganic covalent crystals. Cadmium chloride, for example, consists of

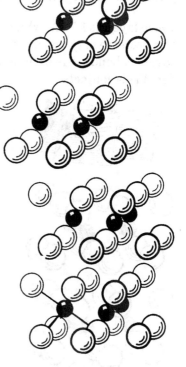

Fig. 7-11.—The arrangement of cadmium atoms (small circles) and chlorine atoms (large circles) in the rhombohedral crystal cadmium chloride, $CdCl_2$. The atoms are arranged in octahedral layers, as represented in Figure 7-10. These layers are superimposed in the manner indicated in the figure. Repetition in the direction of the trigonal axis (vertical) occurs after three layers.

layers of $CdCl_6$ octahedra condensed together by sharing each chlorine atom among three (Figs. 7-10 and 7-11). These layers are superimposed, with only the weak van der Waals forces holding them together. (The crystals show pronounced basal cleavage, resulting from this layer structure.) The distance between chlorine atoms of different layers is 3.76 Å, which is only slightly larger than the ionic value, 3.62 Å. In cadmium iodide, which has a similar layer structure (Fig. 7-12), the distance between iodine atoms of different layers is 4.20 Å, slightly smaller than the ionic value, 4.32 Å.

Other nonmetallic elements also are found to have van der Waals radii approximately equal to their ionic radii. For sulfur, for example, in the layer crystal molybdenite, the van der Waals radius effective be-

[57] R. G. Dickinson and C. Bilicke, *J.A.C.S.* **50**, 764 (1928).

TABLE 7-20.—VAN DER WAALS RADII OF ATOMS

		H	1.2 Å		
N	1.5 Å	O	1.40 Å	F	1.35 Å
P	1.9	S	1.85	Cl	1.80
As	2.0	Se	2.00	Br	1.95
Sb	2.2	Te	2.20	I	2.15

Radius of methyl group, CH_3, 2.0 Å.
Half-thickness of aromatic molecule, 1.70 Å.

tween layers is 1.75 Å, which is slightly less than the ionic radius, 1.85 Å; this decrease may be due to the fact that the sulfur atom, forming three covalent bonds, has only one unshared pair left to take care of its van der Waals contacts.

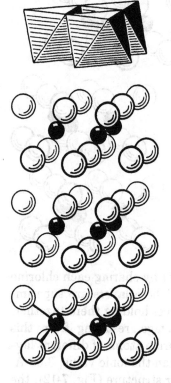

FIG. 7-12.—The arrangement of cadmium atoms (small circles) and iodine atoms (large circles) in the hexagonal crystal cadmium iodide. The atoms are arranged in octahedral layers, as represented in Figure 7-10. The sequence of layers is different from that for cadmium chloride; each layer is directly above the layer below it.

The value of the effective van der Waals radius of an atom in a crystal depends on the strength of the attractive forces holding the molecules together, and also on the orientation of the contact relative to the covalent bond or bonds formed by the atom (as discussed below); it is accordingly much more variable than the corresponding covalent radius. In Table 7-20 there are given the ionic radii of nonmetallic elements for use as van der Waals radii. They have been rounded off to the nearest 0.05 Å, and are to be considered as reliable only to 0.05 or 0.10 Å.[58] For the elements of the nitrogen group values about 0.2 Å less than the ionic radii are included, as indicated by the few available experimental data.

The methyl group as a whole can be assigned the radius 2.0 Å. In metaldehyde,[59] for example, each methyl group is surrounded by eight methyl groups of other molecules, two at 3.90 Å, four at 4.03 Å, and two at 4.11 Å, and in hexamethylbenzene[60] the methyl-methyl dis-

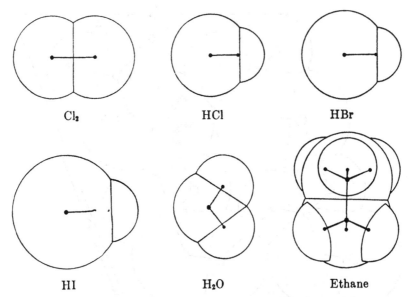

FIG. 7-13.—Drawings of representative molecules, with atoms shown as spheres with radii equal to their van der Waals radii.

tances between molecules lie between 4.0 and 4.1 Å. The methylene group, CH_2, may be assigned the same van der Waals radius as the methyl group, 2.0 Å. This value is substantiated by the following distances representing intermolecular contacts: $CH_2—CH_2 = 3.96$, $CH_2—O = 3.32, 3.33$, $CH_2—N = 3.55, 3.69$ Å in diketopiperazine;[61] and $CH_2—CH_2 = 4.05$, $CH_2—O = 3.38, 3.52$ Å in glycine;[62] and similar values in many other crystals.

[58] Interatomic distances between atoms in different molecules have been discussed briefly by several authors, including S. B. Hendricks, *Chem. Revs.* **7**, 431 (1930); M. L. Huggins, *ibid.* **10**, 427 (1932); and N. V. Sidgwick, *Ann. Repts. Chem. Soc.* **29**, 64 (1933).

[59] L. Pauling and D. C. Carpenter, *J.A.C.S.* **58**, 1274 (1936).

[60] K. Lonsdale, *Proc. Roy. Soc. London* A123, 494 (1929); L. O. Brockway and J. M. Robertson, *J. Chem. Soc.* **1939**, 1324.

[61] R. B. Corey, *J.A.C.S.* **60**, 1598 (1938).

[62] G. Albrecht and R. B. Corey, *J.A.C.S.* **61**, 1087 (1939); Marsh, *loc. cit.* (27).

It has been emphasized by Mack[63] that the dimensions of the methyl group and of other hydrocarbon groups can be accounted for by assigning to the hydrogen atom a van der Waals radius of about 1.29 Å. The effective radius in metaldehyde, diketopiperazine, and glycine varies between 1.06 and 1.34 Å. In Table 7-20 an average value, 1.2 Å, is given.

Drawings showing the effect of van der Waals radii in determining the shapes of molecules are given in Figures 7-13, 7-14, 7-15, and 7-16.

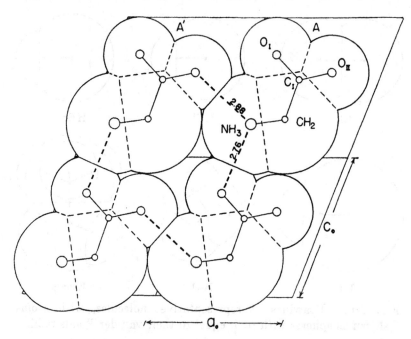

Fig. 7-14.—The arrangement of molecules in a layer of the crystal glycine. The packing of the molecules is determined by the van der Waals radii of the groups and by the N—H···O hydrogen bonds (Chap. 12).

The distances between saturated hydrocarbon molecules in crystals can be calculated by the use of these radii, with consideration also of the possibility of molecular or group rotation. Another factor must be introduced for aromatic molecules.[64] The double bonds in these molecules project above and below the plane of the ring in such a way as to give to the ring an effective thickness of about 3.4 Å, as observed in anthracene, durene, hexamethylbenzene, benzbisanthrene, and many other aromatic hydrocarbons. The same value is also found between the layers of graphite.

[63] E. Mack, Jr., *J.A.C.S.* **54**, 2141 (1932).

[64] Mack, *loc. cit.* (63); *J. Phys. Chem.* **41**, 221 (1937).

It is interesting to note that the van der Waals radii given in Table 7-20 are 0.75 to 0.83 Å greater than the corresponding single-bond co-valent radii; to within their limit of reliability they could be taken as equal to the covalent radius plus 0.80 Å.

The effective radius of an atom in a direction that makes only a small

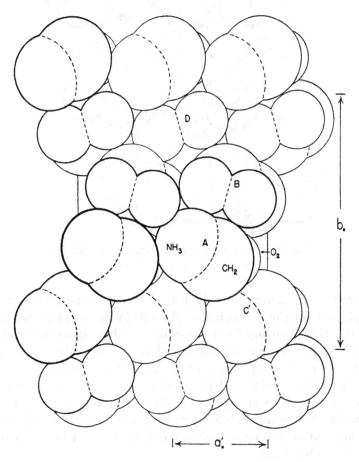

Fig. 7-15.—Another packing drawing of the glycine crystal (Albrecht and Corey).

angle with the direction of a covalent bond formed by the atom is smaller than the van der Waals radius in directions away from the bond. This might well be expected from the fact that the electron pair which would give the chloride ion, $:\!\overset{..}{\underset{..}{Cl}}\!:^-$, for example, its size in the direction toward the left is pulled in to form the bond in methyl chloride, $H_3C\!:\!\overset{..}{\underset{..}{Cl}}\!:$.

In consequence of this, atoms that are bonded to the same atom can approach one another much more closely than the sum of the van der

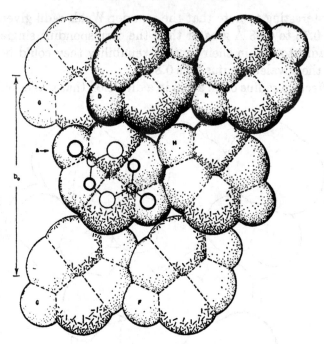

FIG 7-16.—A packing drawing of the diketopiperazine crystal (Corey).

Waals radii.   In carbon tetrachloride the chlorine atoms are only 2.87
Å apart, and yet the properties of the substance indicate that there is
no great strain resulting from the repulsion that should correspond to
the van der Waals diameter 3.6 Å.   Even in methylene chloride and
chloroform, where the strain might be relieved by increasing the bond
angle, the chlorine-chlorine distance is only 2.92 Å.   We conclude that
the nonbonded radius of an atom in directions close to the bond direction (within 35°) is about 0.5 Å less than the van der Waals radius; a
unicovalent atom can be considered as a sphere that is whittled down
on the side of the bond.

CHAPTER 8

Types of Resonance in Molecules

WITH the background of information given in the preceding chapters about the nature of the phenomenon of the resonance of a molecule among several valence-bond structures and its relation to such properties as the energy of the molecule and its interatomic distances, we are now ready to begin the discussion of the structure of molecules to which a single valence-bond formula cannot be assigned. Some of these resonating molecules have already been mentioned as examples; in the selection of others for treatment the effort has been made to illustrate all of the principal types of resonance and to present substantiating evidence in each case. The discussion is not intended to be exhaustive; indeed, it could not be, since once that the nature of the resonance phenomenon has been recognized it is seen that it is of significance in every branch of chemistry and for nearly every class of substances.

The discussions that may be given of resonance in various molecules may seem to the reader to be so far from quantitative in nature as to be without value. It is true that the picture presented of the structure of a resonating molecule is often rather indefinite; but in the years that have gone by since the quantum-mechanical resonance phenomenon was first applied to problems of molecular structure encouraging progress has been made in the formulation of a semiquantitative system, with the aid of both experimental and theoretical methods, and we can hope for further progress in the future.

8-1. THE STRUCTURE OF SIMPLE RESONATING MOLECULES

Carbon Monoxide and Carbon Monosulfide.—It was stated in Chapter 6 that carbon monoxide resonates among the four structures $:C:\overset{..}{\underset{}{O}}:^-$,

$:C::\overset{..}{O}:$, $:C::\overset{-}{O}:$, and $:C:::O:^+$, with the resonance energy and elec-

265

tric dipole moment providing evidence. The value of the dipole moment[1] is 0.112 D, and its direction[2] corresponds to a positive charge on the oxygen atom.

In the discussion of the observed value[3] 1.130 Å of the internuclear distance in the molecule we must make use of the normal covalent structures $:\overset{+}{C}-\overset{..}{\underset{..}{O}}:^-$, $:C=\overset{..}{O}:$, $:C=\overset{..}{\underset{..}{O}}:$, and $:\overset{-}{C}\equiv\overset{+}{O}:$, rather than the extreme covalent structures $:C:\overset{..}{\underset{..}{O}}:$, and so forth, because our empirical system of covalent radii is based upon the normal covalent structures, with the normal amount of ionic character. The expected bond lengths are 1.43 Å for the first structure, 1.21 Å for the second and third (Table 7-5), and 1.07 Å for the fourth.[4] From Table 7-12 we see that the observed bond length corresponds to bond number 2.4, which may be described as representing 10 percent $:\overset{+}{C}-\overset{..}{O}:^-$, 20 percent each $:C=\overset{..}{\underset{..}{O}}:$ and $:C=\overset{..}{O}:$, and 50 per cent $:\overset{-}{C}\equiv\overset{+}{O}:$, as calculated for independent resonance of the xy and xz bonds.

This resonating structure of the molecule would lead to the observed value of the electric dipole moment if each bond (including the bent bonds) had 16 percent partial ionic character. This amount of ionic character is somewhat less than the amount given in Table 3-10 as that applying to a carbon-oxygen bond (22 percent). We conclude that the structure derived from the observed bond length is compatible with the observed electric dipole moment, to within the uncertainty of the calculation.

The bond length in CS has been determined by microwave spectroscopy[5] to be 1.535 Å, which corresponds to bond number 2.4. Another value for the bond number can be derived from the observed value[5] 1.97 D of the electric dipole moment. Carbon and sulfur have the same electronegativity, and accordingly the dipole moment arises entirely from the difference in the contributions of the two structures

---

[1] C. A. Burrus, *J. Chem. Phys.* **28**, 427 (1958).

[2] B. Rosenblum, A. H. Nethercot, Jr., and C. H. Townes, *Phys. Rev.* **109**, 400 (1958).

[3] L. Gerö, G. Herzberg, and R. Schmid, *Phys. Rev.* **52**, 467 (1937).

[4] In earlier editions of this book it was suggested that a correction in radius should be made for atoms with incomplete outer shells. The correction was based on comparison of the C—B bond length in boron trimethyl with the value corresponding to an assumed boron radius (0.88 Å) that is now seen to be too large (correct value about 0.81 Å). Other comparisons (such as 0.971 Å in OH with 0.965 Å in H_2O and 1.037 Å in NH with 1.014 Å in NH_3) indicate that no large correction is needed.

[5] R. C. Mockler and G. R. Bird, *Phys. Rev.* **98**, 1837 (1955).

:C≡S: and :C—S:⁻ to the normal state of the molecule. By dividing the value 1.97 D by the internuclear distance and the charge of the electron we obtain 27 percent for this difference, and 2.27 as the value of the bond number—probably somewhat more reliable than that given by the bond length. We can predict that the direction of the moment will be found to be that from sulfur (positive) to carbon (negative). The bond number 2.27 corresponds with independent resonance of the xy and xz bonds to 13 percent :C—S:⁻, 23.5 percent :C=S:, 23.5 percent :C=S:, and 40 percent :C≡S:⁺.

The measured interaction of the electric quadrupole moment of the sulfur atom and the electric quadrupole field set up by the electrons has also been interpreted[5] as showing that the structure :C≡S: makes a 40-percent contribution.

Carbon Dioxide and Related Molecules.—It is not surprising that so unconventional a molecule as carbon monoxide should have a resonating structure; but recognition of the fact that the carbon dioxide molecule, for which the valence-bond formula O=C=O has been written ever since the development of valence theory, is not well represented by this structure alone and that other valence-bond structures also make important contributions must have come as a surprise to everyone.

The carbon-oxygen distance in this molecule is known[6] to be 1.159 Å. If one structure O=C=O alone represented the molecule the distance should be 1.18 Å, the double-bond length with the adjacent-bent-bond correction. There are, however, two such structures, :O=C=O: and :O=C=O:, differing in the planes of the double bonds. In addition, resonance might occur with two other structures, :O≡C—O: and :O—C≡O:. If there were independent resonance of the xy and xz bonds the bond lengths would be 1.16 Å, as given with use of the correction in Table 7-12 for $n = 2.00$. This value agrees with the observed value, and we conclude that the normal state of the carbon dioxide molecule can be described as composed to the extent of about 25 percent each of the four structures :O=C=O:, :O=C=O:, :O≡C—O:, and :O—C≡O:.

* E. K. Plyler and E. F. Barker, *Phys. Rev.* **38**, 1827 (1931); D. M. Dennison. *Rev. Modern Phys.* **12**, 175 (1940).

The observed resonance energy, relative to the ketonic type of double bond, is 33 kcal/mole.

Resonance of the same type as in carbon dioxide would be expected in carbon oxysulfide and carbon disulfide. The observed interatomic distances,[7] C—O = 1.164 Å and C—S = 1.559 Å in SCO and C—S = 1.553 Å in CS_2, agree with the expected values C—O = 1.16 Å and C—S = 1.56 Å.

The bond lengths in carbon oxyselenide[8] are C—O = 1.159 Å and C—Se = 1.709 Å, and in TeCS are[9] Te—C = 1.904 Å and C—S = 1.557 Å. These values are close to the expected ones.

The carbon suboxide molecule is linear,[10] as expected for the two double-bond structures *A* and *B* and also for structures *C* and *D*:

$$A \quad :\ddot{O}\!=\!C\!=\!C\!=\!C\!=\!\ddot{O}:$$

$$B \quad :O\!=\!C\!=\!C\!=\!C\!=\!\ddot{O}:$$

$$C \quad :\overset{+}{O}\!\equiv\!C\!-\!C\!\equiv\!C\!-\!\overset{-}{\ddot{O}}:$$

$$D \quad :\overset{-}{\ddot{O}}\!-\!C\!\equiv\!C\!-\!C\!\equiv\!\overset{+}{O}:$$

These four structures contain the same number of covalent bonds, but there is a separation of formal charge in *C* and *D* and none in *A* and *B*, so that *C* and *D* would be expected to make a somewhat smaller contribution to the normal state of the molecule than *A* and *B*. The values C—O = 1.18 Å and C—C = 1.274 Å are expected for resonance between *A* and *B*, and C—O = 1.16 Å and C—C = 1.254 Å for equal resonance among the four structures. The observed values[11] C—O = 1.160 Å and C—C = 1.279 Å suggest that there is resonance

[7] M. W. P. Strandberg, T. Wentink, Jr., and R. L. Kyhl, *Phys. Rev.* 75, 270 (1949); H. C. Allen, Jr., E. K. Plyler, and L. R. Blaine, *J.A.C.S.* 78, 4843 (1956).

[8] M. W. P. Strandberg, T. Wentink, Jr., and A. G. Hill, *Phys. Rev.* 75, 827 (1949).

[9] W. A. Hardy and G. Silvey, *Phys. Rev.* 95, 385 (1954).

[10] L. O. Brockway and L. Pauling, *Proc. Nat. Acad. Sci. U. S.* 19, 860 (1933); H. Mackle and L. E. Sutton, *Trans. Faraday Soc.* 47, 937 (1951); R. L. Livingston and C. N. Rao, to be published; O. Bastiansen, to be published. It has been reported on the basis of analysis of infrared and Raman spectra that the molecule is bent, with an angle of 158° at each of the end carbon atoms (H. D. Rix, *J. Chem. Phys.* 22, 429 [1954]). This is probably incorrect.

[11] Brockway and Pauling, Mackle and Sutton, Livingston and Rao, and Bastiansen, *loc. cit.* (10).

between *A* and *B* with some contribution of the four structures of type

$:\overset{+}{O}\!\!\equiv\!\!C\!\!-\!\!\overset{-}{\underset{\cdot\cdot}{C}}\!\!=\!\!C\!\!=\!\!\overset{\cdot\cdot}{O}:$, but the covalent radii and their corrections are

not reliable enough to permit a more precise conclusion to be drawn. The frequencies of vibration of the molecule are compatible with this structure.[12]

The Cyanides and Isocyanides.—For methyl cyanide (and other

alkyl cyanides) the structure

$$H\!\!-\!\!\overset{\displaystyle \overset{H}{\diagdown}}{\underset{\displaystyle \diagup}{\underset{H}{C}}}\!\!-\!\!C\!\!\equiv\!\!N: \text{ may be accepted as a}$$

first approximation. There is evidence that a significant contribution, totaling about 20 percent, is made also by the several structures

$$H\!\!-\!\!\overset{\displaystyle \overset{H^+}{}}{\underset{\displaystyle \diagup}{\underset{H}{C}}}\!\!=\!\!C\!\!=\!\!\overset{\cdot\cdot}{\overset{-}{N}}:. \text{ The differences in electronegativity of the atoms indi-}$$

cate that each of the H—C bonds should have about 4 percent ionic character, with hydrogen positive, and each of the three bent bonds of the C≡N triple bond about 7 percent, with nitrogen negative. If these ionic aspects of the bonds were synchronized, the electron pair on the methyl carbon atom and the freed orbital of the cyanide carbon atom could form the second half of a carbon-carbon double bond, as represented in the structure given above.. The observed electric dipole moment of methyl cyanide is 3.44 D, far larger than the sum of the moments of the methyl group and the cyanide group, about 1.5 D. If the observed moment is attributed to the conjugated structures $H_3^+CCN^-$, their contribution is calculated to total 24 percent. Also, the C—C distance in the molecule[13] is 1.459 Å, which corresponds (Table 7-9) to about 17 percent double-bond character, and hence to a 17-percent contribution of these conjugated structures, in approximate agreement with the less reliable dipole-moment value.

The same value, 1.460 Å, is found also in malononitrile,[14] $CH_2(CN)_2$, indicating that the conjugation[15] is determined essentially by the cya-

[12] H. W. Thompson and J. W. Linnett, *J. Chem. Soc.* **1937**, 1376.

[13] M. Kessler, H. Ring, R. Trambarulo, and W. Gordy, *Phys. Rev.* **79**, 54 (1950); L. F. Thomas, E. I. Sherrard, and J. Sheridan, *Trans. Faraday Soc.* **51**, 619 (1955); M. D. Danford and R. L. Livingston, *J.A.C.S.* **77**, 2944 (1955).

[14] N. Muller and D. E. Pritchard, *J.A.C.S.* **80**, 3483 (1958).

[15] This sort of conjugation, which has been given the name hyperconjugation, is discussed in a later section of this chapter.

nide group.   The value 1.475 Å has been reported for trifluoromethyl cyanide.[16]

For the carbon-nitrogen distance the value 1.157 Å in methyl cyanide has been reported, and it is essentially the same in other molecules. This value is that expected for a triple bond with a small amount of double-bond character.

The interesting question as to the relative importance of the two structures R—$\overset{..}{N}$=C: and R—$\overset{+}{N}$≡C:⁻ for the alkyl isocyanides can be answered.   The atoms C—N—C in methyl isocyanide are observed[17] to have a linear configuration, which is the stable one for the second structure and not for the first one, and accordingly we may assume that the second structure is the more important of the two.   A rough quantitative conclusion can be reached by consideration of the observed value[17] 1.167 Å for the N—C distance.   The expected values for the two structures are 1.262 Å and 1.150 Å, respectively, and the observed value is found, with use of the method discussed in the derivation of Equation 7-4 and with $k_3/k_2 = 2$, to correspond to 74 percent contribution of H_3C—$\overset{+}{N}$≡$\overset{-}{C}$: and 26 percent of H_3C—$\overset{..}{N}$=C:.

The observed length of the bond between the methyl carbon atom and the nitrogen atom[17] is 1.427 Å, nearly equal to the expected value for a pure single bond, 1.432 Å (the adjacent-bent-bond correction − 0.040 Å has been made).   From this small difference we conclude that there is only a small contribution, less than 3 percent, of structures

such as H—$\overset{H^+}{\underset{\diagup}{C}}$=$\overset{+}{N}$=$\overset{--}{\overset{..}{C}}$:⁻ ; these structures are ruled out by the adjacent-

H

charge rule, which is discussed in the following section.

8-2. THE ADJACENT-CHARGE RULE AND THE ELECTRONEUTRALITY RULE

In Section 6-1 nitrous oxide was considered to resonate among three structures *A*, *B*, and *C*, which are so similar in nature as to contribute about equally to the normal state of the molecule.   There is, however, a fourth structure, *D*, that must be discussed:

$$A \qquad \overset{-..}{:N}=\overset{+}{N}=\underset{..}{O}:$$

$$B \qquad \overset{-}{:N}=\overset{+}{N}=\overset{..}{O}:$$

[16] Danford and Livingston, *loc. cit.* (13).

[17] Kessler *et al.. loc. cit.* (13); L. O. Brockway, *J.A.C.S.* **58**, 2516 (1936)

$$C \quad :N\equiv\overset{+}{N}-\overset{..}{\underset{..}{O}}:^{-}$$

$$D \quad \overset{-}{\underset{..}{:}}\overset{-}{N}-\overset{+}{N}\equiv O:^{+}$$

It is analogous to the fourth structure for carbon dioxide, which has the same number of electrons as nitrous oxide, and might be of importance for the latter molecule also. Resonance among the four structures, contributing equally, leads to the values $N—N = 1.15$ Å and $N—O = 1.11$ Å; that is, to an $N—O$ distance smaller than the $N—N$ distance, which is contrary to observation:[18] $N—N = 1.126$ Å, $N—O = 1.186$ Å. The observed values are those expected for resonance among the first three structures.

We can attribute the lack of importance of structure D to the instability resulting from the charge distribution, which gives adjacent atoms electric charges of the same sign. The *adjacent-charge rule*,[19] which states that structures that place electric charges of the same sign on adjacent atoms make little contribution to the normal states of molecules, has been further substantiated by observations on covalent azides and on fluorine nitrate.[20] In the ionic crystals NaN_3 and KN_3, the azide ion is linear and symmetrical, each of the end atoms being 1.15 ± 0.02 Å from the central one.[21] Resonance among the structures A, B, C, and D,

$$A \quad \overset{-}{\underset{..}{:}}\overset{-}{N}=\overset{+}{N}=\overset{-}{N}:^{-}$$

$$B \quad :N=\overset{+}{N}=\overset{..}{\underset{..}{N}}:^{-}$$

$$C \quad \overset{-}{\underset{..}{:}}\overset{-}{N}-\overset{+}{N}\equiv N:$$

$$D \quad :N\equiv\overset{+}{N}-\overset{..}{\underset{..}{N}}:^{--}$$

contributing equally, leads to agreement with the observed value, whereas structure A and B would require the value 1.17 Å. The cova-

[18] Plyler and Barker, *loc. cit.* (6); V. Schomaker and R. Spurr, *J.A.C.S.* **64**, 1184 (1942); A. E. Douglas and C. K. Møller, *J. Chem. Phys.* 22, 275 (1954).

[19] L. Pauling, *Proc. Nat. Acad. Sci. U. S.* 18, 498 (1932).

[20] L. Pauling and L. O. Brockway, *J.A.C.S.* **59**, 13 (1937).

[21] S. B. Hendricks and L. Pauling, *J.A.C.S.* 47, 2904 (1925); L. K. Frevel, *ibid.* 58, 779 (1936). The value $N—N = 1.16$ Å has been reported for ammonium azide (L. K. Frevel, *Z. Krist.* 94, 197 [1936]) and 1.12 Å for strontium azide (F. J. Llewellyn and F. E. Whitmore, *J. Chem. Soc.* 1947, 881).

lent molecule methyl azide, on the other hand, has the configuration[22]

$$\underset{\substack{1.47\ \text{Å} \\ \\ H_3C}}{}\overset{\substack{1.24\ \text{Å} \qquad\qquad 1.10\ \text{Å}}}{N\ \text{---}\ N\ \text{---}\ N}$$

120°

the distances having probable errors of 0.02 Å.   A similar configuration for the covalent azide group has been found in cyanuric triazide,[23] $C_3N_3(N_3)_3$, in which the two N—N distances have the values 1.26 Å and 1.11 Å, and in hydrazoic acid,[24] HN_3, with distances 1.240 Å and 1.134 Å and bond angle H—N—N = 112.7°.   These values of the bond lengths are incompatible with resonance among the structures A, B, C, and D,

$$A \qquad R\text{---}\overset{..}{N}\text{=}\overset{+}{N}\text{=}\overset{-}{\underset{..}{N}}\text{:}$$

$$B \qquad R\text{---}N\text{=}\overset{+}{N}\text{=}\overset{..}{\underset{..}{N}}\text{:}$$

$$C \qquad R\text{---}\overset{-}{\underset{..}{N}}\text{---}\overset{+}{N}\text{≡}N\text{:}$$

$$D \qquad R\text{---}\overset{+}{N}\text{≡}\overset{+}{N}\text{---}\overset{..}{\underset{..}{N}}\text{:}^{-\,-}$$

but agree well with equal resonance between A and C, the calculated values being 1.25 Å and 1.12 Å.   The significance of the adjacent-charge rule is seen from the fact that the elimination of structure D for the covalent azide and not for the azide ion is the result of the positive formal charge given to the nitrogen atom by the formation of a covalent bond.

Another structural feature of importance is the bond angle R—N—N. This angle has the value 116° for structure A (unstrained), 108° for C, and 180° for D (the value 116° applies also to B, but with the plane of the molecule normal to that for A) (Sec. 4-8).   An average value would be expected for resonance among several structures.   The observed values, 120° ± 10° in methyl azide, 114° ± 3° in cyanuric triazide, and 112.7° ± 0.5° in hydrogen azide, are in agreement with the value expected for resonance between A and C, about 112°.

[22] L. Pauling and L. O. Brockway, *J.A.C.S.* **59**, 13 (1937).

[23] E. W. Hughes, *J. Chem. Phys.* **3**, 1 (1935); I. E. Knaggs, *Proc. Roy. Soc. London* A150, 576 (1935).

[24] E. Amble and B. P. Dailey, *J. Chem. Phys.* **18**, 1422 (1950); E. H. Eyster, *ibid.* **8**, 135 (1940).

Another application of the adjacent-charge rule, to the fluorine nitrate molecule, will be discussed in a following section, with mention also of the stability of covalent and ionic azides and nitrates.

The electroneutrality rule may also be applicable in the discussion of these molecules. This rule (Sec. 5-7) states[25] that in general the electronic structure of substances is such as to cause each atom to have essentially zero resultant electric charge. Exceptionally large charges may result from the partial ionic character of bonds between atoms with great difference in electronegativity, if there is no way in which the charges can be reduced. The electroneutrality rule may be said to account for the very small electric dipole moments of CO and NNO, 0.112 D and 0.166 D, respectively. A structure such as D (above) for NNO and the covalent azides would, if it made a considerable contribution, confer a large negative charge on the end nitrogen atom. Moreover, the structure itself, with a double negative formal charge on one atom as well as adjacent charges of like sign on two other atoms, has two features leading to its instability and hence to a decrease in its contribution to the normal state of the molecule.

Cyanates and Thiocyanates.—Three reasonable structures can be written for hydrogen cyanate:

$$N=C=O:$$

$$A \quad H$$

$$:N-C\equiv O:$$

$$B \quad H$$

$$N\equiv C-O:$$

$$C \quad H$$

For all three the cyanate group would be expected to be linear. Expected values of the H—N—C bond angle (unstrained) are 116° for A, 108° for B, and 180° for C (Sec. 4-8). Resonance among the three structures would lead to an averaged value of this angle. (The fourth structure, like A but with the double bond in the alternative planes, is not considered because it does not provide a bond orbital for the N—H bond.)

The observed (microwave) dimensions of the molecule[26] are N—C = 1.21 Å, C—O = 1.17 Å (both ±0.01 Å), and angle

[25] I. Langmuir, *Science* 51, 605 (1920); L. Pauling, *J. Chem. Soc.* 1948, 1461.
[26] L. H. Jones, J. N. Shoolery, R. G. Shulman, and D. M. Yost, *J. Chem. Phys.* 18, 990 (1950).

H—N—C = 128.1° ± 0.5°. These bond lengths correspond to those expected for approximately equal resonance among the three structures. The observed bond angle is 10° less than the average for the three structures, suggesting that C makes a somewhat smaller contribution than the other two structures.

Hydrogen thiocyanate, HNCS, has[27] N—C = 1.218 Å, C—S = 1.557 Å, and angle H—N—C = 136°. These bond lengths are those expected for resonance among the three structures, and the bond angle also has the expected value.

The dimensions[28] (microwave) of H_3CNCS are H_3C—N = 1.47 Å, N—C = 1.22 Å, C—S = 1.56 Å, and angle C—N—C = 142°. These also agree well with those expected for resonance among the three structures. The structure of H_3CSCN is, however, somewhat different,[28] with H_3C—S = 1.81 Å (the normal single-bond value), S—C = 1.61 Å, C—N = 1.21 A, and angle C—S—C = 142°. The bond lengths indicate about 70 percent H_3C—S̈=̈C=N: and 30 percent H_3C—S̈—C≡N:, with no contribution of the structure H_3C—S̈≡C—N̈:⁻⁻, which has an unfavorable distribution of electric charge. The reported value of the bond angle is, however, much larger than the expected value, about 113°.

8-3. THE NITRO AND CARBOXYL GROUPS; ACID AND BASE STRENGTHS

Resonance between the two equivalent structures $R—N$ and $R—N$, with perhaps a small contribution by $R—N$, is expected for the nitro group. This would lead to the tetrahedral value 125°16′ for the O—N=O bond angle and to the value 1.27 Å for the N—O distance, the three atoms of the group and the atom of R attached to nitrogen being coplanar, with the two oxygen atoms sym-

[27] C. I. Beard and B. P. Dailey, *J. Chem. Phys.* **18**, 1437 (1950).
[28] C. I. Beard and B. P. Dailey, *J.A.C.S.* **71**, 927 (1949).

metrically related to the R—N axis.   The observed value of the angle in *p*-dinitrobenzene[29] is 124°, and that of the N—O bond length is 1.23 Å.

In nitryl chloride,[30] NO_2Cl, the values are 130°35′ ± 15′ and 1.202 ± 0.001 Å, respectively (microwave spectroscopy).   Both values indicate that there is a significant contribution of the structure

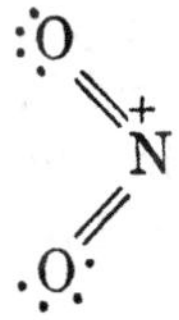

:O
  \\
   N⁺   Cl⁻.   The N—Cl bond length, 1.840 ± 0.002 Å, is 0.11 Å larger
  //
.O:

than that for a single bond, indicating that the contribution of this structure is 24 percent (Equation 7-7).

Whereas in the nitro group the two resonating structures are equivalent, they are made nonequivalent in the carboxyl group and its esters, becoming equivalent again in the corresponding ions:

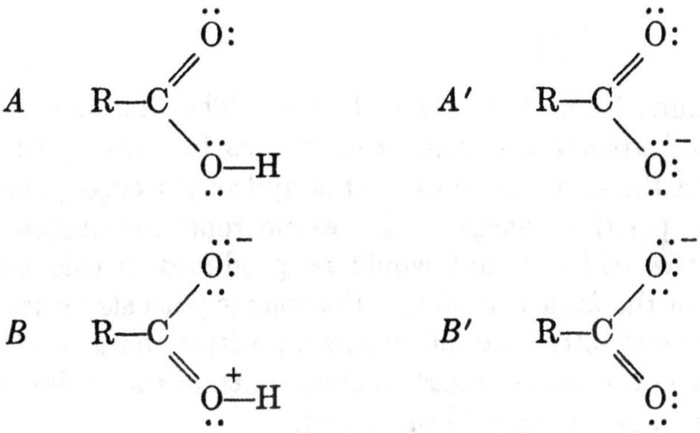

The lack of equivalence of structures *A* and *B* does not inhibit their resonance very thoroughly, however, since the corresponding resonance energy is still large, having the value of 28 kcal/mole for acids and 24 kcal/mole for esters.

The predicted configuration for the carboxylate ion group is that with the angle O—C=O equal to 125°16′ and each C—O bond length equal to 1.27 Å.   (The single-bond and double-bond lengths are 1.41 Å and 1.21 Å, respectively.)   The experimental values lie close to these; for example, for the formates of sodium, calcium, strontium, and barium the average values are 125.5° ± 1° and 1.25 ± 0.01 Å, respectively.[31]

[29] F. J. Llewellyn, *J. Chem. Soc.* 1947, 884.
[30] D. J. Millen and K. M. Sinnott, *J. Chem. Soc.* 1958, 350.
[31] References in Sutton, *Interatomic Distances*.

For formic acid two electron-diffraction studies[32] have given the values $O{=}C{-}OH = 123° \pm 1°$, $C{=}O = 1.22 \pm 0.01$ Å, and $C{-}OH = 1.36 \pm 0.01$ Å, and these values have been supported by several infrared and microwave spectroscopic investigations. Nearly the same values have been reported also for many other carboxylic acids. The bond numbers calculated from these bond lengths are about 1.85 and 1.15, respectively; that is, the presence of the hydrogen atom causes structure *A* to make an 85 percent and structure *B* only a 15 percent contribution. Essentially the same resonance is found in esters: for methyl formate, for example, the bond lengths in the carboxylate group are[33] $C{=}O = 1.22 \pm 0.03$ Å and $C{-}OCH_3 = 1.37 \pm 0.04$ Å.

The concept of resonance provides an obvious explanation of some of the characteristic properties of the carboxyl group, the most striking of which is its acid strength. If the electronic structure of a carboxylic

acid were R—C (with O double-bonded above and OH below), its acid strength would differ only by a rather

small amount from that of an alcohol. The double-bonded oxygen atom would attract electrons from the carbon atom, which in turn would exert the same influence on the hydroxyl oxygen, leaving it with a resultant positive charge. This would repel the proton, and an increase in the acid constant would be produced in this way, through operation of the *inductive effect*. Resonance with structure *B* provides a much more effective way of placing a positive charge on the hydroxyl oxygen, however, and the high acid strength of the carboxyl group can be attributed in the main to this effect.

An alternative point of view about the acid strength of the carboxyl group is the following: There is a certain decrease in free energy accompanying the ionization of the hydroxyl group in a nonresonating molecule, corresponding to the value of the acid constant K_A. The free-energy decrease for ionization of the carboxyl group is larger than this because of the gain in resonance energy of the group during the change from the unsymmetrical configuration, in which resonance is partially inhibited, to the symmetrical configuration of the ion, with complete resonance. The effect of this on the acid constant is given by the equation

$$\text{Change in resonance energy} = RT \ln (K_{A'}/K_A)$$

[32] V. Schomaker and J. M. O'Gorman, *J.A.C.S.* **69**, 2638 (1947); I. L. Karle and J. Karle, *J. Chem. Phys.* **22**, 43 (1954).

[33] J. M. O'Gorman, W. Shand, and V. Schomaker, *J.A.C.S.* **72**, 4222 (1950).

in which $K_{A'}$ is the acid constant of the resonating group.[34] If the resonance energy of the carboxylate ion were 36 kcal/mole, that for the acid being 28 kcal/mole, the acid constant would be raised to the observed value 1.8×10^{-5} (for acetic acid and the following members of the series) from the value 2×10^{-11}. The difference between this and the value for the hydroxyl group in alcohols and water, about 10^{-16}, could be attributed to the inductive effect. It is unfortunate that an experimental value for the resonance energy of the carboxylate ion is not available.[35]

Further increase of the acid constant by substitution of electronegative atoms such as chlorine in the hydrocarbon chain (from $K_A = 1.86 \times 10^{-5}$ for acetic acid to 1.5×10^{-3} for chloroacetic acid, 5×10^{-2} for dichloroacetic acid, and 2×10^{-1} for trichloroacetic acid, for example) is attributed to the inductive effect, the effect of the electronegative atom being transmitted through the chain to the oxygen atom,[36] and to electrostatic interactions.

The acid constant of phenol, 1.7×10^{-10}, is much larger than that of the aliphatic alcohols. This we attribute to resonance with the structures F, G, and H,

which give the oxygen atom a positive formal charge. The inductive effect of the ring is negligible; accordingly the increase in acid constant by a factor of about 10^6 from aliphatic to aromatic alcohols indicates that the resonance energy of the phenolate ion among the structures I to V

is about 8 kcal/mole greater than that for phenol. This is to be ex-

[34] It is likely that the entropy effect of resonance is negligible here.

[35] The value 12 kcal/mole has been reported from an approximate quantum-mechanical calculation by K. Wirtz, *Z. Naturforsch.* 2a, 264 (1947).

[36] See, for example, G. Schwarzenbach and H. Egli, *Helv. Chim. Acta* 17, 1183 (1934).

pected, since these five structures are closely similar in nature, differing
only in the position of the negative charge, whereas for the unionized
molecule the structures F, G, and H, with separated charges, are much
less stable than the conventional structures and contribute only a
small amount (7 kcal/mole—Table 6-2) to the resonance energy.

A nitro group substituted in phenol should increase the acid constant
by virtue of the inductive effect of the electronegative group (with N^+
attached to the ring); moreover, in the ortho and para positions there
would occur an additional resonance effect, due to the contribution of
structures such as the following:

Unionized molecule   Ion

These place a positive charge on the oxygen atom of the unionized molecule, and so cause it to repel the proton.   On analysis of the experimental values for K_A at 25°C it is found that the inductive effect of a
nitro group increases K_A by a factor of about 45, and the resonance effect in the ortho and para positions gives another factor of about 22.
The acid constant of a nitrophenol can be found approximately by multiplying that for phenol, 1.1×10^{-10}, by the factor 45 for every meta
nitro group and 1000 for every ortho or para nitro group in the molecule. The comparison of the values calculated in this way with those
found by experiment is shown in Table 8-1.

The extra factor 22 for resonance in the ortho and para positions
corresponds to an extra resonance energy of 1.8 kcal/mole (= $RT \ln$
22) for the ion relative to the unionized molecule.   This is not unreasonable; there is only one structure of this type involved for each
ortho or para nitro group, and the unfavorable distribution of charge
makes it of little significance for the unionized molecule.[37]

A similar increase in acid strength is produced by other groups such
as cyanide and aldehyde.   Thus K_A for o-hydroxybenzonitrile (in
50-50 by weight ethanol-water solution) is 4.5×10^{-9}, the increase
in acidity over phenol being due to resonance with structures such as

[37] C. M. Judson and M. Kilpatrick, *J.A.C.S.* **71**, 3110, 3115 (1949).

$$\overset{+\,\cdot\cdot}{O}H \qquad \overset{\cdot\cdot\,-}{N}:$$

$$C$$

. Since this resonance involves the double bond in the

1,2 position of the ring, the phenomenon may be used to investigate the amount of double-bond character of bonds. Data bearing on this have been obtained by Arnold and Sprung,[38] who have found for K_A (in ethanol-water) the values 2.2×10^{-7} for 1-hydroxy-2-naphthonitrile

TABLE 8-1.—ACID STRENGTHS OF NITROPHENOLS

Number of nitro groups		Calculated K_A	Observed[a] K_A	Compound
Meta	Ortho, para			
0	0	(1.1×10^{-10})	1.1×10^{-10}	Phenol
1		5.0×10^{-9}	4.5×10^{-9}	m-Nitrophenol
2		2.2×10^{-7}	2.1×10^{-7}	3,5-Dinitrophenol
	1	1.1×10^{-7}	0.6×10^{-7}	o-Nitrophenol
			0.7×10^{-7}	p-Nitrophenol
			12×10^{-6}	2,3-Dinitrophenol
1	1	5.0×10^{-6}	6.1×10^{-6}	2,5-Dinitrophenol
			3.8×10^{-6}	3,4-Dinitrophenol
	2	1.1×10^{-4}	1×10^{-4}	2,4-Dinitrophenol
			2×10^{-4}	2,6-Dinitrophenol
	3	1.1×10^{-1}	1.6×10^{-1}	2,4,6-Trinitrophenol

[a] All data are for 25°C.

and 2.1×10^{-9} for 3-hydroxy-2-naphthonitrile. In our discussion of aromatic molecules (Sec. 6-3) we have attributed to the 1,2 bond of naphthalene two thirds double-bond character, to the bonds in benzene one half, and to the 2,3 bond of naphthalene one third. It is seen that the increase in acid constant runs parallel to the amount of double-bond character of the connecting bond, and that the empirical relation formulated in this way could be used in assigning amounts of double-bond character to bonds on the basis of acidity measurements.

Because the 1,2 and 1,4 interactions in benzene are nearly equivalent, we expect the acidity of 4-hydroxy-1-naphthonitrile to involve an average amount of double-bond character, and hence to lie near the value for o-hydroxybenzonitrile; this is verified, the experimental value being 4.5×10^{-9}.

Similar behavior is shown by the hydroxyaldehydes, with the ob-

[38] R. T. Arnold and J. Sprung, *J.A.C.S.* **61**, 2475 (1939).

served K_A values 1.2×10^{-8} for 1-hydroxy-2-naphthaldehyde, 6×10^{-10} for salicylaldehyde, and 1.3×10^{-10} for 3-hydroxy-2-naph-thaldehyde.

A straightforward treatment of the base strength of aniline can be given. A saturated aliphatic amine such as methylamine has a base constant K_B of about 5×10^{-4}, corresponding to the reaction

$$R-\overset{\overset{\textstyle H}{|}}{\underset{\underset{\textstyle H}{|}}{N}}: + H_2O \rightarrow R-\overset{\overset{\textstyle H}{|}}{\underset{\underset{\textstyle H}{|}}{\overset{+}{N}}}-H + OH^-.$$

In these substances an unshared electron pair is available on the nitrogen atom for forming the bond with the proton. In aniline, on the other hand, this pair of electrons is involved in resonance; and whereas the aniline molecule resonates among the three structures F, G, and H,

as well as the normal structures

, the phenylammonium ion

is restricted to the normal structures $H-\overset{+}{N}-H$. The ion is thus made

unstable relative to the unionized molecule (using the aliphatic compounds for comparison) by a free energy equal to the resonance energy due to the three structures F, G, and H, and its base dissociation constant is reduced very greatly, to the value $K_B = 3.5 \times 10^{-10}$.

Since this change by the factor $1/1.4 \times 10^6$ is due entirely to the complete inhibition of the F, G, H resonance by addition of the proton, the quantity $RT \ln 1.4 \times 10^6 = 8.4$ kcal/mole represents the F, G, H resonance energy in aniline. This value is probably more accurate than that given by thermochemical data, 6 kcal/mole (Table 6-2), and the agreement between the two is satisfactory.

8-4. THE STRUCTURE OF AMIDES AND PEPTIDES

During the last two decades great progress has been made in the investigation of the structure of amides and peptides, because of their importance to the structure of proteins.   Whereas the first and second editions of this book (1939, 1940) contained the statement that no information exists about the configuration and dimensions of the amide group, we may now say that the structural information that has been obtained for this group is more extensive and more reliable than that for any other group of comparable complexity.[39]

The principal resonance structures of amides, A and B,

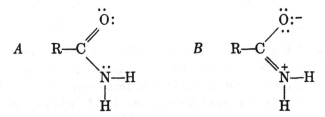

are not equivalent, and it is expected that A contributes somewhat more than B to the structure of the group.   The resonance energy is about 21 kcal/mole (Table 6-2).   The amides are very weak bases; the resonance with structure B is nearly completely inhibited by addition of a proton to the nitrogen atom.   The corresponding calculated value for the base constant, 1×10^{-20}, is so small as to be without significance, except to show that the amides do not form salts with acids by adding a proton to the amino group.

The structure of the amide group can be illustrated by the formamide molecule, which has been subjected to a careful study by microwave spectroscopy.[40]   The molecule is completely planar, as required for resonance of A and B.   Its dimensions are given in the following diagram:

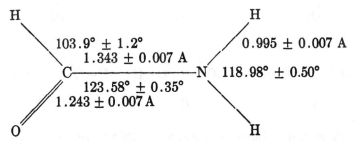

These dimensions are essentially the same as those that had been

[39] Much of the progress has been the result of the work of Professor Robert B. Corey and his collaborators.

[40] R. J. Kurland and E. B. Wilson, Jr., *J. Chem. Phys.* **27,** 585 (1957).

assigned to the peptide group from analysis of the results of a number of careful x-ray structure determinations of crystals of amino acids, simple peptides, and related substances:[41]

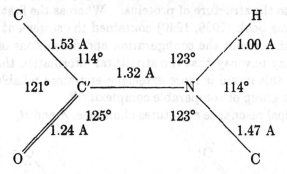

The value of the O=C'—N bond angle is close to the tetrahedral value 125°16'. The bond lengths are about as expected for resonance between structures A and B: the C'—N bond length, 1.32 Å, is 0.13 Å less than would be expected for a single bond adjacent to a double bond, showing that there is a considerable contribution by the resonance structure B. This contribution is evaluated as about 40 percent. For 60 percent A and 40 percent B the bond lengths from Tables 7-5 and 7-9 are C—O = 1.26 Å and C—N = 1.34 Å; there may also be some contribution of multiple bonds involving the p orbitals normal to the plane of the group, correlated with the ionic aspect of the adjacent bond.[42]

8-5. THE CARBONATE, NITRATE, AND BORATE IONS AND RELATED MOLECULES

Carbonic acid and its derivatives resonate among the three structures A, B, and C, the resonance being complete for the ion and somewhat inhibited in the acid

[41] R. B. Corey and J. Donohue, *J.A.C.S.* **72**, 2899 (1950); R. B. Corey and L. Pauling, *Proc. Roy. Soc. London* **B141**, 10 (1953); R. B. Corey, *Fortschr. Chem. org. Naturstoffe* **8**, 310 (1951); R. B. Corey and L. Pauling, *ibid.* **11**, 18 (1954).

[42] Evidence that this contribution amounts to about 20 percent for the C—O bond in amides has been discussed by L. Pauling in *Symposium on Protein Structure*, ed. by A. Neuberger, John Wiley and Sons, New York, 1958. This would shorten the C—O bond to 1.24 Å.

and its esters. (Specific mention need not be made of the structure

$$
\begin{array}{c}
:\overset{\cdot\cdot}{\underset{}{O}}:^{-} \\[2pt]
| \\[2pt]
C^{+} \\[-2pt]
\diagup \quad \diagdown \\[2pt]
H-\overset{\cdot\cdot}{\underset{\cdot\cdot}{O}} \qquad \overset{\cdot\cdot}{\underset{\cdot\cdot}{O}}-H
\end{array}
$$

The contribution of structures of this type, which give partial ionic character to bonds, is to be assumed in all cases.)

The value 42 kcal/mole for the resonance energy is given by the thermochemical data for the dialkyl carbonates. This value for resonance of the double bond among three positions is not unreasonable when it is compared with the corresponding value of 24 kcal/mole for esters of the fatty acids, in which the bond resonates between two positions.

The resonating structure requires that the carbonate ion be planar with bond angles 120°, and three C—O distances equal to 1.32 Å (Tables 7-5 and 7-9). This value presumably should be decreased by about 0.02 Å because of the contribution of $p\pi$ electrons to the bonds. The trigonal planar configuration of the ion was found in the original x-ray study of calcite[43] and has been verified since by examination of other carbonate crystals. The value calculated for the C—O distance is in satisfactory agreement with that found in recent reinvestigations[44] of calcite, 1.30 ± 0.01 Å.

For the nitrate ion, with the same type of structure as the carbonate ion, a similar configuration is expected and observed:

$$
\begin{array}{c}
\overset{\cdot\cdot}{\underset{}{O}}: \\[2pt]
\| \\[2pt]
N^{+} \\[-2pt]
\diagup \quad \diagdown \\[2pt]
^{-}:\overset{}{\underset{\cdot\cdot}{O}} \qquad \overset{}{\underset{\cdot\cdot}{O}}:
\end{array}
$$

The value of the N—O bond length,[44] 1.218 ± 0.004 Å, is much smaller than that given by Tables 7-5 and 7-9, and it is not clear whether or not the difference can be attributed to $p\pi$ bonding. A somewhat larger value, 1.234 ± 0.01 Å, is reported for the nitrate ion[45] in $NO_2^+NO_3^-$.

[43] W. L. Bragg, *Proc. Roy. Soc. London* A89, 468 (1914).

[44] N. Elliott, *J.A.C.S.* 59, 1380 (1937), reported 1.313 ± 0.010 Å, and R. L. Sass, R. Vidale, and J. Donohue, *Acta Cryst.* 10, 567 (1957), reported 1.294 ± 0.004 Å.

[45] E. Grison, K. Eriks, and J. L. de Vries, *Acta Cryst.* 3, 290 (1950).

In molecules in which an atom or group is attached by a covalent bond to one of the oxygen atoms a planar configuration is found, with the N—OR bond length about 1.38 Å and the N—O bond length in the —NO$_2$ group about 1.28 Å. Experimental values are 1.36, 1.39, 1.36, and 1.41 Å for N—OR and 1.26, 1.29, 1.28, and 1.22 Å (all ±0.05 Å) for methyl nitrate,[46] fluorine nitrate,[46] pentaerythritol tetranitrate[47] (C(CH$_2$ONO$_2$)$_4$), and nitric acid,[48] respectively. These bond lengths indicate that the N—OR bond has about 15 percent of double-bond character. The barrier to rotation of the HO group about the O—N bond in HNO$_3$ has been reported as 9.3 ± 1.1 kcal/mole[49] and 9.5 ± 0.5 kcal/mole[50] from spectroscopic studies, which suggests about 25 percent (assuming the energy of the N=O bond to be about 80 kcal/mole).

The borate group in boric acid is observed[51] to have a planar trigonal configuration with the B—O distance 1.360 ± 0.005 Å. Similar BO$_3$ groups occur in many salts of boric acid. In calcium metaborate,[52] CaB$_2$O$_4$, there are infinite chains of BO$_3$ groups joined by shared oxygen atoms, and in potassium metaborate,[53] K$_3$B$_3$O$_6$, these groups are similarly joined into trimeric ions (Fig. 8-1).

In many complex borates there are BO$_4$ tetrahedra as well as BO$_3$ triangles. For example, the ion [B$_3$O$_3$(OH)$_5$]$^{--}$ consists of two tetrahedra and one triangle, each sharing a corner with each of the other two:

The unshared corners are occupied by OH groups. (Note that the

[46] Pauling and Brockway, *loc. cit.* (20).

[47] A. D. Booth and F. J. Llewellyn, *J. Chem. Soc.* 1947, 837.

[48] L. R. Maxwell and V. M. Mosley, *J. Chem. Phys.* 8, 738 (1940).

[49] H. Cohn, C. K. Ingold, and H. Poole, *J. Chem. Phys.* 24, 162 (1956); *J. Chem. Soc.* 1952, 4272.

[50] A. Palm and M. Kilpatrick, *J. Chem. Phys.* 23, 1562 (1955).

[51] W. H. Zachariasen, *Acta Cryst.* 7, 305 (1954).

[52] W. H. Zachariasen, *Proc. Nat. Acad. Sci. U. S.* 17, 617 (1931); W. H. Zachariasen and G. E. Ziegler, *Z. Krist.* 83, 354 (1932).

[53] W. H. Zachariasen, *J. Chem. Phys.* 5, 919 (1937).

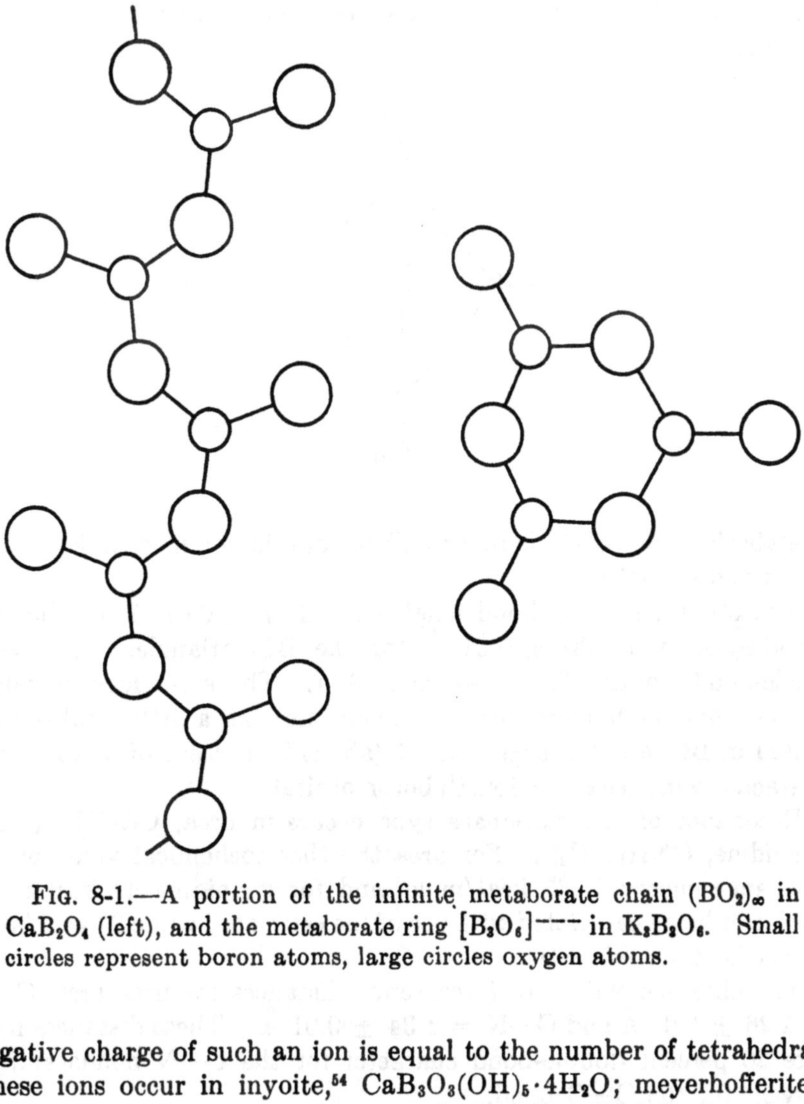

Fig. 8-1.—A portion of the infinite metaborate chain $(BO_2)_\infty$ in CaB_2O_4 (left), and the metaborate ring $[B_3O_6]^{---}$ in $K_3B_3O_6$. Small circles represent boron atoms, large circles oxygen atoms.

negative charge of such an ion is equal to the number of tetrahedra.) These ions occur in inyoite,[54] $CaB_3O_3(OH)_5 \cdot 4H_2O$; meyerhofferite,[55] $CaB_3O_3(OH)_5 \cdot H_2O$; and synthetic $CaB_3O_3(OH)_5 \cdot 2H_2O$.[56] In colemanite,[57] $CaB_3O_4(OH)_3 \cdot H_2O$, there are chains in which the $[B_3O_3(OH)_5]^{--}$ ions have joined together with the replacement of two OH groups by a shared O. Borax,[58] $Na_2B_4O_5(OH)_4 \cdot 8H_2O$, contains $[B_4O_5(OH)]_4^{--}$

[54] J. R. Clark, *Acta Cryst.* 12, 162 (1959).
[55] C. L. Christ and J. R. Clark, *Acta Cryst.* 9, 830 (1956).
[56] J. R. Clark and C. L. Christ, *Acta Cryst.* 10, 776 (1957).
[57] C. L. Christ, J. R. Clark, and H. T. Evans, Jr., *Acta Cryst.* 11, 761 (1958)
[58] N. Morimoto, *Mineral J. Japan* 2, 1 (1956).

ions composed of two tetrahedra and two triangles linked in the following way:

OH
|
B
/ | \
O   O
/ | \
O—B   O   B—O
\ | /
O   O
\ | /
B
|
OH

Metaboric acid, HBO_2, is reported[59] to contain chains of linked tetrahedra and triangles.

The observed B—O bond lengths are 1.47 ± 0.01 Å for the BO_4 tetrahedron and 1.37 ± 0.01 Å for the BO_3 triangle. The B—O single-bond length (Table 7-5) is 1.43 Å. The experimental values suggest that the bonds in BO_4 have bond number a little smaller than 1, and in BO_3 a little larger than 1 (about 20 percent of double-bond character, with use of the fourth boron orbital).

Resonance of the carbonate type occurs in urea, $CO(NH_2)_2$, and guanidine, $CNH(NH_2)_2$. For urea the thermochemical value of the resonance energy is 37 kcal/mole, and for guanidine 47 kcal/mole, the latter being calculated with use of the estimated value 24 kcal/mole for the heat of sublimation.

The observed values of interatomic distances for urea are[60] C—O $= 1.26 \pm 0.01$ Å and C—N $= 1.34 \pm 0.01$ Å. These distances indicate 20 percent double-bond character for the C—N bonds and 60 percent for the C—O bond.

Reliable interatomic-distance data are not available for guanidine or the guanidinium ion.

The base strengths of guanidine and its derivatives present an interesting problem. Guanidine itself is a very strong base, approaching the alkalies in strength. This fact can be accounted for by arguments closely related to those used for the carboxylic acids in the preceding section. The guanidinium ion resonates among the three structures

[59] W. H. Zachariasen, *Acta Cryst.* **5**, 68 (1952).
[60] R. W. G. Wyckoff and R. B. Corey, *Z. Krist.* **89**, 462 (1934); P. A. Vaughan and J. Donohue, *Acta Cryst.* **5**, 530 (1952).

which are all equivalent, whereas guanidine itself resonates among the three structures

which are not equivalent. The difference in resonance energy can be estimated to be of the magnitude of 6 to 8 kcal/mole, which would increase the basic strength very greatly.

The monoalkylguanidines and N,N-dialkylguanidines should be somewhat weaker bases than guanidine itself, for the following reason. The replacement of one or two hydrogens of an $-NH_2$ group by alkyl radicals tends to prevent the double bond from swinging to this group, because carbon is more electronegative than hydrogen and hence tends to cause the adjacent nitrogen atom not to assume a positive charge. In consequence, resonance of the double bond is to some extent restricted to the two other nitrogen atoms. This causes a decrease in the basic strength toward that characteristic of an imidine, the decrease

being about twice as great for $\mathrm{HNC}\big\langle{}^{NH_2}_{NR_2}$ as for $\mathrm{HNC}\big\langle{}^{NH_2}_{NHR}$. A very

much larger effect is expected for the N,N'-dialkylguanidines. The alkyl groups on two of the nitrogen atoms would tend to force the double bond to the third nitrogen atom, the structure R—N

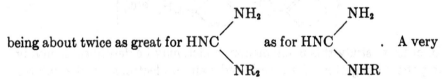

being more important than the other two. This nitrogen atom would hence have little tendency to add a proton, and the substance would be a weak base. On the other hand, the N,N',N''-trialkylguanidines should be about as strong bases as guanidine itself, inasmuch as the conditions for resonance in this molecule and its symmetrical ion are exactly the same as for guanidine and its ion. These conclusions are in agreement with the available data;[61] guanidine, the monoalkylguanidines, N,N-dimethylguanidine, and N,N'N''-trimethylguanidines are strong bases, whereas the N,N'-dialkylguanidines are weak bases.

8-6. THE STRUCTURE AND PROPERTIES OF THE CHLORO-ETHYLENES AND CHLOROBENZENES

The chemical properties of the chlorobenzenes and chloroethylenes differ strikingly from those of saturated aliphatic chlorine compounds and of aromatic compounds with chlorine substituted in a side chain. For example, methyl chloride and benzyl chloride are hydrolyzed by boiling alkali, giving the corresponding alcohols, whereas chlorobenzene is not affected by this treatment. In general there is a pronounced diminution in reactivity of a chlorine atom adjacent to an aromatic nucleus or double bond.

The obvious explanation of this involves resonance of the following type, which gives to the C—Cl bond some double-bond character:

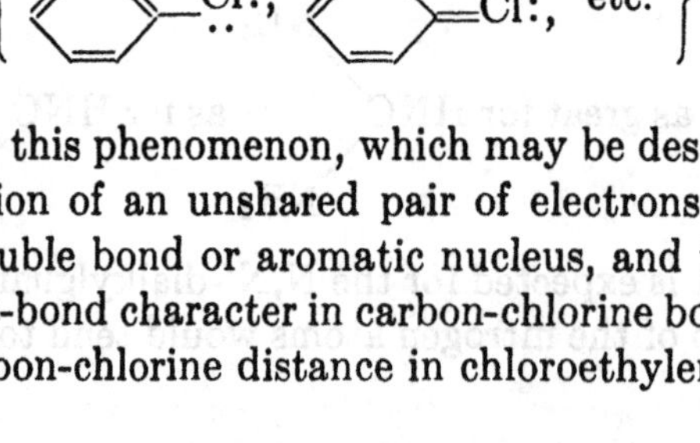

In order to study this phenomenon, which may be described as involving the conjugation of an unshared pair of electrons on the chlorine atom with the double bond or aromatic nucleus, and to determine the amount of double-bond character in carbon-chlorine bonds of this type, values of the carbon-chlorine distance in chloroethylenes[62] and chloro-

---

[61] T. L. Davis and R. C. Elderfield, *J.A.C.S.* **54**, 1499 (1932).

[62] L. O. Brockway, J. Y. Beach, and L. Pauling, *J.A.C.S.* **57**, 2693 (1935). The value C—Cl = 1.72 ± 0.01 Å has been reported for C_2Cl_4 (electron diffraction) by I. L. Karle and J. Karle, *J. Chem. Phys.* **20**, 63 (1952), and 1.68 ± 0.02 Å for $CFClCH_2$ (microwave) by J. K. Bragg, T. C. Madison, and A. H. Sharbaugh. *Phys. Rev.* **77**, 148 (1950).

benzenes[63] have been determined by the electron-diffraction method.

The results of the investigations are given in Table 8-2. Whereas in carbon tetrachloride, methyl chloride, and similar molecules the carbon-chlorine distance is 1.765 Å, in these substances the values found are from 0.03 to 0.09 Å less than this, and, moreover, there is found for the chloroethylenes a reasonable correlation between the

TABLE 8-2.—INTERATOMIC DISTANCES IN CHLORO-
ETHYLENES AND CHLOROBENZENES

Molecule	C—Cl distance[a]	Molecule	C—Cl distance[a]
CH_2CHCl	1.726 Å	C_6H_5Cl	1.69 Å
CH_2CCl_2	1.69	p-$C_6H_4Cl_2$	1.69
cis-CHClCHCl	1.67	m-$C_6H_4Cl_2$	1.69
trans-CHClCHCl	1.69	o-$C_6H_4Cl_2$	1.71
$CHClCCl_2$	1.71	1,3,5-$C_6H_3Cl_3$	1.69
C_2Cl_4	1.72	1,2,4,5-$C_6H_2Cl_4$	1.72
		C_6Cl_6	1.70

[a] These values are reliable to about ± 0.02 Å.

amount of shortening and the number of chlorine atoms conjugated with the double bond. The shortening is about 0.08 Å for substituted ethylenes containing one or two chlorine atoms, which corresponds, when interpreted by the method of Section 7-5, to about 20 percent double-bond character. The shortening observed for trichloroethylene and tetrachloroethylene, 0.05 Å, corresponds to about 10 percent double-bond character. It seems probable that there is competition between the chlorine atoms in these compounds for the double bond, leading to a decrease in double-bond character as compared with vinyl chloride. Corresponding to the decrease in double-bond character is the somewhat greater reactivity observed for trichloroethylene and tetrachloroethylene than for vinyl chloride and the dichloroethylenes.

The shortening observed for all of the chlorobenzenes studied is about 0.06 or 0.07 Å, corresponding to about 15 percent double-bond character. (The variations from the values 1.69 or 1.70 Å reported in the table have little significance, although there is indication of a small increase in distance with increase in the number of chlorine atoms in the molecule.) From these results it can be concluded that

[63] L. O. Brockway and K. J. Palmer, *J.A.C.S.* **59**, 2181 (1937). Two microwave values for C_6H_5Cl have been reported: 1.706 Å (G. Erlandsson, *Arkiv Fysik* **8**, 341 [1954]) and 1.670 $\pm$ 0.003 Å (R. L. Poynter, dissertation, **1954**, quoted in Sutton, *Interatomic Distances*).

the benzene ring has about the same power of conjugation with a chlorine atom as has a double bond, and moreover that its capacity for conjugation is somewhat greater, in that little saturation is indicated for hexachlorobenzene, in contrast to tetrachloroethylene.

Decreases in interatomic distances in the bromine and iodine derivatives of ethylene and benzene have also been reported,[64] with magnitudes about the same as for the chlorine derivatives.   In dichloroacetylene,[65] dibromacetylene, and diiodoacetylene,[66] in which the halogens are conjugated with a triple bond, there are decreases of about 0.13 Å, corresponding to 25 percent double-bond character.

A discussion of nuclear electric quadrupole coupling in the vinyl halides has led to the estimate of about 6 percent double-bond character for the C—Cl bond in vinyl chloride and 3 percent for the C—I bond in vinyl iodide.[67]   Values of electric dipole moments of monohalogenated benzenes have been interpreted as corresponding to 4 percent of double-bond character for the C—X bonds.[68]

Theoretical studies of the phenomenon described above have been made by Sherman and Ketelaar and others.[69]

8-7. RESONANCE IN CONJUGATED SYSTEMS

For a molecule such as butadiene-1,3, $CH_2\!=\!CH\!-\!CH\!=\!CH_2$, it is customary to write one valence-bond formula involving alternating single and double bonds and to take cognizance of the difference in properties from a molecule containing isolated double bonds by saying that here the double bonds are conjugated.   From the new point of view the phenomenon of conjugation is attributed to resonance between the ordinary structure and certain structures involving one less double

---

[64] H. de Laszlo, *Proc. Roy. Soc. London* A146, 690 (1934); J. A. C. Hugill, I. E. Coop, and L. E. Sutton, *Trans. Faraday Soc.* 34, 1518 (1938).

[65] The value C—Cl = 1.632 ± 0.001 Å has been found for HCCCl in a microwave investigation by A. A. Westenberg, J. H. Goldstein, and E. B. Wilson, Jr., *J. Chem. Phys.* 17, 1319 (1949), and 1.637 Å in H₃CCCCl by C. C. Costain, *ibid.* 23, 2037 (1955).

[66] H. de Laszlo, *Nature* 135, 474 (1935).   L. O. Brockway and I. E. Coop, *Trans. Faraday Soc.* 34, 1429 (1938), reported C—Cl = 1.68 ± 0.04 Å in HCCCl and C—Br = 1.80 ±0.03 Å in HCCBr, and J. Y. Beach and A. Turkevich, *J.A.C.S.* 61, 299 (1939), have found values 0.01 Å less than these for ClCN and BrCN, respectively.

[67] J. H. Goldstein, *J. Chem. Phys.* 24, 106 (1956); J. H. Goldstein and J. K. Bragg, *Phys. Rev.* 75, 1453 (1949); 78, 347 (1950).

[68] C. P. Smyth, *J.A.C.S.* 63, 57 (1941).

[69] J. Sherman and J. A. A. Ketelaar, *Physica* 6, 572 (1939); J. E. Lennard-Jones, *Proc. Roy. Soc. London* A158, 280 (1937); W. G. Penney, *ibid.* 306; C. A. Coulson, *ibid.* A169, 413 (1939); *J. Chem. Phys.* 7, 1069 (1939); J. E. Lennard-Jones and C. A. Coulson, *Trans. Faraday Soc.* 35, 811 (1939); many recent papers.

bond; $\overset{|}{C}H_2\!\!-\!\!CH\!\!=\!\!CH\!\!-\!\!\overset{|}{C}H_2$ for butadiene, and to a smaller extent $C^+H_2\!\!-\!\!CH\!\!=\!\!CH\!\!-\!\!\overset{\cdot\cdot}{C}H_2{}^-$, etc. These structures are less stable than the ordinary structure and contribute only a small amount to the normal state of the molecule, giving the 2,3 bond a small amount of double-bond character.

The quantum-mechanical treatment of this problem[70] indicates that the single bonds in a conjugated system have about 20 percent double-bond character and that the extra resonance energy resulting from the conjugation of two double bonds is about 5 to 8 kcal/mole. The calculations also show that a double bond and a benzene nucleus are approximately equivalent in conjugating power.

The thermochemical data for biphenyl, 1,3,5-triphenylbenzene, phenylethylene, and stilbene (Table 6-2) correspond to a value of about 7 kcal/mole for the conjugation energy of a double bond and a benzene ring or of two benzene rings. Somewhat lower values, between 2 and 6 kcal/mole, are given for the conjugation energy in dienes by values of heats of hydrogenation.[71]

In butadiene-1,3 and cyclopentadiene the value found[72] for the carbon-carbon distance for the bond between the conjugated double bonds is 1.46 Å. This, interpreted by use of Table 7-9, corresponds to 15 percent double-bond character. The same amount of double-bond character is also indicated by the following x-ray values for the bond between two benzene rings or a benzene ring and a double bond: stilbene ($C_6H_5\!\!-\!\!CH\!\!=\!\!CH\!\!-\!\!C_6H_5$), 1.44 Å; biphenyl, 1.48 Å; *p*-diphenylbenzene, 1.46 Å.

This amount of double-bond character should give to the bond in some part the properties of a double bond; in particular, the conjugated systems should tend to remain planar. Chemical evidence for cis and

trans isomers of the type and

has not been forthcoming; the restriction of rotation about the central bond is not great enough to prevent easy interconversion of these molecular species. It is great enough, however, to cause the con-

[70] L. Pauling and J. Sherman, *J. Chem. Phys.* **1**, 679 (1933).

[71] G. B. Kistiakowsky, J. R. Ruhoff, H. A. Smith, and W. E. Vaughan, *J.A.C.S.* **58**, 146 (1936).

[72] V. Schomaker and L. Pauling, *J.A.C.S.* **61**, 1769 (1939).

jugated molecules to retain in general the planar equilibrium configuration, and this configuration has been verified by various physical techniques, such as x-ray diffraction, electron diffraction, and spectroscopy. In the gas phase the trans molecules are found in greater number than the cis molecules, and in crystals the configuration is usually trans (about the single bonds).

It was discovered by Gillam and El Ridi[73] that the carotenoids exist in isomeric forms, and these forms were identified by Zechmeister[74] as involving cis-trans isomerism about the double bonds. Steric hindrance between methyl groups and hydrogen atoms restricts the cis configuration largely to certain of the double bonds.[75] The absorption spectra provide information about the location of the cis double bonds in the long conjugation chains, and, moreover, show that in all chains the trans configuration about the single bonds is the stable one.[75]

The explanation of the stability of the trans configuration about the single bonds can be given on the basis of the principles discussed in preceding sections. For conjugation, with resonance between the principal structure $=$ — $=$ and the less important structures $\cdot$— $=$—$\cdot$, $^+$— $=$— $\overset{..}{\cdot}$, and $\overset{..}{\cdot}$ $=$—$^+$, a planar configuration is required by the double-bond character of the central bond. The trans and cis planar configurations $=\diagdown_=$ and $\diagup\diagdown$ are thus stabilized by the resonance energy, several kcal/mole, relative to the nonplanar configurations. Moreover, the bonds (ordinary single bonds and bent bonds) adjacent to the central bond have the stable relative azimuthal orientation for the trans configuration and the unstable orientation for the cis configuration (Secs. 4-7, 4-8); the difference in energy can be estimated from the barrier heights restricting rotation to be about 1.5 kcal/mole.

A surprising prediction is that it is the cis configuration (rather than the trans) about the triple bond (and the two adjacent single bonds) that is the stable one in conjugated systems of double bonds and a triple bond, $\ldots\diagup\overset{\equiv}{}\diagdown\ldots$ Conjugation stabilizes both the cis and the trans planar configurations, and the cis is further stabilized by the favored eclipsed orientation of the two groups adjacent to the triple bond, as discussed in Section 4-8 for dimethylacetylene. The

[73] A. E. Gillam and M. S. El Ridi, *Nature* **136**, 914 (1935).

[74] L. Zechmeister and L. Cholnoky, *Ann. Chem.* **530**, 291 (1937); L. Zechmeister, *Chem. Revs.* **34**, 267 (1944).

[75] L. Pauling, *Fortschr. Chem. org. Naturstoffe* **3**, 203 (1939).

predicted difference in energy of the cis and the trans configuration is about 0.4 kcal/mole. In fact, the trans configuration has been found in crystals of 9,9'-dehydro-β-carotene.[76] However, it seems likely that it is the intermolecular forces that stabilize this configuration of the molecules in the crystal, because spectroscopic studies indicate that both 7,7'-bis-desmethyl-9,9'-dehydro-β-carotene[77] and 9,9'-dehydro-β-carotene[78] in ether solution have the cis configuration as the stable one. Moreover, hydrogenation of each of the substances gives mainly the cis isomer of the corresponding bis-desmethyl carotenoid or β-carotene, with only small amounts of the trans isomer.

Many x-ray investigations showing the planarity of conjugated molecules have been reported. Thus the stilbene molecule is planar,[79] whereas the closely similar unconjugated molecule dibenzyl is not planar;[80] and planarity has been observed also for trans-azobenzene,[81] oxalic acid and oxalate ion, dimethyloxalate,[82] and many other conjugated molecules. Because the amount of double-bond character of the conjugated "single" bond is small, however, the forces that strive toward planarity are not very strong, and may be rather easily overcome by steric effects. This is illustrated by Figure 8-2, showing a molecule of cis-azobenzene drawn to scale, with use of 1.0 Å as the van der Waals radius of hydrogen. It is seen that contact of the ortho hydrogen atoms of the two rings prevents the molecule from assuming the planar configuration, and it has in fact been found by x-ray examination[83] that each of the phenyl groups is rotated through about 50° out of the coplanar orientation.

A similar scale drawing of 1,3,5-triphenylbenzene, Figure 8-3, shows that there is some steric repulsion between hydrogen atoms in this molecule (drawn with nonbonded radius 1.0 Å); the same amount is expected for biphenyl (Fig. 8-4). Triphenylbenzene has been shown by an x-ray study to be nonplanar,[84] with two of the phenyl groups rotated through about 30° in one direction and the third rotated by 27° in the other direction. The gas molecule has been shown by elec-

---

[76] W. G. Sly, Ph.D. thesis, Calif. Inst. Tech., 1957.

[77] H. H. Inhoffen, F. Bohlmann, and G. Rummert, *Ann. Chem.* 569, 226 (1950).

[78] H. H. Inhoffen, F. Bohlmann, K. Bartram, G. Rummert, and H. Pommer, *Ann. Chem.* 570, 54 (1950).

[79] J. M. Robertson, *Proc. Roy. Soc. London* A150, 348 (1935).

[80] J. M. Robertson and I. Woodward, *Proc. Roy. Soc. London* A162, 568 (1937).

[81] J. J. de Lange, J. M. Robertson, and I. Woodward, *Proc. Roy. Soc. London* A171, 398 (1939).

[82] M. W. Dougill and G. A. Jeffrey, *Acta Cryst.* 6, 831 (1953).

[83] J. M. Robertson, *J. Chem. Soc.* 1939, 232.

[84] K. Lonsdale, *Z. Krist.* 97, 91 (1937); M. S. Farag, *Acta Cryst.* 7, 117 (1954).

tron diffraction[85] to have the rings rotated by 45° ± 5°, probably with a statistical distribution between the two directions. The angle between the two rings in biphenyl[86] is 45° ± 10°.

The effect of steric hindrance in giving rise to optical activity of *o,o'*-substituted biphenyls is well known.[87]

An interesting method of determining the amount of conjugation energy of two benzene rings and a nitrogen-nitrogen double bond has been developed, subsequent to the discovery[88] that the cis isomer of

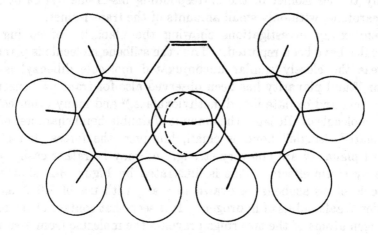

Fig. 8-2.—The planar configuration of cis-azobenzene, drawn to scale using 1.0 Å for the van der Waals radius for hydrogen. Steric interactions of hydrogen atoms prevent the assumption of this configuration.

azobenzene is formed from the ordinary (trans) isomer in solution through the action of light. As mentioned above, steric interaction of the two rings in the cis compound is so great that a planar configuration cannot be achieved, the benzene rings being rotated 50° out of the planar configuration. Since planarity is an essential attribute of double-bond character, we may assume that the amount of conjugation energy in cis-azobenzene and related molecules is small (it is probably not greater than 2 or 3 kcal/mole) and accept the difference in energy of the cis and trans isomers as the conjugation energy for the trans configuration. This difference has been determined by the measurement of heats of fusion of the two crystalline substances to the

[85] O. Bastiansen, *Acta Chem. Scand.* **6**, 205 (1952).

[86] O. Bastiansen, *Acta Chem. Scand.* **3**, 408 (1949).

[87] See, for example, H. Gilman, *Organic Chemistry*, John Wiley and Sons, New York, 1943, p. 343.

[88] G. S. Hartley, *Nature* **140**, 281 (1937).

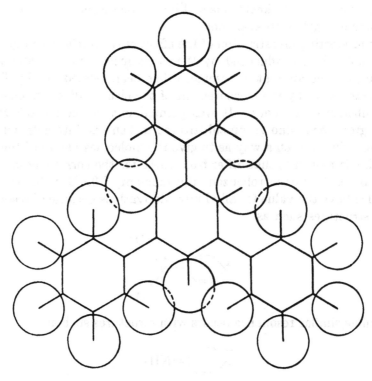

Fig. 8-3.—The planar configuration of 1,3,5-triphenyl-benzene, drawn to scale using 1.0 Å for the van der Waals radius of hydrogen.

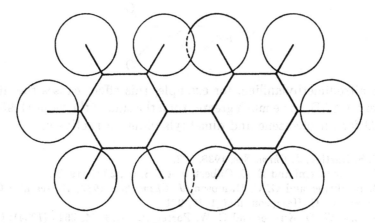

Fig. 8-4.—The planar configuration of biphenyl, drawn to scale using 1.0 Å for the van der Waals radius of hydrogen.

(trans) liquid[89] as 12 kcal/mole and by the measurement of their heats of combustion[90] as 10 kcal/mole.

An interesting investigation of the effect of the nuclear configuration of a molecule, as influenced by the steric interactions of atoms and groups, on the amount of resonance has been carried out by Birtles and Hampson[91] by the measurement of the electric dipole moments of the substituted durenes. It was pointed out in Section 6-3 that a nitro group or amino group substituted in benzene interacts with the benzene ring in such a way as to give the molecules values of the electric dipole moments that differ from those for the corresponding alkyl derivatives.[92] For nitrobenzene the moment, 3.95 D, is increased to 0.64 D above the value for alkyl nitro derivatives because of resonance with structures such as

In aniline similar resonance occurs with structures such as

With an electron-donating group and an electron-receiving group para to one another in the same molecule additional resonance occurs with structures such as

For *p*-nitrodimethylaniline, for example, this effect causes the dipole moment, 6.87 D, to be much greater than the sum of the values 3.95 and 1.58 D for nitrobenzene and dimethylaniline, respectively.

[89] G. S. Hartley, *J. Chem. Soc.* **1938**, 633.

[90] R. J. Corruccini and E. C. Gilbert, *J.A.C.S.* **61**, 2925 (1939).

[91] R. H. Birtles and G. C. Hampson, *J. Chem. Soc.* **1937**, 10; see also C. E. Ingham and G. C. Hampson, *ibid.* **1939**, 981.

[92] See also W. D. Kumler and C. W. Porter, *J.A.C.S.* **56**, 2549 (1934); C. K. Ingold, *Chem. Revs.* **15**, 225 (1934); L. G. Groves and S. Sugden, *J. Chem. Soc.* **1937**, 1992; C. P. Smyth, *J. Phys. Chem.* **41**, 209 (1937); K. B. Everard and L. E. Sutton, *J. Chem. Soc.* **1951**, 2816; J. W. Smith, *ibid.* **1953**, 109; R. C. Cass, H. Spedding, and H. D. Springall, *ibid.* **1957**, 3451.

In order that resonance of these types may occur, giving partial double-bond character to the bonds connecting the benzene ring and the attached groups, the molecule must approximate the planar configuration appropriate to this double-bond character.   This is possible

in nitrobenzene itself; but in nitrodurene,

$$CH_3 \qquad CH_3$$
$$\langle\!\!\langle\ \rangle\!\!\rangle\!\!-NO_2,$$
$$CH_3 \qquad CH_3$$

steric interaction between the oxygen atoms of the nitro group and the ortho methyl groups prevents the assumption of the planar configuration, the nitro group being rotated somewhat about the N—C bond.   Accordingly the resonance with the ring should be less complete than for nitrobenzene.   This prediction is verified by the observed moments: nitrobenzene, 3.93 D; nitrodurene, 3.39 D; alkyl nitro compounds, 3.29 D.   It is seen that the moment in nitrodurene is reduced nearly to the alkyl value.   A similar result is obtained with nitroaminodurene, with moment 4.98 D, 1.12 smaller than that for *p*-nitroaniline, 6.10 D.   In this case there is inhibition also of the resonance involving interaction between the nitro and amino groups.

Steric effects of this sort should be small for the amino group, because of its small size, and nonexistent for bromine and other cylindrically symmetrical groups.   It is found that corresponding durene and benzene derivatives involving these groups differ only slightly in moment, the small differences being attributed to induction in the methyl groups; values found are the following: aminodurene, 1.39 D; aniline, 1.53 D; bromodurene, 1.55 D; bromobenzene, 1.52 D.

Steric interactions are often of significance in the orientation of substituents in aromatic molecules.   In the discussion of the Mills-Nixon effect in Section 6-3 it was mentioned that *ar*-tetrahydro-*β*-naphthol undergoes substitution in the 1 position on reaction with the phenyldiazonium ion, this effect being explained as the result of a somewhat greater contribution by one Kekulé structure than by the other to the normal state of the molecule.   Bromination also takes place, like diazotization, in the 1 position, but sulfonation and nitration are anomalous, leading to 3 derivatives.   This we attribute to the effect of steric repulsion of the sulfonic acid group or nitro group in the 1 position by the adjacent methylene group, leading to an increase in the heat of activation for 1 substitution great enough to overcome the rather small advantage over 3 substitution resulting from the Mills-Nixon effect.

Spectroscopic methods provide valuable information about the degree of planarity of conjugated systems.   Fluorene,

$$CH_2$$

which is planar, absorbs much more strongly than benzene in the 2500 Å region.   Biphenyl is intermediate, indicating some deviation from planarity, with interference with the conjugation of the two benzene rings, and the o,o'-substituted biphenyls absorb only little more strongly than benzene itself, showing a greater deviation from planarity than for biphenyl.[93]   Similarly, the absorption spectra of 9,10-diphenylanthracene, 9,10-di(α-naphthyl)anthracene, and 9,9'-dianthryl are almost identical with the spectrum of anthracene.[94]   Many other investigations of this sort have been reported.[95]

Overcrowded Molecules.—Many condensed aromatic molecules that are forced into nonplanar configurations by steric repulsion have been studied during recent years.   An example is 3,4-5,6-dibenzophenanthrene,

HH

The two carbon atoms shown with the symbol H attached would be only 1.40 Å apart if the molecule were planar, and the two hydrogen atoms would have to occupy the same position in space.   The molecule is found[96] to be deformed into a portion of a flat helix (either right or left handed), so as to separate these two carbon atoms, with their hydrogens, to about 3.0 Å.   Similar results have been reported for several other molecules.[97]

[93] L. W. Pickett, G. F. Walter, and H. France, *J.A.C.S.* **58**, 2182 (1936); M. T. O'Shaughnessy and W. H. Rodebush, *ibid.* **62**, 2906 (1940).

[94] R. N. Jones, *J.A.C.S.* **63**, 1658 (1941).

[95] Styrene, stilbene, etc., R. N. Jones, *J.A.C.S.* **65**, 1815, 1818 (1934); *Chem. Revs.* **32**, 1 (1943); nitro derivatives, W. G. Brown and H. Reagen, *J.A.C.S.* **69**, 1032 (1947); G. N. Lewis and G. T. Seaborg, *ibid.* **62**, 2122 (1940); G. Thomson, *J. Chem. Soc.* **1944**, 404; E. A. Braude, E. R. H. Jones, H. P. Koch, R. W. Richardson, F. Sondheimer, and J. B. Toogood, *ibid.* **1949**, 1890; etc.

[96] A. O. McIntosh, J. M. Robertson, and V. Vand, *J. Chem. Soc.* **1954**, 1661.

[97] G. M. J. Schmidt and others, *J. Chem. Soc.* **1954**, 3288, 3295, 3302, 3314; S. C. Nyburg, *Acta Cryst.* **7**, 779 (1954).

In di-p-xylylene,

$$H_2C \underline{\hspace{3cm}} CH_2$$

$H_2C \underline{\hspace{3cm}} CH_2$

the two benzene rings are nearly parallel to one another. If each were planar and the CH_2 carbon atoms had tetrahedral angles, the rings would be only 2.50 Å apart. Steric repulsion causes the benzene rings to deform so that the two C-substituted atoms of each ring are 0.133 Å from the plane of the other four. The distances between rings are increased in this way to 2.83 Å for the C-substituted carbon atoms and 3.09 Å for the others.[98] A similar deformation has also been found for di-m-xylylene.[99]

Conjugated Systems Involving Triple Bonds.—In diacetylene, $HC{\equiv}C{-}C{\equiv}CH$, resonance occurs with more structures than in butadiene, because the linear configuration permits two kinds of double bonds (in perpendicular planes) to contribute (Sec. 7-7). The observed value of the C—C bond length in this molecule[100] is 1.379 $\pm$ 0.001 Å; the same value has been found also in cyanogen, cyanoacetylene, methyldiacetylene, diacetylene dicarboxylic acid, and several other molecules.[101] This value corresponds to bond number 1.33 (Sec. 7-7); the amount of double-bond character of the central bond, 33 percent, is, as expected, thus about twice as great as for a single bond between two double bonds.

Vinylacetylene,[102] $H_2C{=}CH{-}C{\equiv}CH$, and vinyl cyanide,[103] have a planar bent structure with C=C—C bond angle 123° and C—C bond lengths 1.446 and 1.426 Å, respectively, corresponding to 13 to 20 percent of double-bond character (Table 7-9; note that correction -0.06 Å is made for adjacent double bond and triple bond). This agrees well with the value, 15 percent, found for butadiene.

The single bond between a carbon-carbon triple bond and a carbon-

[98] C. J. Brown, *J. Chem. Soc.* 1953, 3265.

[99] C. J. Brown, *J. Chem. Soc.* 1953, 3278.

[100] G. D. Craine and H. W. Thompson, *Trans. Faraday Soc.* **49**, 1273 (1953), The electron-diffraction value 1.36 $\pm$ 0.03 Å had been reported by L. Pauling. H. D. Springall, and K. J. Palmer, *J.A.C.S.* **61**, 927 (1939).

[101] For references see Sutton, *Interatomic Distances*.

[102] J. R. Morton, quoted in following reference.

[103] C. C. Costain and B. P. Stoicheff, *J. Chem. Phys.* **30**, 777 (1959).

oxygen double bond has been found to have the length 1.445 ± 0.001 Å by a microwave study,[103a] showing that the resonance is about the same as in vinylacetylene.

The chemical properties of conjugated systems, including especially their power of transmitting the effects of groups, can be accounted for in a qualitative way by the resonance concept. For example, an electron-donating group such as $(CH_3)_2\ddot{N}—$ in a molecule $(CH_3)_2\ddot{N}CH=CH—CH=R$ can transmit its electrons to group R by resonance with the structure $(CH_3)_2N^+=CH—CH=CH—\ddot{R}^-$. Phenomena of this type have been discussed especially by Robinson.[104]

8-8. RESONANCE IN HETEROCYCLIC MOLECULES

For pyridine, pyrazine, and related six-membered heterocyclic molecules Kekulé resonance occurs as in benzene, causing the molecules to be planar and stabilizing them by about 40 kcal/mole. The inter-atomic distances observed in these molecules,[105] $C—C = 1.40$ Å, $C—N = 1.33$ Å, and $N—N = 1.32$ Å, are compatible with this structure. The resonance energy found for quinoline, 69 kcal/mole, is about the same as that of naphthalene.

In cyanuric triazide,[106] $C_3N_3(N_3)_3$, and the cyanuric tricyanamide ion,[107] $[C_3N_3(NCN)_3]^{---}$, the cyanuric ring has the configuration and dimensions expected for Kekulé resonance. An interesting type of resonance[108] is shown by the cyameluric nucleus, C_6N_7, in the [yamelurate ion, $[C_6N_7O_3]^{---}$, and the hydromelonate ion, $cC_6N_7(NCN)_3]^{---}$ (Fig. 8-5). The electronic structure of this nucleus corresponds to resonance not only between the two valence-bond

[103a] C. C. Costain and J. R. Morton, *J. Chem. Phys.* **31**, 389 (1959).

[104] R. Robinson, *Outline of an Electrochemical [Electronic] Theory of the Course of Organic Reactions*, Institute of Chemistry of Great Britain and Ireland, London, 1932; *Society of Dyers and Colourists, Jubilee Journal*, **1934**, 65.

[105] For pyridine, $C—C = 1.395 \pm 0.005$ Å and $C—N = 1.340 \pm 0.005$ Å (microwave), B. Bak, L. Hansen, and J. Rastrup-Andersen, *J. Chem. Phys.* **22**, 2013 (1954); for *s*-triazene, $C—N = 1.319 \pm 0.005$ Å (x-ray diffraction), P. J. Wheatley, *Acta Cryst.* **8**, 224 (1955); for *s*-tetrazine, $C—N = 1.334 \pm 0.005$ Å, $N—N = 1.321 \pm 0.005$ Å (x-ray diffraction), F. Bertinotti, G. Giacomello, and A. M. Liquori, *ibid.* **9**, 510 (1956).

[106] Hughes, *loc. cit.* (23), 1, 650; Knaggs, *loc. cit.* (23); *J. Chem. Phys.* **3**, 241 (1935).

[107] J. L. Hoard, *J.A.C.S.* **60**, 1194 (1938).

[108] L. Pauling and J. H. Sturdivant, *Proc. Nat. Acad. Sci. U. S.* **23**, 615 (1937). Chemical evidence regarding the structure of the cyameluric nucleus has been presented by C. E. Redemann and H. J. Lucas, *J.A.C.S.* **61**, 3420 (1939).

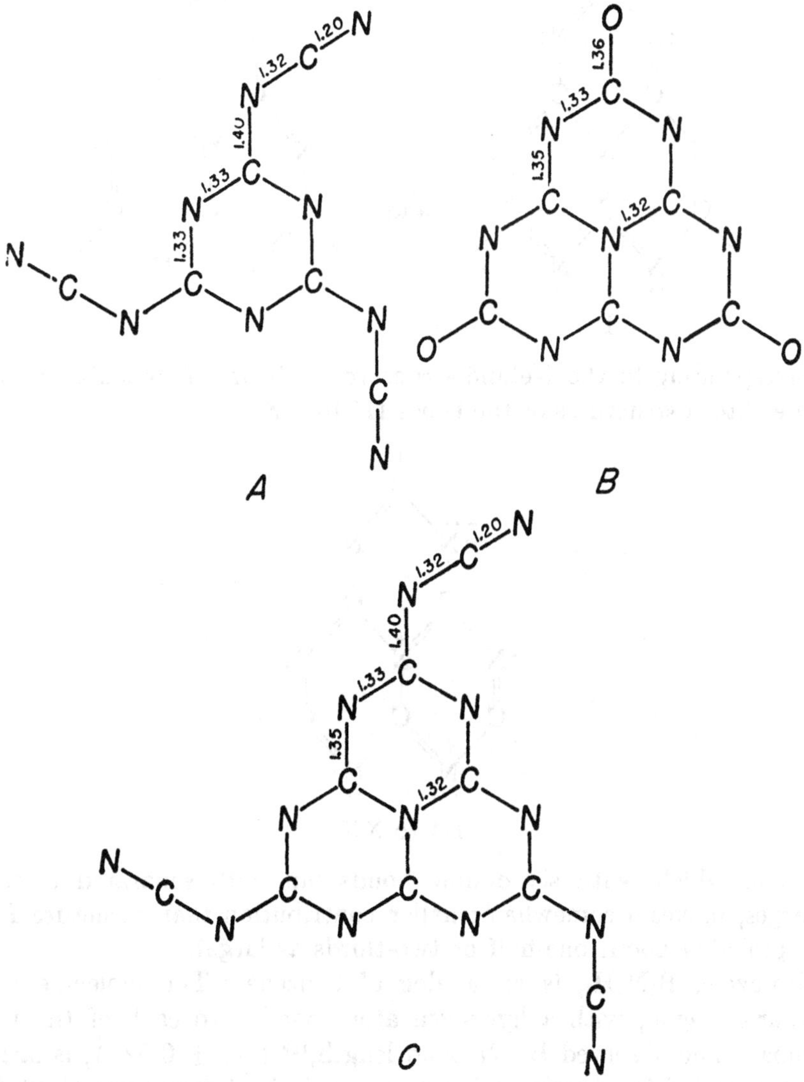

Fig. 8-5.—The structures of (A) the cyanuric tricyanamide ion, [C$_6$N$_9$]$^{---}$, (B) the cyamelurate ion, [C$_6$N$_7$O$_3$]$^{---}$, and (C) the cyameluric tricyanamide (hydromelonate) ion, [C$_9$N$_{13}$]$^{---}$. Predicted values of interatomic distances are shown. The molecules are planar.

structures I and II

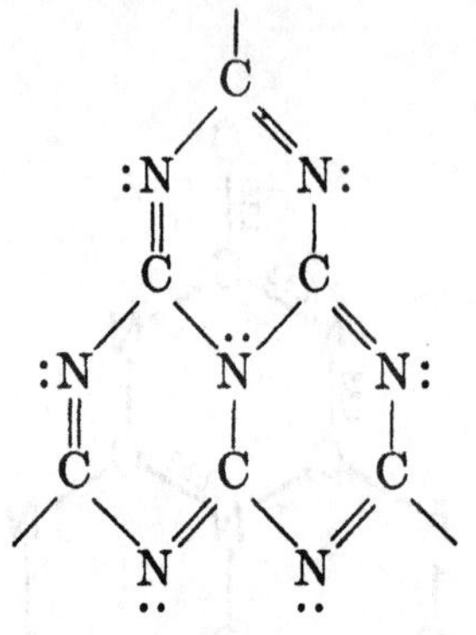

I   and   II

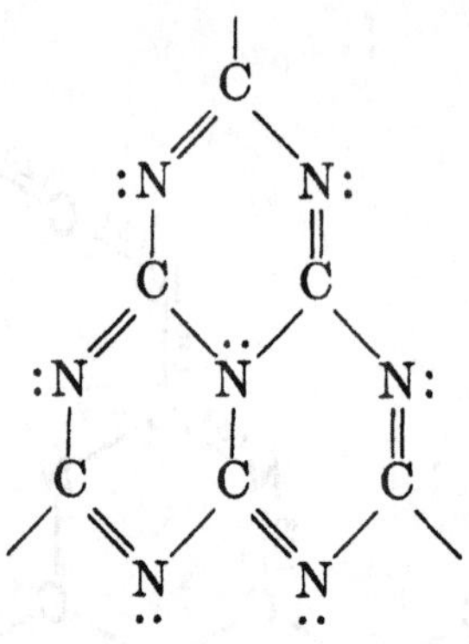

(corresponding to the Kekulé structures of benzene) but also among the eighteen structures of the types III to XX,

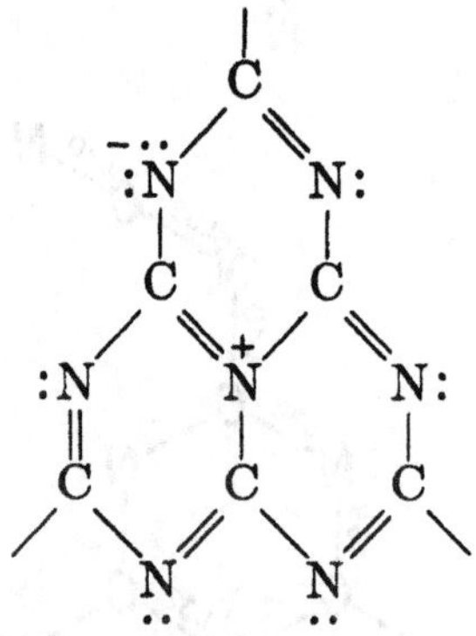

III to XX

each of which, with six double bonds but with separated electric charges, makes a somewhat smaller contribution than structure I or II (probably about one-half or two-thirds as large).

Borazole, $B_3N_3H_6$, is an analog of benzene. The molecule is a planar hexagon, with a hydrogen atom bonded to each of the ring atoms. The observed B—N bond length,[109] 1.44 ± 0.02 Å, is larger than expected for resonance between two Kekulé structures (1.33 Å), indicating that structures with an unshared pair on the nitrogen atom make an important contribution to the normal state. If the Kekulé structures contribute to the extent required for electroneutrality of the atoms, with the bonds having 22 percent of partial ionic character, the B—N bond number would be 1.28; this value agrees with the observed bond length.

[109] S. H. Bauer, *J.A.C.S.* **60**, 524 (1938).

For the five-membered heterocyclic molecules furan, pyrrole, and thiophen, with the conventional structure

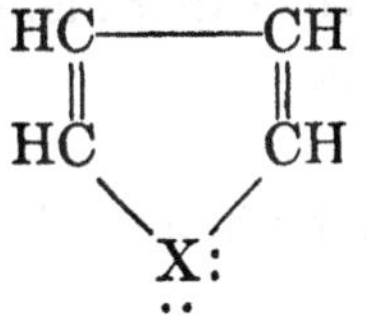

$$
\begin{array}{ccc}
HC & \!\!\!\!\!\!\text{------} & CH \\
\| & & \| \\
HC & & CH \\
& \diagdown \quad \diagup & \\
& \underset{\cdot\cdot}{X}\!\!: &
\end{array}
$$

I

resonance is expected to occur with structures of the types

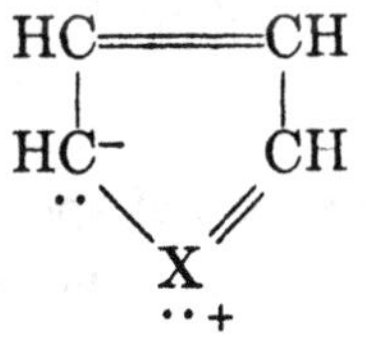

$$
\begin{array}{ccc}
HC & \!\!\!\!\!\!\text{=====} & CH \\
| & & | \\
\overset{\cdot\cdot}{HC}{}^{-} & & CH \\
& \diagdown \quad \diagup\!\!\!\diagup & \\
& \underset{\cdot\cdot\,+}{X} &
\end{array}
$$

II, III

and

$$
\begin{array}{ccc}
HC & \!\!\!\!\!\!\text{------} & \overset{\cdot\cdot}{C}H^{-} \\
\| & & | \\
HC & & CH \\
& \diagdown \quad \diagup\!\!\!\diagup & \\
& \underset{\cdot\cdot\,+}{X} &
\end{array}
$$

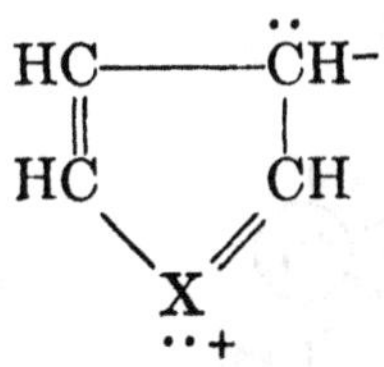

IV, V

The thermochemical data for these substances give the values 23, 31, and 31 kcal/mole, respectively, for the energy of this resonance. It is interesting that the extent of the resonance, as indicated by the magnitude of the resonance energy, increases with decrease in electronegativity of X; the very electronegative oxygen atom has a smaller tendency to assume the positive charge accompanying structures II to V than has the less electronegative nitrogen or sulfur atom. This conclusion is further substantiated by the observed interatomic distances, C—O = 1.37 Å, C—N = 1.42 Å, and C—S = 1.74 Å in furan,[110] pyrrole,[111] and thiophen,[111] respectively; these correspond (with consideration of the electric charge effect) to about 23 percent total contribution of structures II to V for furan, 24 percent for pyrrole, and 28 percent for thiophen.

These values indicate that each of the structures with separated charges makes one-quarter the contribution of the principal structure

[110] B. Bak, L. Hansen, and J. Rastrup-Andersen, *Discussions Faraday Soc.* 19, 30 (1955).

[111] Schomaker and Pauling, *loc. cit.* (72).

I. The corresponding values of the C—C bond lengths (Table 7-9) are 1.439 Å ($n = 1.25$, single bond of I) and 1.377 Å ($n = 1.625$), in satisfactory agreement with the experimental values, 1.440 ± 0.016 Å and 1.354 ± 0.016 Å, respectively.

Similar resonance is shown by indole, (resonance energy 54 kcal/mole), carbazole, (resonance energy 91 kcal/ mole), and related molecules.

The pyrimidines and purines are of special interest because of their presence in the nucleic acids. The dimensions of these molecules are

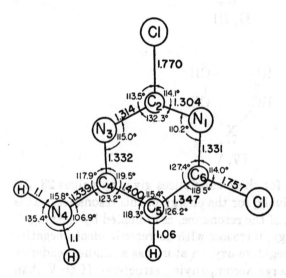

Fig. 8-6.—A drawing showing the dimensions of the molecule of 4-amino-2,6-dichloropyrimidine as determined by x-ray crystal analysis.

illustrated by Figures 8-6, 8-7, 8-8, 8-9, and 8-10, showing molecules of 4-amino-2,6-dichloropyrimidine,[112] 5-bromo-4,6-diaminopyrimidine,[112] uracil,[113] adenine (in adenine hydrochloride hemihydrate[114]), and guanine (in guanine hydrochloride monohydrate[115]), respectively. The conclusion has been drawn[116] that in these rings the C—C bond

[112] C. J. B. Clews and W. Cochran, *Acta Cryst.* **2**, 46 (1949).

[113] G. S. Parry, *Acta Cryst.* **7**, 313 (1954).

[114] W. Cochran, *Acta Cryst.* **4**, 81-92 (1951); J. M. Broomhead, *ibid.* **1**, 324 (1948).

[115] J. M. Broomhead, *Acta Cryst.* **4**, 92 (1951).

[116] L. Pauling and R. B. Corey, *Arch. Biochem. Biophys.* **65**, 164 (1956); M. Spencer, *Acta Cryst.* **12**, 59, 66 (1959).

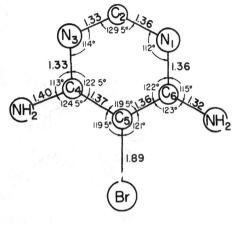

Fig. 8.7—A drawing showing the dimensions of the molecule of 5-bromo-4,6-diaminopyrimidine.

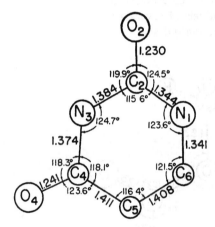

Fig. 8-8.—A drawing showing the dimensions of the molecule of uracil.

Fig. 8-9.—A drawing showing the dimensions of the molecule of adenine as determined by x-ray analysis of crystals of adenine hydrochloride hemihydrate.

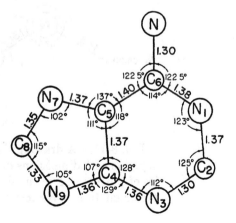

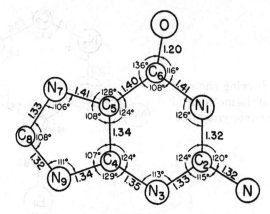

FIG. 8-10.—A drawing showing the dimensions of the molecule of guanine as determined by x-ray analysis of crystals of guanine hydrochloride monohydrate.

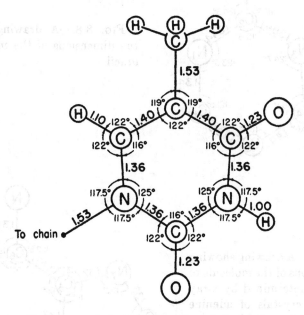

FIG. 8-11.—A drawing showing the molecule of thymine with dimensions derived from x-ray studies of purines and pyrimidines. The point of attachment to a polynucleotide chain is indicated.

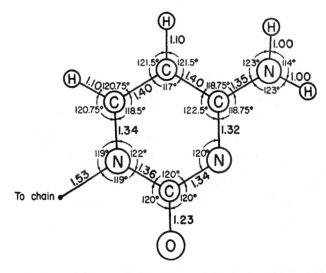

Fig. 8-12.—A drawing showing the molecule of cytosine with dimensions derived from x-ray studies of purines and pyrimidines. The point of attachment to a polynucleotide chain is indicated.

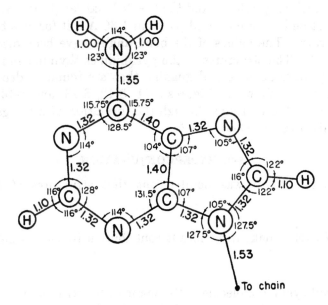

Fig. 8-13.—A drawing showing the molecule of adenine with dimensions derived from x-ray studies of purines and pyrimidines. The point of attachment to a polynucleotide chain is indicated.

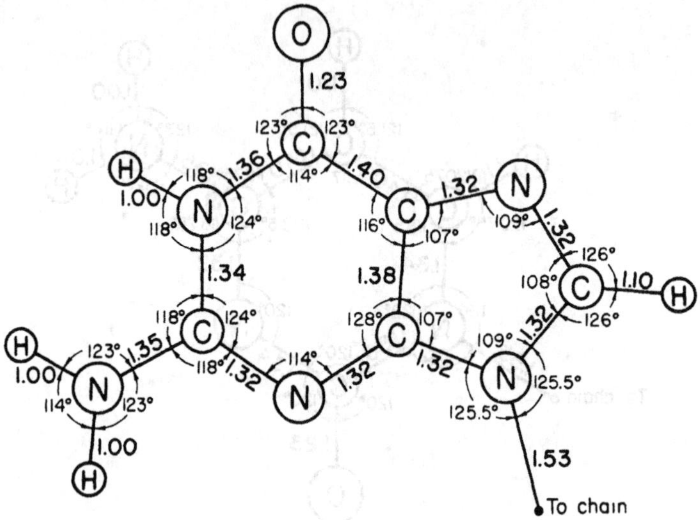

Fig. 8-14.—A drawing showing the molecule of guanine with dimensions derived from x-ray studies of purines and pyrimidines. The point of attachment to a polynucleotide chain is indicated.

length is uniformly 1.40 Å, and the C—N bond length is 1.32 Å if the nitrogen atom has an unshared pair and 1.36 Å if it forms a bond outside the ring. The values of the bond angles have been discussed in Section 4-8. The structures of the pyrimidines thymine and cytosine and the purines adenine and guanine that are found in deoxyribose-nucleic acid are shown in Figures 8-11, 8-12, 8-13, and 8-14.[116] The interaction of these groups through the formation of hydrogen bonds will be discussed in Chapter 12.

8-9. HYPERCONJUGATION

In Section 8-1 it was pointed out that structures of the type

H+

H—C≡C=N̈:⁻   make a significant contribution to the normal state of

/

H

the methyl cyanide molecule. Resonance with structures of this sort, involving the breaking of a single bond (in this case an H—C bond), is called *hyperconjugation*.

Hyperconjugation was first proposed by Wheland, in a discussion of the effect of alkyl groups on the stability of substituted methyl radicals.[117] The name was proposed by Mulliken, Rieke, and Brown, who gave a general theoretical discussion of the phenomenon and pointed out its widespread significance.[118]

For toluene, for example, the value 0 for the electric dipole moment would be expected if the methyl group were attached to the ring by a normal single bond and the C—H bonds of the group and the ring had the same amount of ionic character. The observed value, 0.37 D, indicates that structures of the type

contribute to a total extent of about 2.5 percent. Correspondingly the heats of hydrogenation of alkylolefines are less than those of the corresponding olefines by about 2 kcal/mole per methyl group adjacent to the double bond.

Hyperconjugation affects many properties of hydrocarbons and other molecules. Its effects are in general somewhat smaller than those of conjugation; they correspond to resonance energies of 1 or 2 kcal /mole (instead of 5 or 10), changes in bond length by 0.01 or 0.02 Å, and transfer of 0.01 or 0.02 units of electric charge. These structural changes are, however, great enough to have significant effect on many of the physical and chemical properties of substances.[119]

[117] G. W. Wheland, *J. Chem. Phys.* **2**, 474 (1934).

[118] R. S. Mulliken, C. A. Rieke, and W. G. Brown, *J.A.C.S.* **63**, 41 (1941).

[119] For thorough discussions of hyperconjugation see G. W. Wheland, *Resonance in Organic Chemistry*, John Wiley and Sons, New York, 1955; J. W. Baker, *Hyperconjugation*, Clarendon Press, Oxford, 1952; C. L. Deasy, *Chem. Revs.* **36**, 145 (1945); F. Becker, *Angew. Chem.* **65**, 97 (1953). The theory is discussed by A. Lofthus, *J.A.C.S.* **79**, 24 (1957).

CHAPTER 9

The Structure of Molecules and Complex Ions Involving Bonds with Partial Double-Bond Character

FROM the discussion in the preceding chapters several significant conclusions have been drawn about the structure of molecules and complex ions involving bonds between atoms of the heavier elements—those beyond the first row of the periodic system—and atoms such as the halogens or oxygen and groups such as hydroxyl, amino, carbonyl, cyanide, or nitro. It has been seen that the heavier atoms are not rigorously restricted by the octet rule, but can make use of d orbitals in bond formation; that electron donors such as the halogen and oxygen atoms and the hydroxyl and amino groups held by a single bond are able under certain circumstances to swing another pair of electrons into position for bond formation, thus giving some double-bond character to the bond; and that electron acceptors such as the cyanide and nitro groups held by a single bond are able to provide an orbital for a pair of electrons from the rest of the molecule, to give some double-bond character to the bond. By application of these ideas, with recourse to experimental information to settle doubtful points, a detailed description of the structure of molecules and complex ions of the heavier elements can be formulated, as described in the following sections. One new structural feature of compounds of carbon (and other first-row elements) is also discussed in this chapter—the effect of an electronegative atom attached to a carbon atom in inducing partial double-bond character of other bonds formed by the carbon atom (Sec. 9-3).

9-1. THE STRUCTURE OF SILICON TETRACHLORIDE AND RELATED MOLECULES

For about 75 years the silicon tetrachloride molecule (which we select as an example) was assigned the simple valence-bond structure A. G. N. Lewis introduced the practice of showing the unshared

$$A \quad \begin{matrix} & \text{Cl} \\ & | \\ \text{Cl} & -\text{Si} - \text{Cl} \\ & | \\ & \text{Cl} \end{matrix} \qquad\qquad A' \quad \begin{matrix} & :\ddot{\text{Cl}}: \\ & | \\ :\ddot{\text{Cl}} & -\text{Si} - \ddot{\text{Cl}}: \\ & | \\ & :\ddot{\text{Cl}}: \end{matrix}$$

electron pairs of the valence shells (structure A'), and the recognition of the partial ionic character of covalent bonds (Chap. 3) showed that the Si—Cl single bond can be described as having the amount of ionic character (Si^+Cl^-) that corresponds to the difference in electronegativity of silicon and chlorine (about 30 percent—Sec. 3-9). Structure A is analogous to the structures of carbon tetrachloride and silicon tetramethyl, in which the central atom uses its sp^3 tetrahedral bond orbitals to form four single bonds with its ligands.

When the structure of silicon tetrachloride was determined[1] it was found that the configuration of the molecule is that of the regular tetrahedron, as expected, but that the Si—Cl distance, 2.01 Å ± 0.02 Å, is much smaller than the sum of the covalent radii, 2.16 Å. A part of the 0.15 Å decrease in bond length can be attributed to the partial ionic character of the bond, as discussed in Section 7-2. The bond shortening calculated by Equation 7-1 is 0.08 Å. The remaining shortening, 0.07 Å, is attributed to partial double-bond character of the bond. Since the double bond may be formed with use of either the p_y or the p_z orbitals, the discussion of Section 7-7 is assumed to apply; Table 7-12 then shows that the observed shortening corresponds to about 23 percent of double-bond character.

This amount of double-bond character is to be expected from consideration of the principle of electroneutrality (Sec. 8-2). The 30 percent partial ionic character of the Si—Cl bond that corresponds to the difference in electronegativity of the atoms would place the charge +1.2 on the silicon atom in the $SiCl_4$ molecule. This electric charge would be reduced to zero if each bond had 30 per cent double-bond character, or to +0.2 (a value approximating electroneutrality) if each bond had 25 percent double-bond character. This amount of double-bond character (and the same amount of partial ionic character for each bond) is given by resonance among the six equivalent structures of type B:

$$B \quad \begin{matrix} & :\ddot{\text{Cl}}:^- \\ & | \\ :\ddot{\text{Cl}} & -\text{Si} =\text{Cl}:^+ \\ & | \\ & :\ddot{\text{Cl}}: \end{matrix}$$

[1] By the electron-diffraction method: R. Wierl, *Ann. Physik* **8**, 521 (1931); L. O. Brockway and F. T. Wall, *J.A.C.S.* **56**, 2373 (1934).

It is probably significant that for these structures only the four stable orbitals of the silicon valence shell need to be used.   The bond-angle strain is considerably less than for pure sp^3 orbitals, since d character can be introduced with little promotion energy.   Some contribution may be made also by the six structures of type C:

$$C \qquad \overset{\displaystyle :\!\overset{\cdot\cdot}{\underset{\cdot\cdot}{Cl}}\!:^{-}}{\underset{\displaystyle :\!\overset{\cdot\cdot}{\underset{\cdot\cdot}{Cl}}\!:^{-}}{\overset{+}{:}Cl\!=\!Si\!=\!\overset{\cdot\cdot}{\underset{}{Cl}}\!:^{+}}}$$

A significant dependence of the Si—Cl bond length on the nature of the other ligands in the molecule has been found.   Microwave spectroscopy has provided the values[2] (reliable to 0.002 Å) 2.021 Å in $HSiCl_3$, 2.021 Å in CH_3SiCl_3, and 2.048 Å in H_3SiCl, corresponding, by the method of interpretation given above, to about 18 percent, 18 percent, and 9 percent, respectively, of double-bond character.

It is reasonable to attribute the small amount of double-bond character in H_3SiCl to the failure of the H—Si bonds to release a silicon orbital in sufficient amount; each H—Si bond is expected from the electronegativity difference (0.3) to have 2.3 percent H^-Si^+ ionic character, and the three would hence release 0.07 silicon bond orbital, permitting 7 percent double-bond character to the Si—Cl bond, in close agreement with the value 9 percent given by the interatomic distance.

The normal Si—Cl distance 2.01 Å or 2.02 Å has been reported for several molecules, in addition to $SiCl_4$, $HSiCl_3$, and CH_3SiCl_3, in which two or three chlorine atoms are attached to one silicon atom: $SiCl_3SH$, H_2SiCl_2, Si_2Cl_6, Si_2Cl_6O, and $(CH_3)_2SiCl_2$.   For all of these molecules structures of the type $Cl^-Si\!=\!Cl^+$ can be written, and the same amount of double-bond character as in $SiCl_4$ is to be expected.

The Si—Cl distance in $SiClF_3$ is small—only 1.99 Å, corresponding to 31 percent double-bond character.   The increase over the amount for the other molecules is probably to be attributed to the release of bond orbitals by the largely ionic (70 percent) Si—F bonds, permitting the Si—Cl bond to assume the amount of double-bond character that completely neutralizes the transfer of electric charge corresponding to its normal partial ionic character.

The observed Si—Br bond length in H_3SiBr, 2.209 Å, is 0.10 Å less than the sum of the single-bond radii, 2.31 Å.   The predicted shortening for ionic character is 0.07 Å; the remainder, 0.03 Å, corresponds to 8 percent of double-bond character, approximately equal to the 7 percent bond-orbital release of the three H—Si bonds.   Similarly,

---

[2] For references for these and other values of interatomic distances in this chapter see Sutton, *Interatomic Distances.*

the Si—I bond length in H_3SiI, 2.433 Å is 0.07 Å less than the sum of the radii; of this shortening 0.05 Å is due to partial ionic character and 0.02 Å to double-bond character. The Si—Br distance in $SiBr_4$, $HSiBr_3$, $SiBr_2F_2$, and $SiBrF_3$ is about 2.16 Å, corresponding to about 26 percent double-bond character; the predicted amount of ionic character of the Si—Br bond is 22 percent.

As another example we may discuss H_3GeCl, with the observed (microwave) Ge—Cl distance 2.148 Å, and $HGeCl_3$, with distance 2.114 Å. (The electron-diffraction value for $GeCl_4$ is 2.09 ± 0.02 Å.) The sum of the radii, 2.21 Å, is changed by the correction for partial ionic character (Sec. 7-2) to the predicted single-bond value 2.174 Å. The observed bond length for H_3GeCl hence corresponds (Sec. 7-7) to 7 percent double-bond character, equal to the amount of bond orbital released by the three Ge—H bonds, and the bond lengths for $HGeCl_3$ and $GeCl_4$ correspond to 19 percent and about 28 percent double-bond character, approaching the value 30 percent that would neutralize the charge transfer of the partial ionic character of the single bond.

The bonds between chlorine, bromine, and iodine and the heavier fifth-group and sixth-group atoms seem to have little double-bond character. The observed bond lengths are approximately equal to the calculated single-bond values (with the correction for partial ionic character, as given in Section 7-2); for example, observed for PCl_3, 2.043 Å (calculated 2.03 Å); for $AsCl_3$, 2.161 Å (2.17 Å); for SCl_2, 2.00 Å (2.00 Å).

9-2. SILICON TETRAFLUORIDE AND RELATED MOLECULES

The calculated Si—F single-bond length (Sec. 7-2) is 1.69 Å; the observed Si—F distances are much smaller: 1.54 Å for SiF_4, 1.56 Å for $SiHF_3$, $SiClF_3$, and $SiBrF_3$, and 1.594 Å for SiH_3F. The extra shortening, 0.15 to 0.10 Å, corresponds to 65 to 35 percent of double-bond character (Sec. 7-7).

The difference in electronegativity of silicon and fluorine (2.2) has been interpreted as leading to 70 percent ionic character of the Si—F bond. This amount of ionic character, if uncompensated, would place the charge +2.8 on the silicon atom in SiF_4 and +0.7 in SiH_3F. The charge would be reduced to +0.2 by 65 percent double-bond character in SiF_4 (as given by the above discussion of the observed bond length) and to +0.35 in SiH_3F. These values are in reasonable accord with the electroneutrality principle, and, although the foregoing quantitative considerations are not thoroughly reliable, we conclude that the silicon-fluorine bonds have a large amount of double-bond character.[3]

All other bonds between fluorine and second-row or heavier atoms

---

[3] It is interesting that the internuclear distance of the gas molecule SiF is 1.603 Å; presumably the bond is rather similar to the Si—F bonds in stable molecules of quadrivalent silicon.

have interatomic distances that indicate a large amount of double-bond character. For example, the observed distance in PF_3, 1.535 Å, is 0.115 Å less than the calculated single-bond value, 1.65 Å, corresponding to about 40 percent of double-bond character.

9-3. THE FLUOROCHLOROMETHANES AND RELATED MOLECULES; THE EFFECT OF BOND TYPE ON CHEMICAL REACTIVITY

The structure $H_3C\!\!=\!\!\ddot{F}$: does not make a significant contribution to the normal state of the methyl fluoride molecule; the carbon atom has only four stable orbitals, of which three are occupied in the H—C bonds, leaving only one available for bonding the fluorine atom. If, however, two or more fluorine atoms are substituted into the methane molecule, resonance with structures of the type

$$F^-$$
$$H\!-\!\underset{\underset{H}{|}}{C}\!\!=\!\!F^+$$

may occur. It was discovered by Brockway[4] that the carbon-fluorine bond distances in CF_4, CH_2F_2, CHF_2Cl, and CF_2Cl_2 are significantly less than those in CH_3F, CH_2FCl, and $CHFCl_2$, and he attributed the difference to the effect of the double-bond character of one carbon-fluorine bond induced by the partial ionic character of another carbon-fluorine bond formed by the same carbon atom. This effect of induction of double-bond character by the partial ionic character of another bond is expected to be much more important for fluorine-substituted methanes than for molecules containing chlorine, bromine, or iodine for two reasons: first, the C—F bond has a larger amount of ionic character (43 percent) than the other carbon-halogen bonds (6 percent or less), and, second, multiple bonds with first-row atoms are more stable than those with heavier atoms.

The average of the observed distances 1.385 Å in CH_3F and 1.375 Å in C_2H_5F may be taken to be the normal carbon-fluorine single-bond distance. A decrease of about 0.03 Å is observed for molecules with two fluorine atoms attached to the same carbon atom (1.358 Å in CH_2F_2, 1.36 Å in $CHClF_2$, 1.35 Å in CCl_2F_2, 1.345 Å in CH_3CHF_2), about 0.05 Å for molecules with three fluorine atoms (1.332 Å in CHF_3, 1.328 Å in $CClF_3$, and 1.328 Å in ClF_3), and 0.06 Å for CF_4. These values of the bond shortening correspond (Sec. 7-7) to 8 percent, 15 percent, and 19 percent of double-bond character, respectively.[5]

[4] L. O. Brockway, *J. Phys. Chem.* **41**, 185 (1937).

[5] It is interesting that the observed distance 1.271 Å in the gas molecule CF corresponds to 40 percent of double-bond character, approximately enough to reduce the electric charge on the atoms (caused by the 43 percent of partial ionic character of the single bond) to zero.

For CF_4 this amount of double-bond character (19 percent), with the amount of partial ionic character of the C—F single bond (43 percent), corresponds to the electric charge +0.96 on the carbon atom. Hence we may say that this molecule contains a C^+ atom, with one negative charge resonating among the four fluorine atoms. As a first approximation the structure may be described as a resonance hybrid of the 12 structures of type A:

$$A \quad F^-\ \begin{matrix} F^- \\ | \\ C^+ = F^+ \\ | \\ F \end{matrix}$$

with, of course, contributions from many other structures, such as the four of type B, and the one of type C:

$$B \quad \begin{matrix} F^- \\ | \\ F-C^+-F \\ | \\ F \end{matrix} \qquad C \quad \begin{matrix} F \\ | \\ F-C-F \\ | \\ F \end{matrix}$$

The average carbon-chlorine bond length in substituted alkanes not containing fluorine is 1.767 ± 0.002 Å. Smaller values are observed in molecules in which fluorine atoms also are attached to the carbon atoms: 1.759 ± 0.003 Å in CH_2ClF, 1.751 ± 0.004 Å in $CClF_3$, and 1.74 ± 0.02 Å in CCl_2F_2, $CHClF_2$, $CHCl_2F$, and $CClF_2CClF_2$. Similarly, the normal carbon-bromine bond length is 1.937 ± 0.003 Å, and the value 1.91 Å has been found for $CBrF_3$ and CBr_3F. It is likely that this C—Cl and C—Br bond shortening by fluorine attached to the same carbon atom is the result of partial double-bond formation with use of the carbon bond orbital released by the large ionic character of the C—F bond. The shortening (0.01 to 0.03 Å) corresponds to 3 to 8 percent of double-bond character, as compared with 8 to 19 percent for C—F bonds; the difference probably reflects the smaller tendency of the heavier atoms to form multiple bonds and the smaller ionic character of the single bonds (which produces a charge separation that favors the reverse transfer of charge by double-bond formation).

We have seen (Sec. 8-6) that the partial double-bond character of the carbon-chlorine bond in the chloroethylenes and chlorobenzenes provides an explanation of the great stability of these substances relative to the chloroparaffins. The same explanation[6] applies to the fact[7] that although the substitution of one fluorine atom in an alkane mole-

[6] Brockway, *loc. cit.* (4).
[7] A. L. Henne and T. Midgley, Jr., *J.A.C.S.* **58**, 882 (1926).

cule gives an unstable product, which easily loses hydrogen fluoride to form an olefine or hydrolyzes to the alcohol, molecules containing two fluorine atoms attached to the same carbon atom are much more stable, the stability applying not only to the fluorine atoms but also to other halogen atoms on the same carbon atom.

9-4. PARTIAL DOUBLE-BOND CHARACTER OF BONDS BETWEEN THE HEAVIER NONMETAL ATOMS

The heavier nonmetals may be expected to make some use of the less stable orbitals of the outermost shell ($3d$ for P, S, Cl; $4d$ for As, Se, Br; etc.), as is indicated by the existence of compounds such as PCl_5 and SF_6, in which the central atom forms more bonds than permitted by the use of orbitals occupied by electron pairs in the adjacent noble gas. In our earlier discussion of the structure of PCl_5 it was pointed out that a rough quantum-mechanical treatment leads to the conclusion that the structure in which the phosphorus atom forms five covalent bonds, with use of one $3d$ orbital in addition to the $3s$ and three $3d$ orbitals, makes a significant contribution to the normal state of the molecule (about 8 percent).

It might accordingly be expected that the diatomic chlorine, bromine, and iodine molecules would have some partial double-bond character, corresponding to resonance of the normal structure $:\overset{..}{X}\!-\!\overset{..}{X}:$ with the structures $:\overset{-}{\underset{..}{X}}\!=\!\overset{+}{X}:$, $:\overset{+}{X}\!=\!\overset{-}{\underset{..}{X}}:$, and $:\overset{..}{X}\!\equiv\!\overset{..}{X}:$, in which one or both of the atoms make use of one orbital in addition to those of the noble-gas shell.[8]

An estimate of the amount of double-bond character in the molecules Cl_2, Br_2, and I_2 can be made by the consideration of interatomic distances. Let us evaluate the pure single-bond radii of the halogens by use of the observed carbon-halogen distances in the halogen-substituted alkanes (not containing fluorine). In these molecules the bond orbitals available to the carbon atom do not allow double-bond character to the bonds with the halogen atoms, except for the very small amount (1 percent or less) corresponding to hyperconjugation (Sec. 8-9). (The effect of fluorine atoms in inducing double-bond character of the adjacent bonds has been discussed in the preceding section.) The observed carbon-halogen bond lengths for substituted alkanes are[9] 1.767 ± 0.002 Å for C—Cl, 1.937 ± 0.003 Å for C—Br, and 2.135 ± 0.010 Å for C—I. The correction for electronegativity difference gives the radius-sum values 1.807 Å for C—Cl, 1.961 Å for C—Br, and 2.135 Å for C—I (no electronegativity difference), which on subtrac-

---

[8] R. S. Mulliken, *J.A.C.S.* **77**, 884 (1955).

[9] These are the averaged values given by Sutton, *Interatomic Distances.*

tion of 0.772 Å, the carbon single-bond radius, lead to the values 1.035 Å, 1.189 Å, and 1.363 Å for the pure single-bond radii of chlorine, bromine, and iodine, respectively.

The observed bond lengths in Cl_2 (1.988 Å), Br_2 (2.284 Å), and I_2 (2.667 Å) are therefore shorter than the lengths expected for pure single bonds by 0.082 Å, 0.094 Å, and 0.059 Å. These shortenings correspond (Sec. 7-7) to between 18 percent and 33 percent of double-bond character.

A similar treatment of bond lengths indicates somewhat smaller amounts of double-bond character, 5 to 20 percent, for S—S, Se—Se, P—P, and As—As bonds.

The observed bond lengths for bonds between unlike atoms (not of the first row) agree with those calculated from the average for bonds between like atoms, with the electronegativity correction. For example, the values 2.20 Å for P—P and 1.98 Å for Cl—Cl lead, with the electronegativity correction −0.054 Å, to 2.036 Å for P—Cl, in agreement with the experimental value 2.043 ± 0.003 Å for PCl_3. We conclude that these bonds have about the same amount of double-bond character as those between like atoms.

9-5. THE BORON HALOGENIDES

In the boron trimethyl molecule the boron atom is surrounded by three pairs of valence electrons, which are involved in the formation of single covalent bonds to the three carbon atoms of the methyl groups. An electron-diffraction study[10] has shown the molecule to be planar (except for the hydrogen atoms), as would be expected for sp^2 hybrid orbitals. The B—C distance is 1.56 ± 0.02 Å, which agrees reasonably well with the value 1.54 Å calculated, with the electronegativity correction, by the use of 0.81 Å for the boron single-bond radius.[11]

A similar structure, *A*,

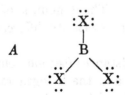

might be assigned to the boron halogenides. However, the boron atom has a fourth stable orbital that can be used for bond formation,

[10] H. A. Levy and L. O. Brockway, *J.A.C.S.* **59**, 2085 (1937).

[11] The gas molecule B_2, to which we may confidently assign the structure :B—B: (with unshared pairs largely 2s in character), has bond length 1.589 Å, corresponding to the radius 0.795 Å for boron.

and the three structures B, C, and D may be expected to be about as stable as A, the

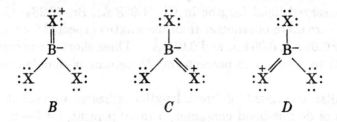

extra bond compensating for the separation of charge.   The molecules are hence expected to resonate equally among the three structures B,

TABLE 9-1.—INTERATOMIC DISTANCES FOR BORON HALOGENIDES

Bond	Distance calculated for single bond	Amount of double-bond character	Distance calculated with double-bond character	Observed in BX₃	Observed in BX
B—F	1.37 Å	63%	1.24 Å		1.262 Å
		(33)	1.29	1.295 Å	
B—Cl	1.77	22	1.71	1.73 ±0.02	1.716
B—Br	1.94	15	1.89	1.87 ±0.02	1.887
B—I	2.13	6	2.11		

C, and D, with A making a contribution such that the transfer of charge due to the partial ionic character of the bonds overcomes that due to the double bonds, making the resultant average charge of the boron atom nearly zero.   The amounts of partial ionic character corresponding to the electronegativity differences are given in the third column of Table 9-1.

The pure single-bond distances (second column) are calculated from the boron radius 0.80 Å and the halogen radii given in Section 9-4 (with 0.72 Å for fluorine), with the electronegativity correction (Sec. 7-2).   The corrections for double-bond character are made in the usual way (Sec. 7-5).   The fifth and sixth columns give the observed bond lengths for BF₃, BCl₃, BBr₃ and the gas molecules BF, BCl, and BBr, respectively.

The observed and calculated bond lengths agree to within the uncertainties of the calculation.   We conclude that when bond orbitals and electrons are available multiple-bond formation occurs to the extent required to make the electric charges of the atoms zero, in accordance with the electroneutrality principle.

The difference in observed bond length in BF (1.262 Å) and BF_3 (1.295 Å) results from the limitation to 33 percent of the double-bond character for each of the three bonds in BF_3; in BF the fourth orbital can be used to the extent needed to achieve electroneutrality.

There are some molecules in which the fourth orbital of boron is used for formation of another bond and is not available to permit double-bond character to be assumed by the B—X bonds. In these molecules the bond lengths should approximate the values calculated for B—X single bonds (Table 9-1 column 2). An example is ammonia-boron trifluoride, with the structure

```
   H           F
    \         /
  H—N——B—F
    /         \
   H           F
```

The observed B—F bond length[12] is 1.38 ± 0.01 Å, in agreement with the calculated value 1.37 Å. The values 1.39 ± 0.01 Å for trimethyl-amine-boron trifluoride[13] and 1.43 ± 0.03 Å for dimethylether-boron trifluoride[14] have also been reported. In trichloroborazole, to which we may assign the structure

```
            Cl
            |
    H       B       H
     \     / \     /
      N         N
      |         ‖
      B         B
     /  \     /  \
   Cl     N     Cl
          |
          H
```

the reported B—Cl bond lengths, 1.78 ± 0.03 Å from electron diffraction of the gas molecule[15] and 1.76 ± 0.01 Å from x-ray diffraction of the crystal,[16] agree with the calculated value, 1.77 Å. Similar agreement is found for the B—Br bond length in bromodiborane, B_2H_5Br:

[12] J. L. Hoard, S. Geller, and W. M. Cashin, *Acta Cryst.* **4,** 396 (1951).

[13] S. Geller and J. L. Hoard, *Acta Cryst.* **4,** 399 (1951).

[14] S. H. Bauer, G. R. Finlay, and A. W. Laubengayer, *J.A.C.S.* **67,** 339 (1945).

[15] K. P. Coffin and S. H. Bauer, *J. Phys. Chem.* **59,** 193 (1955).

[16] D. L. Coursen and J. L. Hoard, *J.A.C.S.* **74,** 1742 (1952).

observed,[17] 1.934 ± 0.010 Å; calculated, 1.94 Å.   (For the structure of diborane see Chap. 10.)

9-6. THE OXIDES AND OXYGEN ACIDS OF THE HEAVIER ELEMENTS

The older conventional valence-bond formulas for an ion such as the sulfate ion,

$$A \qquad O=\!\!\overset{\displaystyle O^-}{\underset{\displaystyle O}{\overset{|}{\underset{|}{S}}}}\!\!-O^-$$

involving single and double covalent bonds from the central atom to the surrounding oxygen atoms in numbers determined by the position of the central atom in the periodic table, have now fallen into general disuse, in consequence of the suggestion, originally made by Lewis in his 1916 paper and accepted by most subsequent investigators, that the octet rule is to be applied to the sulfur atom and other second-row and heavier atoms, and that only four covalent bonds are to be represented in the electronic structures of the sulfate ion and similar ions:

$$B \qquad \overset{\displaystyle :\!\overset{..}{O}\!\!:^-}{\underset{\displaystyle :\!\overset{..}{O}\!\!:^-}{^-:\!\overset{..}{O}\!\!-\!\!\overset{++}{S}\!\!-\!\!\overset{..}{O}\!\!:^-}}$$

On considering the question of the structure of these ions from the resonance point of view, we see that structure B, although it makes some contribution, is not of overwhelming importance, that other structures involving double bonds between the central atom and oxygen are significant, and that the available evidence indicates that the older valence-bond formulas such as A, with the double bonds resonating among the oxygen atoms, making them equivalent, and with the bonds considered to have partial ionic character, represent the ions somewhat more satisfactorily than the extreme structures of the type of B.

The observed values of interatomic distances in the tetrahedral ions of the ortho oxygen acids of the second-row elements are given in Table 9-2.   They are 0.15 to 0.19 Å less than the values calculated for the single-bond lengths with the covalent radii and the correction for

[17] K. Hedberg, M. E. Jones, and V. Schomaker, *2nd Int. Congr. Cryst.*, Stockholm, 1951; C. D. Cornwall, *J. Chem. Phys.* **18**, 1118 (1950).

TABLE 9-2.—INTERATOMIC DISTANCES IN TETRAHEDRAL IONS MO₄

	Si—O in SiO₄----	P—O in PO₄---	S—O in SO₄--	Cl—O in ClO₄-
Observed	1.61[a]	1.54[b]	1.49[c]	1.44 Å[d]
Single-bond value	1.77	1.73	1.70	1.69
Difference	− 0.16	− 0.19	− 0.21	− 0.25
Bond number	1.55	1.70	1.83	2.10
Bond ionic character	0.51	0.39	0.22	0.06
Charge on central atom	+ 0.96	+ 0.85	+ 0.29	− 0.90

[a] The average for many silicates is 1.62 + 0.02 Å. J. V. Smith, *Acta Cryst.* **7**, 479 (1954), has reviewed the best values to that time and proposed 1.60 ±0.01 Å.

[b] The most reliable values for the P—O distance in the phosphate group are 1.538 Å in KH₂PO₄ (neutron diffraction, G. E. Bacon and R. S. Pease, *Proc. Roy. Soc. London* A220, 397 [1953]), 1.54 Å in CaHPO₄ (G. MacLennan and C. A. Beevers, *Acta Cryst.* **8**, 579 [1955]), and 1.54 Å in BPO₄ (G. E. R. Schulze, *Z. physik. Chem.* **B24**, 215 [1934]).

[c] The value 1.49 Å has been reported in orthorhombic hydrazinium sulfate, N₂H₆SO₄, by I. Nitta, K. Sakurai, and Y. Tomiie, *Acta Cryst.* **4**, 289 (1951), and 1.51 Å in sulfohalite, Na₆(SO₄)₂ClF (A. Pabst. *Z. Krist.* **89**, 514 [1934]; T. Watanabe, *Proc. Imp. Acad. Tokyo* **10**, 575 [1934]).

[d] The average of values for LiClO₄, LiClO₄·3H₂O, and KClO₄, R. J. Prosen and K. N. Trueblood, unpublished research, Univ. Calif., Los Angeles, and for OH₃ClO₄, F. S. Lee and G. B. Carpenter, *J. Phys. Chem.*, in press.

partial ionic character (Sec. 7-2), as given in the second row of the table. Hence structure *B* alone is unsatisfactory. In accordance with the discussion in earlier sections of the chapter, we may well expect that the bivalent oxygen atom in these ions will strive to share four valence electrons with the central atom, and that structures of the types *C, D, E,* and *F* will contribute largely to the normal state of the ion:

In these structures the formation of a double covalent bond with one oxygen atom may be correlated with the ionic aspect of the bond to another atom, so that the bond orbitals used by the sulfur atom have the normal sp^3 nature (with only the usual small amount of d and s character—Chap. 4); some use may also be made of the $3d$ orbitals, in sp^3d or sp^3d^2 bond orbitals.

Let us interpret the most accurately determined bond length in Table 9-2, that for the phosphate ion, 1.54 Å, in terms of the assumed double-bond character.[18] The bond shortening 0.19 Å corresponds to bond number 1.70 (Equation 7-6, with $D_1 - D_2 = 0.21$ Å and $D_1 - D_3 = 0.34$ Å). The transfer of charge for 70 percent double-bond character of each of four bonds is -2.80 to the phosphorus atom, which with consideration of the formal charge $+1$ for structure B would leave -1.80. If each bond has the amount of ionic character corresponding to the electronegativity difference of the two atoms, 39 percent (Table 3-10), the charge $+2.65$ is transferred thereby to the phosphorus atom, which thus has a total residual charge of $+0.85$. This value is a reasonable one; in Chapter 13 it is pointed out that the properties of crystals of silicates and phosphates (failure to share edges and faces of XO_4 tetrahedra) indicate that the central atoms are eleccally charged, and a charge not greater than 1 is compatible with the electroneutrality principle.

Similar treatment of the other three ions leads to the values $+1.06$, $+0.29$, and -0.90 for the residual charges on the Si, S, and Cl atoms, respectively. The sequence $+1.06$, $+0.85$, $+0.29$, -0.90 for Si, P, S, Cl is a reasonable one, in comparison with the increasing electronegativity of the atoms, 1.8, 2.1, 2.5, 3.0, with oxygen 3.5, except that the negative value for Cl is unlikely.

It has been shown by neutron diffraction[19] that in orthorhombic KH_2PO_4 the phosphorus atom has two oxygen atoms and two OH groups ligated to it. The corresponding P—O distances, 1.508 Å and 1.583 Å, correspond to 92 percent and 48 percent of double-bond character, respectively. The average is the same as for the four equivalent P—O bonds, 70 percent. A similar distortion (with one short bond, 1.52 Å, and three longer ones, 1.57 $\pm$ 0.02 Å) has been reported in crystalline phosphoric acid.[20]

Very little accurate structural information about esters of the oxygen acids of the heavier elements has been reported. The molecule of tetramethyl orthosilicate, $Si(OCH_3)_4$, has[21] Si—C = 1.64 $\pm$ 0.03 Å, C—O = 1.42 $\pm$ 0.04 Å, and angle Si—O—C = 113° $\pm$ 2°. These values are approximately those expected from the foregoing considerations; the amount of double-bond character of the Si—O bonds is indi-

[18] This discussion closely resembles that given by L. Pauling, *J. Phys. Chem.* **56**, 361 (1952). For alternative discussions, see K. S. Pitzer, *J.A.C.S.* **70**, 2140 (1948); A. F. Wells, *J. Chem. Soc.* **1949**, 55; W. E. Moffitt, *Proc. Roy. Soc. London* A200, 409 (1950).

[19] G. E. Bacon and R. S. Pease, *Proc. Roy. Soc. London* A230, 359 (1955).

[20] J. P. Smith, W. E. Brown, and J. R. Lehr, *J.A.C.S.* **77**, 2728 (1955).

[21] K. Yamasaki, A. Kotera, M. Yokoi, and Y. Ueda, *J. Chem. Phys.* **18**, 1414 (1950).

cated by the bond length to be about the same as for the silicate ion, and the Si—O—C bond angle lies between the value for two single bonds, about 108°, and a double bond and a single bond, 114° (Sec. 4-8).

The only accurate information about the structure of the phosphate di-ester group, which is important because of its occurrence in the nucleic acid chain, is that given by an x-ray investigation of dibenzyl-phosphoric acid.[22] The P—O bonds to the esterified oxygen atoms have length 1.56 ± 0.01 Å and O—P—O angle $104° \pm 2°$. The P—OH bond and the fourth P—O bond have lengths 1.55 ± 0.01 Å and 1.47 ± 0.01 Å, respectively, and their mutual angle is 117°.

The pyro, meta, and other poly acids of the second-row atoms contain MO_4 tetrahedra with shared corners (oxygen atoms). As expected, the M—O bond lengths for the shared oxygen atoms are greater than for the others. Thus in the triphosphate ion of the $Na_5P_3O_{10}$ crystal[23] the P—O bond lengths to the shared oxygen atoms are 1.61 ± 0.03 Å (central phosphorus atom) and 1.68 ± 0.03 Å (outer phosphorus atoms), and those for the eight unshared oxygen atoms are 1.50 ± 0.03 Å. These values correspond to 35 percent, 13 percent, and 98 percent of double-bond character, respectively, and to the charge $+0.94$ on the central phosphorus atom and $+0.69$ on the outer phosphorus atoms, nearly the same as for the orthophosphate ion, $+0.85$ (Table 9-2). The bond angle O—P—O for the bonds to the shared oxygen atoms is 98° and the angle P—O—P is 121°.

In the diphosphate (pyrophosphate) ion in $Na_4P_2O_7 \cdot 10H_2O$ the P—O bond lengths to the shared oxygen atom are 1.63 Å and to the unshared oxygen atoms 1.47 ± 0.02 Å, and the value of the P—O—P angle is 134°.[24]

Many minerals can be described as salts of polysilicic acids, with SiO_4 tetrahedra sharing corners with one another. Their structure is discussed in Chapter 13.

The Chlorate Ion and Related Ions.—The conventional electronic structure for the chlorate ion is

[22] J. D. Dunitz and J. S. Rollett, *Acta Cryst.* **9**, 327 (1956).
[23] D. R. Davies and D. E. C. Corbridge, *Acta Cryst.* **11**, 315 (1958).
[24] D. M. MacArthur and C. A. Beevers, *Acta Cryst.* **10**, 428 (1957).

in which an unshared pair of electrons on the chlorine atom occupies one of its outer orbitals. The Cl—O distance observed[25] in $NaClO_3$ and $KClO_3$ is 1.46 ± 0.01 Å, with bond angle $108.0° \pm 1.0°$; the ion has a trigonal pyramidal configuration. This bond length corresponds to 91 percent of double-bond character and resultant charge -0.73 on the chlorine atom. It might well be expected that the charge on the chlorine atom would be the same as in the perchlorate ion, $+0.35$, which would lead to 64 percent of double-bond character and predicted Cl—O bond length 1.51 Å.

An experimental value[26] for the Cl—O bond length for the chlorite ion, ClO_2^-, is 1.57 ± 0.03 Å for NH_4ClO_2; the bond angle is $110° \pm 2°$. The bond length corresponds to bond number 1.37 and charge $+0.38$ on the chlorine atom, which is reasonable.

It is of interest to note that in these complexes, as in the halogen compounds of the elements of the fifth and sixth groups, the bond angles have values close to those expected for single covalent bonds: $108.0°$ in the chlorate ion and $110°$ in the chlorite ion.

It is probable that in the oxyacids of the heavier atoms, such as H_2CrO_4, H_2MnO_4, $HMnO_4$, and H_2SeO_4, the M—O bonds have a large amount of double-bond character, and that the properties of the acids are influenced by this to some extent.

The Strengths of the Oxygen Acids.—The strengths of acids are, of course, closely related to their molecular structure. It is interesting that the acid constants of acids involving a central atom to which oxygen atoms and hydroxyl groups are ligated are given roughly by two simple rules.[27]

Rule 1. The successive acid constants K_1, K_2, K_3, $\cdots$ of a polyprotic acid are in the ratios $1:10^{-5}:10^{-10} \cdots$. For phosphoric acid, for example, the first ionization constant has the value 0.75×10^{-2}, the second 0.62×10^{-7}, and the third 1×10^{-12}. We see that these three constants are closely in the ratio $1:10^{-5}:10^{-10}$.

For sulfurous acid, H_2SO_3, the first and second constants have the values 1.2×10^{-2} and 1×10^{-7}, which are again in the ratio $1:10^{-5}$. It is found that this rule, that each ionization constant of an acid is 100,000 times smaller than the preceding one, holds well for all of the acids of the class under consideration.

[25] R. G. Dickinson and E. A. Goodhue, *J.A.C.S.* **43**, 2045 (1921); W. H. Zachariasen, *Z. Krist.* **71**, 517 (1929); J. G. Bower, R. A. Sparks, and K. N. Trueblood, unpublished research, Univ. of Calif., Los Angeles.

[26] R. B. Gillespie and K. N. Trueblood, unpublished research, Univ. of Calif., Los Angeles.

[27] L. Pauling, *General Chemistry*, W. H. Freeman and Co., San Francisco, 1947; *School Science and Math.* **1953**, 429.

TABLE 9-3.—STRENGTHS OF OXYGEN ACIDS

First class: Very weak acids, $X(OH)_n$ or H_nXO_n
First acid constant about 10^{-7} or less K_1

	K_1
Hypochlorous acid, HClO	9.6×10^{-7}
Hypobromous acid, HBrO	2×10^{-9}
Hypoiodous acid, HIO	1×10^{-11}
Silicic acid, H_4SiO_4	1×10^{-10}
Germanic acid, H_4GeO_4	3×10^{-9}
Boric acid, H_3BO_3	5.8×10^{-10}
Arsenious acid, H_3AsO_3	6×10^{-10}
Antimonous acid, H_3SbO_3	10^{-11}

Second class: Weak acids, $XO(OH)_n$ or H_nXO_{n+1}
First acid constant about 10^{-2} K_1

	K_1
Chlorous acid, $HClO_2$	1×10^{-2}
Sulfurous acid, H_2SO_3	1.2×10^{-2}
Phosphoric acid, H_3PO_4	0.75×10^{-2}
Phosphorous acid, H_2HPO_3	1.6×10^{-2}
Hypophosphorous acid, HH_2PO_2	1×10^{-2}
Arsenic acid, H_3AsO_4	0.5×10^{-2}
Periodic acid, H_5IO_6	2.3×10^{-2}
Nitrous acid, HNO_2	0.45×10^{-3}
Acetic acid, $HC_2H_3O_2$	1.80×10^{-5}
Carbonic acid, H_2CO_3	0.45×10^{-6}

Third class: Strong acids, $XO_2(OH)_n$ or H_nXO_{n+2}
First acid constant about 10^3
Second acid constant about 10^{-2} K_1 K_2

	K_1	K_2
Chloric acid, $HClO_3$	Large	
Sulfuric acid, H_2SO_4	Large	1.2×10^{-2}
Selenic acid, H_2SeO_4	Large	1×10^{-2}

Fourth class: Very strong acids, $XO_3(OH)_n$ or H_nXO_{n+3}
First acid constant about 10^8

Perchloric acid, $HClO_4$	Very strong
Permanganic acid, $HMnO_4$	Very strong

Rule 2. *The value of the first ionization constant is determined by the formula of the acid, written as* $XO_m(OH)_n$: *if m is zero (no excess of oxygen atoms over hydrogen atoms, as in* $B(OH)_3$) *the acid is very weak, with* $K_1 \leqq 10^{-7}$; *for m = 1 the acid is weak, with* $K_1 \cong 10^{-2}$; *for m = 2 the acid is strong, with* $K_1 \cong 10^3$: *for m = 3 the acid is very strong, with* $K_1 \cong 10^8$.

It is interesting that the same factor, 10^{-5}, occurs in this rule as in Rule 1.

The application of the rule is shown by the constants given in Table 9-3.

The first rule can be understood as reflecting the increase in electric attraction of the negative ion for the positive proton with increase in the degree of ionization. An explanation of the second rule can be given by the following argument: Let us consider the acids HClO, $HClO_2$, $HClO_3$, and $HClO_4$. According to the second rule the first acid, hypochlorous acid, should be a very weak acid, the second acid, chlorous acid, should be a weak acid, the third acid, chloric acid, should be a strong acid, and the fourth acid, perchloric acid, should be a very strong acid. If hypochlorous acid, HClO, ionizes, the negative ion that is formed, ClO^-, has its negative charge concentrated on a single oxygen atom. The force of attraction of the proton to this oxygen atom would be characteristic of the force that leads to the formation of an O—H valence bond. Now let us consider chlorous acid. In the chlorite ion, ClO_2^-, the negative charge is divided between two oxygen atoms, and as the proton approaches one of the oxygen atoms, in the formation of the O—H bond in chlorous acid, the attraction would be expected to be smaller than in the case of hypochlorite ion. The acid constant for chlorous acid would accordingly be expected to be larger than that for hypochlorous acid. Similarly, in the chlorate ion, ClO_3^-, formed by ionization of chloric acid, the total negative charge would be divided among the three oxygen atoms, and the attraction of one of the oxygen atoms for an approaching proton would be still smaller, corresponding roughly to that for one-third of a negative charge, rather than for one-half of a negative charge for chlorite ion and one negative charge for the hypochlorite ion; this would be expected to cause chloric acid to be a still stronger acid than chlorous acid. The same argument leads us to expect perchloric acid to be still stronger than chloric acid.

It is seen that all of the acids listed in the first section of Table 9-3 have one hydrogen atom for every oxygen atom: their formulas are of the types $Cl(OH)$, $As(OH)_3$, and $Si(OH)_4$.

In the second part of Table 9-3, the class of weak acids with first acid constant about 10^{-2}, there are several acids in which the number of hydrogen atoms is one less than the number of oxygen atoms. These include acids such as chlorous acid, $ClO(OH)$; sulfurous acid, $SO(OH)_2$; phosphoric acid, $PO(OH)_3$; and periodic acid, $IO(OH)_5$.

Also given in this class are two acids, phosphorous acid and hypophosphorous acid, which seem to be out of place, inasmuch as their formulas, H_3PO_3 and H_3PO_2, seem not to put them in this class. Their acid constants, 1.6×10^{-2} and 1×10^{-2}, respectively, are, however, appropriate to the class, and an explanation must be sought for the apparent abnormality. The explanation is that one of the hydrogen atoms in phosphorous acid is bonded directly to the phosphorus atom. and two of the hydrogen atoms in hypophosphorous acid are bonded to the phosphorus atom.   The correct structural formula of phosphorous acid is $HPO(OH)_2$; this formula shows that the phosphorus atom has, in addition to a hydrogen atom directly bonded to it, one oxygen atom and two hydroxyl groups bonded to it.   The structural formula for hypophosphorous acid is $H_2PO(OH)$; in this acid the phosphorus atom has two hydrogen atoms and one oxygen atom bonded to it, as well as one hydroxyl group.   There is independent evidence of several sorts, obtained from physical chemical experiments, to show that in these acids there are hydrogen atoms bonded directly to phosphorus. The ions of these acids—phosphite ion, HPO_3^{--}, and hypophosphite ion, $H_2PO_2^{--}$—represent intermediate structures between the phosphate ion, PO_4^{---}, and the phosphonium ion, PH_4^+.   In each of these ions there is a phosphorus atom bonded to four other atoms, hydrogen or oxygen, which surround it tetrahedrally.

Nitrous acid, acetic acid (as well as the other carboxylic acids), and carbonic acid deviate somewhat from the simple rule in the values of their acid constants.   The deviation for nitrous acid and the carboxylic acids can be attributed to their electronic structure—the tendency of first-row atoms to form stable double bonds more easily than heavier atoms.   For carbonic acid the low value of the first acid constant is due in part to the existence of some of the unionized acid in the form of dissolved CO_2 molecules rather than the acid H_2CO_3.   It has been found that the ratio of the concentration of dissolved CO_2 molecules to H_2CO_3 molecules is about 25, so that the true acid constant for the molecular species H_2CO_3 is about 2×10^{-4}.

Oxygen acids that do not contain a single central atom have strengths corresponding to reasonable extensions of the rules, as shown by the following examples:

	K_1	K_2
Very weak acids: $K_1 = 10^{-7}$ or less		
Hydrogen peroxide, HO—OH	2.4×10^{-12}	
Hyponitrous acid, HON—NOH	9×10^{-8}	1×10^{-11}
Weak acids: $K_1 = 10^{-2}$		
Oxalic acid, HOOC—COOH	5.9×10^{-2}	6.4×10^{-5}

For hyponitrous acid and oxalic acid the second ionization constant is only about 10^{-3} times the first one, rather than 10^{-5}, as in the case with acids with a single central atom.   The larger value of the second ionization constant for these acids can be explained as resulting from a smaller effect of the negative charge produced by the first ionization, because of its larger distance from the second hydroxyl group that is undergoing ionization.

The arguments that have been presented to explain the two simple rules that represent reasonably well the observed strengths of the oxygen acids do not in fact account for the simple form of these rules.   We may consider that it is a fortunate circumstance that the successive ionization constants for a single polyprotic acid have the same ratio, 10^{-5}, and also that the first ionization constants for oxygen acids of the various sorts discussed in Table 9-3 have the same ratio, 10^{-5}, which happens also to be equal to the ratio for the first rule.   It is this fact that has made it possible to summarize the acid constants in such a simple way and that makes the two rules easily remembered and easily used, without confusion.

The acid constants of the hydrohalogenic acids are discussed in Appendix XI.

Sulfuryl Fluoride and Related Molecules.—The substitution of halogen for hydroxyl in these acids results in molecules in which the oxygen and halogen atoms are attached to the central atom by bonds similar in character to those in oxygen acids and the halides.   Sulfuryl fluoride, SO_2F_2, which has been carefully studied by microwave spectroscopy,[28] has distances S—O = 1.405 $\pm$ 0.003 Å and S—F = 1.530 $\pm$ 0.003 Å and angles O—S—O = 123°58' $\pm$ 12' and F—S—F = 96°7' $\pm$ 10'.   The S—O distance is 0.105 Å less than that in the sulfate ion, 0.295 Å less than the single-bond value.   It corresponds to bond number 2.38, and hence, with consideration of the 22 percent partial ionic character of the bonds, to the charge -0.24 on each oxygen atom.   Moreover, the S—F distance corresponds to bond number 1.33, and hence, with consideration of the 43 percent partial ionic character, also to the charge -0.24 on each fluorine atom.   The charge on the sulfur atom is $+0.96$. Thus the interatomic distances correspond to a structure agreeing well with the electroneutrality rule and the electronegativities of the atoms. The increase in the charge on the sulfur atom from $+0.70$ in the sulfate ion to $+0.96$ in sulfuryl fluoride reflects in a reasonable way the replacement of two oxygen atoms by atoms of the more electronegative element fluorine.

The value of the O—S—O angle, much larger than that of the F—S—F angle, reflects the large amount of double-bond character in the S—O bonds.

In many other molecules of this sort also, such as POF_3, $POCl_3$, PSF_3, $PSCl_3$, and SOF_2, the bond lengths and bond angles agree to within their experimental uncertainties with the values expected for structures with the amounts of double-bond character indicated by the atomic electronegativities and the electroneutrality principle.   For example, for the molecules $PSCl_3$, PSF_3, $PSBr_2F$, and $PSBr_3$ the ob-

---

[28] D. R. Lide, Jr., D. E. Mann, and R. M. Fristrom, *J. Chem. Phys.* **26**, 734 (1957).

served P—S bond lengths lie in the range 1.85 Å to 1.89 Å; the calculated bond length for bond number 2.08, which makes the sulfur atom electrically neutral, is 1.87 Å.

Oxides of the Heavier Elements.—The structures of the oxides of the heavier nonmetallic elements are similar to those of the oxygen acids. In sulfur dioxide the S—O distance is observed[29] to be 1.432 ± 0.001 Å, which is somewhat less than that in the sulfate ion. The value of the O—S—O bond angle, 119.54°, lies close to that ex-

pected for the structure
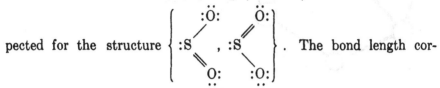
. The bond length cor-

responds to bond number 2.34 and residual charges +0.36 on the sulfur atom and −0.18 on each oxygen atom. This charge distribution leads to 1.25 D for the electric dipole moment of the molecule (with neglect of contributions by unshared electron pairs), somewhat smaller than the observed value,[29] 1.59 ± 0.01 D.

Sulfur trioxide, for which the structure
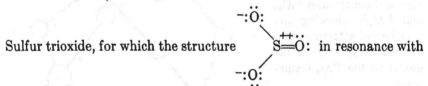
in resonance with

other structures may be written, has, as expected, a planar structure with bond angles 120°. The S—O bond length[30] is 1.43 ± 0.02 Å, equal, to within experimental error, to that in SO_2. The calculated bond number is therefore the same, 2.34, as is the charge on the oxygen atoms, −0.18, that on the sulfur atom being +0.54.

Sulfur trioxide easily polymerizes to form the trimer S_3O_9 and also an infinite asbestos-like polymer. In these polymers each sulfur atom is surrounded by a tetrahedron of four oxygen atoms, two of which are shared with other tetrahedra. The S—O bond lengths for the unshared oxygen atoms are 1.40 Å for the trimer[31] and 1.41 Å for the infinite polymer.[32] These values correspond to bond number about 2.5 and zero charge on the oxygen atoms. For the shared oxygen atoms the bond lengths are between 1.59 Å and 1.63 Å, corresponding to bond number about 1.26 and zero charge on the oxygen atoms.

The sulfur bond angles, about 125° for the two unshared oxygen

[29] D. Kivelson, *J. Chem. Phys.* **22**, 904 (1954); G. F. Crable and W. V. Smith, *ibid.* **19**, 502 (1951); M. H. Sirvetz, *ibid.* 938.

[30] K. J. Palmer, *J.A.C.S.* **60**, 2360 (1938).

[31] H. C. J. De Decker and C. H. MacGillavry, *Rev. trav. chim.* **60**, 153 (1941).

[32] R. Westrik and C. H. MacGillavry, *Acta Cryst.* **7**, 764 (1954).

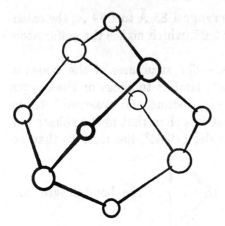

FIG. 9-1.—The structure of the molecules P_4O_6 and As_4O_6. Large circles represent phosphorus or arsenic atoms, small circles oxygen atoms.

FIG. 9-2.—The structure of the molecules P_4O_{10} and $P_4O_6S_4$, showing the positions of attachment of the four oxygen or sulfur atoms to the P_4O_6 framework.

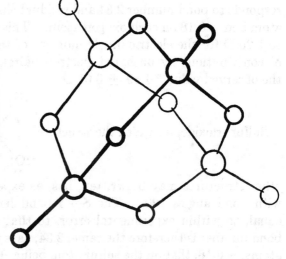

atoms and 100° for the two shared oxygen atoms, reflect the larger amount of multiple-bond character for the bonds to the unshared atoms.

The molecules P_4O_6, P_4O_{10}, and As_4O_6 have interesting configurations.[33] In P_4O_6 and As_4O_6 the four phosphorus or arsenic atoms are at the corners of a tetrahedron, each bonded to three oxygen atoms along the tetrahedron edges (Fig. 9-1). The values of the P—O and As—O distances, 1.65 and 1.74 Å, respectively, indicate 22 percent and 10 percent, respectively, of double-bond character for the bonds. This makes itself evident also in the values of about 126° observed for the P—O—P and As—O—As bond angles. The P_4O_{10} molecule is closely similar to the P_4O_6 molecule in structure, with the addition of an oxy-

[33] L. R. Maxwell, S. B. Hendricks, and L. S. Deming, *J. Chem. Phys.* 5, 626 (1937); G. C. Hampson and A. J. Stosick, *J.A.C.S.* 60, 1814 (1938).

gen atom to each phosphorus atom, completing the PO_4 tetrahedra (Fig. 9-2). These unshared oxygen atoms are at the surprisingly small distance 1.39 Å from the adjacent phosphorus atoms. In the $P_4O_6S_4$ molecule, which has a similar structure,[34] the P—S distance is also very short, having the value 1.85 Å. The P—O distance, 0.34 Å less than the single-bond length, corresponds to a triple bond. The partial ionic character of the bonds, 39 percent, then leads to the charge -0.17 on the unshared oxygen atom. The P—S distance is close to the value 1.87 Å mentioned above (in the discussion of $PSCl_3$, etc.) as corresponding to electrical neutrality for the sulfur atom. The phosphorus bond angles in P_4O_{10}, 116.5° between the triple bond to the unshared oxygen atom and the bond (about 25 percent double-bond character; bond length 1.62 Å) to a shared oxygen atom, and 102.5° for two bonds to shared oxygen atoms, show the expected effect of the difference in character of the bonds.

Similar structural features have been found in many other oxides of the heavier metals.

9-7. THE STRUCTURE AND STABILITY OF CARBONYLS AND OTHER COVALENT COMPLEXES OF THE TRANSITION METALS

The problem of the stability of the complexes of the transition metals was for many years a puzzling one. Why is the cyanide group so facile in the formation of complexes with these elements, whereas the carbon atom in other groups, such as the methyl group, does not form bonds with them? Why do the transition metals and not other metals (beryllium, aluminum, etc.) form cyanide complexes? In the ferrocyanide ion, $[Fe(CN)_6]^{----}$, for example, the iron atom has a formal charge of 4^-, on the assumption that it forms six covalent bonds with the six ligands; how can this large negative charge be made compatible with the tendency of metals to lose electrons and form positive ions?

The answers to these questions and other questions about the transition-metal complexes have been provided by a new idea about their structure, developed in 1935 to account for the bond lengths observed in the nickel tetracarbonyl molecule. This idea is that atoms of the transition groups are not restricted to the formation of single covalent bonds, but can form multiple covalent bonds with electron-accepting ligands by making use of the $3d$ (or $4d$, $5d$) orbitals and electrons of the transition metal.

The structure of nickel tetracarbonyl has been discussed briefly in Section 5-9. It was pointed out that the observed diamagnetism is compatible with the structure A:

[34] A. J. Stosick, *J.A.C.S.* **61**, 1130 (1939).

$$
A \quad \overset{\displaystyle \ddot{O}}{\underset{\displaystyle \underset{\ddot{O}}{\overset{\|}{C}}}{\overset{\|}{\underset{|}{C}}}} \quad \ddot{:}O\!\equiv\!C\!-\!Ni\!-\!C\!\equiv\!O\ddot{:}
$$

With this structure the nickel atom has achieved the krypton electron configuration: its outer shell contains five unshared pairs (in the five $3d$ orbitals) and five shared pairs (occupying the $4s4p^3$ tetrahedral bond orbitals). The Ni—C bond length expected for this structure is about 2.16 Å, as found by use of the tetrahedral radius 1.39 Å obtained by extrapolation from the adjacent values in Table 7-13 (Cu, 1.35 Å; Zn, 1.31 Å).

When an investigation of the structure of the molecule was carried out, by the electron-diffraction method, it was found[35] that the molecule has the tetrahedral configuration predicted for structure A, but that the internuclear distance is surprisingly small, only 1.82 ± 0.03 Å.

The small distance suggests that the bonds have multiple-bond character, corresponding to resonance of structure A with other structures of types B, C, D, and E:

$$
B \quad :O\!\equiv\!C\!-\!Ni\!=\!C\!=\!\ddot{O}: \qquad\qquad C \quad :O\!=\!C\!=\!Ni\!=\!C\!=\!\ddot{O}:
$$

$$
D \quad :O\!=\!C\!=\!Ni\!=\!C\!=\!\ddot{O}: \qquad\qquad E \quad :O\!=\!C\!=\!Ni\!=\!C\!=\!\ddot{O}:
$$

[35] L. O. Brockway and P. C. Cross, *J. Chem. Phys.* **3**, 828 (1935). Spectroscopic evidence verifying the tetrahedral configuration has been reported by

The double bonds Ni=C are formed by use of the $3d$ orbitals and associated electrons of the nickel atom. For example, for the structures of type C the six bond orbitals of the nickel atom are d^2sp^3 hybrids, and the other three $3d$ orbitals are occupied by unshared pairs, and for structure E there are eight d^4sp^3 bond orbitals and one unshared pair.

There is uncertainty in the determination of the amount of double-bond character of the bonds, in part because of the dependence of the nickel radius on the amount of d character of the bond orbitals. For structures C we may use the d^2sp^3 value 1.21 Å (as given for Ni(IV) in Table 7-15); with the correction -0.12 Å for 50 percent double-bond character this gives the Ni—C bond length 1.86 Å. Similarly, for structures D the d^3sp^3 radius 1.15 Å (Table 7-18) with correction -0.17 Å for 75 percent double-bond character gives 1.74 Å. The observed value lies between these two. The electronegativity principle suggests about 78 percent of double-bond character (which gives zero charge for the nickel atom, with the assumption that all bonds have 12 percent of ionic character, corresponding to the C—Ni electronegativity difference 0.7). We conclude that the nickel-carbon bonds in nickel tetracarbonyl have a large amount of double-bond character, and that it is this structural feature that accounts for their stability.[36]

The ion $[\mathrm{Ni(CN)_4}]^{----}$, isoelectronic with $\mathrm{Ni(CO)_4}$, can be made by reduction with potassium of $[\mathrm{Ni(CN)_4}]^{--}$ in liquid ammonia solution.[37] Its infrared spectrum[38] corresponds to the expected tetrahedral structure. The value of the C—N stretching vibrational frequency shown by the spectrum, 2135 cm^{-1}, is considerably greater than the value for $[\mathrm{Ni(CN)_4}]^{--}$, 1985 cm^{-1}. From this comparison it can be concluded[38] that the C—N bond has more double-bond character in $[\mathrm{Ni(CN)_4}]^{----}$ than in $[\mathrm{Ni(CN)_2}]^{--}$. It was pointed out by Amr El-Sayed and Sheline[38] that this conclusion agrees with expectation from the electroneutrality principle. The calculation given above for $\mathrm{Ni(CO)_4}$, leading to 78 percent double-bond character for the Ni—C bonds, applies also to $[\mathrm{Ni(CN)_4}]^{----}$. A similar calculation for $[\mathrm{Ni(CN)_4}]^{--}$ leads to 34 percent double-bond character (for electroneutrality of the nickel atom); we conclude that the structure Ni=C=N̈: makes a larger con-

---

B. L. Crawford, Jr., and P. C. Cross, *ibid.* **6**, 525 (1938), and B. L. Crawford, Jr., and W. Horwitz, *ibid.* **16**, 147 (1948).

[36] Hybrid orbitals for nickel tetracarbonyl have been discussed by G. Giacometti, *J. Chem. Phys.* **23**, 2068 (1955), who reaches the conclusion that with *dsp* hybrid orbitals the bonds can have as much as 75 percent of double-bond character.

[37] J. W. Easter and W. M. Burgess, *J.A.C.S.* **64**, 1187 (1942).

[38] M. F. Amr El-Sayed and R. K. Sheline, *J.A.C.S.* **80**, 2047 (1958).

tribution, and Ni—C≡N: a smaller, for [Ni(CN)₄]⁻⁻⁻⁻ than for [Ni(CN)₄]⁻⁻, as is indicated by the C—N vibrational frequencies.[39]

The molecules Co(CO)₃NO and Fe(CO)₂(NO)₂, which are isoelectronic with nickel tetracarbonyl, have structures of the same type, the interatomic distances being observed[40] to have the values Co—C = 1.83, Fe—C = 1.84, Co—N = 1.76, Fe—N = 1.77, C—O = 1.15, and N—O = 1.11 Å. Similar tetrahedral structures have also been found[41] for the isoelectronic iron and cobalt carbonyl hydrides HCo(CO)₄ and H₂Fe(CO)₄, with interatomic distances Co—C = 1.75 and 1.83 Å, Fe—C = 1.79 and 1.84 Å, and C—O = 1.15 Å. These bond lengths indicate that the bonds are similar to those in nickel tetracarbonyl. There is some question about the positions of the hydrogen atoms in the last two compounds. The electron-diffraction study[41] was made on the assumption that they are bonded to oxygen atoms of the carbonyl groups. However, no absorption corresponding to the O—H stretching vibration has been found in the infrared spectrum. Hieber[42] proposed that the hydrogen atoms are bonded to the metal atom, and Edgell and his coworkers,[43] who found that the C—O vibration in HCo(CO)₄ is split into three bonds in the infrared spectrum, suggested that the proton is bonded not only to the cobalt atom but also to the carbon and oxygen atoms of three carbonyl groups. Splitting of the infrared frequencies of H₂Fe(CO)₄ (as compared with [Fe(CO)₄]⁻⁻) has been reported also.[44] The acid constants of H₂Fe(CO)₄ are $K_1 = 4 \times 10^{-5}$ and $K_2 = 4 \times 10^{-14}$; salts KHFe(CO)₄ and K₂Fe(CO)₄ have been made.[45] The two acid constants differ by such a great amount that it is probable that the two hydrogen atoms are bonded to the same atom, which would surely be the iron atom and not an oxygen atom. This evidence and evidence from the nuclear magnetic resonance spectrum have been advanced[46] in support of the bonding of hydrogen to the metal atoms. One theoretical study[47] has indicated that the Co—H bond length is 2.0 Å, and another[48] has indicated a

[39] In [Ni(CN)₄]⁻⁻⁻⁻ nickel has oxidation-number 0 and in [Ni(CN)₄]⁻⁻ +2; the intermediate oxidation state, +1, is represented in [Ni₂(CN)₆]⁻⁻⁻⁻, the structure of which is discussed in Sec. 11-15.

[40] L. O. Brockway and J. S. Anderson, *Trans. Faraday Soc.* 33, 1233 (1937).
[41] R. V. G. Ewens and M. W. Lister, *Trans. Faraday Soc.* 35, 681 (1939).
[42] W. Hieber. *Angew. Chem.* 49, 463 (1936).
[43] W. F. Edgell, C. Magee, and G. Gallup, *J.A.C.S.* 78, 4185 (1956).
[44] H. Stammreich, reported by F. A. Cotton, *J.A.C.S.* 80, 4425 (1958).
[45] P. Krumholz and H. M. A. Stettiner, *J.A.C.S.* 71, 3035 (1949).
[46] R. A. Friedel, I. Wender, S. L. Shufler, and H. W. Sternberg, *J.A.C.S.* 77, 3391 (1955); F. A. Cotton and G. Wilkinson, *Chem. & Ind.* (London), 1956. 1305.
[47] W. F. Edgell and G. Gallup, *J.A.C.S.* 78, 4188 (1956).
[48] F. A. Cotton, *J.A.C.S.* 80, 4425 (1958).

value less than 1.2 Å. It is probable that the values for Co—H and Fe—H are close to those found in the diatomic gas molecules Co—H and Fe—H, which are 1.54 Å and 1.48 Å, respectively; those values are close to those given by the covalent radii (Chap. 7). If the carbonyl groups were at tetrahedral angles, the hydrogen atom would be only about 2.8 Å from three oxygen atoms and 1.9 Å from three carbon atoms. The interaction with the carbon atoms would be great, probably distorting the molecule to increase the hydrogen-carbon distance. It seems not unlikely that $HCo(CO)_4$ has a configuration approximating the trigonal bipyramid and $H_2Fe(CO)_4$ one approximating the octahedron.

Other substitution products of nickel tetracarbonyl have also been reported; examples[49] are o-phenylene-bisdimethylarsine-dicarbonyl nickel and dipyridyl-dicarbonyl nickel. The infrared spectra indicate that bonds formed by the carbonyl groups are similar to those in nickel tetracarbonyl.

For iron pentacarbonyl, $Fe(CO)_5$, a trigonal bipyramidal structure has been reported.[50] The value found for the Fe—C distances, 1.84 Å, shows that the bonds in this molecule also have considerable double-bond character.

An electron-diffraction investigation of $Cr(CO)_6$, $Mo(CO)_6$, and $W(CO)_6$ has been carried out by Brockway, Ewens, and Lister.[51] The molecules are regular octahedra, with Cr—C = 1.92, Mo—C = 2.08, and W—C = 2.06 Å (all ±0.04 Å). These values are about 0.10 Å less than those for single bonds, indicating that the bonds have some double-bond character.

The discovery that the iron-group elements can form bonds which have in part the character of multiple bonds by making use of the orbitals and electrons of the 3d subshell, while surprising, need not be greeted with skepticism; the natural formula for a compound RCO is that with a double bond from R to C, and the existence of the metal carbonyls might well have been interpreted years ago as evidence for double-bond formation by metals. The double-bond structure for nickel tetracarbonyl (structure E) was in fact first proposed by Langmuir[52] in 1921, on the basis of the electroneutrality principle, but at that time there was little support for the new idea.

The single-bonded structures are not to be ignored; they seem to play a determinative part with respect to the stereochemical properties

[49] R. S. Nyholm and L. N. Short, *J. Chem. Soc.* 1953, 2670.

[50] Ewens and Lister, *loc. cit.* (41).

[51] L. O. Brockway, R. V. G. Ewens, and M. W. Lister, *Trans. Faraday Soc.* 34, 1350 (1938).

[52] I Langmuir. *Science* 54, 59 (1921).

of the central atom, as discussed in Chapter 5.   Nickel tetracarbonyl
and its isosteres, for example, are tetrahedral in configuration, whereas
the nickel cyanide complex ion $[\mathrm{Ni(CN)_4}]^{--}$, in which the nickel-carbon
bonds also have some double-bond character, is square, this difference
being that predicted by discussion of the nature of the orbitals used in
the formation of single bonds.

In many metal carbonyls containing two or more metal atoms there
are metal-metal bonds.   The structures of some of these molecules will
be discussed in Chapter 11.   The metal cyclopentadienyls and similar
molecules involving fractional metal-carbon bonds will be discussed in
Chapter 10.

The Cyanide and Nitro Complexes of the Transition Elements.—The
structural formula usually written for the ferrocyanide ion,

$$
A \quad
\begin{bmatrix}
& & \mathrm{N} & & \\
& & \mathrm{C} & \mathrm{CN} & \\
& & & \diagup & \\
& \mathrm{NC}\!-\!\mathrm{Fe}\!-\!\mathrm{CN} & & \\
& & \diagup\,| & & \\
& \mathrm{NC} & \mathrm{C} & & \\
& & \mathrm{N} & &
\end{bmatrix}^{----}
$$

with single covalent bonds from the iron atom to each of the six carbon
atoms, is seen to be surprising in that it places a charge of 4− on the
iron atoms, whereas iron tends to assume a positive charge, as in the
ferrous ion, and not a negative charge.   Now the cyanide group is an
electronegative group, and the Fe—C bonds accordingly have some
ionic character, which, however, can hardly be great enough to remove
the negative charge completely from the iron atom.   As suggested by
the discussion of the metal carbonyls in the previous section, we assume
that the cyanide group in this complex can function as an acceptor of
electrons, and that the bonds resonate among the following types:

$$A \quad \mathrm{Fe} \quad (\mathrm{CN})^{-}$$

$$B \quad \mathrm{\bar{F}e{:}C{:}{:}{:}N{:}}$$

$$C \quad \mathrm{Fe{:}{:}C{:}{:}\overset{..}{N}{:}^{-}}$$

The first of these represents an electrostatic bond between the iron
atom and the cyanide ion, the second a single covalent bond from iron
to carbon, and the third a double covalent bond, with use of another $3d$
orbital of the iron atom, with its pair of electrons.   The first and the
third of these place a negative charge on the cyanide group, and the
second leaves the group neutral.   Resonance among these with the
second structure contributing only about one-third would make the
iron atom in the complex electrically neutral, the negative charge 4−
being divided among the six cyanide groups.   The magnitude of the

contribution of the third structure could be found from the value of the Fe—C distance, which, however, has not been accurately determined.

It is interesting to note that by using all of the $3d$, $4s$, and $4p$ orbitals of the iron atom the valence-bond structure B

can be written for the ferrocyanide ion. This structure (which is, of course, in resonance with the equivalent structures obtained by redistributing the bonds) gives the iron atom a negative charge of only unity, dividing the residual charge 3— among the six nitrogen atoms, each of which then has the charge $\frac{1}{2}$; and it is probable that the ionic character of the bonds is great enough to transfer further negative charge also to the nitrogen atoms, making the iron atom neutral or even positive. The structures of this type, involving some iron-carbon double bonds, are without doubt of greater significance to the normal state of the complex ion than the conventional structure A written at the beginning of this section; it may well be convenient, however, to continue to represent the complex ion by the conventional structure, just as for convenience the benzene molecule is often represented by a single Kekulé structure.

For other anionic cyanide complexes of the transition elements, such as $[Fe(CN)_6]^{---}$, $[Co(CN)_6]^{---}$, $[Mn(CN)_6]^{4-}$, $[Cr(CN)_6]^{4-}$, $[Ni(CN)_4]^{--}$, $[Zn(CN)_4]^{--}$, and $[Cu(CN)_2]^{-}$, and the analogous complexes of the elements of the palladium and platinum groups, similar structures involving partial double-bond character of the metal-carbon bond can be written.

The nitrosyl group and the nitro group also are able to accept an additional pair of binding electrons, and the bonds in complexes such as $(Fe(CN)_5NO]^{--}$, $[Co(NH_3)_5NO]^{++}$, and $[Co(NO_2)_6]^{---}$ have to a considerable extent the character of the structures

The crystal hexamethylisocyanide-iron(II) chloride trihydrate, $Fe(CNCH_3)_6Cl_2 \cdot 3H_2O$, contains the octahedral complex $[Fe(CNCH_3)_6]^{++}$ with the bond length Fe—C equal to 1.85 Å, corresponding to about 50 percent of double-bond character.[53] The bonds in the complex may be be described as involving resonance between Fe—C≡N⁺—CH₃ and Fe=C=N̈—CH₃. For each of these structures the Fe—C—N angle should be a straight angle, as is observed. The two structures correspond to the values 180° and 114° (Sec. 4-8) for the angle at the nitrogen atom, however, and for the resonance structure an intermediate value would be expected; the value reported is 173°.

For a complex such as $[Co(NH_3)_6]^{+++}$, containing six amino groups, structures involving metal-nitrogen double bonds cannot be formulated. The stability of these complexes is to be attributed to the large amount of ionic character of the single bonds between the metal atom and its ligands, as discussed in Section 5-7. The atoms and groups that occur in the octahedral complexes of cobalt are in the main strongly electronegative; they include NH_3, OH_2, $(OH)^-$, $(O_2)^{--}$ (peroxide), $H_2NCH_2CH_2NH_2$ (ethylenediamine), $(C_2O_4)^{--}$ (oxalate), $(NO_3)^-$, $(SO_4)^{--}$, and others. The atoms bonded to cobalt in all of these groups have about the same electronegativity (that of N^+ in $M—N^+H_3$, for example, being not much different from that of O in M—OH, etc.). The somewhat less electronegative chlorine and bromine atoms can also be introduced, but only to a limited extent (occupying a maximum of two of the six positions), whereas the still less electronegative iodine atom cannot be introduced.

The interatomic distances observed in these complexes are compatible with this structure; the value Co—N = 1.95 Å ± 0.02 Å for the ammonia ligands has been found[54] in the crystals $M[Co(NH_3)_2(NO_2)_4]$, with M = Ag, K, and NH_4, and nearly the same value has been reported for several other crystals. The value 1.96 Å ± 0.02 Å has been found[54] for the nitro ligands in these crystals; we conclude from comparison with the value for the ammonia ligands that the structure

M=N with two O groups makes very little contribution.

To summarize: we attribute the stability of the octahedral complexes

[53] H. M. Powell and G. W. R. Bartindale, *J. Chem. Soc.* 1945, 799.

[54] G. B. Bokii and E. A. Gilinskaya, *Izvest. Akad. Nauk S.S.S.R.* 1953, 238.

of cobalt to the removal of the negative charge assigned to the central atom on the basis of a normal covalent single-bonded structure from it to the surrounding electronegative groups; in the case of cyanide group, and to a smaller extent the nitrosyl and nitro groups, the transfer of charge is also accomplished in part by the formation of double bonds from the cobalt atom to the attached groups.

The iron-group elements are electropositive, tending to form positive ions; and this property is reflected in the nature of the complexes that they form.   The metals of the palladium and platinum groups, on the other hand, have little tendency to form positive ions, but prefer to remain neutral or even to become negative; this characteristic is indicated by their position (2.2) in the electronegativity scale.   In consequence these elements can form covalent octahedral complexes not only with cyanide, ammonia, hydroxide, and related groups but also with chlorine, bromine, and even iodine atoms.   In the hexachloroplatinate ion, $[PtCl_6]^{--}$, the ionic character of the bonds removes some of the negative charge from platinum to chlorine; but in the hexaiodo-osmiate ion, $[OsI_6]^{----}$, the bonds to the weakly electronegative iodine atoms can have little ionic character and a good part of the negative charge would be left on the central atom; some of the negative charge may be removed by the contribution of double-bond structures involving a fifth outer orbital of the halogen atom (Sec. 9-4).

Molybdenum and tungsten are classed with the elements of the palladium and platinum groups rather than with those of the iron group with respect to nobility, and they are to be similarly classed with respect to complex formation.   The stability of the complex ions $[Mo(CN)_8]^{---}$ and $[Mo(CN)_8]^{----}$ and their tungsten analogs cannot be attributed to double-bond formation, because of the small number of $4d$ electrons; these complexes presumably involve eight single covalent bonds with some ionic character, which transfer some of the negative charge from the central atom to the attached groups.   The fact that the large ligancy eight is shown in combination with cyanide and not with chloride has probably a steric explanation.   In the cyanide groups with structure $M:C:::N:$ all of the carbon electrons are concentrated closely about the internuclear axis, and the only unshared pair projects toward the outside of the complex; hence there is little steric repulsion between eight cyanide groups attached to the same atom, whereas eight larger groups could not be fitted in.

A great amount of knowledge about the properties of metal complexes has been gathered; much of it is summarized in the book by Martell and Calvin.[55]

[55] A. E. Martell and M. Calvin, *Chemistry of the Metal Chelate Compounds*, Prentice-Hall, New York, 1952.

CHAPTER 10

The One-Electron Bond and the

Three-Electron Bond; Electron-

deficient Substances

I N a few molecules and crystals it is convenient to describe the interactions between the atoms in terms of the one-electron bond and the
three-electron bond. Each of these bonds is about half as strong as a
shared-electron-pair bond; each might be described as a half-bond.
There are also many other molecules and crystals with structures that
may be described as involving fractional bonds that result from the
resonance of bonds between two or more positions. Most of these
molecules and crystals have a smaller number of valence electrons than
of stable bond orbitals. Substances of this type are called *electron-
deficient substances*. The principal types of electron-deficient substances are discussed in the following sections (and in the next chapter,
on metals).

10-1. THE ONE-ELECTRON BOND

The one-electron bond in the hydrogen molecule-ion is about half as
strong as the electron-pair bond in the hydrogen molecule ($D_0 = 60.95$
kcal/mole for H_2^+, 102.62 kcal/mole for H_2—Secs. 1–4, 1–5); and, since
the same number of atomic orbitals is needed for a one-electron bond
as for an electron-pair bond, it is to be expected that in general molecules containing one-electron bonds will be less stable than those in
which all the stable bond orbitals are used in electron-pair-bond formation. Moreover, there is a significant condition that must be satisfied
in order for a stable one-electron bond to be formed between two atoms;
namely, that the two atoms be identical or closely similar (Sec. 1–4).
For these reasons one-electron bonds are rare—much rarer, indeed,
than three-electron bonds, to which similar restrictions apply.

Of the 33 excited states of the H_2 molecules that have been reported,[1] 20 have internuclear distance within ± 0.03 Å of the value for H_2^+, 1.06 Å; it is therefore probable that these states can be described as corresponding to an H_2^+ ion, with a one-electron bond, and a second electron in an outer orbit, with only a small bonding or antibonding effect.

Similar excited states have been observed for diatomic molecules of the alkali metals. They may be interpreted as involving a molecule-ion, such as Li_2^+, with a one-electron bond, plus a loosely-bound outer electron. The internuclear distances are about 0.3 Å greater than for the corresponding normal states:[2] 2.94 Å for Li_2^+ (2.672 Å for Li_2), 3.41 Å for Na_2^+ (3.079 Å for Na_2), and 4.24 Å for K_2^+ (3.923 Å for K_2). The values of the bond energies for the one-electron bonds, as indicated by the vibrational levels, are about 60 percent of those for the corresponding electron-pair bonds.

The only other substances[3] in which one-electron bonds are important are the ferromagnetic metals. It will be pointed out in the following chapter that in these substances the interaction of the spins of atomic electrons and those of bonding electrons causes some of the pairs of bonding electrons to be split into unpaired bonding electrons with parallel spins.

10-2. THE THREE-ELECTRON BOND

Lewis in his 1916 paper and in his book on valence emphasized the fact that there exist only a few stable molecules and complex ions (other than those containing atoms of the transition elements) for which the total number of electrons is odd. He pointed out that in general an "odd molecule," such as nitric oxide or nitrogen dioxide, would be expected to use its unpaired electron to form a bond with another such molecule, and that the monomeric substance should accordingly be very much less stable than its dimer; and he stated that the method by which the unpaired electron is firmly held in the stable odd molecule was not at that time understood. Since then the explanation of the phenomenon has been found, as the result of the

---

[1] G. Herzberg, *Molecular Spectra and Molecular Structure*, vol. I, "Diatomic Molecules," D. Van Nostrand Company, Princeton, N.J., 1950.

[2] It is interesting that these differences as well as the difference for H_2^+ and H_2, 0.32 Å, are considerably greater than the value 0.18 Å given by Equation 7-7.

[3] In earlier editions of this book the boranes were discussed as examples of substances containing one-electron bonds. New structural information has shown that they are electron-deficient substances containing fractional bonds (Sec. 10-7).

application of quantum mechanics to the problem; the stability of odd molecules is the result of the power of certain pairs of atoms to form a new type of bond, the *three-electron bond*.[4]

The Conditions for Formation of a Stable Three-Electron Bond.— Let us consider the normal state of a system of three electrons and two nuclei or kernels, A and B, each with one stable bond orbital.   There are only two essentially different ways of introducing the three electrons into the two available orbitals, I and II:

$$\text{I} \qquad \text{A:} \qquad \cdot \text{B}$$

$$\text{II} \qquad \text{A} \cdot \qquad :\text{B}$$

The exclusion principle permits only two electrons, which must have opposed spins, to occupy either one of the orbitals; the third electron must occupy the other orbital.[5]

It is found on carrying out the energy calculations that structure I alone does not correspond to the formation of a stable bond; it leads instead to repulsion or at best to only a very weak attraction between the atoms.   Structure II alone also leads to a similar type of interaction.   If, however, the atoms A and B are identical or are closely similar, so that the two structures have nearly the same energy, then resonance will occur between them, which will stabilize the molecule and lead to an interaction between the atoms corresponding to the formation of a stable bond.[6]   This bond, corresponding to resonance of the type {A:·B, A·:B}, may be called the *three-electron bond* and represented by the symbol A···B.   It is found by calculation and by experiment to be about one-half as strong as an electron-pair bond (that is, to have half as great a value of the bond energy).   The system of two molecules A⋯B, each containing a stable three-electron bond in addition to another bond between A and B, has accordingly about the same energy as A—B—B—A, involving an additional covalent bond; and we can expect that in some cases the heat of formation of the dimer will be positive and in others it will be negative, with corresponding differences in stability of the two forms.   This is in accord with the results of observation: nitric oxide, to which we assign a structure involving a three-electron bond, does not form a stable dimer,

[4] L. Pauling, *J.A.C.S.* **53**, 3225 (1931).

[5] The spin of the unpaired electron can be either positive or negative.   The structures with positive spin and those with negative spin have (with the very small spin-orbit interactions neglected) the same energy; together they correspond to a doublet state of the molecule.

[6] Notice the close similarity of this argument to that given for the one-electron bond in Sec. 1-4.

whereas the similar substance nitrogen dioxide does form its dimer, dinitrogen tetroxide.

In order that there may be resonance between structures I and II and the formation of a stable three-electron bond the atoms A and B must be identical or similar; the conditions for formation of the bond are thus the same as those of the one-electron bond discussed in Section 1-4, and the two bonds show the same dependence of bond energy on the energy difference of the resonating structures. It is found on examination of the energy quantities that a stable three-electron bond might be formed between unlike atoms which differ by not much more than 0.5 in electronegativity, that is, between oxygen and fluorine, nitrogen and oxygen, nitrogen and chlorine, chlorine and oxygen, etc. Three-electron bonds between oxygen and fluorine, oxygen and oxygen, nitrogen and oxygen, and chlorine and oxygen have been recognized in stable molecules, and others are indicated also by spectroscopic data.

It may be pointed out that the one-electron bond, the electron-pair bond, and the three-electron bond use one stable bond orbital of each of two atoms, and one, two, and three electrons, respectively.

The Helium Molecule-Ion.—The simplest molecule in which the three-electron bond can occur is the helium molecule-ion, He_2^+, consisting of two nuclei, each with one stable $1s$ orbital, and three electrons. The theoretical treatment[7] of this system has shown that the bond is strong, with bond energy about 55 kcal/mole and with equilibrium internuclear distance about 1.09 Å. The experimental values for these qualities, determined from spectroscopic data for excited states of the helium molecule, are about 58 kcal/mole and 1.080 Å, respectively, which agree well with the theoretical values. It is seen that the bond energy in $He \cdots He^+$ is about the same as that in $H \cdot H^+$, and a little more than half as great as that of the electron-pair bond in $H:H$.

10-3. THE OXIDES OF NITROGEN AND THEIR DERIVATIVES

Nitric Oxide.—Nitric oxide is the most stable of the odd molecules. For the first of the two structures I and II

$$\text{I} \quad :\!\overset{\cdot\cdot}{\underset{}{N}}\!\!=\!\!\overset{\cdot\cdot}{\underset{}{O}}: \qquad \text{II} \quad :\!\overset{-\ \cdot\cdot}{\underset{}{N}}\!\!=\!\!\overset{\cdot\ +}{\underset{}{O}}:,$$

we would expect great ease of polymerization to stable molecules of the type

$$\begin{array}{c} :\!\overset{\cdot\cdot}{O}\!\!=\!\!N: \\ | \\ :\!N\!\!=\!\!\underset{\cdot\cdot}{O}:, \end{array}$$

[7] E. Majorana, *Nuovo cimento* **8**, 22 (1931); L. Pauling, *J. Chem. Phys.* **1**, 56 (1933); S. Weinbaum, *ibid.* **3**, 547 (1935).

and structure II, because of its unfavorable charge distribution, should be somewhat less stable than I.   The unfavorable charge distribution is, however, partially neutralized by the ionic character of the double bond, and the difference in stability of I and II is small enough to permit nearly complete resonance between them.   We accordingly assign to the molecule the structure : $N \overset{\cdots}{=} O$:, involving a double bond and a three-electron bond between the two atoms.   Of the four valence orbitals of each atom one is used for the unshared pair of electrons, two for the double bond, and the fourth for the three-electron bond.

The properties of the molecule are accounted for by this structure. The extra energy of the three-electron bond stabilizes the molecule relative to structure I to such extent that the heat of the reaction $2NO \rightarrow N_2O_2$ is small,[8] and the substance does not polymerize in the gas phase.

This structure, which may be described as involving a $2\frac{1}{2}$ bond, is expected to lead to a bond length intermediate between that for a double bond and that for a triple bond.   The N—O single-bond length is 1.44 Å (Sec. 7-2), and the double-bond and triple-bond lengths may be taken as 0.04 A less than for $N{=}N$ and $N{\equiv}N$, and hence equal to 1.20 Å and 1.06 Å, respectively.   This triple-bond value agrees well with the experimental value 1.062 Å for NO^+, which has the structure :$\overset{+}{N}{\equiv}O$:.   The observed distance for NO is 1.151 Å, somewhat larger than that expected for a $2\frac{1}{2}$ bond; it corresponds, when interpreted with use of an equation similar in form to Equation 7-7, to the bond number 2.31.   We conclude that the difference in electronegativity of the two atoms decreases the contribution of structure II to such an extent that the three-electron bond is about a one-third bond, rather than a one-half bond.

A study of the hyperfine structure of the electron spin magnetic resonance spectrum, resulting from the interaction with the nuclear spins, has led to the conclusion[9] that structure I contributes 65 percent and structure II 35 percent, and that the odd electron occupies a $2p\pi$ orbital with 2.5 percent s character.

The electric dipole moment of the molecule is small, about 0.16 D. Structure I would lead to a moment with the oxygen atom negative, because of the partial ionic character of the bonds; this moment is neutralized by structure II.

Dinitrogen Dioxide.—Crystals of nitric oxide[10] contain its dimer,

[8] The enthalpy of formation of N_2O_2 is 3.7 kcal/mole: A. L. Smith and H. L. Johnston, *J.A.C.S.* **74**, 4696 (1952).

[9] G. C. Dousmanis, *Phys. Rev.* **97**, 967 (1955); see also M. Mizushima, *ibid* **105**, 1262 (1957).

[10] W. J. Dulmage, E. A. Meyers, and W. N. Lipscomb, *Acta Cryst.* **6**, 760 (1953).

with the form of a rectangle with short edge 1.12 ± 0.02 Å and long edges 2.40 Å. The long edges represent very weak bonds, with bond number about 0.06 (Equation 7-7). It is not unlikely that there is some mobility of an electron from one NO to the other in the dimer, and that its structure can be represented by the resonance structures *A*, *B*, and *C*:

$$:\overset{\cdots}{N}\!\!=\!\!O: \qquad\qquad :N\!\!\equiv\!\!\overset{+}{O}: \qquad\qquad \overset{-}{:}N\!\!=\!\!\overset{\cdots}{O}: $$

$$ \qquad A \qquad\qquad\qquad\qquad B \qquad\qquad\qquad\qquad C $$

$$:\overset{\cdots}{O}\!\!=\!\!N: \qquad\qquad :O\!\!=\!\!\overset{\cdots\,-}{N}: \qquad\qquad :\overset{+}{O}\!\!\equiv\!\!N: $$

The structures *B* and *C* have no odd electrons, and for resonance with them *A* must have the spins of the two odd electrons opposed. Hence the substance should be diamagnetic—as it has been observed to be.[11] The residual entropy at low temperature,[12] approximately $R\ln2$ per mole of N_2O_2, can be explained as resulting from a disorder in the crystal, each N_2O_2 rectangle having two possible orientations.

Klinkenberg and Ketelaar[13] have shown that $NOClO_4$, $NOBF_4$, and $(NO)_2SnCl_6$ (the last usually being written as $2NOCl \cdot SnCl_4$) are similar in structure to NH_4ClO_4, NH_4BF_4, and $(NH_4)_2SnCl_6$, and hence contain the nitrosyl cation, $(NO)^+$.

In NaNO, which is diamagnetic,[14] there probably exist $(NO)^-$ anions with the structure $[:\overset{\cdots}{N}\!\!=\!\!\overset{\cdots}{O}:]^-$. It is interesting that this anion, which is isoelectronic with molecular oxygen, does not have the same $^3\Sigma$ structure, which would lead to paramagnetism.

The Nitrosyl Halogenides.—Nitrosyl fluoride, chloride, and bromide, ONF, ONCl, and ONBr, have been studied by the electron-diffraction method[15] and by microwave spectroscopy.[16] Their configuration is nonlinear. The N—O distance is 1.14 ± 0.02 Å. Although a reasonable electronic structure, I,

I

[11] E. Lips, *Helv. Phys. Acta* **8**, 247 (1935).

[12] H. L. Johnston and W. F. Giauque, *J.A.C.S.* **51**, 3194 (1929).

[13] L. J. Klinkenberg, *Rec. trav. chim.* **56**, 749 (1937); L. J. Klinkenberg and J. A. A. Ketelaar, personal communication.

[14] J. H. Frazer and N. O. Long, *J. Chem. Phys.* **6**, 462 (1938).

[15] J. A. A. Ketelaar and K. J. Palmer, *J.A.C.S.* **59**, 2629 (1937).

[16] ONCl by J. D. Rogers, W. J. Pietenpol, and D. Williams, *Phys. Rev.* **83**, 431 (1951); ONF by D. W. Magnuson, *J. Chem. Phys.* **19**, 1071 (1951).

can be formulated for these molecules, this seems not to be correct, for the observed N—F, N—Cl, and N—Br distances, 1.52 ± 0.03, 1.96 ± 0.01, and 2.14 ± 0.02 Å, are very much greater than the expected single-bond values (Sec. 7-2), 1.38, 1.73, and 1.86 Å, respectively. It seems probable that these molecules resonate between structure I and the ionic structure II (with a small contribution also from structure III, representing conjugation of the double bond and an electron pair of the halogen atom),

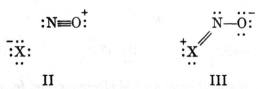

| II | III |

and that the ionic bond gives rise to the increase in the N—X distance.

A study of BrNO by microwave spectroscopy[17] has yielded values of the bond lengths and bond angle in agreement with the electron-diffraction values (N—O = 1.15 ± 0.06 Å, N—Br = 2.14 ± 0.06 Å, angle BrNO = $114° \pm 5°$). The fine structure due to coupling of the electric quadrupole moment of the bromine nucleus with the electrons has been interpreted as showing that structure II contributes 39 percent to the normal state of the molecule, structure I 49 percent, and structure III, representing conjugation, 12 percent. These values are reasonable, except that the observed Br—N distance indicates that the contribution of the ionic structure II is larger than 39 percent. The increase by 0.28 Å over the single-bond value interpreted by Equation 7-7 leads to bond number 0.34, corresponding to at least 66 percent contribution of structure II. A similar treatment of ClNO (increase 0.23 Å) and FNO (increase 0.14 Å) leads to bond numbers 0.42 and 0.58, respectively.

The contribution of structure III would be expected to amount to a few percent for BrNO and ClNO, and much more for FNO, because of the greater stability of the fluorine-nitrogen double covalent bond. A resonating structure with 50 percent contribution of II, 25 percent of I, and 25 percent of III is compatible with both the observed F—N distance in FNO and the observed value[18] 1.81 D of the electric dipole moment.

We may ask why the no-bond or ionic structure II makes a large contribution for the nitrosyl halogenides and not for other substances. The answer must be that the nitrosyl group tends to lose an electron— it is far less electronegative than the nitrogen atom or oxygen atom. This greatly decreased electronegativity would result from an unusually

[17] T. L. Weatherly and H. Williams, *J. Chem. Phys.* **25**, 717 (1956); D. F. Eagle, T. L. Weatherly, and H. Williams, *ibid.* **30**, 603 (1959).
[18] Magnuson, *loc. cit.* (16).

great stability of the N≡O triple bond relative to the double bond. We conclude that the extra stability of the triple bond in the nitrogen molecule, amounting to 79 kcal/mole (Sec. 6-2), applies also to the nitrosonium ion.

Nitrosyl-Metal Complexes.—Reasonable structures of metal-nitrosyl complexes are M=N⁺=Ö: and M—N=Ö:. The structure M—N≡Ö: is to be rejected, in that it involves similar electric charges on adjacent atoms. The substances $Na_2[Fe(CN)_5NO]\cdot 2H_2O$, $[Ru(NH_3)_4NO \cdot H_2O]Cl_3$, and $[Ru(NH_3)_5NO]Br_3$ are diamagnetic, as would be expected in case that the structure M=N=Ö: made a significant contribution; seven of the nine outer orbitals of the metal atom are used in forming bonds (two with the nitrosyl group), and the other two are occupied by unshared electron pairs.

The "brown-ring" test for nitrates involves the formation of a dark-brown unstable complex of nitric oxide and hydrated ferrous ion. The composition of the complex[19] is probably $[Fe(OH_2)_5NO]^{++}$. Let us assume that the iron-nitrosyl bond corresponds to the structure Fe=N=Ö: and that the bond orbitals have the same character (50 percent d character) as in nickel tetracarbonyl and the carbonyl-nitrosyls (Sec. 9-7). One d orbital of the iron atom would then be used in forming the two bonds to the nitrogen atom, leaving four d orbitals for occupancy by the unshared electrons. (The water molecules are assumed to be held without the use of any $3d$ orbitals, as shown [Chap. 5] by the observed magnetic moments of the hydrated ferrous and ferric ions.) The iron atom has five unshared electrons to be placed in the $3d$ orbitals. With four $3d$ orbitals available for these electrons, three electrons would remain unpaired. This structure, which is rare for octahedral complexes (Chap. 5), is supported by the observation[19] that the magnetic moment is 3.9 magnetons, corresponding to three unpaired electrons. The complex $[Fe(C_2H_5OH)_5NO]^{+++}$, containing an iron atom with one less electron (four in the four $3d$ orbitals), has a similar structure; its magnetic moment[19] is 5.0 magnetons, corresponding to four unpaired electrons.

To the compound[20] $Fe(NO)_3Cl$ we assign the structure

$$:\overset{\cdot\cdot}{\underset{\cdot\cdot}{Cl}}:$$
$$:\overset{\cdot\cdot}{O}=N=\overset{|}{Fe}=N=\overset{\cdot\cdot}{O}:$$
$$\overset{\parallel}{N}$$
$$\overset{\parallel}{O}:$$

[19] W. P. Griffith, J. Lewis, and G. Wilkinson, *J. Chem. Soc.* 1958, 3993.
[20] W. Hieber and R. Nast, *Z. anorg. Chem.* 244, 23 (1940).

In this structure all nine outer orbitals of the iron atom are used in forming bonds or holding unshared electron pairs. The compound $Fe(NO)_4$ has also been made.[21] It is interesting that the structure with four $Fe{=}\overset{+}{N}{=}\overset{..}{O}:$ groups cannot be assigned to it, because this structure would require ten stable outer orbitals for the iron atom; instead, we assign to the compound the structure

There are two possibilities with this structure: (1) that the single bond Fe—N resonates with the three double bonds, and (2) that the bonds remain fixed, permitting one group to assume the angle 113° that is unstrained for $-\overset{..}{N}{=}$ and the other three to have the angle 180°. The infrared absorption spectrum[22] has been interpreted as supporting the second structure.

Nitrogen Dioxide.—To nitrogen dioxide we assign the resonating structure

in which one oxygen atom is held to nitrogen by a double bond and one by a single bond plus a three-electron bond. In the first edition of this book the prediction was made that the N—O distance is about 1.18 Å and the ONO bond angle about 140°. The prediction of the value of the bond angle was based on the argument that the configuration

[21] W. Manchot and H. Gall, *Ann. Chem.* **470**, 271 (1929).

[22] Griffith, Lewis, and Wilkinson, *loc. cit.* (19). It may be mentioned that the very low volatility of the solid substance suggests that bonds are formed linking the molecules into larger complexes.

$\overset{..}{:}\overset{..}{O}=\overset{..}{N}\overset{..}{\cdots}\overset{..}{O}:$ is intermediate between $:\overset{..}{O}=N^+=\overset{..}{O}:$ and $:\overset{..}{O}=\overset{..}{N}-\overset{..}{O}:^-$, with angles 180° and 113°, respectively, and that the angle should have an intermediate value. The N—O distance is intermediate between that in NO_2^+, 1.154 Å, and that in NO_2^-, 1.236 Å. An infrared study[23] of NO_2 has now yielded the values 1.188 ± 0.004 Å for N—O and 134.1° ± 0.25° for the angle ONO, in agreement with the less accurate electron-diffraction values.[24]

The nitronium cation, NO_2^+, with a structure resembling that of carbon dioxide (linear, N—O = 1.154 Å), was discovered by spectroscopic methods.[25] Its linear configuration has been verified for NO_2ClO_4 by spectroscopic methods[26] and for $NO_2^+NO_3^-$ (crystalline dinitrogen pentoxide) by x-ray diffraction,[27] and also for several other crystals.[28]

Dinitrogen Tetroxide.—The dimer of nitrogen dioxide is not very stable: its enthalpy of formation from the monomer is 13.873 kcal/mole. The molecule is found in both the crystal (x-ray diffraction)[29] and the gas (electron diffraction)[30] to be planar, with orthorhombic symmetry. The N—N bond length reported for the crystal is 1.64 Å. The value for the gas molecule, 1.75 Å, is probably somewhat more accurate.[31] This value is 0.28 Å greater than the single-bond value found for hydrazine; the bond number of the bond is accordingly about 0.34 (Equation 7-7).

The explanation of this weak bond is provided by the stability of the three-electron bonds in the NO_2 molecules that compose the dimer, which strive to prevent the two odd electrons from settling down on the

[23] G. E. Moore, *J. Opt. Soc. Am.* **43**, 1045 (1953).

[24] S. Claesson, J. Donohue, and V. Schomaker, *J.A.C.S.* **16**, 207 (1948).

[25] C. K. Ingold, D. J. Millen, and H. G. Poole, *Nature* **158**, 480 (1946); D. R. Goddard, E. D. Hughes, and C. K. Ingold, *ibid.*; E. D. Hughes, C. K. Ingold, and R. I. Reed, *ibid.* 448; F. H. Westheimer and M. S. Kharasch, *J.A.C.S.* **68**, 1871 (1946); G. M. Bennett, J. C. D. Brand, and G. Williams, *J. Chem. Soc.* **1946**, 869. The existence of the cation was first suggested by H. von Euler, *Angew. Chem.* **35**, 580 (1922).

[26] W. E. Gordon and J. W. T. Spinks, *Can. J. Res.* **A18**, 358 (1940).

[27] E. Grison, K. Eriks, and J. L. de Vries, *Acta Cryst.* **3**, 290 (1950).

[28] In $NO_2HS_2O_7$ by J. W. M. Steeman and C. H. MacGillavry, *Acta Cryst.* **7**, 402 (1954); in $(NO_2)_2S_3O_{10}$ by K. Eriks and C. H. MacGillavry, *ibid.* 430.

[29] J. S. Broadley and J. M. Robertson, *Nature* **164**, 915 (1949).

[30] D. W. Smith and K. Hedberg, *J. Chem. Phys.* **25**, 1282 (1956).

[31] The ONO angle is reported in the x-ray study as 108°. It is highly probable, as discussed above, that the electron-diffraction value 133.7° is correct. If the atoms are moved from their reported positions in the crystal by minimum amounts (weighted by inverse of scattering power for x-rays) to achieve this value of the angle, the N—N distance becomes 1.74 Å, agreeing with the electron-diffraction value.

nitrogen atoms to form the N—N bond in the structure

(There is, of course, resonance of the double bonds between the alternative positions.)   In the foregoing discussion of NO it was concluded that the occupancy of the nitrogen atom by the odd electron of the N···O bond is about 65 percent.   If the three-electron bond in NO_2 is similar and the resonance for the two NO_2 molecules is unsynchronized, the odd electrons would have 42 percent simultaneous location on the nitrogen atoms, and hence the bond number 0.42 would be expected. This result agrees satisfactorily with the value 0.34 given by the bond length.

About 4 percent of conjugation of the two N=O bonds would be expected, giving the N—N bond enough double-bond character to require the observed planarity.   Some contribution to the potential function restraining the molecule to the planar configuration is made by the fanning out of the orbitals for the N—N bond in the molecular plane consequent to the resonance of the double bond.

The above argument about the structure of N_2O_4 is based upon the assumption[32] that the odd electron of NO_2 occupies a σ orbital (symmetric in the plane of the nuclei) rather than a π orbital (antisymmetric).   The alternative assumption has been proposed;[33] with it the N—N bond would be described as a π bond without an associated σ bond.   It is easy to see, however, that this alternative assumption is incorrect.   Let us consider O=$\overset{+}{N}$=O (linear) and the nitrite ion

(bent).   In the nitrite ion (angle ONO = 115.4°) the nitrogen atom can be described as a tetrahedron with two corners defining an edge shared with one oxygen atom and a third corner occupied by the other oxygen atom; the double bond hence

---

[32] The wave function for a molecule with a nuclear configuration that has a plane of symmetry must be either symmetric or antisymmetric in the plane. In the simple molecular-orbital treatment an antisymmetric wave function for the molecule results from occupancy of antisymmetric orbitals by an odd number of electrons.

[33] D. W. Smith and K. Hedberg, *loc. cit.* (30).

lies in a plane perpendicular to the plane of the nuclei. The NO_2 molecule is intermediate in structure between these two ions, and accordingly it has an intermediate configuration—this is a configuration with the double bond still perpendicular to the plane of the nuclei and with bond angle intermediate between 115.4° and 180°, and an odd electron occupying a tetrahedron corner in the nuclear plane, i.e., occupying a σ orbital.

10-4. THE SUPEROXIDE ION AND THE OXYGEN MOLECULE

On oxidation the alkali metals are converted into oxides to which the formula R_2O_4 and the name alkali tetroxide were assigned until recently, in the belief that the substances were analogous to the tetrasulfides and contained the O_4^{--} anion with structure

With the discovery of the three-electron bond it was seen that these alkali oxides might contain the ion O_2^-, with the structure

$$\left[:\overset{..}{O}\text{---}\overset{..}{O}:\right]^-,$$

involving a single bond and a three-electron bond between the two identical atoms. This suggestion was verified by the measurement of the magnetic susceptibility of the potassium compound.[34] The superoxide ion O_2^- contains one unpaired electron, corresponding to the observed paramagnetism, which gives $\mu = 2.04$ Bohr magnetons, the theoretical value for the $^2\Pi$ state being about 1.85; whereas the O_4^{--} ion would be diamagnetic. Paramagnetism has been verified[35] also for $Ca(O_2)_2$, and one crystalline form of NaO_2 has been reported to be antiferromagnetic.[36]

The existence of the superoxide ion in crystalline KO_2 and NaO_2 has been verified also by x-ray examination.[37] The interatomic distance,

[34] E. W. Neuman, *J. Chem. Phys.* 2, 31 (1934); W. Klemm and H. Sodomann, *Z. anorg. Chem.* 225, 273 (1935).

[35] P. Ehrlich, *Z. anorg. Chem.* 252, 370 (1940).

[36] G. S. Zhdanov and Z. V. Zvonkova, *Doklady Akad. Nauk S.S.S.R.* 82, 743 (1952).

[37] For KO_2, W. Kassatochkin and W. Kotow, *J. Chem. Phys.* 4, 458 (1936); S. C. Abrahams and J. Kalnajs, *Acta Cryst.* 8, 503 (1955); for NaO_2, Zhdanov and Zvonkova, *loc. cit.* (36); G. F. Carter and D. H. Templeton, *J.A.C.S.* 75, 5247 (1953).

reported as 1.28 ± 0.01 Å, is in satisfactory agreement with that expected for the single bond plus the three-electron bond.

On being heated rubidium superoxide loses one-fourth of its oxygen, to form a substance with the stoichiometric formula Rb_2O_3. This substance, originally supposed to contain the ion O_3^{--}, with the structure

$$\text{:O:}^{-}$$
$$|$$
$$\text{:O——O:}^{-},$$

probably has the formula $Rb_2O_2 \cdot 2RbO_2$; that is, it contains both the peroxide ion, with structure $^{-}\text{:O——O:}^{-}$, and the superoxide ion.[38]

The structure of potassium perchromate, K_3CrO_8, presents an interesting problem. The x-ray structure determination[39] shows that four O_2 groups surround the chromium atom. The configuration is roughly that of a tetragonal antiprism with opposite edges of the two square faces shortened to the O—O bond length 1.34 Å. Four Cr—O distances are 1.93 Å and the other four are 2.02 Å. The same structure is shown by K_3NbO_8, K_3TaO_8, Rb_3TaO_8, and Cs_3TaO_8. For these substances the normal oxidation number $+5$ for niobium and tantalum corresponds to having the O_2 groups peroxide groups. However, the oxidation number $+5$ is an unusual one for chromium, and, moreover, the observed O—O bond length for the chromium complex, 1.34 Å, does not agree with the peroxide single-bond length (1.49 ± 0.01 Å in BaO_2,[40] 1.48 ± 0.01 Å in H_2O_2, nearly the same in other peroxides). The structure with four superoxide ions, expected bond length 1.28 Å, is probably acceptable so far as the experimental value 1.34 Å is concerned, but the oxidation number $+1$ of chromium is unusual. It seems likely that the complex contains chromium with the common oxidation number $+3$, two peroxide groups, and two superoxide groups, with resonance of the electrons such that each O_2 group is midway between a peroxide and a superoxide. The chromium atom would be made electrically neutral by forming six bonds with 50 percent ionic character, resonating among the eight Cr—O positions.

For the normal state of the oxygen molecule we would expect the structure

$$A \qquad \text{:O}{=}\text{O:}$$

---

[38] This structure has been verified by magnetic and x-ray data for both Rb_4O_6 and Cs_4O_6 by A. Helms and W. Klemm, *Z. anorg. Chem.* **242**, 201 (1939).

[39] I. A. Wilson, *Arkiv. Kemi, Mineral., Geol.* **15B**, 1 (1941).

[40] S. C. Abrahams and J. Kalnajs, *Acta Cryst.* **7**, 838 (1954).

with a double bond. The normal molecule has, however, the term symbol $^3\Sigma$, showing it to contain two unpaired electrons; in consequence of this, the substance is strongly paramagnetic. It seems probable[41] that the first excited state of the molecule, the $^1\Delta$ state, is represented by this double-bonded structure, and that the normal state, which is more stable by 22.4 kcal/mole, corresponds to a structure in which the two atoms are held together by a single bond and two three-electron bonds.[42] The numbers of electrons and orbitals permit this structure, B,

$$B \qquad :O\!::\!:O:$$

to be formed, each oxygen atom using one of its four valence orbitals for an unshared pair, one for a single bond, and two for the two three-electron bonds.

Since the bond energy of a three-electron bond is about one-half that of a single bond, structure B would be expected to have about the same stability as structure A. There is another interaction to be considered, however—the coupling of the two three-electron bonds. Each of these involves one unpaired electron spin. The two unpaired spins can combine to give either a singlet state, by opposition, or a triplet state, by remaining parallel; one of these will be stabilized by the corresponding interaction energy and the other destabilized. Theoretical arguments have been given[43] which lead to the conclusion that the triplet state should be the more stable, as observed. If the two odd electrons are somewhat unsynchronized they will be during one phase of their motion on the same oxygen atom, and their interaction will then be larger than when they are on different atoms. By Hund's first rule (Sec. 2-7) the interaction is such as to make the triplet more stable than the singlet. Strong support for these ideas is provided by the existence of a $^1\Sigma$ state 37.8 kcal/mole less stable than the normal state; this is to be identified as the state with structure B with unfavorable mutual interaction of the two three-electron bonds (opposed spins of the two odd electrons). The average energy of the normal state and this state is close to that of the double-bonded state.

It is probably the presence of unpaired electron spins in the normal oxygen molecules that gives rise to an interaction between them, somewhat stronger and more definitely directed than ordinary van der Waals forces, that leads to the formation of O_4 (or $(O_2)_2$) molecules. These double molecules were discovered by Lewis,[44] by the analysis of

[41] G. W. Wheland, *Trans. Faraday Soc.* **33**, 1499 (1937).

[42] Pauling, *loc. cit.* (4).

[43] Wheland, *loc. cit.* (41).

[44] G. N. Lewis, *J.A.C.S.* **46**, 2027 (1924).

the data on the magnetic susceptibility of solutions of liquid oxygen in liquid nitrogen. The enthalpy of formation of O_4 from $2O_2$ is very small, 0.16 kcal/mole, so that the O_4 molecules exist in air only in very low concentration. This is, however, large enough to give rise to absorption spectra,[45] analysis of which has verified the existence of the molecule.

The magnetic data show that the spins of two oxygen molecules combined in O_4 are paired together, to give a normal O_4 molecule containing no unpaired electrons. This does not have the structure

$$\ddot{\text{O}}\text{---}\ddot{\text{O}}$$
$$\text{(structure)}$$

(Sec. 3-5), but consists instead of two O_2 molecules, with nearly the same configuration and structure as when free, held together by bonds much weaker than ordinary covalent bonds. Whether the molecule has a planar rectangular configuration or that of a tetragonal bisphenoid is not known. It has been reported[46] that one form of crystalline oxygen contains rotating O_4 molecules in a cubic close-packed arrangement.

The Ozonide Ion.—The red crystalline substance potassium ozonide, KO_3, is obtained by recrystallizing from liquid ammonia the product of reaction of ozone and potassium hydroxide.[47] The corresponding ozonides NaO_3 and CsO_3 have been shown[48] to have magnetic susceptibility corresponding to the presence of the O_3^- ion with one odd electron. The electronic structure of the ozonide ion is

$$\text{(structures)}$$

The bond length is predicted to be 1.35 Å and the bond angle 108°.

[45] O. R. Wulf, *Proc. Nat. Acad. Sci. U.S.* **14**, 609 (1938); see also W. Finkelnburg and W. Steiner, *Z. Physik* **79**, 69 (1932); J. W. Ellis and H. O. Kneser, *ibid.* **86**, 583 (1943); *Phys. Rev.* **44**, 420 (1933); H. Salow and W. Steiner, *Z. Physik.* **99**, 137 (1936).

[46] L. Vegard, *Nature* **136**, 720 (1935).

[47] I. A. Kazarnovskii, G. P. Nikolskii, and T. A. Abletsova, *Doklady Akad. Nauk S.S.S.R.* **64**, 69 (1949).

[48] T. P. Whaley and J. Kleinberg, *J.A.C.S.* **73**, 79 (1951).

10-5. OTHER MOLECULES CONTAINING THE THREE-ELECTRON BOND

A few molecules in addition to those discussed in the preceding sections can be assigned structures involving one or two three-electron bonds. The normal states of the molecules SO, S_2, Se_2, and Te_2 are $^3\Sigma$ states, like that of the normal oxygen molecule, and it is probable that the electronic structures with a single bond plus two three-electron bonds are satisfactory for these molecules. The observed values of interatomic distances, 1.493, 1.888, 2.152, and 2.82 Å, respectively, are about those expected.

It was reported some time ago[49] that the substance OF exists, but the evidence is weak.[50] It might well be possible for the substance to be stable, however, and to be formed to some extent by the dissociation of O_2F_2, inasmuch as the conditions for resonance of the type $\{:\!\ddot{\text{O}}\!-\!\ddot{\ddot{\text{F}}}:, \; :\!\ddot{\text{O}}\!-\!\dot{\ddot{\text{F}}}:\}$, corresponding to the structure $:\!\text{O}\!\cdots\!\text{F}:$ with a single bond plus a three-electron bond, are satisfied for the atoms oxygen and fluorine, which differ in electronegativity by only 0.5. This structure for OF is closely similar to that of the NO molecule.

The anion Cl_2^- has been reported in potassium chloride crystals irradiated with x-rays at the temperature of liquid nitrogen.[51] It may be assigned the structure $:\!\ddot{\text{Cl}}\!\cdots\!\ddot{\text{Cl}}:^-$, and the bond length may be predicted to have the value 2.16 Å.

The cation Cl_2^+, to which we assign the structure $:\!\ddot{\text{Cl}}\!\cdots\!\ddot{\text{Cl}}:^+$, has been studied spectroscopically. It has bond length 1.891 Å; this value is slightly larger than the value expected[52] (Equation 7-5) for a single bond plus a three-electron bond, 1.863 Å.

[49] O. Ruff and W. Menzel, *Z. anorg. Chem.* **211,** 204 (1933); **217,** 85 (1934).

[50] See P. Frisch and H. J. Schumacher, *Z. anorg. Chem.* **229,** 423 (1936).

[51] W. Känzig, *Phys. Rev.* **99,** 1890 (1955); T. Castner and W. Känzig, *J. Phys. Chem. Solids* **3,** 178 (1957); C. J. Delbecq, B. Smaller, and P. H. Yuster, *Phys. Rev.* **111,** 1235 (1958).

[52] A more refined calculation of the expected bond length might be justified. In Section 9-4 it was pointed out that the bond length 1.988 Å in Cl_2 probably reflects some double-bond character, and that the length for a pure single bond is about 0.082 Å greater. For Cl_2^+ one half of this amount of double-bond character would be expected, inasmuch as one orbital for each atom is used for the formation of the three-electron bond. Hence the expected value for Cl_2^+ is 1.904 Å, in good agreement with the observed value. In a similar way we use Equation 7-5 to calculate the expected value 2.242 Å for an excited state with a single bond plus a three-electron antibond. The reported values for the observed excited state are 2.28 Å and 2.30 Å (two levels of a doublet).

The ion F_2^+ has also been studied spectroscopically.[53] Its bond energy is 76 kcal/mole, which is not unreasonable for the structure $[:F\!\cdots\!F:]^+$, in comparison with the values for F_2 and O_2^-.

The ion Ne_2^+ has also been found in mass-spectrographic studies.[54] Its bond length is rather uncertain: 1.7 to 2.1 Å. We would predict, for the structure $[:Ne\!\cdots\!Ne:]^+$, the value 1.69 Å. The corresponding experimental value of the bond energy is 17 kcal/mole, which is about the expected value for a three-electron bond (one-half that for the F—F bond).

The Cl—O distance in the odd molecule ClO_2 has been found[55] to be 1.491 ± 0.014 Å. This value is compatible with the structure

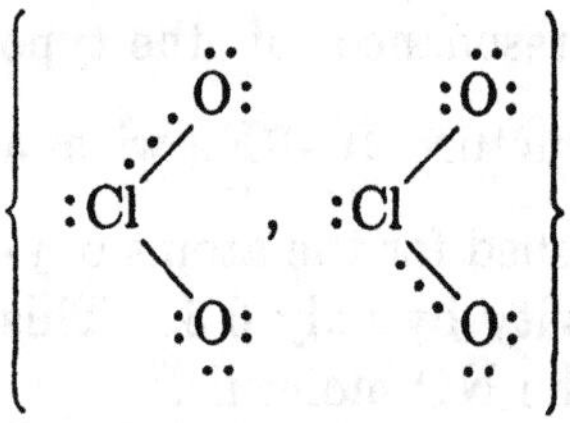

, involving resonance of the three-electron bond

between the Cl—O positions. The OClO bond angle is $116.5° \pm 2.5°$.

No structural studies have as yet been made for other simple odd molecules, NO_3, ClO_4, IO_4, which may contain three-electron bonds. The nitrosodisulfonate ion, $[ON(SO_3)_2]^{--}$, which has been shown by magnetic measurements[56] of the potassium salt to be an odd ion, prob-

ably has the structure

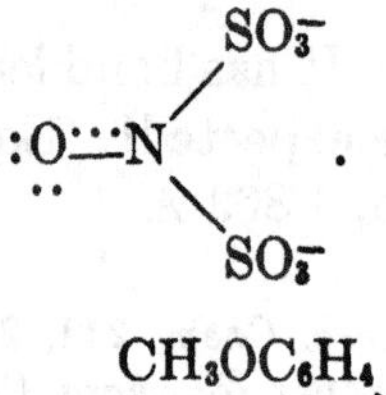

. To di-*p*-anisyl nitric oxide we

assign the similar structure

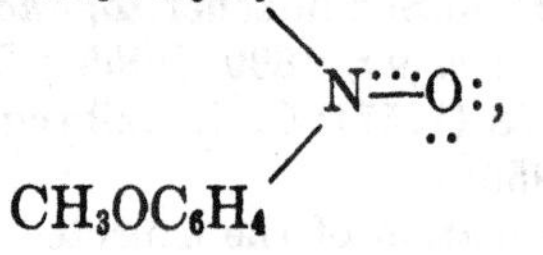

, and to the tetra-*p*-

tolylhydrazinium ion[56] the structure

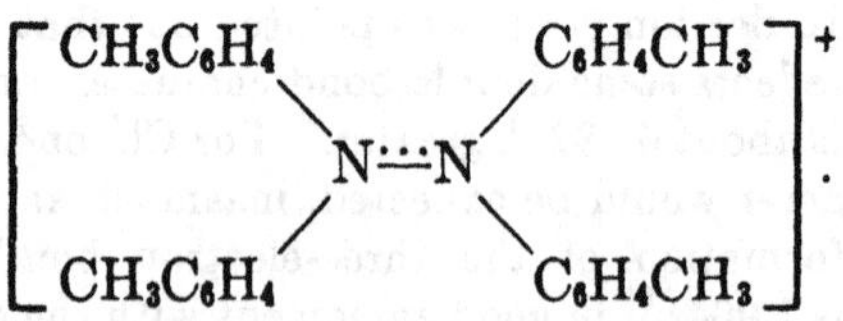

.

[53] R. P. Iczkowski and J. L. Margrave, *J. Chem. Phys.* **30**, 403 (1959).
[54] E. A. Mason and J. T. Vanderslice, *J. Chem. Phys.* **30**, 599 (1959).
[55] J. D. Dunitz and K. Hedberg, *J.A.C.S.* **72**, 3108 (1950).
[56] H. Katz, *Z. Physik* **87**, 238 (1933)

It is probable that in these compounds there is also some resonance of the type described in the following section.

The Structure of the Semiquinones and Related Substances.—The reduction of a quinone, such as *p*-benzoquinone, (I), leads in general to the corresponding hydroquinone, (II). The molecule corresponding to the intermediate stage of reduction, (III), is not expected to be stable; although it is intermediate between I and II in regard to number of electrons, it is rendered unstable by the fact that the loss in bond energy from I with four double bonds to II with three occurs completely on the addition of the first hydrogen atom (at III). It is for this reason that odd molecules are in general of little importance.

There is, however, a way of stabilizing a semiquinone—the molecule intermediate between a quinone and a hydroquinone. In basic solution the semiquinone will exist as the ion, . The structure written (with Kekulé resonance in the benzene ring, of course) is not the only one for the molecule; there is an equivalent structure obtained by interchanging the odd electron of the bottom oxygen atom with a pair of electrons of the other oxygen atom. The semiquinone ion accordingly has the following resonating structure:

with contributions also from the less important structures of the type

It is seen that the resonance indicated in IV is closely analogous to that of the three-electron bond; in each case there is interchange of a single electron and an electron pair. In He_2^+, NO, etc. this interchange takes place directly between adjacent atoms, whereas in the semi-quinone ion it takes place by way of a conjugated system. We may accordingly expect the resonance energy of IV to be about one-half the energy of a single covalent bond; and this is just enough to permit the intermediate stage of reduction from quinone to hydroquinone to be observable.

The condition for stabilization of the semiquinone by resonance is that the two structures IV be equivalent. This condition is satisfied for the semiquinone anion, but not for the semiquinone III itself, in which the presence of the hydrogen atom destroys the equivalence of the two structures. We thus expect the semiquinone to be stable only in the form of the anion. This is verified by experiment. Michaelis and his collaborators[57] have shown that the semiquinone of phenanthrene-3-sulfonate is stable in alkaline solution as the semiquinone ion,

; proof of the existence of the monomeric ion by

[57] L. Michaelis and M. P. Schubert, *J. Biol. Chem.* **119**, 133 (1937); L. Michaelis

measurement of its paramagnetic contribution to the magnetic susceptibility of the solution (due to the spin magnetic moment of the unpaired electron) has been obtained by these investigators. The semiquinone ion is in equilibrium with a dimeric form, perhaps involving an O—O bond; and in acid solution only the dimer is present, in accordance with the foregoing discussion of the expected instability of the unsymmetrical semiquinones.

Many substances containing nitrogen have been shown to exist in intermediate reduction states corresponding to the semiquinone state, and for these substances it is found in general that the conditions are satisfied for resonance of the extended three-electron-bond type. The tetramethyl-*p*-phenylenediaminium ion, shown by its paramagnetism to be monomeric,[58] resonates between the two structures of the type

$$CH_3—\overset{+}{N}—CH_3$$

. Similar resonance occurs for the semiquinone cat-

$$CH_3—\overset{..}{N}—CH_3$$

ions of *p*-naphthophenazine, $\overset{.\,+}{NH}$ ... $\overset{..}{NH}$, and of pyocyanine,

$\overset{.\,+}{NH}$ OH ... $\overset{..}{NCH_3}$ It is interesting to note that in the pyocyanine semi-

quinone the NH and NCH$_3$ groups are sufficiently alike to permit resonance complete enough to effect stabilization.

---

and E. S. Fetcher, Jr., *J.A.C.S.* **59**, 2460 (1937); L. Michaelis, G. F. Boeker, and R. K. Reber, *ibid.* **60**, 202 (1938); L. Michaelis, R. K. Reber, and J. A. Kuck *ibid.* 214; L. Michaelis, M. P. Schubert, R. K. Reber, J. A. Kuck, and S. Granick, *ibid.* 1678; G. Schwarzenbach and L. Michaelis, *ibid.* 1667.

[58] Katz, *loc. cit.* (56); R. Kuhn and K. Schön, *Ber.* **68B**, 1537 (1935).

Resonance between structures

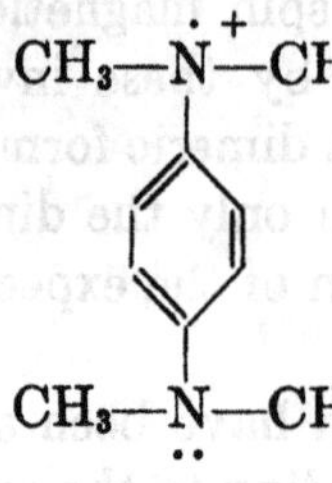

$$CH_3-\overset{..}{\underset{}{N}}{}^{+}-CH_3 \qquad CH_3-\overset{..}{N}-CH_3$$

and

$$CH_3-\underset{..}{N}-CH_3 \qquad H_3C-\underset{+}{N}-CH_3$$

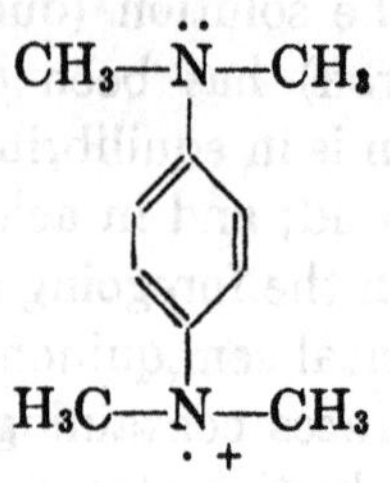

takes place through structures such as

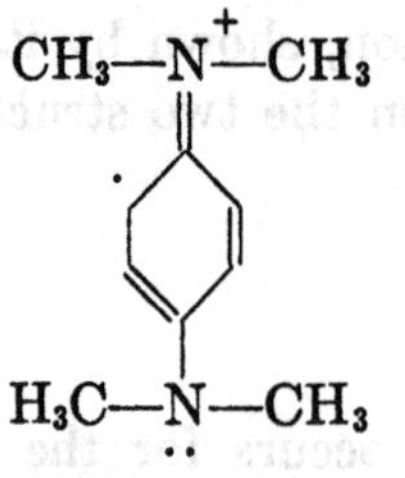

$$CH_3-\overset{.}{\underset{}{N}}{}^{+}-CH_3 \qquad H_3C-\overset{..}{\underset{+}{N}}-CH_3$$

and

$$H_3C-\underset{..}{N}-CH_3 \qquad H_3C-\underset{..}{N}-CH_3$$

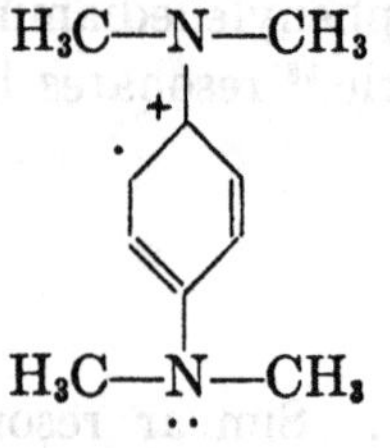

for which the odd electron is attached to a ring carbon atom. Small contributions are also made by structures with the odd electron on the methyl carbon atoms and the hydrogen atoms. Studies of the proton-spin fine structure of the electron-spin magnetic resonance spectrum of the substance[59] and its ring-deuterated derivative[60] give for the spin densities at methyl protons and ring protons the values 0.0148 and 0.0042, respectively. These values show that the odd electron is usually in the neighborhood of one or the other nitrogen atom, and hence that the first two structures are the most important.

Among the many other substances[61] showing resonance of the semi-quinone type mention may be made of those of the dipyridyl group.

On the reduction of γ,γ'-dipyridyl, 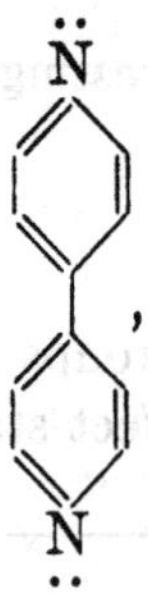   , in acid solution a substance

[59] S. I. Weissman, J. Townsend, D. E. Paul, and G. E. Pake, *J. Chem. Phys.* 21, 2227 (1953).

[60] T. R. Tuttle, Jr., *J. Chem. Phys.* 30, 331 (1959).

[61] Semiquinones of oxazines, thiazines, and selenazines are discussed by S. Granick, L. Michaelis, and M. P. Schubert, *J.A.C.S.* 62, 204, 1802 (1940).

with a deep violet color is obtained, and similar violet substances

are obtained by reduction of the biquaternary dipyridyl bases

(called "viologens") in either acid or alkaline solution. By analogy with the semiquinones resonance of the type

may be suggested, with contributions also from structures of the types

and

These violet substances, like semiquinones in general, are deeply colored. The color is correlated with resonance involving the transfer of electric charge from one end to another of a large molecule, as in the triphenylmethane dyes and other deeply colored substances.[62]

[62] C. R. Bury, *J.A.C.S.* **57**, 2115 (1935); E. Q. Adams and L. Rosenstein, *ibid.*

It has been found[63] that the stability of free radicals of the type A

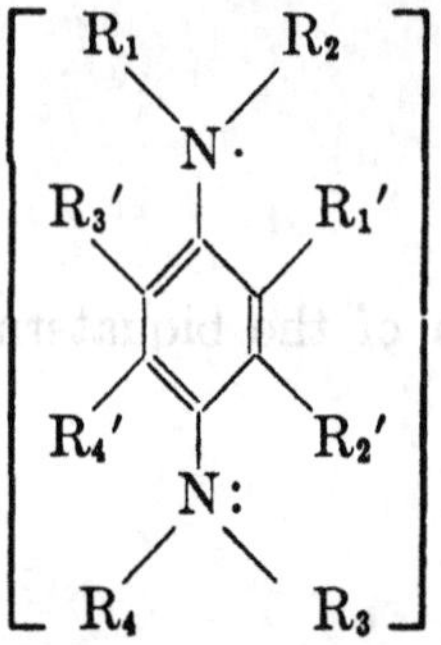

$$
\left[
\begin{array}{c}
R_1 \quad\quad R_2 \\
N\cdot \\
R_3{}' \quad\quad R_1{}' \\
\\
R_4{}' \quad\quad R_2{}' \\
N: \\
R_4 \quad\quad R_3
\end{array}
\right]
\qquad A
$$

is dependent on the nature of the groups R and R' in a way that can be interpreted in terms of steric inhibition of resonance of the type adduced in Section 6-3 in explanation of the results on the electric dipole moments of substituted durenes.

The radical obtained from diaminodurene, $NH_2C_6(CH_3)_4NH_2$, is comparable in stability with that obtained from p-phenylenediamine; the ortho methyl groups apparently do not come into pronounced contact with the hydrogen atoms of the amino groups. (This conclusion is compatible with the values of the van der Waals radii of the groups.) On the other hand, although the phenylene diamine radical with four methyl groups attached to nitrogen, $[(CH_3)_2NC_6H_4N(CH_3)_2]^+$, is stable, the corresponding durene radical, $[(CH_3)_2NC_6(CH_3)_4N(CH_3)_2]^+$, is very unstable; no detectable concentration of it has ever been obtained. It is clear that this instability is the result of the action of steric repulsion between methyl groups. The planar configuration required for resonance with structures such as B

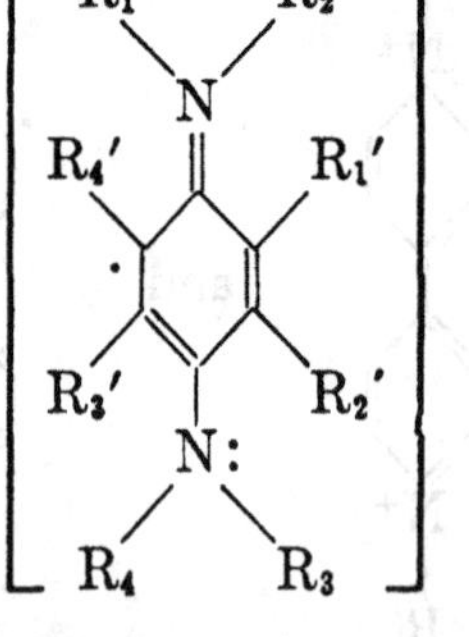

$$
\left[
\begin{array}{c}
R_1 \quad\quad R_2 \\
N \\
R_4{}' \quad\quad R_1{}' \\
\\
R_3{}' \quad\quad R_2{}' \\
N: \\
R_4 \quad\quad R_3
\end{array}
\right]^+
\qquad B
$$

would place pairs of methyl groups such as R_2 and $R_1{}'$ only 2.4 Å apart; this configuration is very unstable, the distance for van der Waals con-

---

36, 1472 (1914); A. Baeyer, *Ann. Chem.* **354**, 152 (1907); L. Pauling, *Proc. Nat. Acad. Sci. U.S.* **25**, 577 (1939).

[63] S. Granick and L Michaelis, *J.A.C.S.* **65**, 1747 (1943).

tact being 4.0 Å. In consequence, the molecule must assume a non-planar configuration, with inhibition of the resonance of this type and with decrease in stability of the radical by the amount of the corresponding resonance energy.

Semiquinone formation is undoubtedly of great significance in physiological processes. Thus it has been found[64] that whereas diaminodurene increases the respiration of erythrocytes to about the same extent as methylene blue, tetramethyldiaminodurene has no catalytic effect at all.

10-6. ELECTRON-DEFICIENT SUBSTANCES

Electron-deficient substances are substances in which the atoms have more stable orbitals than electrons in the valence shell.[65] An example is boron. The boron atom has four orbitals in its valence shell, and three valence electrons.

A characteristic feature of the structure of most electron-deficient substances is that the atoms have ligancy that is not only greater than the number of valence electrons but is even greater than the number of stable orbitals.[66] Thus most of the boron atoms in the tetragonal form of crystalline boron have ligancy 6. Also, lithium and beryllium, with four stable orbitals and only one and two valence electrons, respectively, have structures in which the atoms have ligancy 8 or 12. All metals can be considered to be electron-deficient substances (Chap. 11).

Another generalization that may deserve to be called a structural principle[67] is that an electron-deficient atom causes adjacent atoms to increase their ligancy to a value greater than the orbital number. For example, in the boranes, discussed in the following section, some of the hydrogen atoms, adjacent to the electron-deficient atoms of boron, have ligancy 2.

The structure of tetragonal boron has been determined with care.[68] There are 50 boron atoms in the unit of structure. All but two of them are in icosahedral groups of 12, as shown in Figure 10-1. In the B_{12} icosahedron each boron atom forms five bonds with adjacent atoms. The icosahedra and the two additional boron atoms per unit (interstitial atoms) are arranged in such relative positions that each icosahedral boron atom also forms one more bond, extending in the direction directly out from the center of the icosahedron. Thus each of the

[64] S. Granick, L. Michaelis, and M. P. Schubert, *Science* **90**, 422 (1939).

[65] R. E. Rundle, *J.A.C.S.* **69**, 1327 (1947); *J. Chem. Phys.* **17**, 671 (1949).

[66] This principle was first pointed out to me, in a conversation, by Prof. V. Schomaker.

[67] This principle has not previously been published.

[68] J. L. Hoard, R. E. Hughes, and D. E. Sands, *J.A.C.S.* **80**, 4507 (1958).

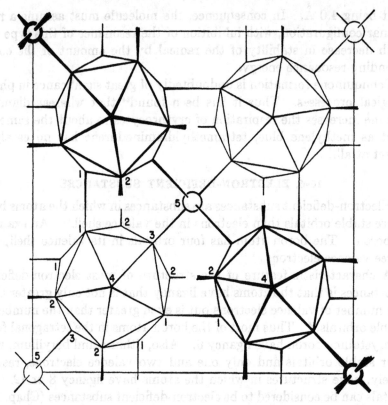

Fig. 10-1.—Structure of tetragonal boron as viewed in the direction of the c axis. One unit cell is shown. Two of the icosahedral groups (light lines) are centered at $z = \frac{1}{4}$, and the other two (heavy lines) at $z = \frac{3}{4}$. The interstitial boron atoms (open circles) are at $(0, 0, 0)$ and $(\frac{1}{2}, \frac{1}{2}, \frac{1}{4})$. Numbers identify the various structurally nonequivalent boron atoms. All of the extra-icosahedral bonds are shown with the exception of B_4—B_4, which is formed parallel to the c axis from each icosahedron to the icosahedra in cells directly above and below.

icosahedral boron atoms has ligancy 6; the interstitial atoms have ligancy 4. The bonds (8 per unit) formed by the interstitial atoms are expected to have bond number[69] 0.89, and the others (140 per unit) to have bond number slightly less than $\frac{1}{2}$ (0.485, when correction is made for the slightly stronger bonds formed with the interstitial atoms).

[69] This value for the bond number is the one given by the statistical theory of resonating bonds described in the following section. A more detailed discussion of boron is given by L. Pauling and B. Kamb, *Laue Festschrift, Z. Krist.*, 112, 472 (1959).

The icosahedral boron atoms may be described as forming three single bonds each of which resonates between two positions.

The expected half-bond boron-boron distance is twice the single-bond radius 0.81 Å (Table 7-18) plus the half-bond correction 0.18 Å (Equation 7-7). This value, 1.80 Å, agrees well with the experimental value, 1.797 $\pm$ 0.015 Å. For the other bonds the calculated distance is 1.65 Å and the observed value is 1.62 $\pm$ 0.02 Å.

Similar icosahedral B_{12} groups are found in other modifications of boron and also in the compound $B_{12}C_3$, in which there are linear C_3 groups.[70]

Crystalline boron is expected to be stabilized to some extent by the resonance energy of the bonds among the alternative positions. The amount of resonance stabilization can be estimated in the following way: The enthalpy of formation of $(-\Delta H^\circ)$ for $B(CH_3)_3(g)$ is 25.7 kcal/mole and that of ethane is 16.5 kcal/mole. Hence the enthalpy of formation of $B(CH_3)_3(g)$ from elementary boron and ethane is 0.9 kcal/mole. If elementary boron contained normal B—B single bonds, the reaction would involve only breaking some of these bonds and some C—C bonds in ethane and forming three B—C bonds, and the value of $-\Delta H^\circ$ would be expected to be 17.3 kcal/mole, as given by Equation 3-12, which relates bond energy to electronegativity difference. We conclude that the difference between these two values, 16.4 kcal/mole, is the resonance energy of elementary boron.[71] Two thirds of this quantity, 10.9 kcal/mole, is the resonance energy of a boron-boron bond between two positions.

As another example we may discuss[72] the crystals MB_6, in which M represents Ca, Sr, Ba, Y, La, Ce, Pr, Nd, Gd, Er, Yt, or Th. The structure, shown in Figure 10-2, involves a boron framework with atoms M in the interstices. Each boron atom forms bonds with five other boron atoms, four in its B_6 octahedron and the fifth in an adjacent octahedron. The five bonds have the same length, which for CaB_6 has

[70] G. S. Zhdanov and N. G. Sevast'yanov, *Doklady Akad. Nauk S.S.S.R.* **32**, 432 (1941); H. K. Clark and J. L. Hoard, *J.A.C.S.* **65**, 2115 (1943).

[71] Rough agreement is found with the result of another calculation. From a consideration of values of enthalpy change accompanying addition of $N(CH_3)_3$ to $B(CH_3)_4$, $B_2H_2(CH_3)_4$, and B_2H_4, S. H. Bauer, A. Shepp, and R. E. McCoy, *J.A.C.S.* **75**, 1003 (1953), concluded that the enthalpy of dimerization of BH_3 is 32 ± 2 kcal/mole. The enthalpy of formation of $B_2H_6(g)$ is -7.5 kcal/mole, and hence that of BH_3 is -19.8 kcal/mole. If elementary boron had no resonance energy, the value would be 0.7 kcal/mole (Equation 3-12). Hence the resonance energy of boron is 20.5 ± 1 kcal/mole. This value is probably less reliable than the value 16.4 given above.

[72] M. von Stackelberg and F. Neumann, *Z. physik. Chem.* **B19**, 314 (1932); G. Allard, *Bull. Soc. chim. France* **51**, 1213 (1932).

the experimental value[73] 1.72 ± 0.01 Å.   Nearly the same value is found in CeB$_6$, and similar values are found in the other compounds, ranging from 1.69 Å for YB$_6$ to 1.77 Å for BaB$_6$.   These values reflect the sizes of the atoms M (the radius for ligancy 12 is 1.797 for Y, 1.970 for Ca, 2.215 Å for Ba—Table 11-1).

It is probable that the atoms M form bonds with one another, and that the boron atoms use their valence electrons in forming boron-boron bonds.   The bond number of these bonds would then be 0.60,

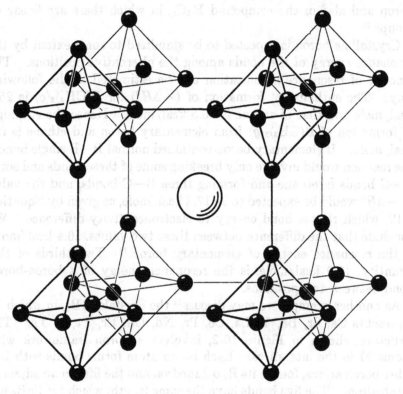

FIG. 10-2.—The atomic arrangement in the cubic
crystal calcium hexaboride, CaB$_6$

with the expected B—B bond length 1.75 Å.   It is also not improbable that there is some transfer of valence electrons from M to the boron framework.   If one electron were transferred, the B—B bond number would be 0.633, with expected bond length 1.74 Å.

There is reason to expect the bond numbers in the resonating system of an electron-deficient molecule to be approximately ½.   Let us consider N bond positions and M electron pairs.   There are $N!/(N - M)!M!$ ways of distributing the M pairs among the N positions.   For given N, this function has its maximum value when M is

[73] L. Pauling and S. Weinbaum, *Z. Krist.* **87**, 181 (1934).

equal to $N/2$. The resonance energy increases with increase in the number of resonance structures, and hence would have its maximum for this value of M. (A factor tending to keep the ligancy small is the repulsion of adjacent nonbonded atoms.) Hence we may expect to find often that the number of bond positions in the resonating system is approximately twice the number of resonating electron pairs, and the bond numbers are about $\frac{1}{2}$. The boranes (Sec. 10-7) provide good examples of this rule.

The following sections contain discussion of the boranes (Sec. 10-7), related substances (Sec. 10-8), ferrocene and related substances (Sec 10-9), and other electron-deficient compounds (Sec. 10-10).

10-7. THE STRUCTURE OF THE BORANES[74]

Boron forms a series of hydrides of surprising composition.[75] The simple substance BH_3 has not been prepared; instead, hydrides of various compositions occur, including B_2H_6, B_4H_{10}, B_5H_9, B_5H_{11}, B_6H_{10}, B_9H_{15}, and $B_{10}H_{14}$.

The structural problem presented by these substances is not a simple one; the fundamental difficulty is that there are not enough valence electrons in the molecules to bind the atoms together with electron-pair bonds. In B_2H_6, for example, there are twelve valence electrons; all twelve would be needed to hold the six hydrogen atoms to boron by covalent bonds, leaving none for the boron-boron bond.

It was suggested by Sidgwick[76] that electron pairs are used for the boron-boron bond and four of the boron-hydrogen bonds and that one-electron bonds are formed between boron atoms and the two remaining hydrogen atoms. Structures based upon this suggestion were discussed in the previous editions of this book.

It was then discovered that the configurations of the molecules correspond to an increase in ligancy of the boron atoms to 5 or 6 and of some of the hydrogen atoms to 2. These configurations provide strong support of the suggestion made by Lewis[77] that the electron pairs reso-

---

[74] The discussion in this section resembles that given by L. Pauling and B. Kamb, *Proc. Nat. Acad. Sci. U.S.* **45**, in press (1959); see also K. Hedberg, *J.A.C.S.* **74**, 3486 (1952); W. N. Lipscomb, *J. Chem. Phys.* **22**, 985 (1954); W. H. Eberhardt, B. Crawford, Jr., and W. N. Lipscomb, *ibid.* 989; W. C. Hamilton, *Proc. Roy. Soc. London* **A235**, 295 (1956); *J. Chem. Phys.* **29**, 460 (1958); M. Yamazaki, *ibid.* **27**, 1401 (1957).

[75] See, for a review of this subject, A. Stock, *Hydrides of Boron and Silicon,* Cornell University Press, 1933; H. I. Schlesinger and A. B. Burg, *Chem. Revs.* **31**, 1 (1942).

[76] N. V. Sidgwick, *The Electronic Theory of Valency*, Clarendon Press, Oxford, 1927, p. 103.

[77] G. N. Lewis, *J. Chem Phys.* **1**, 17 (1933).

nate among the several interatomic positions in such a way as to produce fractional bonds, with resonance stabilization of the molecules.

Diborane, B_2H_6, has the configuration shown in Figure 10-3. This configuration was proposed long ago;[78] it has been verified by spectroscopic and other physical evidence.[79] The molecular dimensions were obtained by a careful electron-diffraction study.[80]

Four of the hydrogen atoms have ligancy 1. Their H—B distance is 1.187 ± 0.030 Å. This value agrees moderately well with the single-bond value 1.13 Å; it corresponds to bond number 0.80 ± 0.08, which shows that the bonds are not pure single bonds. The bridging hydrogen atoms, with ligancy 2, have B—H bond length 1.334 ± 0.027 Å, corresponding to bond number 0.46 ± 0.05, and the B—B distance is

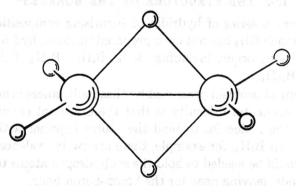

Fɪɢ. 10-3.—The configuration of atoms in diborane, B_2H_6.

1.770 ± 0.013 Å, corresponding to bond number 0.56 ± 0.03. (The bond numbers are calculated with radius 0.81 Å for boron and 0.32 for hydrogen.)

The sum of the bond numbers of the nine bonds is 5.60 ± 0.55. It should be six, because there are twelve valence electrons in the molecule, and therefore the values given above need to be increased by an average of 0.04.

A simple theoretical treatment that can be applied to electron-deficient substances in general can be given for the diborane molecule.[81] Let us consider the various valence-bond structures that can be written for the molecule[82] (with its known configuration) with use of the six

[78] W. Dilthey, *Z. angew. Chem.* **34,** 596 (1921).

[79] F. Stitt, *J. Chem. Phys.* **8,** 981 (1940); **9,** 780 (1941); H. C. Longuet-Higgins and R. P. Bell, *J. Chem. Soc.* **1943,** 250; K. S. Pitzer, *J.A.C.S.* **67,** 1126 (1946); W. C. Price, *J. Chem. Phys.* **15,** 614 (1947); **16,** 894 (1948).

[80] K. Hedberg and V. Schomaker, *J.A.C.S.* **73,** 1482 (1951).

[81] Pauling and Kamb, *loc. cit.* (74).

[82] In general the condition may be made, from the electroneutrality principle, that atoms have formal charge 0, +1, or −1, only.

electron pairs and the stable valence orbitals (one for hydrogen, four for boron). Twenty structures may be written: they are of type *A* (2 structures), *B* (2), *C* (4), *D* (8), and *E* (4).

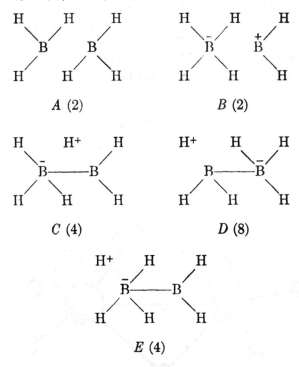

A (2) *B* (2)

C (4) *D* (8)

E (4)

We may, as a first approximation, assume that these structures contribute equally to the normal state of the molecule. The bond numbers calculated in this way are 0.85 for nonbridging B—H, 0.45 for bridging B—H, and 0.80 for B—B. The two B—H values agree well with those obtained from the interatomic distances, but the B—B value does not. Possibly the structures with neutral atoms should be given greater weight than those with separated charges. If the structures of type *A* are given triple weight the calculated bond numbers become 0.875 for nonbridging B—H, 0.459 for bridging B—H, and 0.667 for B—B.

The resonance energy for diborane may be found in the following way: From the electronegativities of boron and hydrogen we estimate (Equation 3-12) that the enthalpy of formation of BH_3 from hydrogen and single-bonded boron would be 0.69 kcal/mole. Elementary boron is more stable than single-bonded boron by 16.4 kcal/mole (Sec. 10-6); hence the enthalpy of formation of BH_3 from the elements in their standard states is estimated to be −15.7 kcal/mole. The observed enthalpy of formation of B_2H_6 from the elements is −7.5 kcal/mole, and accordingly that from $2BH_3$ is 23.9 kcal/mole. Since $2BH_3$ and

B_2H_6 contain the same bonds, except for their resonance in B_2H_6, we may take this quantity, 23.9 kcal/mole, to be the resonance energy of B_2H_6. As a first approximation the structure may be described as involving the resonance of each of two B—H bonds between two positions (to the bridge hydrogen atoms), with a contribution also of the B—B bond resonance. The value 23.9 kcal/mole is seen to be reasonable in comparison with the value 10.9 kcal/mole obtained in Section 10-6 for the resonance energy of a B—B bond between two positions.

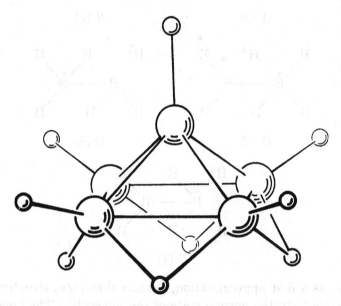

FIG. 10-4.—Arrangement of atoms in the molecule of pentaborane, B_5H_9.

Pentaborane, B_5H_9, has the structure[83] shown in Figure 10-4. In this molecule the B—H bonds for nonbridging hydrogen atoms have the length 1.22 ± 0.07 Å, corresponding to the bond number 0.68 ± 0.20, and the bridge hydrogens have the length 1.35 ± 0.02 Å, corresponding to the bond number 0.43 ± 0.04. It may be pointed out that the experimental values of the electric dipole moment, 2.13 D for pentaborane[84] and 3.52 D for decaborane,[85] require a separation of

[83] Electron diffraction: K. Hedberg, M. E. Jones, and V. Schomaker, *J.A.C.S.* **73**, 3538 (1951); *Proc. Nat. Acad. Sci. U.S.* **38**, 680 (1952); x-ray diffraction: W. J. Dulmage and W. N. Lipscomb, *J.A.C.S.* **73**, 3539 (1951); *Acta Cryst.* **5**, 260 (1952); microwave spectroscopy: H. J. Hrostowski and R. J. Myers, *J. Chem. Phys.* **22**, 262 (1954).

[84] H. J. Hrostowski, R. J. Myers, and G. C. Pimentel, *J. Chem. Phys.* **20**, 518 (1952).

[85] A. W. Laubengayer and R. Bottei, *J.A.C.S.* **74**, 1618 (1952).

charge in the molecules corresponding approximately to these bond numbers.[86]

The B—B bond length for the sides of the base of the pyramid is 1.800 ± 0.005 Å and that for the other four bonds is 1.690 ± 0.005 Å. These values correspond to bond numbers 0.50 and 0.75, respectively. The sum of the bond numbers for all of the bonds, which should be 12 (there are 24 valence electrons), is 11.84 ± 1.32.

The method of averaging for all valence-bond structures, as described above for diborane, is extremely laborious for any except very simple molecules. A statistical theory of resonating valence bonds that can be easily applied to complex as well as simple molecules has been developed.[87] It can be illustrated by application to B_5H_9. Let us begin by assigning the probability 1 to the nonbridging B—H bonds and $\frac{1}{2}$ to the other bonds in the molecule:

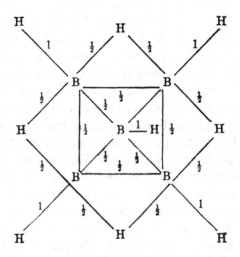

We assume that a hydrogen atom can form no bond or one bond and that a boron atom can form not more than four bonds (corresponding to the number of bond orbitals). For each bond position in the molecule we calculate the probability that the other bonds are so arranged as to permit the bond to be formed, assuming that the occupancy of the positions is unsynchronized. For example, let us consider the apical B—H position. The hydrogen atom, which is not a bridging hydrogen, has probability factor 1 for forming its one bond (the correspond-

[86] A quantum-mechanical discussion of the structure of these molecules in relation to the dipole moments has been reported by W. N. Lipscomb, *J. Chem. Phys.* **25**, 38 (1956).

[87] Pauling and Kamb, *loc. cit.* (74).

ing factor for a bridging hydrogen is $\frac{1}{2}$). The apical boron atom has four $\frac{1}{2}$ bonds, and the probability that at least one of its four orbitals is free is found by a simple calculation to be 15/16. The sum of the probabilities for all 21 bonds is 11.24. The calculated bond numbers for B—H (apex), B—H (base), B—H (bridging), B—B (base), and B—B (slant) are 1.00, 0.87, 0.37, 0.50, and 0.64, respectively. These correspond to bond lengths 1.13 Å (1.22 ± 0.07), 1.17 Å (1.22 ± 0.07), 1.39 Å (1.35 ± 0.02), 1.80 Å (1.800 ± 0.005), and 1.73 Å (1.690 ± 0.005), respectively; the values in parentheses are the experimental values. The statistical theory of resonating bonds is seen to account for the major features of the pentaborane structure.[88]

The enthalpy of formation of $B_5H_9(l)$ is −7.8 kcal/mole relative to the elements in their standard states, 76 kcal/mole relative to single-bonded boron. With the small correction (0.23 kcal/mole per hydrogen atom) for the partial ionic resonance energy of the B—H bonds, this value gives 74 kcal/mole as the resonance energy stabilizing the B_5H_9 molecule. This value is reasonable, in comparison with the values for tetragonal boron and B_2H_6 discussed above; it corresponds to about 10 kcal/mole for each of four bonds resonating between two positions for the bridging hydrogen atoms, two B—B bonds resonating between two positions for the base boron atoms, and three bonds resonating among four positions (that is, one no-bond resonating among the four positions) for the bonds to the apical boron atom.

A great contribution to the structural chemistry of the boranes was made in 1950 when Kasper, Lucht, and Harker[89] reported their determination of the structure of decaborane, $B_{10}H_{14}$. The structure, shown in Figure 10-5, involves a group of ten boron atoms in positions corresponding to a B_{12} icosahedron with two adjacent boron atoms removed. Each of these missing boron atoms is replaced by two hydrogen atoms, which serve as bridges. The other ten hydrogen atoms are attached separately by single bonds to the ten boron atoms. Each boron atom has ligancy 6, with the arrangement of the bonds similar to that in the tetragonal boron crystal. In addition to the bond for each boron that extends out from the dodecahedron center to its nonbridging hydrogen atom, two boron atoms (at the top in Fig. 10-5) form two bonds to bridging hydrogen atoms and three to other borons, four form one bond to a bridging hydrogen and four to other borons, and the remaining four form five bonds to boron atoms.

[88] A refinement of the statistical theory can be made by repeating the calculation with use of the bond numbers given by the preceding calculation as probabilities, until a self-consistent set is obtained. This refinement leads to small changes only in the bond numbers.

[89] J. S. Kasper, C. M. Lucht, and D. Harker, *Acta Cryst.* **3**, 436 (1950); C. M. Lucht, *J.A.C.S.* **73**, 2373 (1951).

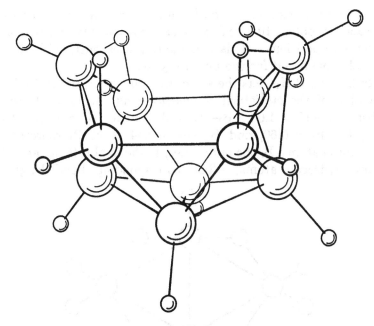

Fig. 10-5.—The atomic arrangement in the molecule of decaborane, $B_{10}H_{14}$. Note that the 10 boron atoms are approximately at corners of a regular icosahedron. Each of the two remaining corners of the icosahedron may be considered to have been replaced by two bridging hydrogen atoms. The other 12 hydrogen atoms are bonded to boron atoms in such a way that the B—H bonds extend out from the center of the icosahedron.

The statistical resonating-bond theory described above can be easily applied to decaborane. Each boron atom forms one bond to an attached hydrogen atom and five bonds that are taken to have probability $\frac{1}{2}$ (to five borons or four borons and one bridging hydrogen). The 21 B—B bonds are hence all alike, as are the 10 B—H bonds and the 8 B—H′ bonds (H′ represents a bridging hydrogen atom). Their bond numbers are calculated to be 0.50, 0.86, and 0.36. These values correspond to the bond lengths B—B = 1.80 Å, B—H = 1.17 Å, and B—H′ = 1.40 Å. These values agree roughly with the experimental values B—B = 1.79 Å (average of values between 1.73 and 2.01), B—H = 1.28 Å, and B—H′ = 1.37 Å.

The enthalpy of formation of $B_{10}H_{14}(c)$, −8 kcal/mole, leads by a calculation similar to that made above for pentaborane to the value 153 kcal/mole for the resonance energy of decaborane. The molecule contains 29 bonds (21 B—B and 8 B—H) that are essentially half-bonds (bond numbers 0.50 and 0.36). If we describe it, as an approximation, as involving fourteen half-bonds resonating between two positions each, we obtain 10.8 kcal/mole as the resonance energy of a bond between two positions. This value agrees excellently with the value 10.9 kcal/mole found for the boron crystal.

Tetraborane, B_4H_{10}, has the structure[90] shown in Figure 10-6. The application of the statistical resonating-bond theory leads to bond numbers B_1—H = 1.00, B_2—H = 0.88, B_1—H' = 0.44, B_2—H' = 0.32, B_1—B_1 = 0.60, and B_1—B_2 = 0.44 (here B_1 represents the central boron atoms, B_2 the outer ones, H the nonbridging, and H' the bridging hydrogen atoms). The corresponding bond lengths are

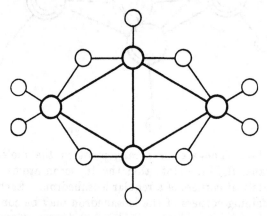

Fig. 10-6.—The atomic arrangement in tetraborane, B_4H_{10}, represented in a plane. The arrangement of the four boron atoms in space is that of the bottom part of the decaborane, Figure 10-5. The four bridging hydrogen atoms and two of the hydrogen atoms attached only to the side boron atoms (right and left) occupy six other icosahedral positions; four B—H bonds extend out from the center of the icosahedron.

B_1—H = 1.13 Å, B_2—H = 1.17 Å, B_1—H' = 1.34 Å, B_2—H' = 1.43 Å, B_1—B_1 = 1.75 Å, and B_1—B_2 = 1.84 Å. These values agree excellently (mean deviation 0.015 Å) with those selected as most probable in an analysis[91] of the x-ray and electron-diffraction results: B_1—H = 1.19 Å, B_2—H = 1.19 Å, B_1—H' = 1.33 Å, B_2—H' = 1.43 Å, B_1—B_1 = 1.75 Å, and B_1—B_2 = 1.85 Å. It is especially interesting that the resonating-bond theory leads to an explanation of the observed differences in the two B—H' bond lengths and in the B—B bond lengths.

[90] Electron diffraction: M. E. Jones, K. Hedberg, and V. Schomaker, *J.A.C.S.* **75**, 4116 (1953); x-ray diffraction: C. E. Nordman and W. N. Lipscomb, *ibid.*; *J. Chem. Phys.* **21**, 1856 (1953).

[91] Lipscomb, *loc. cit.* (74).

Dihydropentaborane, B_5H_{11}, is less stable than pentaborane. Its structure,[92] shown in Figure 10-7, may be described as obtained by breaking one of the base B—B bonds of the B_5 pyramid and introducing two hydrogen atoms. The statistical resonating-bond theory leads to the following bond numbers and bond lengths: B_1—B_2, 0.42, 1.85 Å; B_1—B_3, 0.39, 1.87 Å, B_2—B_3, 0.58, 1.76 Å; B_3—B_3, 0.53, 1.79 Å; B_1—H, 0.76, 1.20 Å; B_2—H, 0.98, 1.13 Å; B_3—H, 0.91, 1.15 Å; B_2—H', 0.42, 1.36 Å; B_3—H', 0.39, 1.38 Å. Here B_1 represents the central boron atom, B_2 the boron atoms forming one hydrogen bridge bond, B_3 those forming two hydrogen bridge bonds, H the nonbridging hydrogen

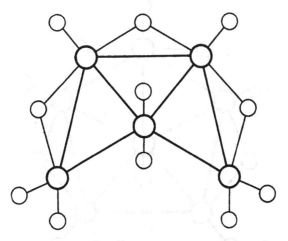

FIG. 10-7.—A diagram representing the structure of dihydropentaborane, B_5H_{11}. Several of the atoms are approximately at icosahedral corners.

atoms, and H' the bridging hydrogen atoms. The x-ray study gave the bond lengths B—H = 1.07 Å, B—H' = 1.24 Å, B_1—B_2 = 1.87 Å, B_1—B_3 = 1.72 Å, B_2—B_3 = 1.75 Å, and B_3—B_3 = 1.77 Å, and the electron-diffraction study gave the average B—B value 1.81 Å, all in rough agreement with the calculated values.

Hexaborane, B_6H_{10}, has the structure[93] shown in Figure 10-8. The six boron atoms and four bridging hydrogen atoms lie approximately at ten corners of an icosahedron.

The statistical resonating-bond calculation leads to the following

[92] Electron diffraction: K. Hedberg, M. E. Jones, and V. Schomaker, *2nd Int. Congr. Cryst.*, Stockholm, 1951; x-ray diffraction: L. R. Lavine and W. N. Lipscomb, *J. Chem. Phys.* 22, 614 (1954).

[93] F. L. Hirshfeld, K. Eriks, R. E. Dickerson, E. L. Lippert, Jr., and W. N. Lipscomb, *J. Chem. Phys.* 28, 56 (1958).

bond numbers and interatomic distances: B_2—H, 0.97, 1.14 Å; other B—H, 0.84, 1.19 Å; B_2—H′, 0.45, 1.34 Å; other B—H′, 0.36, 1.40 Å; B_1—B_2, 0.62, 1.74 Å; B_1—B_3, 0.49, 1.81 Å; B_1—B_4, 0.49, 1.81 Å; B_2—B_2, 0.79, 1.68 Å; B_2—B_3, 0.62, 1.74 Å; B_3—B_4, 0.49, 1.81 Å (H′ is the bridging hydrogen, B_1 boron at apex, B_2 base boron with one hydrogen bridge only, B_3 adjacent to it). The values reported from the x-ray study are B—H, average 1.22 ± 0.06 Å; B—H′, average 1.38 ± 0.08 Å; B_1—B_2, 1.79 ± 0.01 Å; B_1—B_3, 1.75 ± 0.01 Å; B_1—B_4, 1.74 ± 0.01 A, B_2—B_2, 1.60 ± 0.01 Å; B_2—B_3, 1.74 ± 0.01 Å; B_3—B_4, 1.79 ± 0.01 Å.

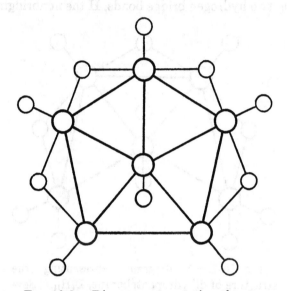

Fig. 10-8.—Diagram representing the structure of the hexaborane molecule, B_6H_{10}. Several of the atoms are at icosahedral positions.

The structure found[94] for enneaborane, B_9H_{15}, is shown in Figure 10-9. There are five bridging hydrogen atoms. Each boron atom has ligancy 6. The boron atoms are approximately at nine corners of an icosahedron. The boron atom at the top (B_1) has two nonbridging hydrogen atoms attached to it, and each of the others has one. The statistical resonating-bond theory leads to the following bond numbers and bond lengths: B_1—H, 0.76, 1.20 Å; other B—H, 0.90, 1.16 Å; B_1—H′, 0.28, 1.46 Å; other B—H′, 0.38, 1.38 Å; B_1—B, 0.38, 1.87 Å; other B—B, 0.52, 1.79 Å. The observed values of the bond lengths agree moderately well: B—H, average 1.15 ± 0.10 Å; B_1—H′, 1.45

[94] R. E. Dickerson, P. J. Wheatley, P. A. Howell, and W. N. Lipscomb, *J. Chem. Phys.* **27**, 200 (1957). The composition of the substance was determined by the x-ray study.

±0.10 Å; other B—H′, 1.36 ± 0.10 Å; B₁—B, 1.86 ± 0.05 Å; other B—B, 1.81 ± 0.05 Å.

The structures of B_2H_6, B_4H_{10}, B_5H_9, B_5H_{11}, B_6H_{10}, B_9H_{15}, and $B_{10}H_{14}$ have common features. In all except B_5H_9 the atoms are arranged in a way closely related to the icosahedron; in B_5H_9 there is a close relation to the octahedron. In some cases a hydrogen atom occupies a corner of the icosahedron; in others the corner is replaced by

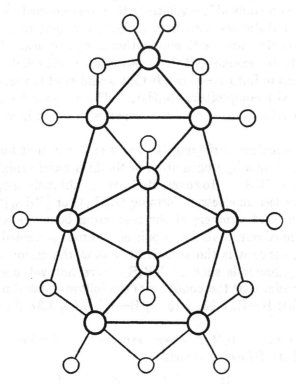

Fig. 10-9.—A diagram representing the structure of enneaborane, B_9H_{15}. All of the boron atoms and some of the hydrogen atoms are approximately at icosahedral positions.

two hydrogen atoms, or even three (B_5H_{11}). There is one hydrogen atom attached to each boron atom in such a way that the B—H bond points outward from the center of the molecule. The ligancy of boron is either 5 (tetragonal pyramid) or 6 (pentagonal pyramid): the average ligancy is 5 in B_2H_6, 5.6 in B_5H_{11}, 5.67 in B_6H_{10}, 5.8 in B_5H_9, and 6 in B_4H_{10}, B_9H_{15}, and $B_{10}H_{14}$.

The most stable boranes are $B_{10}H_{14}$ and B_5H_9. In each of these two molecules every boron atom has one nonbridging H attached, and all ($B_{10}H_{14}$) or most (B_5H_9) of the boron atoms form five half-bonds (in

B_5H_9 the apex boron forms four B—B bonds with bond number 0.64). The conclusion may be drawn that the structural feature leading to greatest stability is ligancy 6 with one nonbridging B—H bond (or B—B bond, as in elementary boron).

We may ask if there are any possible boranes other than $B_{10}H_{14}$ that are based entirely on this structural feature. One is $B_{12}H_{12}$, with a B_{12} icosahedron. It is likely that this molecule is stable, but that the conditions for its preparation are essentially the same as for the colorless or yellow compounds of low volatility that are obtained by heating the boranes.[95] Substances with high molecular weight may result from condensation of boranes with elimination of hydrogen and formation of B—B bonds; for example, $B_{12}H_{12}$ icosahedra might eliminate four hydrogen atoms to form B—B bonds that would hold the icosahedra in a framework with composition $(B_{12}H_8)_\infty$. The yellow color observed for some of the solid boranes would be expected for such large conjugated systems.

Another structure satisfying the requirement is that for B_6H_{11}, obtained by adding a hydrogen atom in the fifth basal bridging position of B_6H_{10} (Fig. 10-8). However, B_6H_{11} is an odd molecule, and would either add or lose an electron, forming $[B_6H_{11}]^-$ or $[B_6H_{11}]^+$. In order to keep the bond numbers of the resonating bonds close to $\frac{1}{2}$, as is needed for maximum stability, a pair of electrons is needed for the two half-bonds; hence it is the anion rather than the cation that will be stable.[96] Compounds such as KB_6H_{11} have not yet been reported. We may predict that the bonds have the following bond numbers and bond lengths: B—H, 0.93, 1.15 Å; B—H', 0.40, 1.37 Å; B—B, 0.54, 1.78 Å.

The substance NaB_3H_8 has been reported.[97] The ion $[B_3H_8]^-$ may be assigned the following structure:

[95] A chrome-yellow $(BH)_x$ has been reported by A. Stock and W. Mathing, *Ber.* **69**, 1456 (1936).

[96] The stability of $[B_6H_{11}]^+$ has been suggested by W. N. Lipscomb, *J. Chem. Phys.* **28**, 170 (1958), on the basis of a molecular-orbital treatment. It seems likely, however, that the argument given above is valid, and that the anion rather than the cation is stable. Lipscomb also suggested that the $[B_4H_7]^-$ ion with a B_4 tetrahedron and three bridging hydrogen atoms about one triangular face should be stable. This structure, with one boron atom forming only a B—H bond plus three fractional B—B bonds, would be very unstable, according to the argument given above. However, the $[B_4H_6Cl]^-$ ion might be stable.

[97] W. V. Hough, L. J. Edwards, and A. B. McElroy, *J.A.C.S.* **78**, 689 (1956).

(It is deficient in electrons; $[B_3H_8]^{---}$ is analogous to propane.) The predicted values of bond numbers and bond lengths are B_2—H, 0.89, 1.16 Å; B_2—H′, 0.38, 1.38 Å; B_1—H′, 0.41, 1.36 Å; B—B, 0.62, 1.75 Å. In other reported salts, such as $Na_2B_4H_{10}$,[98] the anion with a double negative charge may be expected to have the same configuration as for the corresponding borane, except for the decrease in all bond lengths (by 0.03 Å for $[B_4H_{10}]^{--}$) corresponding to the increase by one in the number of resonating electron pairs.

10-8. SUBSTANCES RELATED TO THE BORANES

Bromodiborane, B_2H_5Br, has a structure essentially identical with that of diborane except that a bromine atom replaces one of the nonbridging hydrogen atoms.[99] The B—Br bond length, 1.934 ± 0.010 Å, corresponds to bond number 0.80 (the B—Br single-bond length is

1.894 Å). In 1,1-dimethyldiborane,[99] $(CH_3)_2B \diagup\!\!\!\!\!\overset{\text{H}}{\underset{\text{H}}{\diagdown}}\!\!\!\!\!\diagup BH_2$, the B—C

bond length, 1.61 Å, corresponds to the bond number 0.77. In aminodiborane, $B_2H_5NH_2$, and dimethylaminodiborane, $B_2H_5N(CH_3)_2$, one of the bridging hydrogen atoms is replaced by the amino or dimethylamino group.[100] The B—N bond length, 1.53 ± 0.04 Å, corresponds to bond number 0.80.

A number of other substituted boranes are known. An interesting one is $B_{10}H_{12}(NCCH_3)_2$, the product of reaction of decaborane with acetonitrile.[101] Its structure has been determined[102] by x-ray study of the crystal. The two hydrogen atoms projecting out from the two top boron atoms of Figure 10-5 are replaced by acetonitrile groups. These groups are linear, as would be expected for the reasonable structure B—N≡C—CH_3.

Tetraboron tetrachloride, B_4Cl_4, differs from the boranes in that the ligancy of boron is only 4. The boron atoms lie at the corners of a regular tetrahedron.[103] Each boron atom forms three B—B bonds, with the other atoms of the tetrahedron, and a B—Cl bond, directed out from the center of the molecule. The B—Cl bond length, 1.70 Å,

[98] A. Stock and E. Kuss, *Ber.* **59**, 2210 (1926).

[99] Hedberg, Jones, and Schomaker, *loc. cit.* (92).

[100] K. Hedberg and A. J Stosick, *J.A.C.S.* **74**, 954 (1952).

[101] R. Schaeffer, *J.A.C.S.* **79**, 1006 (1957).

[102] J. van der Maas Reddy and W. N. Lipscomb, *J.A.C.S.* **81**, 754 (1959).

[103] M. Atoji and W. N. Lipscomb, *Acta Cryst.* **6**, 547 (1953). The substance was first prepared by G. Urry, T. Wartik, and H. I. Schlesinger, *J.A.C.S.* **74**, 5809 (1952).

is approximately the single-bond value, 1.72 Å; the bond may have a small amount of double-bond character. The B—B bond length, 1.70 Å, corresponds to bond number 0.74. This value is somewhat larger than the value 0.67 that would apply if each boron used one of its valence electrons completely in forming a B—Cl bond. It suggests that the B—Cl bond has bond number 0.89 (plus some double-bond character, using electron pairs of the chlorine atom).[104]

The corresponding borane B_4H_4 involves boron with ligancy 4, equal to the orbital number (whereas in B_4Cl_4 the partial double-bond character of the B—Cl bonds may be said to correspond to an increased ligancy for boron); accordingly we would not expect B_4H_4 to be stable. Similarly, as discussed in the preceding section, B_6H_6 (octahedral B_6 group, ligancy 5 for boron) would be expected to be unstable and $B_{12}H_{12}$ (icosahedral B_{12}, ligancy 6) to be stable.[105]

Beryllium borohydride, BeB_2H_8, has the structure[106]

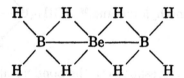

in which each boron or beryllium atom is surrounded tetrahedrally by four hydrogen atoms. The bond lengths and corresponding bond

[104] In tetrachlorodiborine, B_2Cl_4, and tetrafluorodiborine, B_2F_4, the ligancy of boron is 3. These substances are not electron-deficient substances; the halogen atoms have extra electron pairs that can make use of the fourth orbital of boron for double-bond formation. B_2F_4 is planar, with bond lengths B—B = 1.67 ± 0.05 Å and B—F = 1.32 ± 0.04Å and bond angle FBF = 120° ± 2.5° (x-ray investigation of the crystal: L. Trefonas and W. N. Lipscomb, *J. Chem. Phys.* **28**, 54 [1958]). The x-ray investigation of the B_2Cl_4 crystal has led to a similar planar structure, with B—B = 1.80 ± 0.05 Å, B—Cl = 1.72 ± 0.05 Å, and angle ClBCl = 121.5° ± 3° (M. Atoji, W. N. Lipscomb, and P. J. Wheatley, *ibid.* **23**, 1176 [1955]). The gas molecule has been reported to have the same dimensions but with a bisphenoidal (nonplanar) configuration (electron diffraction: Hedberg, Jones, and Schomaker, *loc. cit.* (92); Infrared and Raman spectroscopy: M. J. Linevsky, E. R. Shull, D. E. Mann, and T. Wartik, *J.A.C.S.* **75**, 3287 [1953]). It seems likely that there is largely unrestricted rotation about the B—B bond.

[105] Quantum mechanical calculations have been reported to show instability of these substances: Eberhardt, Crawford, and Lipscomb, *loc. cit.* (74); H. C. Longuet-Higgins and M. de V. Roberts, *Proc. Roy. Soc. London* A230, 110 (1955).

[106] S. H. Bauer, *J.A.C.S.* **72**, 622 (1950). This electron-diffraction study is supported by a spectroscopic investigation: W. C. Price, *J. Chem. Phys.* **17**, 1044 (1949).

numbers are B—H (nonbridging) 1.22 Å, $n = 0.71$; B—H (bridging) 1.28 Å, $n = 0.61$; B—Be 1.74, $n = 0.74$; Be—H 1.63 Å, $n = 0.20$. The sum of these bond numbers is 7.56, a little smaller than the number of bonding electron pairs; the bond numbers should accordingly be increased (each by 0.03). Boron has ligancy 5 and beryllium has ligancy 6.

Aluminum borohydride has the structure[107]

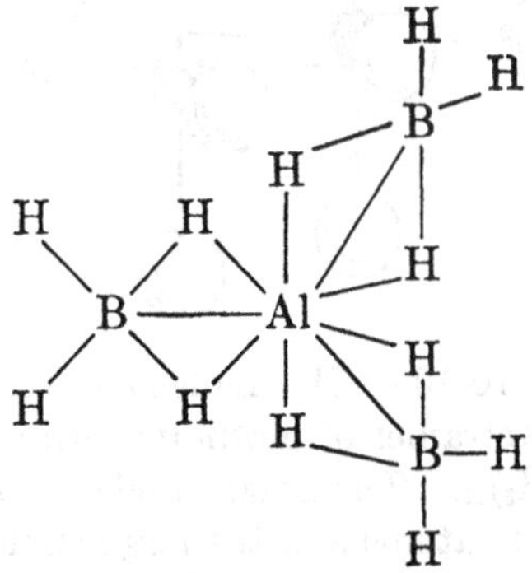

$$
\begin{array}{c}
\text{H} \\
| \\
\text{H} \qquad \text{B---H} \\
| \\
\text{H---B---Al} \\
\text{H---B---H} \\
| \\
\text{H}
\end{array}
$$

in which aluminum is surrounded octahedrally and boron tetrahedrally by hydrogen atoms. The bond lengths and bond angles are B—H (nonbridging) 1.21 Å, $n = 0.74$; B—H (bridging) 1.28 Å, $n = 0.61$; B—Al 2.15 Å, $n = 0.61$; Al—H 2.1 Å, $n = 0.20$. The sum of the bond numbers, 11.13, is 0.87 less than the number of bonding electron pairs, 12, indicating that the bond numbers should be increased by an average of 0.04.

10-9. SUBSTANCES CONTAINING BRIDGING METHYL GROUPS

Several electron-deficient compounds in which there are bridging methyl groups, with ligancy 5 or 6 for carbon, are known. The first one to be recognized was the tetramer of platinum tetramethyl, $Pt_4(CH_3)_{16}$. This substance was subjected to x-ray investigation by Rundle and Sturdivant,[108] who found it to have the structure shown in Figure 10-10. Each carbon atom has ligancy 6; it is bonded to its three hydrogen atoms and also to the three neighboring platinum atoms. The bond lengths were not accurately determined (the platinum-platinum distance was found to be 3.44 Å); however, it is likely that the Pt—C bridging bonds are approximately half-bonds (with

---

[107] Bauer, also Price, *loc. cit.* (106).

[108] R. E. Rundle and J. H. Sturdivant, *J.A.C.S.* **69**, 1561 (1947); the structure was reported in the first edition of this book, 1939.

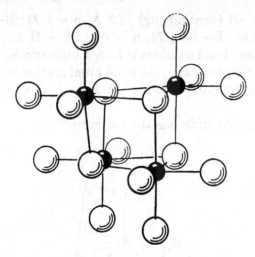

Fɪɢ. 10-10.—The molecular structure
of the tetramer of platinum tetramethyl,
Pt₄(CH₃)₁₆. The small circles represent
platinum atoms and the large circles carbon atoms of the methyl groups.

bond number about 0.83 for the C—H bonds), permitting the value
2.25 Å for the Pt—C bridging bond length to be predicted.[109]

The increase in ligancy of the carbon atom illustrates the principle,
mentioned in Section 10-6, that an electron-deficient atom causes adjacent atoms to increase their ligancy. The platinum atom in the
monomer, $Pt(CH_3)_4$, would make use of only seven of its nine valence
orbitals: four for bonds to the four carbon atoms and three for the
three unshared pairs of $5d$ electrons. The electron deficiency of this
atom then permits the increase in ligancy of carbon.

The trimethylaluminum dimer, $Al_2(CH_3)_6$, has a structure closely
resembling that of diborane (Fig. 10-3), with methyl groups replacing
the hydrogen atoms.[110] The nonbridging and bridging Al—C bond
lengths, 2.00 Å and 2.24 Å, and the Al—Al bond length, 2.56 Å, correspond to the bond numbers 0.74, 0.30, and 0.80, respectively, with sum
5.96, in good agreement with the number of bonding electron pairs (not
including the C—H bonds), 6.

The crystal dimethyl beryllium, $Be(CH_3)_2$, has a structure[111] similar

---

[109] Rundle and Sturdivant reported also the structure of Pt₄(CH₃)₁₂Cl₄, in
which there are bridging chlorine atoms. The Pt—Cl bond length, 2.48 Å, is
just the half-bond value.

[110] X-ray investigation of the crystal: D. N. Lewis and R. E. Rundle, *J. Chem.
Phys.* **21,** 986 (1953).

[111] A. I. Snow and R. E. Rundle, *Acta Cryst.* **4,** 348 (1951).

to that of silicon disulfide (Fig. 11-19). There are infinite polymers,

$$
\begin{array}{ccc}
\text{H}_3 & \text{H}_3 \\
\text{C} & \text{C} \\
\diagup \diagdown & \diagup \diagdown & \diagup \\
\cdots \quad \text{Be} & \text{Be} \quad \cdots \\
\diagdown & \diagup \diagdown & \diagdown \\
\text{C} & \text{C} \\
\text{H}_3 & \text{H}_3
\end{array}
$$

in which each beryllium atom is tetrahedrally surrounded by four bridging methyls. The Be—C bond length, 1.93 Å, and the Be—Be bond length, 2.10 Å, correspond to the bond numbers 0.29 and 0.26, respectively.[112] The sum of these bond numbers per $Be(CH_3)_2$ is 1.42, somewhat less than the number of bonding electron pairs, 2. It is likely that the bond number for each of the five bonds (four Be—C, one Be—Be) per $Be(CH_3)_2$ is about 0.4.

Carbonium Ions as Reaction Intermediates.—The properties of electron-deficient substances may be expected to be of great importance in the theory of chemical reactions. For example, a positively charged (and hence electron-deficient) carbon atom in a complex carbonium ion would be expected to cause adjacent atoms to increase their ligancies, as by the formation of a three-membered ring and by the use of bridging hydrogen atoms. The analysis of the mechanisms of chemical reactions may in the course of time permit much more precise principles to be formulated than are now at hand.

Among the numerous examples in the recent literature of organic chemistry are the discussions of the reactions of norbornyl derivatives with use of a structure for the norbornium ion[113] involving the formation of a three-membered ring through increase in ligancy of the carbon atoms to 5 and similar treatment of reactions of cyclopropylcarbinyl derivatives.[114] In the investigation of carbonium-ion reactions of cyclopropylcarbinyl derivatives[114] it was shown by use of carbon-14 as

a tracer that the three methylene groups of the ion,
$$
\begin{array}{c}
\text{H}_2\text{C} \\
\diagdown \\
\quad \text{CH}\!-\!\overset{+}{\text{C}}\text{H}_2, \\
\diagup \\
\text{H}_2\text{C}
\end{array}
$$
are essentially equivalent, and the suggestion was made that the four

[112] Calculated with single-bond radius 0.889 Å for beryllium (Table 7-18).

[113] T. P. Nevell, E. de Salas, and C. L. Wilson, *J. Chem. Soc.* 1939, 1188; S. Winstein and D. Trifan, *J.A.C.S.* 71, 2953 (1949); 74, 1147, 1154 (1952); J. D. Roberts, C. C. Lee, and W. H. Saunders, Jr., *ibid.* 76, 4501 (1954).

[114] R. H. Mazur, W. N. White, D. A. Semenow, C. C. Lee, M. S. Silver, and J. D. Roberts, *J.A.C.S.* 81, 4390 (1959).

carbon atoms assume a tetrahedral configuration. From the foregoing discussion of the boranes we are led to suggest that three hydrogen atoms serve as bridging atoms and that the ion has the structure

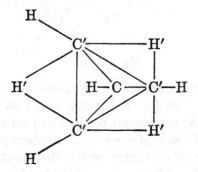

The bond numbers given by the resonating-bond theory are $C-H = 1$, $C'-H = 1$, $C'-C = 1$, $C'-C' = 0.53$, and $C'-H' = 0.40$. The charge of the cation is distributed over the three H' atoms, each $+0.20$, and the three C' atoms, each $+0.13$.

Complexes of Olefines and Silver Ion.—Much work has been done on the interaction of the silver ion, Ag^+, with unsaturated and aromatic hydrocarbons. (Mercuric ion and some other metal ions also react with carbon-carbon double bonds.) The structure proposed by Winstein and Lucas[115] is probably essentially correct. Let us consider a silver ion and ethylene. The system is an electron-deficient one: there are 12 valence electrons and 13 valence orbitals (including one orbital for the silver ion). We may write three structures for the complex:

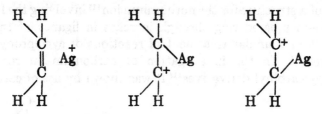

If the three structures contribute equally, the $C-Ag$ bonds have bond number $\frac{1}{3}$ and the $C-C$ bond has bond number $1\frac{1}{3}$.

The crystal $AgClO_4 \cdot C_6H_6$ has been studied with x-rays by Smith and Rundle.[116] Each silver atom was found to be ligated to four carbon atoms: C_1 and C_2 of the benzene molecule on one side of it and C_4 and C_5 of that on the other side. In this way chains of alternating silver atoms and benzene molecules are formed. The bond lengths $Ag-C_1$

[115] S. Winstein and H. J. Lucas, *J.A.C.S.* **60**, 836 (1939).
[116] H. G. Smith and R. E. Rundle, *J.A.C.S.* **80**, 5075 (1958).

and Ag—C_4 are 2.50 Å and Ag—C_2 and Ag—C_5 are 2.63 Å. With the silver radius taken as the bicovalent value, 1.39 Å, these bond lengths correspond to the bond numbers 0.22 and 0.13, respectively. Their expected value, for equal resonance among the two Kekulé structures

and two structures of type 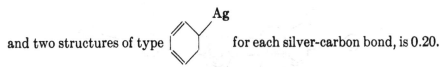 for each silver-carbon bond, is 0.20.

A somewhat similar structure has been found for the crystal containing silver nitrate and cyclooctatetraene.[117] The silver atom is near four carbon atoms of the ring, C_1, C_2, C_5, and C_6. The bond lengths Ag—C_1 and the like are 2.46, 2.51, 2.78, and 2.84 Å, corresponding to the bond numbers 0.26, 0.22, 0.08, and 0.06, respectively.

These two examples indicate that the Ag—C bond is essentially equivalent to half of a C=C double bond in weighting the resonance structures.

10-10. FERROCENE AND RELATED SUBSTANCES

A few years ago the synthesis of a substance of a new type, dicyclopentadienyl iron, $Fe(C_5H_5)_2$ (commonly called ferrocene), was reported almost simultaneously by two groups of investigators.[118] Ferrocene forms orange crystals, which vaporize without decomposition. It can be oxidized to the blue ferricinium cation $[Fe(C_5H_5)_2]^+$. Many similar substances have been reported: ruthenocene, $Ru(C_5H_5)_2$; ruthenicinium ion, $[Ru(C_5H_5)_2]^+$; corresponding compounds of titanium, vanadium, chromium, manganese, and cobalt and of their congeners; corresponding compounds with indenyl replacing dicyclopentadienyl; and corresponding benzene compounds, such as dibenzenechromium, $Cr(C_6H_6)_2$, and its cation, $[Cr(C_6H_6)_2]^+$.

It was shown by Fischer and his collaborators[119] that crystals of ferrocene and the similar compounds of vanadium, chromium, cobalt, nickel, and magnesium are isomorphous, and a determination of the structure of ferrocene was made by Dunitz, Orgel, and Rich.[120] The molecule was found to have the configuration shown in Figure 10-11. Ruthenocene forms crystals that are not isomorphous with those of ferrocene; the molecules have a related structure,[121] differing from that

[117] F. S. Mathews and W. N. Lipscomb, *J.A.C.S.* **80**, 4745 (1958).

[118] T. J. Kealy and P. L. Pauson, *Nature*, **168**, 1039 (1951); S. A. Miller, J. A. Tebboth, and J. F. Tremaine, *J. Chem. Soc.* **1952**, 632.

[119] W. P. Pfab and E. O. Fischer, *Z. anorg. Chem.* **274**, 317 (1953); E. Weiss and E. O. Fischer, *ibid.* **278**, 219 (1955).

[120] J. D. Dunitz, L. E. Orgel, and A. Rich, *Acta Cryst.* **9**, 373 (1956).

[121] G. L. Hardgrove and D. H. Templeton, *Acta Cryst.* **12**, 28 (1959).

of ferrocene only in that the two cyclopentadienyl rings are in the eclipsed rather than the staggered relative orientation.

The C—C bond length found for ferrocene (by electron diffraction of gas molecules)[122] is 1.435 ± 0.015 Å. The C—C distance in rutheno-

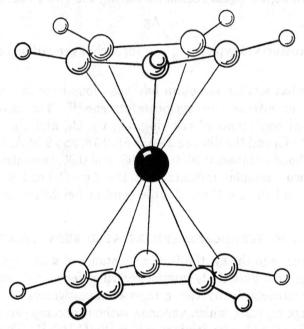

Fig. 10-11.—The structure of ferrocene, $Fe(C_5H_5)_2$.

cene[123] is 1.43 ± 0.02 Å. The experimental values of the Fe—C and Ru—C bond lengths are 2.05 ± 0.01 Å and 2.21 ± 0.02 Å, respectively.

A Resonating-Bond Treatment of Ferrocene.—The electronic structure of ferrocene has been treated by many investigators.[124] We shall discuss it in terms of resonating covalent bonds.[125]

The iron atom might be expected, in this complex as in others (Chap.

[122] The values C—C = 1.440 ± 0.015 Å and Fe—C = 2.064 ± 0.010 Å were reported by K. Hedberg, W. C. Hamilton, and A. F. Berndt (A. F. Berndt, Ph.D. thesis, Calif. Inst. Tech., 1957), and the values 1.43 ± 0.03 Å and 2.03 ± 0.02 Å by E. A. Seibold and L. E. Sutton, *J. Chem. Phys.* 23, 1967 (1955). The crystal values are 1.41 ± 0.03 Å and 2.05 ± 0.03 Å: Dunitz, Orgel, and Rich, *loc. cit.* (120).

[123] Hardgrove and Templeton, *loc. cit.* (121).

[124] G. Wilkinson, M. Rosenblum, M. C. Whiting, and R. B. Woodward, *J.A.C.S.* 74, 2225 (1952); J. D. Dunitz and L. E. Orgel, *Nature* 171, 121 (1953) *J. Chem. Phys.* 23, 954 (1955); E. O. Fischer and R. Jira, *Z. Naturforsch.* 8b, 217 (1953); 9b, 618 (1954); 10b, 354 (1955); W. Moffitt, *J.A.C.S.* 76, 3386 (1954); J. W. Linnett, *Trans. Faraday Soc.* 52, 904 (1956); D. A. Brown, *J. Chem. Phys.* 29, 1086 (1958).

[125] Pauling and Kamb, *loc. cit.* (74).

5), to use all of its nine valence orbitals for bond formation or for occupancy by unshared electrons or electron pairs. There are 18 valence electrons in the molecule (in addition to those required for the single C—H bonds and the single C—C bonds of the two rings), that is, a pair for each of the nine orbitals of the iron atom. The electron pairs may resonate into carbon-carbon positions, conferring some double-bond character on these bonds. The structure should be such as to satisfy the electroneutrality principle.[126]

Let us consider the ways in which the nine valence electron pairs of ferrocene (apart from those used in forming five C—H and five C—C bonds in each cyclopentadienyl ring) may be distributed among the ten C—C and ten Fe—C bond positions with use only of the stable orbitals of the atoms (one orbital of each carbon atom is available, and 9 of the iron atom). The Fe—C bonds have a small amount (12 percent) of ionic character, so that a single negative formal charge may be placed on the iron atom, or it may have zero formal charge; other charge distributions may be rejected on the basis of the electroneutrality principle.

There are 560 structures compatible with these limitations. They are represented by the following diagrams:

A 25   B 25   C 50   D 10   E 10

F 150   G 150   H 30   I 50   J 50   K 10

The number below each diagram is the number of structures of the type indicated. The horizontal bars represent C=C bonds and the vertical bars Fe—C bonds; for example, each of the 25 structures of type A involves two rings, each ring having one carbon atom bonded to the iron atom.

[126] The cyclopentadienyl rings (and the iron atom) of ferrocene are shown to have no electric charge by the identity of the first acid constant of ferrocene dicarboxylic acid with that of benzoic acid (R. B. Woodward, M. Rosenblum, and M. C. Whiting, *J.A.C.S.* **74**, 3458 [1952]).

As an approximation the different structures may be assumed to have the same energy, and hence to contribute equally to the normal state of the molecule. In going from structure A to structure C, for example, one $3d$ electron is promoted to a $4p$ orbital, and an additional bond is formed (two Fe—C bonds formed, one C=C converted to C—C); the bond energy roughly cancels the promotion energy. We may therefore evaluate the properties of the molecule by averaging over the 560 structures A to K, with equal weights.

This average gives 1.225 for the carbon-carbon bond number and 0.471 for the iron-carbon (or ruthenium-carbon) bond number. The calculated average number of unshared pairs on the iron atom is 2.03. Its average formal charge is -0.79. The partial ionic character of the 4.71 Fe—C bonds (12 percent) reduces the charge to -0.22.

It is interesting that the 150 structures G (five Fe—C bonds, two unshared pairs) represent the molecule quite well.

The nature of the bond orbitals of the iron atom can be calculated. If the assumptions are made that the unshared pairs occupy $3d$ orbitals and that the $3d$ orbitals not occupied by them are distributed equally among all the other orbitals (all of which are used in some of the resonance structures of the normal state), the amount of d character in these orbitals is 42.6 percent. The corresponding value of the single-bond radius of iron is 1.135 Å (Sec. 11-8), and the calculated Fe—C bond length for bond number 0.472 is 2.05 Å. This agrees perfectly with the observed value 2.05 $\pm$ 0.01 Å. Similarly for ruthenium the single-bond radius is 1.304 Å and the calculated Ru—C bond length is 2.22 Å, in agreement with the observed value 2.21 $\pm$ 0.02 Å. The calculated bond length for the carbon-carbon bond (bond number 1.225) is 1.445 Å (Table 7-9), in satisfactory agreement with the experimental values 1.435 $\pm$ 0.015 Å for ferrocene and 1.43 $\pm$ 0.02 Å for ruthenocene. We conclude that the resonating-bond theory for these substances is in complete agreement with the experimental results.

Nickelocene, $Ni(C_5H_5)_2$, has the same structure[127] as ferrocene. It is observed to have the paramagnetism corresponding to two unpaired electron spins.

We may apply the resonating-bond theory to nickelocene. There are 20 valence electrons in the molecule, in addition to those of the C—H and C—C single bonds. Altogether 4100 structures compatible with the electroneutrality principle (formal charge 0 or -1 for the nickel atom) may be written. They are of various types, some of which are represented by the following diagrams:

[127] Electron diffraction, Ni—C = 2.20 $\pm$ 0.02 Å, C—C = 1.44 $\pm$ 0.02 Å: K. Hedberg, unpublished work, Calif. Inst. Tech.

$$
\begin{array}{ccccc}
\overline{}\ \overline{}\ | & \overline{}\ |\!|\!| & \overline{}\ |\!|^{+} & \overline{}\ |\!|\uparrow & \overline{}\ |\!|\uparrow \\
:::\text{Ni}\!\uparrow\uparrow & ::\text{Ni}\!\uparrow\uparrow & :::\text{Ni}\!\uparrow\uparrow & :::\text{Ni}\!\uparrow & :::\text{Ni} \\
\underline{}\ \underline{}\ | & \underline{}\ \underline{}\ | & \underline{}\ \underline{}\ | & \underline{}\ \underline{}\ | & \underline{}\ |\!|\uparrow \\
A\ 25 & B\ 50 & C\ 150 & D\ 150 & E\ 225
\end{array}
$$

Here the vertical arrow represents an unpaired electron occupying an atomic orbital, and the other symbols have the same meanings as in the diagrams for ferrocene. For example, for each of the 25 structures A there are two carbon-carbon double bonds in each ring, two nickel-carbon bonds, three unshared pairs occupying $3d$ orbitals of the nickel atom, and two unpaired electrons occupying orbitals of the nickel atom (not pure $3d$ orbitals, because a large fraction of the two remaining $3d$ orbitals is used in forming the nickel-carbon bonds). For these structures A seven of the nine valence orbitals of the nickel atom are used; for others eight or nine. An example is F:

$$
\begin{array}{c}
\overline{}\ |\!|\uparrow \\
:::\text{Ni}\!\uparrow \\
\underline{}\ |\!|^{+} \\
F\ 450
\end{array}
$$

With the assumption that the 4100 structures have equal weights, the carbon-carbon bonds in the ring are calculated to have bond number $n = 1.173$ and the nickel-carbon bonds to have $n = 0.439$, with 34.6 percent d character for the nickel bond orbitals. The number of unshared pairs on the nickel atom is 2.89. The formal charge on the nickel atom is -0.64; of this, the 4.39 Ni—C bonds, with 12 percent ionic character, would provide the opposite charge $+0.53$, leaving the nickel atom essentially neutral (charge -0.11).

The calculated values of bond lengths are 1.456 Å for carbon-carbon and 2.12 Å for nickel-carbon. They agree only roughly with the reported values, 1.44 Å and 2.20 Å, respectively.[128]

The study of proton nuclear magnetic resonance spectra has shown that a shift in frequency occurs that can be accounted for by a positive electron spin density of 0.14 on each carbon atom;[129] leaving 0.6 on the nickel atom. The value of the spin density on the carbon atoms given by the resonating-bond calculation is 0.152, in good agreement with the observed value.

It is interesting that the structures of type F alone provide a good approximation for nickelocene: bond numbers 1.20 for C—C and 0.40

[128] K. Hedberg, reported by Berndt, *op. cit.* (122).
[129] H. M. McConnell and C. H. Holm, *J. Chem. Phys.* **27**, 314 (1957).

for Ni—C, bond lengths 1.45 Å and 2.14 Å, three unshared pairs on the nickel atom, spin density on carbon atoms +0.10.

The observed values of the paramagnetism correspond to 2, 3, 2, 5, 0, 1, and 2 unpaired electron spins for the dicyclopentadienyl compounds of Ti, V, Cr, Mn, Fe, Co, and Ni, respectively. This sequence of values cannot be accounted for in any simple way. It is the result of the interaction of several factors that contribute to the energy of the normal state of each molecule.

One of these factors is the bond energy. For $Ti(C_5H_5)_2$, for example, there are 425 structures for the case of no unpaired electrons and 175 with two unpaired electrons (restricted to the titanium atom, because they are in pure $3d$ orbitals). The numbers of bonds are 4.00 Ti—C and 2.59 C=C for the first case and 2.86 Ti—C and 3.14 C=C for the second case. The extent to which the bond energy favors the first case depends on the relative bond energies of Ti—C and C=C bonds and upon the contributions of resonance energy and promotion energy.

The resonance energy of $3d$ atomic electrons favors states with a large number of unpaired $3d$ electrons. The resonance stabilization is found from the spectroscopic values of atomic energy levels to be $\epsilon N(N-1)/2$, where N is the number of unpaired $3d$ electrons and ϵ varies uniformly from 11 kcal/mole for titanium to 15 kcal/mole for nickel.[130] Hence for titanocene, with two unpaired $3d$ electrons, this factor provides the stabilization energy 11 kcal/mole, and for manganocene, with five unpaired $3d$ electrons, it provides the stabilization energy 130 kcal/mole. The striking difference between manganese and its two neighbors, chromium and iron (found also in other compounds—Sec. 7-9) lacks a convincing explanation. It may be determined by the difference in energy of a $3d$ and a $4s$ or $4p$ electron, which is rapidly changing with atomic number in this region (Sec. 2-7).

Proton magnetic resonance studies[131] of $V(C_5H_5)_2$ and $Cr(C_5H_5)_2$ show negative electron spin densities of -0.06 and -0.12, respectively, on the carbon atoms. These negative spin densities probably arise in a different way from the positive spin density in $Ni(C_5H_5)_2$, discussed above. The unpaired electrons in $3d$ orbitals are restricted to the metal atom. They interact with the shared electron pairs of the M—C bonds in such a way as to make the distribution of the electrons of the shared pair unsymmetrical: the electron with spin parallel to those of the unshared electrons on the metal tends to remain on the metal and the other one on the carbon atom. The observed negative spin densities can be accounted for in this way, with the values of the $3d - 3d$

[130] L. Pauling, *Proc. Nat. Acad. Sci. U.S.* **39**, 551 (1953).

[131] H. M. McConnell and C. H. Holm, *J. Chem. Phys.* **28**, 749 (1958); H. M McConnell, W. M. Porterfield, R. E. Robertson, and T. Cole, *ibid.* **30**, (1959)

interaction energy given above, by use of a reasonable value for the bond energy, about 50 kcal/mole.

Cyclopentadienyl nickel nitrosyl, $(C_5H_5)NiNO$, is a dark-red liquid made by treating nickelocene with nitric oxide.[132] Its structure has been determined by the electron-diffraction method.[133] The observed C—C bond length, 1.434 ± 0.005 Å, corresponds to bond number 1.275 ± 0.025, essentially the same as in ferrocene and nickelocene. The five Ni—C bonds have length 2.144 ± 0.006 Å, corresponding to bond number 0.35 ± 0.01. The nitrosyl group projects from the side of the nickel atom opposite that occupied by the cyclopentadienyl residue. The nickel-nitrogen distance is 1.64 ± 0.02 Å, which corresponds to bond number 1.7. It is likely that there is resonance among the structures Ni—N̈=Ö:, Ni̇=N̈=Ö:, and Ni≡N̈—Ö:⁻ . The N—O distance, 1.154 ± 0.009 Å, indicates some contribution also of the structure Ni :N≡O:. There is some evidence that the NiNO angle is about 160°, as would be expected if the first structure made a significant contribution.

A less accurate structure determination, by the x-ray method, has been reported[133] for cyclopentadienyl manganese tricarbonyl, $(C_5H_5)Mn(CO)_3$. The C—C distance, 1.40 ± 0.06 Å, is about the same as in other compounds. The bonds from the manganese atom to the ring carbon atoms have length 2.15 ± 0.02 Å, corresponding to bond number 0.37 ± 0.03, and those to the carbon atoms of the three carbonyl groups have length 1.77 ± 0.03 Å, corresponding to bond number 1.6 ± 0.2. We conclude that two electron pairs are involved in bond formation with the ring and five in bond formation with the carbonyl groups.[134]

Cyclopentadienyl thallium, $Tl(C_5H_5)$, has been investigated recently by microwave spectroscopy.[135] The cyclopentadienyl group is nearly planar and has pentagonal symmetry. It lies to one side of the thallium atom.

We may discuss the structure that would be expected for this molecule. The thallium atom has three electrons in its outer shell. However, in most of the compounds of thallium two of these electrons form an unshared pair in the 6s orbital, and the remaining electron serves as the valence electron, with the corresponding bond orbital having largely

[132] T. S. Piper, F. A. Cotton, and G. Wilkinson, *J. Inorg. & Nuclear Chem.* 1, 165 (1955).

[133] Berndt, *op. cit.* (122).

[134] Somewhat similar compounds of molybdenum and other metals are discussed in Sec. 11-15.

[135] J. K. Tyler, A. P. Cox, and J. Sheridan, *Nature* 183, 1182 (1959).

p character. The structure that would represent the normal state of the molecule as a first approximation is that in which the thallium atom forms one bond, resonating among the positions to the five carbon atoms. The thallium-carbon bond number would then be 0.20, and the carbon-carbon bond number for the bonds in the ring would be 1.40. The thallium radius 1.570 Å (Table 11-3) leads, with the correction −0.056 Å for partial ionic character, to the value 2.29 Å for the thallium-carbon single-bond distance, and accordingly to the value 2.72 Å for the length of a bond with $n = 0.20$. The corresponding carbon-carbon bond length, for $n = 1.40$, is 1.410 Å.

We would, however, expect some contribution of the structures in which there are a positive charge on one of the carbon atoms in the ring, a negative charge on the thallium atom, and two thallium-carbon bonds, in order to compensate the separation of charge resulting from the partial ionic character of the bonds. From the electronegativity of thallium (1.8) the partial ionic character of the bond is estimated to be 12 percent, so that structures of this sort should make a 12 percent contribution. The corresponding bond numbers are then 0.22 for thallium-carbon and 1.37 for carbon-carbon, corresponding to the bond lengths 2.67 Å and 1.415 Å, respectively.

These expected values of bond lengths are in reasonably good agreement with those reported from the microwave investigation: 2.70 ± 0.01 Å for thallium-carbon and 1.43 ± 0.02 Å for carbon-carbon.

The discussion of electron-deficient substances is continued in the following chapter.

CHAPTER 11

The Metallic Bond

11-1. THE PROPERTIES OF METALS

THE elements that are classed as metals display to a larger or smaller extent certain characteristic properties, including high thermal conductivity and electric conductivity, metallic luster, ductility and malleability, power to replace hydrogen in acids, etc. These properties are shown most strikingly by the elements in the lower left region of the periodic table; in fact, metallic character is closely associated with electropositive character, and in general a small value of the electronegativity of an element, as given by the bond-energy method, the electromotive-force series, or any similar treatment, corresponds to pronounced metallic properties of the elementary substance.

Lorentz[1] advanced a theory of metals that accounts in a qualitative way for some of their characteristic properties and that has been extensively developed in recent years by the application of quantum mechanics. He thought of a metal as a crystalline arrangement of hard spheres (the metal cations), with free electrons moving in the interstices. This free-electron theory provides a simple explanation of metallic luster and other optical properties, of high thermal and electric conductivity, of high values of heat capacity and entropy, and of certain other properties.

One of the most interesting of these properties is the small temperature-independent paramagnetism shown by many metals, including the alkali metals. It was the discussion of this phenomenon by Pauli[2] in 1927 that initiated the development of the modern electronic theory of metals. The fundamental concept is that there exists in a metal a continuous or partially continuous set of energy levels for the "free" electrons. At the absolute zero the electrons (N in number) would

---

[1] H. A. Lorentz, *The Theory of Electrons*, Teubner, Leipzig, 1916.
[2] W. Pauli, Jr., *Z. Physik* **41**, 81 (1927).

occupy the $N/2$ most stable levels in pairs, and, as required by the Pauli exclusion principle, the spins of the electrons of each pair would be opposed, so that the spin magnetic moments of the electrons would not be available for orientation in an applied magnetic field. At higher temperatures some of these pairs are broken, as one of the electrons of a pair is raised to a higher energy level, and the spin moments of these unpaired electrons then make a contribution to the paramagnetic susceptibility of the metal. The number of unpaired electrons increases with increasing temperature; the contribution of one unpaired electron spin to the paramagnetic susceptibility decreases with increasing temperature, however (App. X), and the two effects are found on quantitative discussion to lead to an approximately temperature-independent paramagnetic susceptibility of the observed order of magnitude.

The quantum-mechanical theory of metals has been extensively developed by Sommerfeld and many other investigators.[3] Discussion of it is beyond the scope of this book, however, and instead we shall consider the problem of the structure of metals from a more chemical point of view. The treatment given in the following sections is not to be interpreted as being a rival of the quantum-mechanical theory, but rather as offering an alternative avenue of approach toward the same goal as that of the theoretical physicists.

11-2. METALLIC VALENCE

The great field of chemistry comprising the compounds of metals with one another has been largely neglected by chemists in the past. Approximately three-quarters of the elements are metals. This means that the binary systems of pairs of metals constitute about $\frac{9}{16}$—well over one-half—of all binary systems, and one might conclude that the chemistry of the metals should be the major part of chemistry, more extensive than the chemistry of metals in combination with nonmetals, or of nonmetals and nonmetals. In fact, textbooks of general chemistry may devote only one or two pages out of several hundred to a discussion of the compounds of metals with one another. Perhaps part of the reason for the neglect by chemists of this branch of chemistry is that many compounds of metals with one another show a range of composition, so that the chemist, who likes the precision of compounds

[3] A. Sommerfeld, W. V. Houston, and C. Eckart, *Z. Physik* **47**, 1 (1928); J. Frenkel, *ibid.* 819; W. V. Houston, *ibid.* **48**, 449 (1928); F. Bloch, *ibid.* **52**, 555 (1928); etc. For summarizing discussions and further references, see A. Sommerfeld and N. H. Frank, *Rev. Modern Phys.* **3**, 1 (1931); J. C. Slater, *Rev. Modern Phys.* **6**, 209 (1934); N. F. Mott and H. Jones, *The Theory of the Properties of Metals and Alloys*, Clarendon Press, Oxford, 1936; A. H. Wilson, *The Theory of Metals*, Cambridge University Press, 1936; H. Fröhlich, *Elektronentheorie der Metalle*, J. Springer, Berlin, 1936.

with definite composition, daltonides, rather than the imprecision of the berthollides, turns away from them. Moreover, there is the non-existence of good solvents for metallic systems; both the inorganic chemist and the organic chemist are accustomed to beginning their work on a solid substance by purifying it by recrystallization from a suitable solvent. This process can only occasionally be used with intermetallic compounds—one can, for example, obtain beautiful big crystals of KHg_{13} by cooling a solution of potassium in mercury, but purification by recrystallization is in general not a feasible process for intermetallic compounds. I think, however, that the most important reason for the neglect of this branch of chemistry during the past century is that a theory of valence and structure for intermetallic compounds was not developed at the same time as for other compounds.

In the assignment of valences to metals in intermetallic compounds the chemist would have faced a problem similar to that faced by Frankland, Cowper, Kekulé, and other chemists in their attack on the valence theory of organic chemistry. The compound KHg_{13} might be compared with naphthalene, $C_{10}H_8$. In KHg_{13} the valence of potassium might well be taken as 1, corresponding to its position in the periodic table—presumably the valences of the first-group elements and the second-group elements can be taken as 1 and 2, respectively. But the formula KHg_{13} should not then be taken to require that the valence of mercury be $\frac{1}{13}$. The organic chemist did not conclude from the formula $C_{10}H_8$ for naphthalene that the valence of carbon should be taken as $\frac{4}{5}$, because as soon as the idea of the carbon-carbon bond was developed he accepted for naphthalene a structure in which carbon atoms are bonded to one another, as well as to hydrogen atoms, so as to permit carbon to retain its valence of 4. In the same way we might assume that in KHg_{13} mercury atoms are bonded to one another, as well as to potassium atoms; but the formula does not tell us what the metallic valence of mercury is.

The organic chemist was successful in developing valence theory and discovering the quadrivalence of carbon because he could make many simple compounds, such as CH_4 and CH_3Cl, in which the carbon atom was indicated to be attached to four other (univalent) atoms. If a similar attack on intermetallic compounds could have been made valence theory would probably have been extended to cover this field of chemistry. In fact, however, we see that it fails. In addition to KHg_{13}, potassium forms with mercury the compounds KHg_5, KHg_3, KHg_2, and KHg. If we were to assume that the last of these, with the highest potassium content, involves only bonds between mercury and potassium, we would assign to mercury the metallic valence 1. The compound of sodium with mercury with the highest sodium content is

Na₃Hg, which we should similarly interpret as indicating metallic valence 3 for mercury; and the compound Li₃Hg indicates the same valence. The compound of magnesium with mercury with highest magnesium content is Mg₃Hg; this indicates metallic valence 6 for mercury, if the same assumption, that this compound contains only mercury-magnesium bonds, is made. It is evident that this procedure, analogous to that used by the organic chemist in discovering the quadrivalence of carbon, fails to disclose the metallic valence of mercury.

The properties of the metals themselves can be used to indicate, at least approximately, the values of the metallic valence.[4] If, in the sequence of elements beginning with potassium, we assume that the metallic valence of potassium is 1 and that of calcium is 2, we recognize the expected correlation between valence and properties: the metal calcium is harder, stronger, and denser than potassium, it has a higher melting point, boiling point, enthalpy of fusion, and enthalpy of vaporization than calcium, and in general its properties correlate well with the assumption that the bonds holding the atoms together are twice as strong in calcium as in potassium, corresponding to the respective valences 2 and 1. Similarly, there is a further increase in hardness, density, melting point, and some other properties from calcium to scandium, permitting us to conclude reasonably that scandium has the valence 3 that corresponds to its position in the periodic table. The change in these properties continues from scandium to titanium, titanium to vanadium, and vanadium to chromium, and we may well feel justified in concluding that the metallic valences of titanium, vanadium, and chromium are 4, 5, and 6, respectively. These are just the maximum valences shown by these elements in inorganic compounds—maximum oxidation numbers for titanium, vanadium, and chromium are +4, +5, and +6, respectively, as in the oxides TiO₂, V₂O₅, and CrO₃.

The properties of the following transition metals do not suggest a further increase in metallic valence, to 7 for manganese, 8 for iron, and 9 for cobalt. Instead, the properties mentioned above, hardness, density, melting point, and so on, indicate that the metallic valence remains roughly constant from chromium to nickel, then decreases somewhat from nickel to copper, and shows a further decrease from copper to zinc. I now think that it is reasonable to assign the metallic valence 6 as the normal metallic valence for the elements manganese, iron, cobalt, and nickel, and about 5½ for copper, 4½ for zinc, 3½ for gallium, 2½ for germanium (when functioning as a metal), and 1½ for arsenic.

_____
[4] L. Pauling, *Phys. Rev.* **54**, 899 (1938).

In 1938 I assigned to iron the metallic valence 5.78 (rather than 6), on the basis of an argument that I now consider to be wrong. Iron, with atomic number 26, has eight electrons outside the argon shell, and might use all of them in the formation of chemical bonds. If the bonds were shared-electron-pair bonds the spins of the electrons would have to be paired, and would not contribute to the magnetic moment of the metal. The saturation magnetization of iron corresponds to magnetic moment 2.22 Bohr magnetons per atom, requiring that not more than 5.78 electrons per atom be involved in the formation of electron pairs. I concluded that the metallic valence of iron is accordingly 5.78. There is, however, the possibility that some of the unpaired electrons are involved in the bonding between atoms—in the language of chemical valence theory they would be described as forming one-electron bonds in the metal, and in the language of the ordinary electron theory of metals they would be described as occupying a conduction band in which the electron spins are uncoupled, so that there is only one electron per energy level, instead of two. A simple calculation[5] making use of spectroscopic values of the interaction energy of electrons in the iron atom has led to the value 0.26 electron per atom in this band with uncoupled spins, indicating for the iron atom the total valence 6.04, of which 5.78 represents electrons involved in the formation of electron-pair bonds and 0.26 represents electrons involved in the formation of one-electron bonds. (This calculation was made in the course of working out a simple quantitative theory of ferromagnetism, on the basis of Zener's ideas[6] about the uncoupling of the spins of conduction electrons by interaction with atomic electrons.) It seems accordingly not unreasonable to accept the integral value 6 for the valence of iron, and, for the reasons given above, to accept this value also for the related elements chromium, manganese, cobalt, and nickel.

The mechanical properties of copper and zinc indicate valences less than 6 for these elements. The magnetic properties of alloys of nickel and copper can be used to derive a value for the valence of copper, by the following argument. The saturation magnetic moment of the ferromagnetic alloys of nickel and copper decreases linearly with increase in the atomic fraction of copper in the alloy in such a way as to reach the value 0 for the alloy with 56 atomic percent copper (Fig. 11-1). This alloy has 10.56 electrons per atom. If we assume that the metallic valence remains 6 for both nickel and copper in this alloy, 6 orbitals are required for the valence electrons, and 2.28 orbitals for the remaining 4.56 electrons, which are present as shared electron pairs. Accordingly 8.28 orbitals are occupied outside the argon shell; the remaining 0.72

[5] L. Pauling, *Proc. Nat. Acad. Sci. U.S.* **39**, 551 (1953).
[6] C. Zener, *Phys. Rev.* **81**, 440 (1951).

orbital per atom, of the complement of 9 stable orbitals (five $3d$ orbitals, one $4s$ orbital, and three $4p$ orbitals) is the metallic orbital, which will be discussed later. We now assume that in pure copper the same amount of metallic orbital is required to be present, 0.72 per atom, leaving 8.28 orbitals for occupancy by valence electrons and unshared electron pairs. The 11 electrons of the copper atom outside the argon shell can be introduced into these 8.28 orbitals as $11 - 8.28 = 2.72$ unshared pairs, and $11 - 2 \times 2.72 = 5.56$ unpaired electrons. Accordingly we conclude that the metallic valence of copper is approxi-

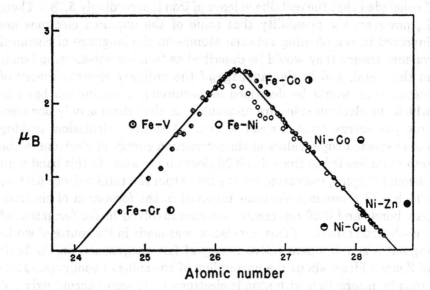

Fig. 11-1.—Observed values of saturation magnetic moments per atom of transition elements of the iron group and their alloys. Values for some alloys, which deviate from the curve, are not shown; these alloys probably involve ferrimagnetism.

mately 5.56. The same argument leads to 4.56 for zinc, 3.56 for gallium, 2.56 for germanium, and 1.56 for arsenic.

11-3. THE METALLIC ORBITAL

The argument given above about the properties of the transition metals suggests that the number of valence electrons involved in bonds between the atoms of the metal is 6 through the sequence from chromium to nickel. Inasmuch as there are nine reasonably stable orbitals available outside the argon shell, we might expect that additional electrons, two for iron and three for cobalt, could occupy two or three of the remaining orbitals individually, producing the saturation magnetic moment of 2 Bohr magnetons for iron and 3 Bohr magnetons for cobalt, but that in nickel, with four additional electrons to be fitted into the three orbitals, two of the electrons would have to occupy an orbital together with opposed spins, causing the magnetic moment to drop again to 2 Bohr magnetons.   The saturation magnetic moment would accordingly be expected to rise from 0 for chromium to a maximum of 3 for cobalt, and then to drop to 2 for nickel and 0 for zinc.   In fact, as shown in Figure 11-1, the experimental values of the saturation magnetic moment for these metals and their alloys with one another rises to a maximum at a point about a quarter of the way between iron and cobalt and drops to 0 at a point intermediate between nickel and copper.   (Some other alloys, which may be antiferromagnetic, do not fit this curve.)   These facts were interpreted[7] as indicating that not all the nine orbitals $3d^54s4p^3$ are available for occupancy by electrons in the transition metals, but instead a somewhat smaller number of orbitals, about 8.3.   The position of the maximum in the saturation magnetic moment curve of Figure 11-1 provides information about the number of orbitals available for occupancy by bonding electrons or atomic electrons.   The experimental points for the iron-cobalt alloys with the body-centered structure indicate a maximum at 26.34 electrons per atom; those for the iron-nickel alloys, for which the number of points is smaller, indicate 26.18; and the two straight lines drawn to approximate all of the values reasonably closely intersect, when extrapolated, at 26.28 electrons per atom.   We may accept this value as the most reliable, and assume that 8.28 electrons per atom outside the argon shell can be occupied in individual orbitals; with a larger number of electrons some unshared electron pairs (two electrons occupying one atomic orbital) are formed.   The value 8.28 as the sum of the bonding orbitals and occupied atomic orbitals leaves 0.72 for the number of metallic orbitals per atom.   This is exactly equal to the number of metallic orbitals given by the position of the foot of the curve, between nickel and copper, and the assumption that 6 electrons per atom are valence electrons, as discussed above.

A reasonable interpretation of the 0.72 metallic orbital per atom was not formulated until ten years later.[8]   It was then suggested that the metallic orbital permits the unsynchronized resonance of electron-pair bonds from one interatomic position to another by the jump of one electron from one atom to an adjacent atom, leading to great stabilization of the metal by resonance energy, and to the characteristic properties of metals.

It is known that lithium, for example, forms some diatomic molecules in the gas phase; these molecules are described as consisting of two atoms held together by a single covalent bond.   In lithium metal each

[7] Pauling, *loc. cit.* (4).
[8] L. Pauling, *Nature* 161, 1019 (1948).

atom has eight nearest neighbors and one valence electron, which permits the formation of an electron-pair bond for each pair of atoms. The bonds may be considered to resonate among alternative positions, mainly the eight positions between each atom and its nearest neighbors (and, to a smaller extent, the six positions between each atom and the six next nearest neighbors). If each atom were to remain electrically neutral, by retaining its valence electron, the stabilization through the permitted synchronized bond resonance,

$$
\begin{array}{ccc}
\text{Li—Li} & \qquad & \text{Li}\ \ \text{Li} \\
& & |\quad\ | \\
\text{Li—Li} & \qquad & \text{Li}\ \ \text{Li}
\end{array}
$$

analogous to that in the benzene molecule, would be relatively small. Much greater stabilization could result from unsynchronized resonance,

$$
\begin{array}{ccc}
\text{Li—Li} & \qquad & \text{Li—Li}^{-} \\
& & |\quad \\
\text{Li—Li} & \qquad & \text{Li}^{+}\ \text{Li} \qquad \text{etc.}
\end{array}
$$

This unsynchronized resonance would require the use of an additional orbital on the atom receiving an extra bond. It is assumed that this additional orbital is the metallic orbital.

A discussion of the nonintegral value, 0.72, of metallic orbitals per atom will be given in the following section, in connection with the discussion of interatomic distances in the allotropic forms of tin.

11-4. INTERATOMIC DISTANCES AND BOND NUMBERS IN METALS

In Chapter 7 a brief discussion was given of interatomic distances for bonds with bond number n less than 1, and the following equation for the relation between the corresponding bond distance $D(n)$ and the bond distance for $n = 1$, $D(1)$, was proposed:

$$D(n) = D(1) - 0.600 \log n \qquad (11\text{-}1)$$

An extensive system of metallic radii has been formulated on the basis of this equation. It is evident that there is some uncertainty about this empirical equation; in particular, the value 0.60 Å for the factor of the logarithmic term is somewhat uncertain; but, in fact, the conclusions about electronic structure, bond numbers, and valence in metals and intermetallic compounds that have been reached through use of the equation would not be significantly changed by some change in the value of this factor.

Experience has shown that Equation 11-1 can usually be applied in the interpretation of observed interatomic distances with considerable confidence. Some intermetallic compounds, however, have structures such as to make it possible that some of the interatomic distances represent bonds in tension and others represent bonds in compression. The application of Equation 11-1 in the interpretation of these interatomic distances would lead to error, and it is necessary in using the equation to keep this possibility in mind.

Let us consider the element tin as the first example of the application of the equation. From the argument given above about the necessity of having 0.72 metallic orbital per atom in order for a substance to be a metal we predict the metallic valence 2.56 for tin. The calculation is made in the following way (analogous to that for copper given above): The tin atom has 14 electrons outside its krypton shell. There are 9 stable orbitals outside this shell, $4d^5 5s 5p^3$. Of these 9 orbitals, 0.72 is allocated as metallic orbital, leaving 8.28 for occupancy by bonding electrons and unshared electron pairs. This requires that there be $14 - 8.28 = 5.72$ unshared electron pairs, leaving 2.56 orbitals for occupancy by bonding electrons; hence the metallic value of tin is predicted to be 2.56.

Now let us consider the interatomic distances in the two allotropic forms of tin, gray tin and white tin. Gray tin has the diamond structure: each tin atom is surrounded by four other tin atoms, at the distance 2.80 Å. It is known that in the molecule tin tetramethyl, $Sn(CH_3)_4$, in which tin is surely quadrivalent, forming single bonds with each of the carbon atoms, the tin-carbon distance is 2.17 Å; and, inasmuch as the single-bond radius of carbon is 0.77 Å, the single-bond radius of tin can be taken as 1.40 Å. The observed distance 2.80 Å in gray tin is, then, just that expected for a single covalent bond, and we are led to the conclusion that tin in this form has valence 4. The tin atom can achieve valence 4 by using all of its nine outer orbitals, either for occupancy by unshared pairs (5 orbitals holding 10 electrons) or by bonding electrons (4 orbitals, holding the remaining 4 of the 14 outer electrons of the tin atom). Accordingly there is no metallic orbital in gray tin; all nine outer orbitals are used either in bond formation or for occupancy by unshared pairs (the electron configuration is $4d^{10} 5s 5p^3$); but gray tin is not a metal—it is a metalloid, and it does not have the characteristic properties of a metal: high electric conductivity, negative temperature coefficient of electric conductivity, malleability, etc.

White tin, on the other hand, has the characteristic properties of a metal. In white tin each tin atom has four nearest neighbors at 3.016 Å and two others at 3.175 Å, its structure being that shown in Figure 11-2. We may apply Equation 11-1 to calculate the valence of tin in white tin, assuming the single-bond distance to be 2.80 Å. The bond numbers for the bonds with length 3.016 Å and 3.175 Å are respectively, 0.44 and 0.24; these values correspond to the valence 2.24. It will be pointed out later that it is likely that the single-bond radius for tin, as for many other metals, depends somewhat on the nature of the bond orbitals and has the value 1.424 Å, rather than 1.40 Å, for valence 2.56. With this value the bond numbers corresponding to the distances 3.016 Å and 3.175 Å are calculated to be 0.52 and 0.28, respectively, leading to the valence 2.64 for tin in the metallic form of the element, in good agreement with the predicted value 2.56 Å.

We may now ask why the valence of tin in the metallic form of the element is not 2, corresponding to one metallic orbital per atom and the electron configuration $4d^{10}5s^25p^2$, but is 2.56. The answer is, I think, given by the quantum-mechanical principle that the actual structure for the normal state of a system is that structure, from among all

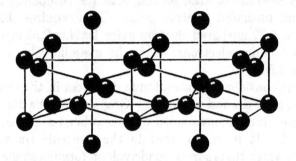

FIG. 11-2.—The arrangement of atoms in the tetragonal crystal white tin (metallic). The tetragonal axis is vertical.

conceivable ones, that minimizes the energy of the system. Let us consider a hypothetical form of tin, resembling white tin, but with valence 2 and one metallic orbital per atom. This structure would be stabilized by the essentially completely unsynchronized resonance of the two valence bonds per atom among the alternative positions, six around each atom. Now let us introduce a single quadrivalent tin atom. The number of bonds would be increased, which would further stabilize the crystal. There would be, it is true, a small interference with resonance of the bonds, because this one atom would not have a metallic orbital; but the amount of interference would be so small as not to diminish the resonance energy comparable to the increase in stabilization caused by the extra bond. A second quadrivalent tin atom would stabilize the metal still further, but finally, as the number of quadrivalent tin atoms became significant, the interference with resonance of the bonds would become so great as to cause a decrease in resonance energy equal to the increase in bond energy accompanying the increase in the number of quadrivalent atoms. At this point the minimum in energy (the maximum stability) of the system would be

TABLE 11-1.—METALLIC VALENCES AND RADII OF THE ELEMENTS

Element	ν	R(L 12)	R₁
Li	1	1.549	1.225
Be	2	1.123	0.889
B	3	0.98	.80
Na	1	1.896	1.572
Mg	2	1.598	1.364
Al	3	1.429	1.248
Si	2.56	1.375	1.173
P	(3)	1.28	1.10
S	(2)	1.27	1.04
K	1	2.349	2.025
Ca	2	1.970	1.736
Sc	3	1.620	1.439
Ti	4	1.467	1.324
V	5	1.338	1.224
Cr	6	1.276	1.186
Mn	6	1.268	1.178
Fe	6	1.260	1.170
Co	6	1.252	1.162
Ni	6	1.244	1.154
Cu	5.56	1.276	1.176
Zn	4.56	1.339	1.213
Ga	3.56	1.404	1.246
Ge	2.56	1.444	1.242
As	1.56	1.476	1.210
Se	(2)	1.40	1.17
Rb	1	2.48	2.16
Sr	2	2.148	1.914
Y	3	1.797	1.616
Zr	4	1.597	1.454
Nb	5	1.456	1.342
Mo	6	1.386	1.296
Tc	6	1.361	1.271
Ru	6	1.336	1.246
Rh	6	1.342	1.252
Pd	6	1.373	1.283
Ag	5.56	1.442	1.342
Cd	4.56	1.508	1.382
In	3.56	1.579	1.421
Sn	2.56	1.623	1.421
Sb	1.56	1.657	1.391
Te	(2)	1.60	1.37
Cs	1	2.67	2.35
Ba	2	2.215	1.981
La*	3	1.871	1.690
Hf	4	1.585	1.442
Ta	5	1.457	1.343
W	6	1.394	1.304
Re	6	1.373	1.283
Os	6	1.350	1.260
Ir	6	1.355	1.265
Pt	6	1.385	1.295
Au	5.56	1.439	1.339
Hg	4.56	1.512	1.386
Tl	3.56	1.595	1.437
Pb	2.56	1.704	1.502
Bi	1.56	1.776	1.510
Th	4	1.795	1.652
U	6	1.516	1.426
*Ce	3.2	1.818	1.646
Pr	3	1.824	1.643
Nd	3	1.818	1.637
Pm	3	1.834	1.633
Sm	3	2.804	1.623
Eu	3	2.084	1.850
Gd	3	1.804	1.623
Tb	3.5	1.773	1.613
Dy	3	1.781	1.600
Ho	3	1.762	1.581
Er	3	1.761	1.580
Tm	3	1.759	1.578
Yb	2	1.933	1.699
Lu	3	1.738	1.557

reached, and this would represent the actual structure of the crystal of white tin.   I have not found it possible to carry through a theoretical argument leading in a reliable way to the value 0.72 as the number of metallic orbitals per atom at which the maximum stability of a metallic crystal is achieved, but a simple theoretical treatment can be carried through that suggests that this value is not unreasonable.[9]

With use of the observed values of interatomic distances in metals, discussed in the following sections, and Equation 11-1, values of R_1, the metallic radius, have been derived.   These values are given in Table 11-1, together with the assumed values of the metallic valence v and the values of the metallic radius for ligancy 12, $R(L12)$.

11-5. THE CLOSEST PACKING OF SPHERES

It is not surprising that often a crystalline substance is a rather closely packed aggregate of atoms or ions, since the van der Waals interactions, Coulomb interactions, and interactions involving metallic valence tend to stabilize structures in which the atoms have large ligancies.   It has been found that the structures of many crystals can be profitably discussed in terms of the packing of spheres, to which we now direct our attention.

Cubic and Hexagonal Closest Packing of Equivalent Spheres.—The problem of packing spheres in ways that leave the minimum of interstitial space has interested many investigators.   Over 75 years ago W. Barlow[10] discovered that there are two ways of arranging equivalent spheres in closest packing, one with cubic and one with hexagonal symmetry.

There is only one way of arranging spheres in a single closest-packed layer.   This is the familiar arrangement in which each sphere is in contact with six others, as in Figure 11-3.   A second similar layer can be superimposed on this layer in such a way that each sphere is in contact with three spheres of the adjacent layer, as shown in Figure

---

[9] L. Pauling, *Proc. Roy. Soc. London* **A196**, 343 (1949).

[10] W. Barlow, *Nature* **29**, 186, 205, 404 (1883); *Z. Krist.* **23**, 1 (1894); **29**, 433 (1898).   In the first of these papers Barlow suggested five "very symmetrical" structures, the sodium chloride, cesium chloride, and nickel arsenide arrangements, and cubic and hexagonal closest packing.   L. Sohncke, *Nature* **29**, 383 (1883), criticized his selection as arbitrary; he also said that an alkali halogenide such as NaCl could not have the sodium chloride arrangement, because it does not show the existence of discrete molecules!   Lord Kelvin, *Proc. Roy. Soc. Edinburgh* **16**, 693 (1889), in discussing the packing of spheres, required that they be not only equivalent but also oriented similarly, and showed that cubic closest packing is the only closest-packed structure satisfying this condition. His additional requirement has no physical significance; hexagonal closest packing is as important an arrangement as cubic closest packing.

Fig. 11-3.—The arrangement of spheres in a closest-packed layer.

11-4. A third layer can then be added in either one of two possible positions, with its spheres either directly above those of the first layer, as in Figure 11-4, or over the holes in the first layer not occupied by the second layer. Once either choice is made, the structure is determined,

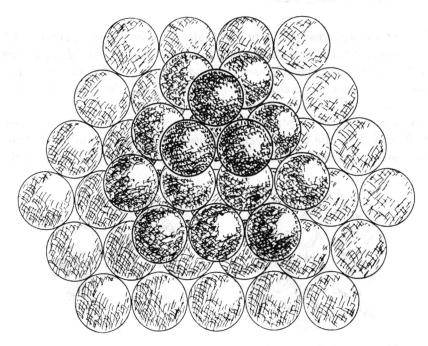

Fig. 11-4.—The arrangement of spheres in hexagonal closest packing.

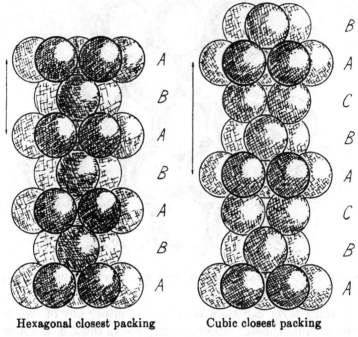

Hexagonal closest packing   Cubic closest packing

Fig. 11-5.—The modes of superposition of closest-packed layers of spheres in hexagonal closest packing (left) and cubic closest packing (right).

if all of the spheres are to be equivalent. The first structure, with hexagonal symmetry, is shown in Figure 11-4 and at the left in Figure 11-5; it is called *hexagonal closest packing*. The second structure, called *cubic closest packing*, is shown in Figures 11-5 (at the right) and 11-6, the latter being taken from one of Barlow's papers.

A convenient description of these structures utilizes the symbols *A*, *B*, and *C* for the three layers of close-packed spheres differing from one another in position. Hexagonal closest packing corresponds to the

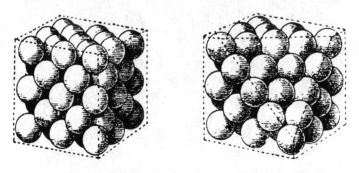

Fig. 11-6.—Cubic closest packing of spheres (after Barlow).

sequence of layers $ABABAB\cdots$ (or $BCBC\cdots$ or $ACAC\cdots$), and cubic closest packing to $ABCABCABC\cdots$. The structure repeats itself after two layers in hexagonal closest packing and after three layers in cubic closest packing.

In each of the closest-packed structures each sphere is in contact with twelve others, a hexagon of six in the same plane and two triangles of three above and below. In hexagonal closest packing the upper triangle has the same orientation as the lower triangle and in cubic closest packing it is rotated through 60°.

The suggestion that in metal crystals the atoms are arranged in closest packing was made by Barlow before the development of the x-ray technique, in order to account for the observations that many metals crystallize with cubic or hexagonal symmetry and that in the latter case many of the observed values of the axial ratio lie near the ideal value $2\sqrt{2}/\sqrt{3} = 1.633$ for hexagonal closest packing.

For crystals involving spherical or nearly spherical molecules that are attracted to one another by van der Waals forces, closest-packed structures, giving the maximum number of intermolecular contacts, can be expected to be stable. It has been shown that all the noble gases (He, Ne, Ar, Kr, and Xe) crystallize in cubic or hexagonal closest packing. Moreover, it has been found that in many molecular crystals of simple gases the molecules are rotating with considerable freedom, permitting them to simulate spheres in their interactions with neighboring molecules;[11] these crystals are usually closest-packed. Thus crystals of molecular hydrogen consist of rotating H_2 molecules in hexagonal closest packing, and in crystalline HCl, HBr, HI, H_2S, H_2Se, CH_4, and SiH_4 the molecules are arranged in cubic closest packing.[12]

Closest-packed Structures Containing Nonequivalent Spheres.— Although the cubic and hexagonal arrangements described above are the only closest-packed arrangements of equivalent spheres, there is an infinite number of other arrangements that do not differ greatly from them. These are the closest-packed arrangements of spheres of the same size which, however, are not all crystallographically equivalent. Any sequence of closest packed layers, such as $ABCBACBC\cdots$, is just as close packed as these two; each sphere is in contact with twelve others, which are arranged about it either as in cubic or as in hexagonal closest packing. These arrangements, which differ so little from the other two, are found to have some importance.

[11] L. Pauling, *Phys. Rev.* **36**, 430 (1930).

[12] Some of these substances have low-temperature modifications in which the molecules do not rotate. In the high-temperature forms the rotation of the molecules is not completely free, but is somewhat hindered; in some cases it may be described as rapid change among alternative orientations.

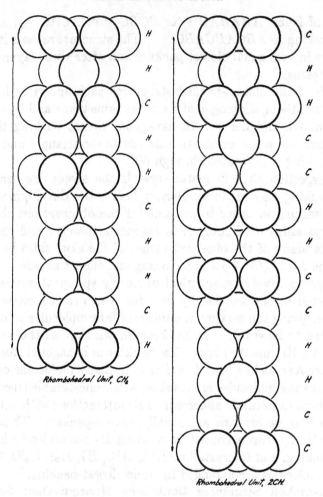

FIG. 11-7.—The four closest-packed structures involving

Figure 11-7 shows the four possible ways of superimposing closest-packed layers so that only two kinds of nonequivalent layers are present.[13] The simplest of these, called double-hexagonal closest packing, corresponds to the sequence $ABACABAC\cdots$, repeating after four layers. It involves alternation of layers (A) with adjacent layers arranged as in cubic closest packing and of layers (B and C) with adjacent layers arranged as in hexagonal closest packing.[14]

[13] L. Pauling, *Chem. Bull.*, *Chicago* 19, 35 (1932).
[14] The structures can accordingly be described by the sequence $c\,h\,c\,h\cdots$, indicating alternation of cubic and hexagonal closest packing. The three other structures have the similar sequences $h\,c\,c\,h\,c\,c\cdots$, $h\,h\,c\,h\,h\,c\cdots$, and $h\,h\,c\,c\,h\,h\,c\,c\cdots$.

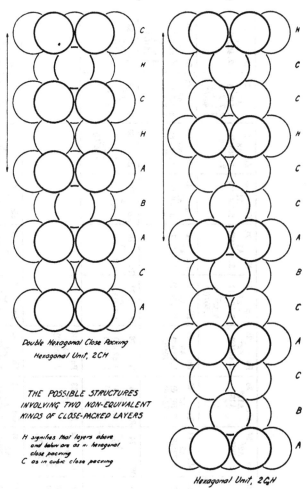

Double Hexagonal Close Packing

Hexagonal Unit, 2CH

THE POSSIBLE STRUCTURES
INVOLVING TWO NON-EQUIVALENT
KINDS OF CLOSE-PACKED LAYERS

H signifies that layers above
and below are as in hexagonal
close packing
C as in cubic close packing

Hexagonal Unit, 2CH

two nonequivalent kinds of spheres with the same radius.

11-6. THE ATOMIC ARRANGEMENTS IN CRYSTALS OF METALLIC ELEMENTS

The structures reported for the metals, as determined by x-ray diffraction, are listed in Table 11-2.

Closest-packed Structures.—If the stability of a metal crystal were determined by the number of bonds formed at a minimum interatomic distance, with no contribution of longer bonds, the structures with closest packing would be the most stable for the metallic elements. These structures, which have been described in the preceding section, involve contact between each atom and the 12 nearest neighbors. (The next interatomic distances are 41 percent larger and presumably have little significance.)

It is noteworthy that 46 of the 58 metallic elements listed in Table

TABLE 11-2.—THE STRUCTURE OF CRYSTALS OF METALLIC ELEMENTS[a]

Li 3 A2[b] 3.039(8)	Be 4 A3[c] 2.226(6) 2.286(6)												
Na 11 A2 3.716(8)	Mg 12 A3 3.197(6) 3.209(6)											Al 13 A1 2.864(12)	Si 14 A4 2.353(4)
K 19 A2 4.544(8)	Ca 20 A1[d] 3.947(12) A3 3.940(6) 3.955(6)	Sc 21 A3 3.256(6) 3.309(6) A1 3.212(12)	Ti 22 A3[e] 2.896(6) 2.951(6)	V 23 A2 2.622(8)	Cr 24 A2[f] 2.498(8)	Mn 25 A12[g] A13 A6 A1 A2	Fe 26 A2[h] 2.482(8)	Co 27 A1 2.506(12) A2 2.501(6) 2.507(6)	Ni 28 A1 2.492(12)	Cu 29 A1 2.556(12)	Zn 30 A3 2.665(6) 2.913(6)	Ga 31 A11 2.442(1) 2.712(2) 2.742(2) 2.801(2)	Ge 32 A4 2.450(4)
Rb 37 A2 4.95(8)	Sr 38 A1[i] 4.303(12)	Y 39 A3 3.551(6) 3.647(6)	Zr 40 A3[j] 3.179(6) 3.231(6)	Nb 41 A2 2.858(8)	Mo 42 A2 2.725(8)	Tc 43 A3 2.703(6) 2.735(6)	Ru 44 A3 2.650(6) 2.706(6)	Rh 45 A1 2.690(12)	Pd 46 A1 2.751(12)	Ag 47 A1 2.889(12)	Cd 48 A3 2.979(6) 3.293(6)	In 49 A6 3.251(4) 3.373(8)	Sn 50 A4 2.810(4) 3.022(4) 3.181(2) A5
Cs 55 A2 5.324(8)	Ba 56 A2 4.347(8)	La 57 A1 3.745(12) A3 3.739(6) 3.770(6)	Hf 72 A3 3.127(6) 3.195(6)	Ta 73 A2 2.860(8)	W 74 A2 2.741(8)	Re 75 A3 2.741(6) 2.760(6)	Os 76 A3 2.675(6) 2.735(6)	Ir 77 A1 2.714(12)	Pt 78 A1 2.775(12)	Au 79 A1 2.884(12)	Hg 80 A10 3.000(6) 3.466(6) (−46°C)	Tl 81 A3[k] 3.408(6) 3.457(6)	Pb 82 A1 3.500(12)
Fr 87	Ra 88	Ac 89 A1 3.756(12)	Th 90 A1[l] 3.595(12)	Pa 91 A6 3.212(8) 3.238(2)	U 92 α[m] 2.77(2) 2.86(2) 3.28(4) 3.37(4)	Np 93 α[n]	Pu 94 A1[o] 3.285(12)						

*Ce 58 A1[p] 3.650(12) A3 3.620(6) *3.652(6)	Pr 59 A1 3.649(12) A3 3.640(6) 3.673(6)	Nd 60 A3 3.628(6) 3.658(6)	Pm 61	Sm 62 A 3.587(6) 3.629(6)	Eu 63 A2 3.989(8)	Gd 64 A3 3.573(6) 3.636(6)	Tb 65 A3 3.525(6) 3.601(6)	Dy 66 A3 3.503(6) 3.590(6)	Ho 67 A3 3.486(6) 3.577(6)	Er 68 A3 3.468(6) 3.559(6)	Tm 69 A3 3.447(6) 3.538(6)	Yb 70 A1 3.880(12)	Lu 71 A3 3.453(6) 3.503(6)

See opposite page for footnotes.

11-2 crystallize with either the cubic closest-packed or the hexagonal closest-packed arrangement or with both.

The close approximation of metal atoms in these crystals to mutually attracting spheres is further shown by the values observed for the axial ratio c/a of the hexagonal closest-packed structures, as tabulated below.

Li	1.637	Sc	1.594	Ho	1.570	Zr	1.589	Ru	1.583
Na	1.634	Y	1.572	Er	1.572	Hf	1.587	Re	1.615
Be	1.585	Gd	1.592	Tm	1.570	Cr	1.626	Os	1.579
Mg	1.624	Tb	1.581	Lu	1.585	Co	1.624	Zn	1.856
Ca	1.640	Dy	1.573	Ti	1.601	Tc	1.605	Cd	1.886

Of the 25 values, 23 lie within 4 percent of the theoretical value $2\sqrt{2}/\sqrt{3} = 1.633$ that makes the twelve smallest interatomic distances equal; the two exceptional substances, zinc and cadmium, are discussed below, together with several other metals that crystallize with structures obtained from closest-packed arrangements by a deformation that shortens some of the twelve small interatomic distances at the expense of others.

The small difference in nature of the cubic and hexagonal closest-packed arrangements is verified by the approximate equality of interatomic distances for the two structures of metals that exist in the corresponding allotropic forms, as given in Table 11-2.

A few metals crystallize with close-packed structures with non-

---

[a] This table is taken in the main from the summary by M. C. Neuberger, *Z. Krist.* 93, 1 (1936), with more recent values of interatomic distances from Sutton, *Interatomic Distances*, and the references given in the following notes. The symbols A1 (cubic closest-packed arrangement), A2 (cubic body-centered arrangement), A3 (hexagonal closest-packed arrangement), A4 (diamond arrangement), etc., are those used in the *Strukturbericht* and *Structure Reports*. The numbers below these symbols are the smallest interatomic distances in A and the number of corresponding neighboring atoms (in parentheses). The values in the table are those for 20°C or 25°C.

[b] An A1 modification of Li, 3.12 (12), has been reported to be formed under shear at 77°K: C. S. Barrett, *Phys. Rev.* 72, 245 (1947), and an A3 modification, 3.111 (6), 3.116 (6), at 78°K: C. S. Barrett, *Acta Cryst.* 9, 671 (1956).

[c] Another modification of Be, with unknown structure, has been reported: F. M. Jaeger and J. E. Zanstra, *Koninkl. Ned. Akad. Wetenschap. Proc.* 36, 636 (1933).

[d] A third modification of Ca, A2, has 3.877 (8) at 500°C.

[e] A2 of Ti has 2.864 (8) at 900°C.

[f] A1 of Cr has 2.61 (12) at 1850°C.

[g] The complex structures A12, A13, and A6 of Mn are discussed in the text. A1 has 2.731 (12) at 1095°C and A2 has 2.668 (8) at 1134°C.

[h] A1 of Fe has 2.578 (12) at 916°C and A2 has 2.539 (8) at 1394°C.

[i] The A3 modification of Sr has 4.32 (6), 4.32 (6) at 248°C, and the A2 modification has 4.20 (8) at 614°C.

A2 of Zr has 3.125 (8) at 862°C.

[k] A2 of Tl has 3.362 (8) at 262°C.

[l] A2 of Th has 3.56 (8) at 1450°C.

[m] A2 of U has 3.058 (8) at 805°C.

[n] A2 of Np has 3.05 (8) at 600°C, extrapolated to 2.97 (8) at 20°C. In each of two low-temperature forms each atom has four nearest neighbors at 2.20 to 2.72 Å and others at 3.06 Å or more.

[o] The value in the table is calculated from the A1 value 3.279 Å at 320°C by use of the linear expansion coefficient − 21 × 10⁻⁶ deg⁻¹ (this δ form of Pu is the only metal known with a negative expansion coefficient). Other forms are α, stable below 117°C, structure unknown; β, 117° to 200°C, structure unknown; γ, 200° to 300°C, a face-centered orthorhombic structure with 3.026 (4), 3.159 (2), 3.287 (4) at 235°C; ε, above 475°C, A2 with 3.150 (8) at 500°C.

[p] A high-density A1 modification has been made by cooling Ce; it has 3.41 (12) at 80°K: A. F. Schuch and J. H. Sturdivant, *J. Chem. Phys.* 18, 145 (1950). A similar A1 modification has also been made by pressure: A. W. Lawson and T.-Y. Tang, *Phys. Rev.* 76, 301 (1949). It has 3.42 (12) at 15,000 atm, room temperature.

equivalent atoms. Americium,[15] cerium,[16] lanthanum, praseodymium, and neodymium[17] have the structure represented by the symbol *c h c h*⋯, as described in the preceding section (double hexagonal closest packing). Samarium[18] has the rhombohedral structure *h h c h h c h h c*⋯. In each case the axial ratio is within 1 percent of the value for closest packing of spheres.

Several metals have been found to crystallize in closest-packed structures with more or less randomness in the sequence of layers with *h* and *c* environment. Lithium and sodium (but not potassium, rubidium, and cesium) are partially transformed to hexagonal closest packing with some packing faults (occasional *c* layers) by cooling and cold work.[19]

The values of the metallic radius in Table 11-1 have been obtained for those metals with closest packing by making the change from the observed distance (the average of the two, for hexagonal closest packing with axial ratio differing by a small amount from 1.633) to the distance for bond number $n = 1$. This correction, as given by Equation 11-1 for metallic valence v and ligancy 12, is $0.600 \log (v/12)$.

Metal Structures Related to Closest-packed Structures.—Zinc and cadmium crystallize with a structure that is identical with hexagonal closest packing except for extension in the direction of the hexagonal axis, the axial ratio c/a having the values 1.856 and 1.886, respectively, which are about 15 percent greater than the value for closest packing of spheres. In consequence, the interatomic distances for the six contacts between each atom and its nearest neighbors in the basal plane are appreciably smaller than those for the other six significant contacts, with three atoms in the plane above and three atoms in the plane below. The interatomic distances are 2.660 and 2.907 Å for zinc and 2.973 and 3.287 Å for cadmium.

From the interatomic distances the conclusion is to be drawn that the bonds in the hexagonal layers of atoms in these metals are stronger than those between layers. This conclusion is substantiated by the properties of the crystals, which show basal cleavage and have larger values of the compressibility, coefficient of thermal expansion, and electrical resistance in the direction perpendicular to the basal plane than in this plane. Moreover, measurements of the intensities of

[15] P. Graf, B. B. Cunningham, C. H. Dauben, J. C. Wallmann, D. H. Templeton, and H. Ruben, *J.A.C.S.* **78**, 2340 (1956).

[16] C. J. McHargue, H. L. Yakel, Jr., and L. K. Jetter, *Acta Cryst.* **10**, 832 (1957).

[17] F. H. Spedding, A. H. Daane, and K. W. Herrmann, *Acta Cryst.* **9**, 559 (1956).

[18] F. H. Ellinger and W. H. Zachariasen, *J.A.C.S.* **75**, 5650 (1953).

[19] C. S. Barrett, *Acta Cryst.* **9**, 671 (1956).

x-ray reflections have shown that the restoring forces of oscillation of the atoms in the basal plane are greater than those for oscillation out of this plane.

The structure of mercury (A10) is closely related to that of zinc and cadmium; it is obtained from cubic closest packing by compression along one of the threefold axes of the cube, the hexagonal layers of atoms thus being brought closer to one another than in a closest-packed structure. Each mercury atom has three neighbors in the layer above and three in the layer below at the interatomic distance 2.999 Å, and six in the same layer at the somewhat larger distance 3.463 Å. Thus mercury, like zinc and cadmium, forms six strong bonds and six weaker bonds, but it differs from its congeners in the directions of the bonds.

Equation 11-1 can be applied to obtain the ratio of bond numbers for the two kinds of bonds, on the assumption that the bond orbitals, and hence $D(1)$, are the same. For bond numbers n' and n'' and corresponding bond distances D' and D'' we obtain the equation

$$D'' - D' = 0.600 \log (n'/n'')$$

For zinc the bond distances given above lead to $n'/n'' = 2.58$. With the assumed valence 4.56 the bond numbers and the metallic radius are calculated to be $n' = 0.55$, $n'' = 0.21$, and $R_1 = 1.252$ Å. This value is somewhat larger than the value given in Table 11-1 (1.213 Å); the difference will be discussed in the following section.

The structures of selenium and tellurium, of arsenic, antimony, and bismuth, and of silicon, germanium, and gray tin involve two, three, and four nearest neighbors, respectively. These structures are interpreted reasonably as involving covalent bonds between each atom and its nearest neighbors. These bonds are formed to the number indicated by the usual valence of the element and also by the octet rule. It was suggested by Hume-Rothery[20] that the sequence of number of nearest neighbors could be continued further to the left in the periodic table, leading to the expected values five for gallium, indium, and thallium (and perhaps boron and aluminum), six for zinc, cadmium, and mercury, seven for copper, silver, and gold, and so on. The rule does not have general validity, but it is given significance by its compatibility with the zinc-cadmium and mercury structures.

Indium has a tetragonal structure (A6) that can be described as cubic closest packing somewhat extended along one axis. Each atom has four neighbors at 3.242 Å and eight at 3.370 Å. One modification of manganese has a similar structure, differing in that there is compres-

[20] W. Hume-Rothery, *Phil. Mag.* 9, 65 (1930); 11, 649 (1931).

sion, rather than extension, along one axis; each atom has eight neighbors at 2.582 Å and four at 2.669 Å

The Cubic Body-centered Arrangement.—In the cubic body-centered arrangement (A2) each atom has eight neighbors at the distance $a_0\sqrt{3}/2$ and six neighbors at the 15 percent larger distance a_0. If the valence of the atom were used only for bonds to the eight nearest neighbors, the effective radius would be that for ligancy 8, which by Equation 11-1 is 0.053 Å less than for ligancy 12. But the observed differences for all those elements that have both an A2 and a closest-packed modification are all less than 0.053 Å, usually 0.03 to 0.04 Å, which supports the description of these metals as involving the formation of eight strong bonds and six weaker bonds between each atom and its neighbors. It has been pointed out by Thewlis[21] that, when the interatomic distances are all corrected to room temperature, the observed differences agree with those predicted by Equation 11-1, with consideration of the six weaker bonds and the assumption that the metallic valence is the same for the A2 structure and the closest-packed structures. Some of the single-bond radii of Table 11-1 have been obtained with use of Equation 11-1 from observed distances for A2 crystals.

The A2 structure is seen from Table 11-2 to be the preferred one for the alkali metals, barium, the fifth-group metals, and the sixth-group metals; it is also observed as one allotropic form for titanium, zirconium, iron, and thallium. The factors determining the choice of the A2 structure by certain elements are not known.

11-7. THE ELECTRONIC STRUCTURE OF THE TRANSITION METALS

In most of the quantum-mechanical treatments that have been carried out for the transition metals the assumption has been made that the 3d shell is full or nearly full of electrons. For example, the treatments of copper by Fuchs[22] and Krutter[23] correspond closely to the configuration $3d^{10}4s$, with the binding due almost entirely to the 4s electron (hybridized somewhat with 4p). For nickel the magnetic moment of 0.6 Bohr magnetons per atom has been assumed to indicate that the electron configuration of the atoms in the metal is approximately $3d^{9.4}4s^{0.6}$, and corresponding configurations with all but about 0.6 electron in the 3d subshell have been assumed also for cobalt and iron. On the other hand, the arguments presented above suggest that the electronic configurations of these elements in the metallic state

[21] J. Thewlis, *J.A.C.S.* **75**, 2279 (1953).

[22] K. Fuchs, *Proc. Roy. Soc. London* **A151**, 585 (1935).

[23] H. M. Krutter, *Phys. Rev.* **48**, 664 (1935).

involve promotion of several electrons to higher orbitals. The configuration $3d^84s4p^2$ for copper would correspond to the valence 5, and $3d^74s4p^3$ (with no metallic orbital) to the valence 7; the valence 5.56 might be expected to result from the contribution of these two configurations in the ratio 72 to 28 percent. For iron, with valence 6 and two unpaired electrons occupying separate orbitals, we must assume the configuration $3d^54s4p^2$.

There is, in fact, no reason for us not to accept configurations that involve a considerable amount of promotion of electrons. Figure 11-8

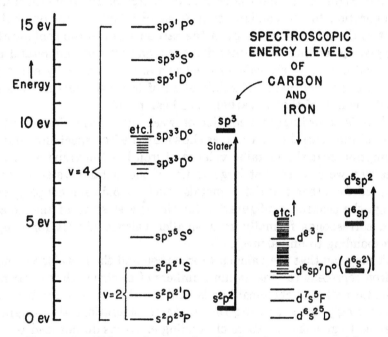

Fig. 11-8.—Values of energy levels of carbon atoms and iron atoms, as determined by spectroscopic methods. The unit of energy used in the scale at the left is the electron volt.

shows the low-lying spectroscopic energy levels of the carbon atom and the iron atom. The three lowest levels of the carbon atom correspond to the configuration $2s^22p^2$, which is the basis of the bivalent state of the carbon atom.

It was pointed out in Section 4-2 that the configuration sp^3, which has promotion energy about 200 kcal/mole relative to the ground configuration $2s^22p^2$, is the basis of the quadrivalent state of the carbon atom and is shown by quantum-mechanical calculations for methane to contribute about 49 percent to this valence state. Now let us consider the iron atom, for which spectroscopic energy levels are shown on the right side of Figure 11-8. The normal state, ^5D, is based on the configuration $3d^64s^2$. The first excited state, ^5S, is based on the configuration $3d^74s$. Among other rather low-lying states (under 100 kcal/mole) are $3d^64s4p$ $^7D^0$ and $3d^8$ ^3F. Analysis of the energy levels suggests that a $4s$ electron in the neutral iron atom is more stable than a $3d$ electron by about 28 kcal/mole, and more stable than a $4p$ electron by 60 kcal/mole. The configuration $3d^54s4p^2$ accordingly lies about 92 kcal/mole above the normal configuration, $3d^54s^2$, and only 36 kcal/mole above the configuration $3d^8$. It is seen from this argument, as indicated in Figure 11-8, that the promotion energy of iron from the normal configuration to the configuration $3d^54s4p^2$ is only about one-half as great as the promotion energy of the carbon atom to the configuration $2s2p^3$ corresponding to its quadrivalence, and we can understand that the bond energy of six valence electrons in iron can easily effect this promotion; the enthalpy of sublimation of iron (97 kcal/mole) is also about one-half of that of carbon (170 kcal/mole).

There is accordingly no conflict between the information about the states of the neutral iron atom, as shown by the iron spectrum, and our assumption that the metallic valence of iron has the value 6. In the same way we can see that large values of the metallic valence can be accepted for other transition metals (such as 5.56 for copper), even though the electron configuration for the lowest state of the isolated atom corresponds formally to a smaller valence ($3d^{10}4s$ for copper, corresponding to univalence).

We may say that the valence 4 for carbon and the metallic valence 6 for iron represent the maximum numbers of electrons that contribute to the formation of chemical bonds, and not the average numbers. It is accordingly not surprising that some considerations that might be expected to give the numbers of bonding electrons do not lead to such large values. An example is a simple theory of the equation of state of metals based on the virial theorem.[24] This theory, in which the energy of formation of the metal from the gaseous metal ions and free valence electrons occurs as a parameter that is given its experimental value, leads to calculated values of the compressibility in good agreement with experiment for the alkali metals (valence 1) and the alkaline-earth metals (valence 2). For the transition metals titanium to nickel agreement is obtained if it is assumed that there are 3 valence electrons per atom, rather than 4, 5, or 6, and for copper, silver, and gold 2.5 rather than 4.5. It is significant that for carbon (diamond) agreement is obtained with the value 2, rather than 4.

[24] W. G. McMillan and A. L. Latter, *J. Chem. Phys.* **29**, 15 (1958).

11-8. METALLIC RADII AND HYBRID BOND ORBITALS

Values of the single-bond radius are shown in Figures 11-9 and 11-10 for the elements of the first and second long periods and the first very long period.

There are some features of the course of the values in the sequences that merit comment. First, what is the cause of the rapid decrease in single-bond radius from potassium to chromium, and the corresponding decreases for the other sequences? We may be sure that this

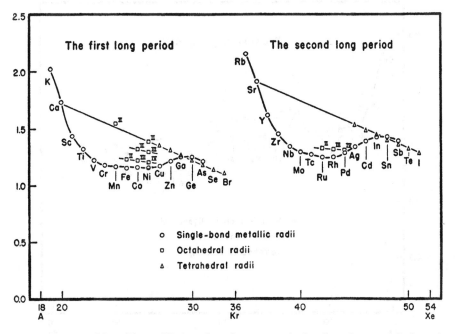

Fig. 11-9.—Metallic radii for the elements of the first long period and the second long period. Values of octahedral radii and tetrahedral radii are also represented.

decrease is not simply the result of increase in atomic number. The decrease corresponding simply to increase in atomic number, without change in bond type, is much smaller in magnitude; it is indicated for the elements of the first long period by the straight line that is drawn through the point for calcium and the points (represented by triangles) for the tetrahedral radius of germanium, arsenic, etc. All of these radii represent bond orbitals of approximately the sp^3 type. For example, the square for manganese is the radius for bipositive manganese found in MnS_2 (hauerite), for which the magnetic properties show that there are five $3d$ orbitals occupied by atomic electrons, leaving only the $4s$ and $4p$ orbitals for use in bonding. The slope of this line corresponds to a decrease in radius by 0.043 Å for increase in atomic number by 1.   The same slope, −0.043 Å, is found for the second long period, and a smaller value, −0.030 Å, for the very long period (the line passing through the points for barium, europium, ytterbium, and the tetrahedral radii in Figure 11-10).

We conclude instead that the rapid decrease in radius from potassium to chromium is due to the nature of the bonding orbitals.   It has been

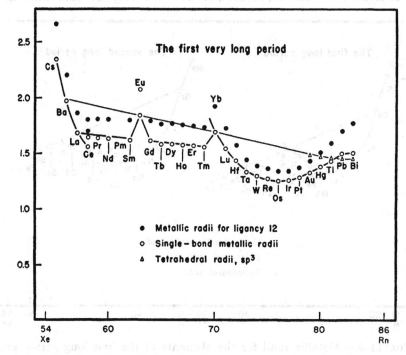

Fig. 11-10.—Metallic radii for the elements of the first very long period.

suggested[25] that in potassium the bond orbitals are mainly s, with 26 percent p character, and that the amount of p character and d character increases to 39 percent d character for chromium, remaining approximately constant at 39 or 40 percent through the sequence to nickel.

That this interpretation is reasonable is indicated by a comparison with the octahedral radii of iron, cobalt, and nickel, indicated in Figure 11-9 by squares, with the oxidation numbers also shown.   The smaller octahedral radii correspond to d^2sp^3 orbitals, with 33 percent d character, and the radii that are about 0.10 greater correspond to dsp^3 orbitals, with 20 percent d character.   It is evident that there is a rapid decrease in single-bond radius with increase in the amount of d character

[25] Pauling, *loc. cit.* (9).

of the bond orbital, and that the value 40 percent for the *d* character of the metallic orbitals is reasonable.

The single-bond radii for the lanthanons (La to Lu, Fig. 11-10) show some interesting features. The magnetic properties[26] require that most of the lanthanons have metallic valence 3, but that two of these metals, europium and ytterbium, have metallic valence 2. The interatomic distances reflect these valences clearly, as shown in Figure 11-10. The stability of metallic valence 2, rather than 3, for these two metals may be attributed to the special stability of a half-filled or completely filled 4*f* subshell.

The smaller of the two single-bond radii indicated for cerium in Figure 11-10 relates to a modification of the metal that was discovered[27] after a search for it had been initiated as a result of the consideration of Figure 11-10. The ordinary, less dense modification of cerium corresponds to valence about 3.2, and the new, high-density modification to valence about 4.

A striking abnormality in behavior is shown by manganese, which crystallizes in three modifications, no one of which contains only atoms with the size expected for metallic valence 6. Gamma-manganese, a tetragonal structure representing a small distortion from cubic closest packing, has interatomic distances corresponding to $R(L\,12) = 1.306$ Å. If we take the single-bond radius of manganese as 1.171 Å, a reliable interpolation between the values for chromium and iron, we calculate for manganese in this modification the valence 4.25. This suggests that 3 of the 7 outer electrons in the manganese atom are atomic electrons, occupying 3*d* orbitals, and 4 are valence electrons. In the more complex modifications of manganese, β-manganese, with 20 atoms in the unit cube, and α-manganese, with 58 atoms in the unit cube, the interatomic distances show clearly that both the low-valent manganese, with valence 4, and normal manganese, with valence 6, are present. In β-manganese there are two crystallographic sorts of atoms, 8 corresponding to valence 6 and 12 to valence 4; in α-manganese there are 4 crystallographic sorts, one, 24 atoms, corresponding to valence 6, and the other three, 34 atoms, to valence 4. No explanation has as yet been proposed for the stability of the state with valence 4 for manganese.

A set of empirical equations representing the single-bond radius of a transition element as a function of atomic number and degree of hybridization of the bond orbitals has been formulated.[28] These

[26] W. Klemm and H. Bommer, *Z. anorg. Chem.* **231**, 138 (1937); **241**, 264 (1939); H. Bommer, *ibid.* **242**, 277 (1939).

[27] Lawson and Tang, also Schuch and Sturdivant, *loc. cit.* (T11-2).

[28] Pauling, *loc. cit.* (9).

equations are the following:
Iron-transition elements:

$$R_1(p) = 1.855 - 0.043z$$
$$R_1(sp^3) = 1.825 - 0.043z$$
$$R_1(\delta, z) = 1.825 - 0.043z - (1.600 - 0.100z)\delta$$

(z is the number of electrons outside the noble-gas shell and δ is the amount of d character in the bond orbitals)
Palladium-transition metals:

$$R_1(p) = 2.036 - 0.043z$$
$$R_1(sp)^3 = 2.001 - 0.043z$$
$$(R_1(\delta, z) = 2.001 - 0.043z - (1.627 - 0.100z)\delta$$

Platinum-transition metals:

$$R_1(p) = 1.960 - 0.030z$$
$$R_1(sp^3) = 1.850 - 0.030z$$
$$R_1(\delta, z) = 1.850 - 0.030z - (1.276 - 0.070z)\delta$$

The nature of the bond orbitals (p, sp^3, sp^3 with some contribution of d) for the hyperelectronic atoms (following nickel, palladium, and platinum) is a function of the valence. Values of the single-bond radius for some values of the valence are given for these elements in Table 11-3. The use of these values will be illustrated later.

It is interesting to note that bond numbers approximately equal

TABLE 11-3.—SINGLE-BOND METALLIC RADII IN
DEPENDENCE ON VALENCE

Cu	Zn	Ga	Ge
(7) 1.138	(6) 1.176	(5) 1.206	(4) 1.223
(5) 1.185	(4) 1.229	(3) 1.266	(2) 1.253
(3) 1.227	(2) 1.309	(1) 1.296	
(1) 1.352			
Ag	Cd	In	Sn
(7) 1.303	(6) 1.343	(5) 1.377	(4) 1.399
(5) 1.353	(4) 1.400	(3) 1.442	(2) 1.434
(3) 1.396	(2) 1.485	(1) 1.477	
(1) 1.528			
Au	Hg	Tl	Pb
(7) 1.303	(6) 1.345	(5) 1.387	(4) 1.430
(5) 1.351	(4) 1.403	(3) 1.460	(2) 1.540
(3) 1.393	(2) 1.490	(1) 1.570	
(1) 1.520			

to the ratios of small integers, especially $\frac{1}{2}$, $\frac{1}{3}$, $\frac{2}{3}$, and $\frac{1}{4}$, occur rather often, and there may be special stability associated with these bond numbers; the importance of half-bonds (with $n = \frac{1}{2}$) has been emphasized by Rundle.[29]

The use of these values may be illustrated by a discussion of tin. It was pointed out in Section 11-4 that application of Equation 11-1 with D_1 taken as 2.80 Å leads to valence 2.24 for the tin atoms in white tin. However, 1.40 Å is not the value for R_1 for tin with this valence; from Table 11-3 we see that a somewhat larger value, about 1.43 Å, should be used. It is found by trial that for four bond distances 3.016 Å and two 3.175 Å, a consistent solution is obtained for R_1 = 1.423 Å, interpolated linearly between the values given in Table 11-3. The valence is 2.64, and the bond numbers are 0.52 and 0.28.

A similar treatment of zinc and cadmium leads to valences 3.93 and 3.98, respectively. The six strong bonds have bond numbers 0.48 and 0.51, and the six weak bonds have bond numbers 0.18 and 0.15.

For all three metals the strong bonds have bond numbers equal to $\frac{1}{2}$ to within the reliability of their evaluation (0.48 is found for the six strong bonds in mercury, too). Bond numbers approximately equal to the ratios of small integers occur rather often, and special stability may be associated with them.

11-9. BOND LENGTHS IN INTERMETALLIC COMPOUNDS

Any intermetallic compound for which interatomic distances are known might be used as an example of the application of Equation 11-1 and the set of single-bond metallic radii. Cementite, Fe_3C, is an important compound which shows some interesting features. In this orthorhombic crystal the iron atoms are in reasonably close packing, each having either 12 iron ligands at the average distance 2.62 Å or 11 at the average distance 2.58 A. Each carbon atom is at the center of a trigonal prism of six iron atoms, with the Fe—C distance 2.01 Å.

We may predict the distances from the structure and radii. Carbon without doubt is quadrivalent, so that the iron-carbon bonds must have bond number $\frac{2}{3}$. The sum of the iron radius 1.167 Å and the carbon radius 0.772 Å is 1.939 Å, and with the correction of Equation 11-1 we obtain 2.04 Å as the predicted Fe—C distance. This is in reasonably good agreement with the experimental value 2.01 Å. Each iron atom forms a bond with bond number $\frac{2}{3}$ with each of two carbon atoms, using up $1\frac{1}{3}$ of its total valence of 6 and leaving $4\frac{2}{3}$ for the Fe—Fe bonds. The predicted bond numbers for ligancy 12 and 11 are 0.39 and 0.42, respectively, and the predicted Fe—Fe distances are 2.58 Å and 2.56 Å, which are approximately equal to the corresponding observed values,

[29] R. E. Rundle, *J.A.C.S.* **69**, 1327 (1947); *J. Chem. Phys.* **17**, 671 (1949).

2.62 Å and 2.58 Å. Possibly the structure is under some strain, which compresses the Fe—C bonds by 0.03 Å and stretches the Fe—Fe bonds by 0.03 Å; an alternative explanation is that the orbitals of the iron atoms have undergone a change in hybridization, those bonding iron to carbon having an increased amount of d character, decreasing the iron radius by 0.03 Å, with a corresponding decrease in amount of d character and increase in radius of the orbitals involved in forming Fe—Fe bonds.

We can understand the increased hardness and strength of cementite over elementary iron by the consideration of the bonds   The volume of cementite per iron atom is only a few percent greater than that of elementary iron, and within this volume there is an increased amount of bonding, involving the valence electrons of the carbon atoms. We may use the heat of sublimation of cementite per iron atom, compared with that of the element, as a measure of the increased strength of the bonds.   The heat of sublimation of carbon is about 78 percent greater than that of iron, and accordingly the heat of sublimation of Fe_3C per iron atom is about 60 percent greater than that of elementary iron. For a reticular structure, such as that shown by cementite, the strength may be taken as roughly proportional to the heat of sublimation of the material in unit volume; accordingly we understand why a small amount of carbon can produce a great amount of change in mechanical strength.

In cementite the carbon atom, although it has its normal covalence 4, has increased its ligancy to 6.   It is interesting to note that in order for an atom to increase its ligancy beyond its covalency it is not necessary that this atom have an extra orbital; it is instead sufficient for the atoms that surround it to have extra orbitals.   The valence bonds of the central atom may then resonate among their alternative positions by pivoting about the central atom.   This sort of pivoting valence-bond resonance is shown by carbon in cementite and by atoms of other nonmetals and metalloids in many compounds.

The compounds with the nickel arsenide structure are especially interesting in this respect.   Very many substances, such as NiAs, FeS, FeSb, and AuSn, crystallize with this structure (Fig. 11-11).   We may discuss AuSn as an example.   Each tin atom is surrounded by six gold atoms at the corners of a trigonal prism, with the distance Au—Sn = 2.847 Å, and each gold atom is surrounded by six tin atoms at the corners of a flattened octahedron, and also by two gold atoms, at 2.756 Å, in the opposed directions through the centers of the two large faces of the octahedron.   The tin atoms are arranged in the positions corresponding to hexagonal closest packing, but the axial ratio c/a has the value 1.278, instead of the normal value 1.633. We predict for gold the valence 5.56, and for tin either the metallic valence 2.56 or the covalence 4. The Au—Sn distance is much too small to correspond to valence 2.56 for tin, but it agrees well with the valence 4. With this valence the Au—Sn bonds have bond numbers $\frac{2}{3}$, and the interatomic distance predicted with use of the single-bond radii 1.338 Å for gold and 1.399 Å for tin is 2.843 Å, in almost exact agreement with the experimental value 2.847 Å. Accordingly the tin atom is quadrivalent, without a metallic orbital, and its four valence bonds resonate among the six positions connecting it with the ligated gold atoms. These bonds use up 4 of the total of 5.56 valences of gold. If the

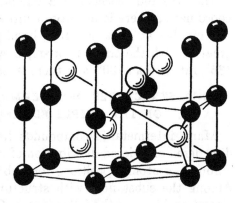

FIG. 11-11.—The arrangement of atoms in the hexagonal crystal nickel arsenide, NiAs. Nickel atoms are represented by black circles; they have ligancy 6, with octahedral coordination of arsenic atoms about themselves. Arsenic atoms are represented by open circles; they have ligancy 6, with coordination of the nickel atoms in a trigonal-prism arrangement.

axial ratio of the crystal had the value 1.633 the remaining 1.56 valences of gold would not be utilized; however, by compression along the c axis, keeping the Au—Sn distance constant, the gold atoms that succeed one another along the c axis can be brought into a suitably small distance from one another to permit Au—Au bonds to be formed. The bond number predicted for these bonds is 0.78, and the Au—Au distance that is predicted is 2.741 Å, in close approximation to the observed distance 2.756 Å. The system of metallic valences and radii accordingly provides an explanation of the interatomic distances, including the compression along the hexagonal axis to give the abnormally small axial ratio.

Some other aspects of these structures may also be discussed. In cementite the carbon atoms, with ligancy 6, coordinate six iron atoms about themselves at the corners of a trigonal pyramid. The arrangement of iron atoms is such that octahedral coordination about carbon would be an alternative possibility, and we may ask why the carbon atom with ligancy 6 prefers the trigonal prism as its coordination polyhedron, and also why the tin atom in AuSn assumes this coordination polyhedron. The choice by the compound AuSn of the nickel arsenide structure rather than the sodium chloride structure can be accounted for in part, inasmuch as there is no easy way of distorting the sodium chloride structure to permit the Au—Au bonds to be formed, utilizing the gold valencies not involved in bonding with the tin atoms; but alternative structures with octahedral coordination or some other type of coordination about tin might well be accepted in place of the nickel arsenide structure. It seems likely that the trigonal prism is preferred to the octahedron by a quadricovalent atom with ligancy 6 because it decreases the bond-angle strain. The four valence bonds of carbon or of quadrivalent tin have the greatest stability when they extend toward the four corners of a regular tetrahedron. With octahedral coordination there is no way of introducing the four bonds that does not involve at least one 180° bond angle; the strain is accordingly great. With the trigonal prism, however, the largest angle is about 135°, rather than 180°, and there is less bond-angle strain.

11-10. STRUCTURES OF INTERMETALLIC COMPOUNDS BASED ON THE SIMPLE ELEMENTARY STRUCTURES

Many intermetallic compounds have structures that involve the disordered or ordered distribution of metal atoms of two or more kinds among the atomic positions for cubic or hexagonal closest packing. Among the substances with structures of this type based on cubic closest packing are the following: $AuCu$, $PtCu$, $AuCu_3$, $PdCu_3$, $PtCu_3$, $CaPb_3$, $CaTl_3$, $CaSn_3$, $CePb_3$, $CeSn_3$, $LaPb_3$, $LaSn_3$, $PrPb_3$, $PrSn_3$. In general there is small difference in radius of the atoms in a compound of this type.

A more complex structure[30] is that of $PuAl_3$. This crystal is based on the sequence of hexagonal layers $c\,c\,h\,c\,c\,h\,c\,c\,h\cdots$, each layer containing plutonium atoms and aluminum atoms in the $\frac{1}{3}$ ratio.

Among the intermetallic compounds that crystallize with structures based on the cubic body-centered structure A2 are the binary compounds $CuPd$, $CuBe$, $CuZn$, $AgMg$, $FeAl$, $AgZn$, $AgCd$, $AuZn$, $AuCd$, $NiAl$, $NdAl$, $SrCd$, $SrHg$, $BaCd$, $BaHg$, and $LaCd$, with the cesium chloride structure. In these crystals each atom has as nearest neighbors, at the corners of a cube about it, eight unlike atoms, whereas in $NaTl$, $LiZn$, $LiCd$, $LiGa$, $LiIn$, $NaIn$, and $LiAl$, which represent another type of structure based on A2 (*B32*), each atom has as nearest neighbors four like atoms and four unlike atoms. Other compounds with structures related to A2 are $LaMg_3$, $CeMg_3$, $PrMg_3$, Fe_3Al, Fe_3Si, Cu_2AlMn, Cu_3Al, Cu_5Sn, and the γ-alloys, discussed below.

[30] A. C. Larson, D. T. Cromer, and C. K. Stanbaugh, *Acta Cryst.* **10**, 443 (1957).

11-11. ICOSAHEDRAL STRUCTURES

The maximum number of rigid spheres that can be brought into contact with another sphere with the same radius is twelve. The corresponding coordination polyhedra, seen in the cubic and hexagonal closest-packed structures, have eight triangular faces and six square faces.

It is possible to retain ligancy 12 with a central sphere as much as 10 percent smaller than the surrounding spheres. These spheres are then arranged at the corners of a regular icosahedron, which has 20 triangular faces (Figure 10-1).

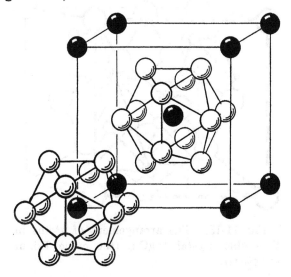

FIG. 11-12.—The atomic arrangement in the cubic crystal $MoAl_{12}$. There is a molybdenum atom at the corner and the center of each cube. It is surrounded by 12 aluminum atoms, at icosahedral corners.

There are many known structures of intermetallic compounds that involve icosahedral coordination about the smaller atoms. Usually these structures are complex, with 20, 52, 58, 162, 184, or more atoms in a cubic unit of structure. Many of the crystals are cubic. The icosahedron has 12 fivefold axes of symmetry, 20 threefold axes, and 30 twofold axes; the fivefold axes cannot be retained in the crystal, but some of the others can be (a maximum of four threefold axes in a cubic crystal).

A simple icosahedral structure[31] is that of $MoAl_{12}$, WAl_{12}, and $(Mn,-Cr)Al_{12}$. In this structure, based on a body-centered cubic lattice,

[31] J. Adam and J. B. Rich, *Acta Cryst.* 7, 813 (1954).

there is at each lattice point a nearly regular icosahedron of twelve aluminum atoms about the smaller central atom (Fig. 11-12).

The cubic Friauf structure is shown by $MgCu_2$ and many other compounds. In this structure, represented in Figure 11-13, each copper atom is surrounded by an icosahedron formed by six magnesium atoms and six copper atoms. The ratio of radii (for ligancy 12) of copper and magnesium is 0.80; hence the ratio for the central atom (copper) and the average for the surrounding atoms (magnesium and copper) is that corresponding to stability of the icosahedron.

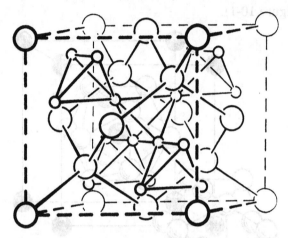

Fig. 11-13.—The arrangement of atoms in the cubic crystal $MgCu_2$ (the cubic Friauf structure).

The larger atoms, magnesium, have ligancy 16 (12 Cu and 4 Mg), as seen in Figure 11-13. This increase in ligancy for magnesium, relative to that (12) for the element, and the assumption of icosahedral coordination by copper cause a decrease in volume of the compound relative to the elements. The volume decrease is 6.7 percent. A part of it may be the result of electron transfer (Sec. 11-12), but similar decreases are found in general for icosahedral structures.

The element manganese has four reported allotropic forms, of which two are icosahedral. The allotrope β-manganese is cubic with 20 atoms in the unit cube, and α-manganese is cubic with 58 atoms in the unit cube. In each structure one of the kinds of atoms has an effective size less than that of the others and shows icosahedral coordination: 8 of the 20 atoms per unit for β-manganese and 24 of the 58 for α-manganese. The larger atoms have ligancy 14 for β-manganese and 13 and 16 for α-manganese. The effective radii correspond to metallic valence close to 6 for the smaller atoms and 4.5 for the larger atoms.

Two other allotropic forms have been reported by Basinski and

Christian,[32] one with structure A1, stable from 1100° to 1130°C, and one with structure A2, stable from 1130° to 1240°C. The lattice constants, corrected to room temperature, correspond to $R(L\,12)$ about 1.30 Å. Another form, obtained by quenching, has been assigned the A6 structure, with 8 bonds at 2.582 Å and 4 at 2.669 Å, corresponding to $R(L\,12) = 1.306$ Å. The calculated valence for the manganese atoms in these three forms is 4.5; hence these atoms may be considered to be similar to the larger atoms in α-manganese and β-manganese.

No satisfactory theory of this small metallic valence of manganese has as yet been proposed.

The compound $Mg_{34}Al_{24}$ crystallizes with the α-manganese structure, and several compounds (Ag_3Al, Cu_5Si) have the β-manganese structure, with some randomness of distribution of atoms of different kinds over the two kinds of atomic positions.

One of the most complex structures known is that of the compound $Mg_{32}(Zn, Al)_{49}$. The unit of structure is a cube containing 162 atoms. The structure is based on icosahedral coordination of larger atoms about somewhat smaller ones.[33] It is characteristic of the icosahedron that groups of four contiguous atoms occur only at corners of a tetrahedron; every triangle formed by three contiguous atoms in the icosahedron has a fourth atom lying approximately above its center. Accordingly a structure involving icosahedral packing may be built up by placing atoms out from the centers of the triangular faces of an inner polyhedron. The metrical nature of the icosahedron is such that the distances from the atom at the center to its twelve ligands is 5 percent smaller than the distances between these ligands; hence the retention of icosahedral packing through successive ligation spheres requires a continued steady increase in average size of the atoms in these spheres. This increase in size can be achieved by placing the smaller atoms, zinc and aluminum, alone in the inner spheres, and introducing the larger atoms, magnesium, as part of the outer spheres.

The structure is based on a body-centered lattice. At each lattice point there is a small atom (Zn, Al). It is surrounded by an icosahedron of twelve atoms (Fig. 11-14). This group is then surrounded by 20 atoms, at the corners of a pentagonal dodecahedron, each atom lying directly out from the center of one of the 20 faces of the icosahedron. The next 12 atoms lie out from the centers of the pentagonal faces of the dodecahedron; this gives a complex of 45 atoms, the outer 32 of which lie at the corners of a rhombic triacontahedron. The next shell consists of 60 atoms, each directly above the center of a

[32] Z. S. Basinski and J. W. Christian, *Proc. Roy. Soc. London* A223, 554 (1954).
[33] G. Bergman, J. L. T. Waugh, and L. Pauling, *Nature* 169, 1057 (1952); *Acta Cryst.* 10, 254 (1957).

triangle that forms one-half of one of the 30 rhombs bounding the
rhombic triacontahedron; these 60 atoms lie at the corners of a truncated icosahedron, which has 20 hexagonal faces and 12 pentagonal
faces.   Twelve additional atoms are then located out from the centers
of 12 of the 20 hexagonal faces.   The very large complexes, shown in
Figure 11-14, are then condensed together in such a way that each of
the 72 outer atoms is shared between two complexes; each outer atom

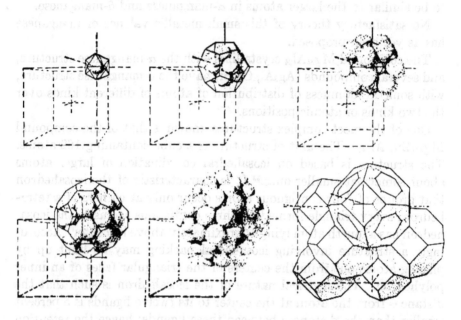

Fig. 11-14.—The atomic arrangement in the cubic crystal $Mg_{32}(Zn, Al)_{49}$.
The six drawings, from left to right in the top row and then left to right in
the bottom row, have the following significance: A central atom surrounded
by 12 atoms at the points of a nearly regular icosahedron; the icosahedral
group of 13 atoms surrounded by 20 atoms at the points of a pentagonal
dodecahedron; the complex of 33 atoms surrounded by 12 atoms at the
corners of an icosahedron; the outermost shell of 60 atoms at the corners of a
truncated icosahedron, plus 12 atoms out from the centers of 12 of the hexagons of this polyhedron; packing drawing showing the outer shell of 72 atoms
surrounding the central complex of 45 atoms; the structure of the crystal, in
which these complexes located about the points of a body-centered cubic
lattice share all of the 72 atoms of the outermost shell with neighboring
complexes.

then contributes 36 atoms per lattice point, which with inner complexes
of 45 atoms gives 81 atoms per lattice point, 162 in the unit cube.
All of the smaller atoms (Al, Zn) have icosahedral coordination; the
larger ones (Mg) have ligancy 14, 15, or 16.

11-12. THE γ-ALLOYS; BRILLOUIN POLYHEDRA

It was pointed out by Hume-Rothery[24] in 1926 that certain intermetallic compounds with closely related structures but apparently unrelated stoichiometric composition can be considered to have the same ratio of number of valence electrons to number of atoms. For example, the β phases of the systems Cu—Zn, Cu—Al, and Cu—Sn are analogous in structure, all being based on the $A2$ arrangement; their compositions correspond closely to the formulas $CuZn$, Cu_3Al, and Cu_5Sn. Considering copper to be univalent, zinc bivalent, aluminum trivalent, and tin quadrivalent, we see that the ratio of valence electrons to atoms has the value $\frac{3}{2}$ for each of these compounds:

$$CuZn: \quad (1 + 2)/2 = 3/2;$$

$$Cu_3Al: \quad (3 + 3)/4 = 3/2;$$

$$Cu_5Sn: \quad (5 + 4)/6 = 3/2.$$

Other alloys that may be placed in this class are $CuBe$, $AgZn$, $AgCd$, $AgMg$, $AuZn$, and Ag_3Al.

A more striking example is provided by the γ-alloys, the principal representatives of which are Cu_5Zn_8, Cu_9Al_4, $Cu_{31}Sn_8$, and Fe_5Zn_{21}. The corresponding phases of several other systems are also known, with ideal compositions Cu_5Cd_8, Ag_5Zn_8, Ag_5Cd_8, Au_5Zn_8, Ag_9Al_4, Cu_9Ga_4, $Ag_{31}Sn_8$, Co_5Zn_{21}, Ni_5Zn_{21}, Rh_5Zn_{21}, Pd_5Zn_{21}, Pt_5Zn_{21}, etc.; in some of these systems the γ phase shows a wide range of composition about the ideal value. These crystals are cubic, with 52 atoms in the unit of structure (27 × 52 for $Cu_{31}Sn_8$, corresponding to tripling the value of a_0). The structure is an icosahedral one. It is obtained from the body-centered structure $A2$ by taking a cube, with edge three times that of the unit for $A2$, which contains $3^3 \times 2 = 54$ atoms, removing two atoms, and displacing the others by small amounts. The atoms of different kinds are distributed among the atomic positions in different ways for the different stoichiometric compositions given above.

For the γ-alloys the ratio of valence electrons to atoms has the surprising value $\frac{21}{13}$:

$$Cu_5Zn_8: \quad (5 + 16)/13 = 21/13;$$

$$Cu_9Al_4: \quad (9 + 12)/13 = 21/13;$$

$$Cu_{31}Sn_8: (31 + 32)/39 = 21/13;$$

$$Fe_5Zn_{21}: \quad (0 + 42)/26 = 21/13.$$

[24] W. Hume-Rothery, *J. Inst. Metals* **35** 295 (1926); see also A. F. Westgren and G. Phragmén, *Z. Metallk.* **18**, 279 (1926); *Metallwirtschaft* **7**, 700 (1928); *Trans. Faraday Soc.* **25**, 379 (1929).

It is to be noted that Fe, Co, Ni, Rh, Pd, and Pt are assigned zero valence electrons in order that the ratio retain its value.

It seems likely that the Hume-Rothery rule is to be explained as resulting from a perturbation of the energy of the valence electrons by their diffraction by the crystal lattice. The distribution in energy (kinetic energy) of free electrons in volume V can be calculated. There is one quantized state (orbital) per volume h^3 in phase space, and two electrons, with opposed spins, can occupy each orbital; in consequence, the number of electrons n with energy not greater than E is $16\sqrt{2}\pi m^{3/2}E^{3/2}/3h^3$. It was pointed out by Brillouin[35] that the distribution is perturbed when an electron has such a wave length ($\lambda = h/\sqrt{2mE}$) and direction as to permit Bragg reflection from an important crystallographic plane (one with large scattering factor for electrons). The perturbation is of such a nature as to stabilize electrons with energy just equal to or less than that corresponding to Bragg reflection and to destabilize electrons with a larger energy. Hence special stability would be expected for metals with just the right number of electrons to correspond to the Brillouin perturbation. This number is proportional to a volume of a polyhedron (the Brillouin polyhedron) in reciprocal space, corresponding to the crystallographic planes giving rise to the perturbation.[36]

It was pointed out by Jones[37] that the first important Brillouin polyhedron for the γ-alloys (bounded by the forms $\{330\}$ and $\{411\}$) contains 22.5 electrons per 13 atoms, and he proposed that some effect determined by the shape of the Brillouin polyhedron could reduce this number to 21 electrons per 13. atoms and thus explain the stability of the γ-alloys.

This consideration ignores the difference between the ordinary valence and the metallic valence described in this chapter. For Cu_5Zn_8, for example, the valences 5.56 for copper and 4.56 for zinc given in Table 11-1 lead to 64.28 valence electrons per 13 atoms, and the same ratio is obtained also for Cu_9Ga_4, $Cu_{31}Sn_8$, etc.

There is, in fact, one other important Brillouin polyhedron for these crystals.[38] It is bounded by the only other strongly reflecting crystallographic forms, $\{600\}$ and $\{442\}$, and its volume is 63.90 electrons per 13 atoms, very nearly equal to that given by the metallic valences of Table 11-1. The number 63.90 corresponds exactly to valences

[35] L. Brillouin, *Compt. rend.* **191**, 198, 292 (1930); *J. de phys. radium* **1**, 377 (1930); **3**, 565 (1932); **4**, 1, 333 (1933); **7**, 401 (1936).

[36] Electron numbers for cubic Brillouin polyhedra are given by D. P. Shoemaker and T. C. Huang, *Acta Cryst.* **7**, 249 (1954).

[37] H. Jones, *Proc. Roy. Soc. London* A144, 225 (1934); A147, 396 (1934).

[38] L. Pauling and F. J. Ewing, *Rev. Modern Phys.* **20**, 112 (1948).

5.53 for copper, 4.53 for zinc, and so on, which agree with those of the table to within the reliability of their determination from the saturation magnetic moments of ferromagnetic metals.   Similar agreement between Brillouin polyhedra and the metallic valences of Table 11-1 (or those given above in the descriptions of the structures) has been found for β-manganese and some other substances.

11-13. ELECTRON TRANSFER IN INTERMETALLIC COMPOUNDS

The consideration of interatomic distances shows that electron transfer from atoms of one element to those of another takes place in many interatomic compounds, and that the numbers of electrons involved are reasonable in relation to the changes in valence resulting from loss or gain of electrons and to the partial ionic character of the bonds between unlike atoms and the striving of atoms toward electroneutrality.[39]

Let us divide atoms into three classes: hypoelectronic (electron-deficient) atoms, hyperelectronic (electron-superfluent) atoms, and buffer atoms.   Hypoelectronic atoms are atoms that can increase their valence by adding electrons.   The hypoelectronic elements include the first three elements of each short period and the first five elements of each long period, as shown in Table 11-4.   Atoms of these elements have more bond orbitals than valence electrons (in the uncharged state), and they can accordingly increase their valence by one unit by accepting an electron.   Hyperelectronic atoms are atoms that can increase their valence by giving up an electron.   The hyperelectronic elements with respect to metallic compounds include the last three elements (before the noble gases) of each short period and the last seven elements of each long period.   Atoms of these elements have more valence electrons than bond orbitals, and they can increase their valence by one unit by giving up one electron of a pair occupying a bond orbital, thus leaving a valence electron in the orbital.   Buffer atoms are atoms that can give up or accept an electron without change in valence.   The five elements Cr, Mn, Fe, Co, and Ni and their congeners in the other two long periods are buffer elements with respect to metallic compounds; they can give up a nonbonding d electron or introduce an electron into the incomplete nonbonding d subshell without change in metallic valence (Cr, Mo, and W are buffer atoms with respect only to addition of an electron).

Carbon and silicon are placed in a separate class in Table 11-4. Carbon is an element with stable valence, 4; either the addition of an electron to a carbon atom or the removal of an electron from it causes a decrease in its valence.   Silicon also has the stable valence 4, except

[39] L. Pauling, *Proc. Nat. Acad. Sci. U.S.* **36**, 533 (1950).

TABLE 11-4.—CLASSIFICATION OF ATOMS WITH RESPECT TO EFFECT OF CHANGE
OF ELECTRON NUMBER ON METALLIC VALENCE

Hypoelectronic Atoms					Atoms with stable valence					Hyperelectronic atoms						
Li	Be	B			C					N	O	F				
Na	Mg	Al			Si					P	S	Cl				
					Buffer atoms											
K	Ca	Sc	Ti	V	Cr*	Mn	Fe	Co	Ni	Cu	Zn	Ga	Ge	As	Se	Br
Rb	Sr	Y	Zr	Nb	Mo*	Tc	Ru	Rh	Pd	Ag	Cd	In	Sn	Sb	Te	I
Cs	Ba	La	Ce^b													
		Lu	Hf	Ta	W*	Re	Os	Ir	Pt	Au	Hg	Tl	Pb	Bi	Po	At

* These three atoms can accept electrons but not give up electrons without change in valence.
^b The rare-earth metals may have some buffering power.

that it may under certain circumstances make use of outer orbitals ($3d$, $4s$, $4p$) and achieve some increase in valence through electron transfer. This effect is less important in alloys of silicon than in compounds of the hypoelectronic atoms.

Let us consider the ways in which an intermetallic compound AB might be stabilized by the transfer of an electron from atom B to atom A.

First, an increase in the number of valence bonds and a corresponding increase in stability would result from electron transfer from B to A if A were hypoelectronic and B were hyperelectronic, or if A were hypoelectronic and B were a buffer, or if A were a buffer and B were hyperelectronic.

Second, according to the electroneutrality principle (Sec. 8-2) an increase in stability would follow from a transfer of electrons if it were to result in a decrease in the electric charges on the atoms. Let B be more electronegative than A. The covalent bonds between A and B would then have some ionic character, of such a nature as to give A a positive electric charge and B a negative charge. Transfer of an electron from B to A reduces the charges on the atoms, and the substance can thus be stabilized. It is interesting that this effect involves the transfer of electrons to the more electropositive atoms (the stronger metals); that is, in the opposite direction to the transfer of electrons that takes place in the formation of ions in electrolytic solutions.

These two stabilizing effects usually operate together, because the electronegativity increases in the sequence hypoelectronic elements, buffers, hyperelectronic elements. Both effects are stronger for compounds of hypoelectronic elements with hyperelectronic elements than for compounds of elements of either of these two classes with buffer elements. Thus we expect electron transfer to be especially important for compounds such as $NaZn_{13}$, less important for compounds such as Al_9Co_2 and Fe_5Zn_{21}, and of little significance for compounds such as Na_2K, FeCr, and Cu_5Zn_8.

In special cases electron transfer may take place even in compounds of two metals in the same class. Stabilizing factors that might operate to this end include the filling of Brillouin zones, the stabilizing of partially filled nonbonding subshells through increase in multiplicity (approach to half-filling) or through completion of the subshell, and the relief of strain resulting from geometric constraints on ratios of interatomic distances through change in bond numbers.

The compound AlP may be taken as a simple example. It has the sphalerite structure, in which each atom is surrounded tetrahedrally by four unlike atoms. Aluminum is a hypoelectronic atom, with normal valence 3 and with single-bond radius 1.248 Å. Phosphorus is a hyperelectronic atom, with normal valence 3 (resulting from occupancy of four orbitals by five electrons) and single-bond radius 1.10 Å. The calculated Al—P single-bond length (including the electronegativity correction, Sec. 7-2) is 2.31 Å, and the value for valence 3 and bond number n equal to $\frac{3}{4}$ is 2.38 Å. The observed distance, 2.35 Å, lies about midway between these values. It corresponds to bond number 0.86, which indicates transfer of 0.44 electron from phosphorus to aluminum, increasing the valence of each atom to 3.44. The difference in electronegativity of phosphorus and aluminum, 0.6, corresponds to 9 percent ionic character of the bonds; accordingly the electron transfer is approximately that required to neutralize the charge of the atoms resulting from the partial ionic character of the bonds.

As another example we may discuss the striking purple alloy Al_2Au, which has the fluorite structure, with $a_0 = 5.99$ Å. Each gold atom has eight aluminum ligates, at 2.59 Å. If gold retained its usual metallic valence, 5.56, the eight Au—Al bonds would have bond number 0.70, and the corresponding correction $-0.600 \log n = 0.093$, plus the single-bond radii 1.342 for gold and 1.248 for aluminum, with the electronegativity correction, would give the predicted Au—Al distance 2.665 Å, which is so much greater than the observed value as to eliminate the assumed valences. Agreement is obtained by assuming gold to have the valence 6.60; the corresponding radius (Table 11-3) is 1.313 Å, and the bond-number correction, for $n = 0.82$, is 0.051, leading to 2.574 Å for the Au—Al bond length.

The valence 6.60 can be achieved by a neutral gold atom (without a metallic orbital). However, in order for gold to have this valence, aluminum must have valence 3.30, or greater if significant Al—Al bonds are formed; and hence at least 0.6 electron per gold atom must have been transferred to the aluminum atoms. Indeed, the observed Al—Al distance 3.00 Å for the six aluminum ligates about each aluminum atom corresponds to $n = 0.15$, and indicates that a significant amount of valence of the aluminum atoms is used in these bonds. It is likely that about 1.5 electrons are removed from each gold atom, which would liberate the customary 0.72 metallic orbital; 0.75 electron added to each aluminum atom would increase the aluminum valence to 3.75, of which 3.30 would be used in bonds to the four gold ligates and the remainder in Al—Al bonds.

This large amount of electron transfer is not incompatible with the electroneutrality principle. The electronegativity of aluminum is 1.5, and that of gold is 2.4. The difference corresponds to 18 percent ionic character of the Au—Al bonds, which with valence 6.60 for gold would lead to the charge -1.19 on the gold atom. To restore it to neutrality 1.19 electrons would have to be transferred to two aluminum atoms.

The proposed structure provides an explanation of the very high melting point (1060°C) and large heat of formation of the compound.[40] Coffinberry and Hultgren[41] pointed out that the properties of the Al—Au alloys indicate the operation of an unusually strong attraction between aluminum atoms and gold atoms.

As an example of a compound in which electron transfer is relatively unimportant we may discuss $PtSn_2$, which also has the fluorite structure, a_0 being 6.41 Å. The normal metallic valences 6 for platinum and 4 for tin permit the formation of Pt—Sn bonds with $n = \frac{3}{4}$ and Sn—Sn bonds with $n = \frac{1}{6}$. The predicted Pt—Sn distance 2.770 Å is only slightly low, the observed distance being 2.78 Å. The predicted Sn—Sn distance for $n = \frac{1}{6}$, 3.27 Å, is somewhat larger than the observed distance, 3.205 Å, and a small amount of strain is accordingly indicated. The strain would be expected to cause a lengthening of the Pt—Sn bonds and shortening of the Sn—Sn bonds in inverse ratio to their total strengths, 6 to 1, and hence it would lead to the bond lengths 2.78 Å and 3.21 Å, in excellent agreement with experiment.

Electron transfer is especially important in the alloys of the alkali and alkaline-earth metals with hyperelectronic elements and buffer elements. In the formation of many of these alloys from the elements a very large volume contraction is observed, resulting in part from the bond-number correction of interatomic distances due to the increase in valence and in part from the decrease in single-bond radius of the hypoelectronic atom with increase in valence. Thus, although the normal radius of sodium for ligancy 12, 1.896 Å, is greater than that of lead, 1.746 Å, the replacement of one-fourth of the lead atoms in pure lead by sodium atoms, to form the phase $NaPb_3$, leads to a contraction, the bond distance decreasing from 3.492 Å to 3.446 Å. This decrease

[40] W. C. Roberts-Austen, *Proc. Roy. Soc. London* **49**, 347 (1891): **50.** 367 (1892).

[41] A. S. Coffinberry and R. Hultgren, *Am. Inst. Mining Met. Engrs.*, Tech. Publ. No. 885, **1938**.

is explained in part by the electronegativity correction and in part by electron transfer, with a little less than one electron transferred to the sodium atom.   In many other intermetallic compounds of the alkali and alkaline-earth metals the interatomic distances similarly indicate that electron transfer occurs to such an extent as to increase the valence by about one unit.

11-14. COMPOUNDS OF METALS WITH BORON, CARBON, AND NITROGEN

Some of the compounds of the metals with boron, carbon, and nitrogen have structures that can be described in a simple way as involving the arrangement of the metal atoms in closest packing or in some other simple structure and the insertion of the small nonmetal atoms into interstices of the lattice of metal atoms.[42] AlN, with the wurtzite structure, can be described in this way, the aluminum atoms having a hexagonal closest-packed arrangement with the nitrogen atoms in tetrahedral positions; this crystal can be described as involving covalent bonds between the nitrogen atom and its four aluminum neighbors. ScN, TiN, ZrN, VN, NbN, TiC, ZrC, VC, NbC, and TaC, which have the sodium chloride arrangement, contain metal atoms in cubic closest packing with nitrogen or carbon atoms in octahedral positions.   Since these first-row atoms are restricted to a maximum of four covalent bonds, it is probable that the octahedral coordination of six metal atoms about each light atom involves resonance of covalent bonds among the six positions.   The structure of Fe_4N is similar in nature; the iron atoms are in cubic closest packing, with nitrogen atoms at the centers of octahedra of six iron atoms (cubic unit, 4Fe at 000, $\frac{1}{2}\frac{1}{2}0$, $\frac{1}{2}0\frac{1}{2}$, $0\frac{1}{2}\frac{1}{2}$; N at $\frac{1}{2}\frac{1}{2}\frac{1}{2}$).

The bond lengths in these substances are the expected ones, as may be illustrated by the discussion of a representative substance with the sodium chloride arrangement, VN.   Each vanadium atom has 6 nitrogen neighbors at 2.06 Å and 12 vanadium neighbors at 2.92 Å.   The calculated values for quinquevalent vanadium and tervalent nitrogen are 2.03 Å and 2.92 Å, respectively.   In this crystal the nitrogen retains its unshared electron pair and its valence 3, whereas in others (such as AlN, analogous to AlP, discussed in the preceding section) there is electron transfer, leading to increase in its valence.

Cementite, Fe_3C, has an interesting structure, involving both octahedral and trigonal-prismatic arrangements of six iron atoms about a carbon atom (*Strukturbericht*, II, 33).   The iron boride FeB (*Strukturbericht*, III, p. 12) contains trigonal prisms of iron atoms about the boron atoms, the Fe—B distance being about 2.15 Å, which is approxi-

[42] G. Hägg, *Z. physik. Chem.* **B6**, 221 (1929); **B12**, 33 (1931).

mately equal to the sum of the covalent radii. However, each boron atom also has two boron atoms close to it, at 1.77 Å, so that B—B covalent bonds are present in the structure.

The process of forming boron-boron bonds is carried on further in aluminum boride, AlB_2, which has a very simple hexagonal structure, consisting of hexagonal layers of boron atoms, like the layers of carbon atoms in graphite, with aluminum atoms in the spaces between the layers (Fig. 11-15). The B—B bond length is 1.73 Å, corresponding to $n = 0.66$; that is, two valence electrons per boron atom are used in the B—B bonds, which are two-thirds bonds.

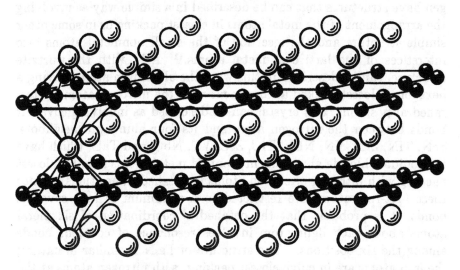

FIG. 11-15.—The structure of the hexagonal crystal AlB_2. Small circles represent boron atoms. They form hexagonal layers resembling the layers of carbon atoms in graphite. Large circles represent aluminum atoms.

The boride UB_{12} has an interesting structure,[48] related to the structures discussed in Section 10-6. It is face-centered cubic, with a_0 = 7.473, the unit cube containing 4 UB_{12}. There are B_{12} groups with cubo-octahedral structure. Each uranium atom is surrounded by 24 boron atoms, which form a regular polyhedron with six square faces and eight hexagonal faces. Each boron atom has five boron neighbors, at 1.76 Å, and two uranium neighbors, at 2.79 Å.

11-15. MOLECULES AND CRYSTALS CONTAINING METAL-METAL BONDS

The existence of the mercury-mercury bond in the mercurous ion, Hg_2^{++}, and in molecules such as mercurous chloride, Cl—Hg—Hg—Cl,

[48] F. Bertaut and P. Blum, *Compt. rend.* 330, 666 (1949).

has been recognized for decades, but until recently other examples of molecules containing metal-metal bonds had not been reported. Now a great many are known.

The complex ion $[W_2Cl_9]^{---}$ was discovered to have such a structure through the determination of the structure[44] of $K_3W_2Cl_9$. The complex ion consists of two WCl_6 octahedra sharing a face. The W—Cl bond lengths are 2.40 Å (unshared Cl) and 2.48 Å (shared Cl). The tungsten atoms are closer to the plane of the shared face than to the plane of the peripheral chlorine. The W—W distance, 2.409 Å, is less than in metallic tungsten, and about equal to the value 2.40 Å expected for a double bond between the tungsten atoms. Each tungsten atom has three valence electrons in addition to those involved in the bonds to the chlorine atoms. The tungsten-tungsten bond can be described as involving resonance among the structures $\overset{..}{W}$—$\overset{..}{W}$, $\overset{..}{W}$=$\overset{..}{W}$, W=$\overset{..}{W}$, and W≡W.

Whereas $K_3W_2Cl_9$ is diamagnetic, the closely similar substance $K_3Cr_2Cl_9$ is paramagnetic, with three unpaired electrons per chromium atom. In the $[Cr_2Cl_9]^{---}$ ion the chromium atoms are 3.12 Å apart,[45] corresponding to bond number 0.05 (that is, there is no Cr—Cr bond).

A bond between two lead atoms is found[46] in hexamethyl dilead, $Pb_2(CH_3)_6$. The Pb—Pb bond length, 2.88 ± 0.03 Å, is about that expected from the tetrahedral radius of lead, as is the Pb—C bond length, 2.25 ± 0.06 Å. This substance is, of course, similar to hexamethylethane and the corresponding compounds of silicon, germanium, and tin.

The structure reported for crystals of molybdenum dioxide and tungsten dioxide by Magnéli[47] also shows the presence of bonds between the metal atoms. These crystals have a distorted rutile structure, in which each metal atom is surrounded by an octahedron of oxygen atoms. The distortion from the ideal structure is of such a nature as to bring two molybdenum or tungsten atoms very close together to form a pair of atoms 2.48 Å apart, the corresponding edge shared by the two octahedra being greatly lengthened. The bond number calculated from this interatomic distance is 1.47, suggesting that there is an effort by each quadrivalent molybdenum or tungsten atom to use its two remaining valence electrons for the formation of a double bond with another atom of molybdenum or tungsten. The distance from

[44] C. Brosset, *Arkiv. Kemi, Mineral., Geol.* **12A**, No. 4 (1935); W. H. Watson, Jr., and J. Waser, *Acta Cryst.* **11**, 689 (1958).

[45] G. J. Wessel and D. J. W. IJdo, *Acta Cryst.* **10**, 466 (1957).

[46] H. A. Skinner and L. E. Sutton, *Trans. Faraday Soc.* **36**, 1209 (1940).

[47] A. Magnéli *Arkiv. Kemi, Mineral., Geol.* **24A**, No. 2 (1946).

the metal atom to the oxygen atom suggests resonance of about four covalent bonds among the six positions, causing the total valence of the molybdenum or tungsten atom to be approximately 6. A similar distorted rutile structure is found[48] for VO_2; the V—V distance is 2.68 Å. In the corresponding crystals molybdenite, MoS_2, and tungstenite, WS_2, however, the metal atoms are so far apart that the bond between them is weak; the Mo—Mo or W—W bond number (for six bonds formed by each atom) is only 0.12. However, the black color and metallic luster of these substances, which resemble graphite in appearance, show that there is a significant interaction between the metal atoms.

Many essentially nonmetallic crystals are known in which metal atoms approach one another to within such distances as to correspond to significantly large fractional bond numbers, and there is little doubt that many of the physical and optical properties of the crystals are essentially determined by this closeness of approach. For example, the oxygen compounds containing iron seem to have a color that is correlated with the distance between iron atoms: pseudobrookite, Fe_2TiO_5, and hematite, Fe_2O_3, with iron-iron distance 2.88 Å, are red, whereas hydrated iron oxides such as lepidocrocite, goethite, limonite, and xanthosiderite tend to be lighter in color. The mineral cubanite, $CuFe_2S_3$, contains pairs of iron-sulfur tetrahedra in which the iron-iron distance, approximately 2.5 Å, corresponds to a bond number of 0.3. It was suggested by the investigator of the crystal, Buerger,[49] that this closeness of approach of iron atoms might be related to the unusual ferromagnetism shown by this sulfide mineral.

The crystal structure of cupric acetate hydrate, $Cu_2(CH_3COO)_4 \cdot 2H_2O$, shows that the pairs of copper atoms are only 2.64 Å apart.[50] This distance corresponds to a bond with $n = 0.33$. The substance has anomalous magnetic properties that have been interpreted as representing a weak bond.[51] Similar bonds have been reported for several crystals containing Ni, Pd, and other metal atoms.

The theory of the color of dyes and other complex organic molecules has been rather well developed in recent years, and the color of these substances is reasonably well understood. However, little progress has been made in the development of a systematizing or correlating theory of the color of inorganic complexes. There is one set of substances that show especially striking coloration. This is the set of substances containing the same element in two different valence states. Substances

[48] G. Anderssen, *Acta Chem. Scand.* 10, 623 (1956).
[49] M. J. Buerger, *J.A.C.S.* 67, 2056 (1945).
[50] J. N. van Niekerk and F. R. L. Schoening, *Acta Cryst.* 6, 227 (1953).
[51] B. N. Figgis and R. L. Martin, *J. Chem. Soc.* 1956, 3837.

of this sort have been recognized for many years as having abnormally deep and intense coloration. For example, the complexes of cuprous copper with chloride ion in solution in concentrated hydrochloric acid are colorless, as is cuprous chloride itself, and the complexes of cupric copper with chloride ion are green. However, if cuprous and cupric solutions are mixed, an intensely colored brown or black solution is obtained, apparently due to complexes containing both cuprous and cupric copper. Similarly, tervalent antimony chloride and quinquevalent antimony chloride are colorless, but a mixture of the two has a deep brown or black color. Crystals of $(NH_4)_2SbCl_6$, a black substance, have been investigated by x-rays by Elliott[52] and shown to have a structure indistinguishable from that of potassium chlorostannate. Moreover, the crystals are diamagnetic, so that the complexes cannot be $[SbCl_6]^{--}$, which would necessarily be paramagnetic because of the presence of an odd number of electrons, but must be alternately $[SbCl_6]^{---}$ and $[SbCl_6]^{-}$. Crystals of cesium aurous auric chloride, $Cs_2AuAuCl_6$, are also intensely black in color.

Another example of the phenomenon is often observed in the chemical laboratory when a solution containing ferrous ion is precipitated with alkali. Ferrous hydroxide is white, and ferric hydroxide is brown. When a ferrous solution is precipitated, however, the initially white precipitate is immediately partially oxidized by atmospheric oxygen to form a ferrous ferric hydroxide, which is black in color (or deep green when finely divided).

A few years ago it was pointed out to me by Sterling Hendricks that ordinary black mica, biotite, which has an intensely black color, owes this color to the presence of iron in both the ferrous and ferric oxidation state. Black tourmalines also usually contain both ferrous and ferric iron. Another intensely black mineral, with black streak, is ilvaite, with composition $Ca(Fe^{++})_2Fe^{+++}(SiO_4)_2OH$.

Molybdenum blue and tungsten blue, which have an intense deep-blue coloration, have the formulas $MoO_{2.5-3}$ and $WO_{2.5-3}$. The tungsten bronzes also contain tungsten in an intermediate valence state; their formulas lie between the limits $Na_2W_2O_6$ and $Na_2W_3O_9$. Many metal oxides, such as Fe_3O_4, U_3O_8, and Pr_6O_{11}, may owe their black color to this phenomenon.

The ion $[Mo_6Cl_8]^{++++}$ is found in solutions of molybdenum dichloride, Mo_6Cl_{12}, and in the crystals[53] $[Mo_6Cl_8](OH)_4 \cdot 14H_2O$, $[Mo_6Cl_8]Cl_4 \cdot 8H_2O$, and $(NH_4)_2[Mo_6Cl_8]Cl_6 \cdot 2H_2O$. Its structure is shown in Figure 11-16. Each molybdenum atom uses two of its six valence

[52] N. Elliott, *J. Chem. Phys.* **2**, 298 (1934).

[53] C. Brosset, *Arkiv. Kemi, Mineral., Geol.* A20, (1945); A?2 (1946); P. A. Vaughan, *Proc. Nat. Acad. Sci. U.S.* **36**, 461 (1950).

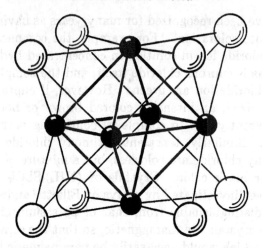

FIG. 11-16.—The structure of the complex ion $[Mo_6Cl_8]^{++++}$.

electrons in the bonds to the chlorine atoms and the other four to form single Mo—Mo bonds along the edges of the Mo_6 octahedron. The Mo—Mo bond length, 2.63 Å, is close to the single-bond value 2.592 Å derived from the metal (Table 11-1).

The ions $[Nb_6Cl_{12}]^{++}$, $[Ta_6Cl_{12}]^{++}$, and $[Ta_6Br_{12}]^{++}$ have a related structure,[54] shown in Figure 11-17. The number of valence electrons is such that the bonds along the edges of the Nb_6 and Ta_6 octahedra have bond number $\frac{2}{3}$. The observed Nb—Nb and Ta—Ta bond lengths, 2.85 Å and 2.90 Å, respectively, agree moderately well with the expected value 2.79 Å.

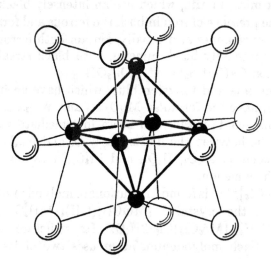

FIG. 11-17.—The structure of the complex ion $[Ta_6Cl_{12}]^{++}$.

[54] P. A. Vaughan, J. H. Sturdivant, and L. Pauling, *J.A.C.S.* **72**, 5477 (1950).

An interesting discovery was made by Powell and Ewens,[55] who determined the crystal structure of iron enneacarbonyl, $Fe_2(CO)_9$, finding the configuration of the molecule to be that shown in Figure 11-18, with a threefold axis of symmetry. Six carbonyl groups are attached to one or the other of the iron atoms; the other three are bonded to both iron atoms, and thus have a structure similar to that in ketones. The iron atoms can be considered to be trivalent. The observed diamagnetism of the substance shows that the spins of the odd electrons of the two iron atoms are opposed; this suggests strongly that there is a covalent bond between the two iron atoms. The Fe—Fe distance, 2.46 Å, is compatible with this idea. Hence in this substance each iron atom forms seven bonds (d^3sp^3), six with carbon atoms and one with the other iron atom; two unshared electron pairs occupy the two remaining $3d$ orbitals of each iron atom.

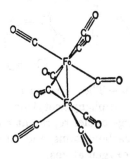

Fig. 11-18.—The structure of di-iron enneacarbonyl, $Fe_2(CO)_9$.

X-ray structure determinations[56] have shown that the molecules $Mn_2(CO)_{10}$ and $Re_2(CO)_{10}$ contain metal-metal bonds with Mn—Mn = 2.93 Å and Re—Re = 3.02 Å, these values being about 0.6 or 0.5 Å longer than the single-bond distances. There are no bridging carbonyls. The other five octahedral positions about each metal atom are occupied by the carbonyl groups, on a straight line out from the metal atom. The two octahedra are twisted into the staggered configuration. These are the first authentic structures of this type in which the two halves of the molecule are held together only by metal-metal bonds. The diamagnetism that is observed is compatible with the structures.

The molecule of diphenylacetylene dicobalt hexacarbonyl has been shown by an x-ray investigation[57] to have the structure shown in Figure 11-19. Each cobalt atom forms six bonds, directed toward the corners

[55] H. M. Powell and R. V. G. Ewens, *J. Chem. Soc.* **1939**, 286.

[56] L. F. Dahl, E. E. Ishishi, and R. E. Rundle, *J. Chem. Phys.* **26**, 1750 (1957).

[57] W. G. Sly, Ph.D. thesis, Calif. Inst. Tech., 1957.

of a distorted octahedron. Each of the two acetylenic carbon atoms forms four single bonds, three of which are in the central tetrahedron: one with the other acetylenic carbon and two with the two cobalt atoms. All bond lengths are reasonable; in the tetrahedron they are C—C = 1.46 Å, Co—C = 1.95 ± 0.06 Å, Co—Co = 2.47 Å. If the Co—CO bonds are taken to be double bonds (their length is 1.75 ± 0.05 Å), all nine orbitals of each cobalt atom are used in bond formation, and also all nine valence electrons.

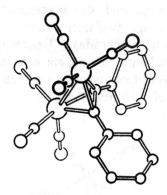

Fig. 11-19.—The structure of dicobalt hexacarbonyl diphenylacetylene, $Co_2(CO)_6C_2(C_6H_5)_2$. Large circles represent cobalt atoms, small circles carbon atoms, and circles of intermediate size oxygen atoms.

We may anticipate that in the next decade many more substances will be found in which metal-metal bonds play an important part.

11-16. THE STRUCTURES OF SULFIDE MINERALS

The sulfide minerals have structures based largely upon covalent bonds between the sulfur atoms and other atoms in the substances. In some of the minerals the bonds resonate among alternative positions, and in some there are also metal-metal bonds, conferring metallic properties, especially metallic luster, upon them.

Sphalerite and wurtzite, the two common forms of zinc sulfide,[58] have the tetrahedral structures shown in Figures 7-5 and 7-6. Pure zinc sulfide is colorless. The minerals are usually yellow, brown, or black, the color probably being due to imperfections and impurities. The luster is not metallic, but resinous or adamantine.

Galena, PbS, is an example of a mineral with metallic luster. The atoms are ordered in the sodium chloride arrangement. Each lead

[58] Other forms, corresponding to more complex arrangements of the tetrahedral layers, also occur in nature: C. Frondel and C. Palache, *Am. Mineralogist* **35**, 29 (1950).

atom has six sulfur neighbors at 2.96 Å and twelve lead neighbors at 4.19 Å. The bond numbers, calculated with use of the metallic single-bond radius for lead, are 0.23 and 0.10, respectively. Hence the lead atom forms a total of 1.38 covalent bonds to sulfur and 1.20 to other lead atoms, its covalence being 2.58. The metallic luster may be attributed to the lead-lead bonds.

Alabandite, MnS, has the same atomic arrangement. Its luster is

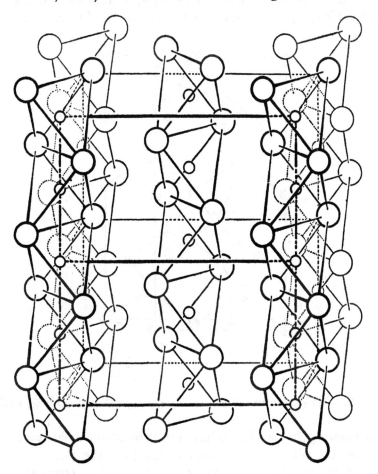

Fig. 11-20.—The structure of the crystal SiS_2. Small circles represent silicon atoms, large circles sulfur atoms (*Strukturbericht*).

not metallic, but dull. The Mn—S bond length, 2.61 Å, shows that the manganese atom has the ^6S structure described in Section 7-9.

The structure of the fibrous synthetic substance silicon disulfide,[59] SiS_2, is shown in Figure 11-20. This structure illustrates a difference

[59] A. Zintl and K. Loosen, *Z. physik. Chem.* A174, 301 (1935); W. Büssem, H. Fischer, and E. Gruner, *Naturwissenschaften* 23, 740 (1935).

between sulfides and oxides that can be explained by the smaller amount of partial ionic character for M—S bonds than for M—O bonds. It will be pointed out in Chapter 13 that the repulsion of the positive charges of the silicon atoms in silicon dioxide makes structures in which the SiO_4 tetrahedra share edges or faces less stable than those in which they share only corners with one another. In SiS_2, on the other hand, the SiS_4 tetrahedra share edges, to form long chains. The Si—S bond length, 2.16 Å, agrees with the calculated single-bond value,

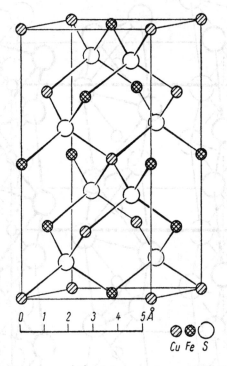

Cu Fe S

Fig. 11-21.—The structure of the tetragonal crystal chalcopyrite, $CuFeS_2$.

2.17 Å, showing that the bonds have little double-bond character. The charge on the silicon atoms is calculated from the electronegativity difference of the atoms, assuming no double-bond character, to be +0.44. The repulsion of the charges causes a distortion of the SiS_4 tetrahedra such that the shared edges are somewhat shorter than the unshared edges (shared, 3.32 Å; unshared, 3.56 and 3.70 Å).

Many sulfide minerals have structures closely related to those of sphalerite and wurtzite. Chalcopyrite, $CuFeS_2$, is an example (Fig. 11-21). Its structure[60] is a tetragonal superstructure of sphalerite, with the copper and iron atoms in the zinc positions of sphalerite.

[60] L. Pauling and L. O. Brockway, *Z. Krist.* **82,** 188 (1932).

Enargite, Cu_3AsS_4, has a structure that is a superstructure of the wurtzite arrangement.[61]   The sulfur atoms are in the same positions as in wurtzite, and the atoms of copper and arsenic replace those of zinc in an ordered way, so as to give discrete AsS_4 groups (Fig. 11-22). The observed As—S bond length, 2.22 Å, agrees exactly with the calculated value for a single bond, 2.22 Å (from the covalent radii, with the correction for electronegativity difference).   The Cu—S bond length, 2.32 Å, corresponds to bond number about 0.7 (the appropriate

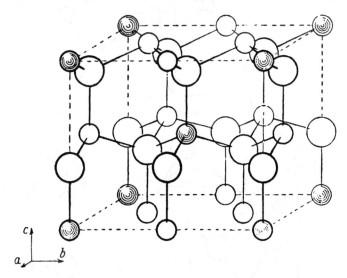

Fig. 11-22.—The structure of the orthorhombic crystal enargite, Cu_3AsS_4. Large circles represent sulfur atoms, small open circles copper atoms, and small shaded circles arsenic atoms. The structure is a superstructure of wurtzite.

single-bond radius of copper is 1.23 Å). Approximately the same Cu—S bond length is found in other copper sulfide minerals. The copper-sulfur bonds have only a small amount of ionic character, and the conclusion may be drawn that the electric charge of the copper atom is negative, probably close to -1.

The mineral sulvanite, Cu_3VS_4, has been found to have a surprising structure.[62]   The crystal is cubic, with 1 Cu_3VS_4 per unit cube, with edge $a_0 = 5.37$ Å, and it was expected that the structure would be a superstructure of sphalerite, which has 4 ZnS in a cubic unit with $a_0 = 5.41$ Å. In fact, the four sulfur atoms and the three copper atoms occupy the corresponding positions in the structure (Fig. 11-23),

[61] L. Pauling and S. Weinbaum, *Z. Krist.* **88**, 48 (1934).
[62] L. Pauling and R. Hultgren, *Z. Krist.* **84**, 204 (1933).

so that each sulfur atom is bonded to three copper atoms at three of the corners of a nearly regular tetrahedron, but the vanadium atom is not at the fourth corner of the tetrahedron; it is in the negative position to this corner. The V—S bond length, 2.19 Å, is that for a single bond, and the Cu—S bond length, 2.29 Å, corresponds to bond number about 0.7.

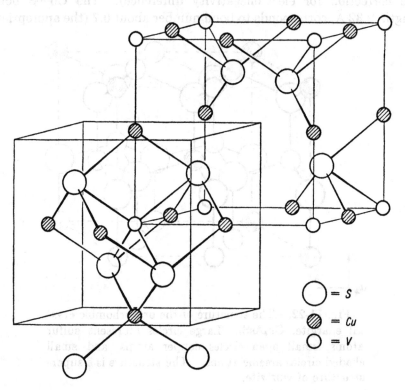

$\bigcirc = S$
$\oslash = Cu$
$\bigcirc = V$

FIG. 11-23.—The structure of the cubic crystal sulvanite, Cu_3VS_4.

The selection of this structure, in which the V—S—Cu bond angle has the value 70°32′, rather than the sphalerite superstructure with tetrahedral angles, is surprising. It is likely that in this sulfide mineral, as well as in the others, the sulfur atom is to be described as

$:\overset{+}{S}{\diagup\!\!\!\!—}$, with one unshared electron pair and three bonds. In sulvanite

one bond is formed with the vanadium atom and the other two resonate among the positions to the three copper atoms. The unshared pair projects into the hole in the crystal, where the fourth metal atom would be in sphalerite. The bond orbitals of sulfur may have enough *d* and *f* character to permit the small bond angle to be assumed without much strain.

The structure is also stabilized by the formation of metal-metal bonds. Each vanadium atom has six copper neighbors at 2.685 Å, corresponding to bond number 0.3. Hence each vanadium atom forms four single bonds with the four adjacent sulfur atoms and six one-third bonds with the six adjacent copper atoms. One electron has been

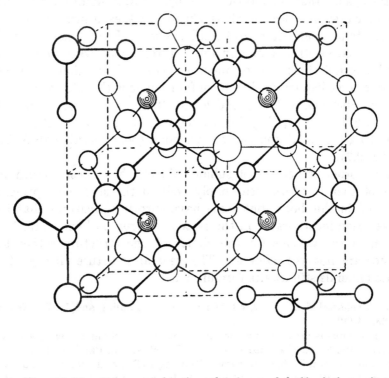

Fig. 11-24.—Diagram showing the forward half of the unit cube of binnite. Large circles represent sulfur atoms, small open circles copper atoms, and small shaded circles arsenic atoms. Bonds between adjacent atoms are indicated. Note that there are two kinds of copper atoms and two kinds of sulfur atoms.

transferred to it, a hypoelectronic atom, from the hyperelectronic copper atoms (Sec. 11-13). The amounts of ionic character of the bonds (V—S, 18 percent; Cu—S, 9 percent; V—Cu, 3 percent) are such that the electric charges of the atoms are changed from the formal values −1 for vanadium, −1 for copper, and +1 for sulfur to −0.22 for vanadium, −0.78 for copper, and +0.64 for sulfur.

A somewhat more complex structure is that of the tetrahedrite minerals. Tetrahedrite and binnite (tennantite) have compositions approximating the formulas $Cu_{10}Zn_2Sb_4S_{13}$ and $Cu_{10}Fe_2As_4S_{13}$, respectively. Their structure[63] is shown in Figure 11-24. It is closely related to the sphalerite structure. In a large cube ($a_0 = 10.19$ Å in binnite) containing 32 ZnS, 8 Zn are replaced by arsenic or antimony atoms and the other 24 by copper atoms (with zinc and iron in apparently random substitution for copper; two bipositive atoms are needed per formula). The sulfur atoms occupy only 24 of the 32 sphalerite positions, such that As and Sb have ligancy 3; in addition, two sulfur atoms are present in the positions 000 and $\frac{1}{2} \frac{1}{2} \frac{1}{2}$, each surrounded by six copper atoms in an octahedral configuration. Each arsenic atom has an unshared electron pair and forms three bonds with sulfur atoms (bond length 2.21 Å). The copper atoms are of two kinds. Those of one kind have ligancy 4; they form four bonds with sulfur atoms, with bond number about 0.75 (bond length 2.28 Å). Those of the other kind have ligancy 3; they form two single bonds with sulfur atoms (bond length 2.23 Å) and one weaker bond (bond length 2.29 Å, bond number 0.7).

Many other structures of sulfide minerals have been determined. Most of them conform reasonably well to the structural principles described in this book, but some have surprising features that have not yet been incorporated in the system of structural chemistry,[64] and in general the reasons for the choice of one structure rather than another are not yet evident. The general structure theory of the sulfide minerals still awaits formulation.

[63] F. Machatschki, *Z. Krist.* **68**, 204 (1928); L. Pauling and E. W. Neuman. *ibid.* **88**, 54 (1934).

[64] An example is provided by the related structures of the minerals sylvanite, calaverite, and krennerite, discussed by G. Tunell and L. Pauling, *Acta Cryst.* **5**, 375 (1952). Sylvanite has the composition $AgAuTe_4$, and calaverite and krennerite $AuTe_2$, with some substitution of Ag for Au. In each of the three structures the gold and silver atoms ligate six tellurium atoms octahedrally. The bond lengths are not equal, however; two bonds are single bonds, and the other four are weaker (in calaverite their bond number is 0.35). As a first approximation the ligation of the gold atoms can be described as involving the square quadricovalent dsp^2 bonds of tripositive gold (Chap. 5) with two bonds extending toward two of the octahedral positions and two resonating among the other four.

CHAPTER 12

The Hydrogen Bond

12-1. THE NATURE OF THE HYDROGEN BOND

IT was recognized some decades ago that under certain conditions
an atom of hydrogen is attracted by rather strong forces to two atoms,
instead of only one, so that it may be considered to be acting as a bond
between them. This is called the *hydrogen bond*.[1] The bond was for
some time thought to result from the formation of two covalent bonds
by the hydrogen atom, the hydrogen fluoride ion $[HF_2]^-$ being assigned the structure $[:\ddot{F}:H:\ddot{F}:]^-$. It is now recognized that the hydrogen atom, with only one stable orbital (the $1s$ orbital), can form
only one covalent bond, that the hydrogen bond is largely ionic in
character, and that it is formed only between the most electronegative
atoms. A detailed discussion of its nature is given in the following
sections.

Although the hydrogen bond is not a strong bond (its bond energy,
that is, the energy of the reaction $XH + Y \rightarrow XHY$, lying in most
cases in the range 2 to 10 kcal/mole), it has great significance in determining the properties of substances. Because of its small bond energy
and the small activation energy involved in its formation and rupture,
the hydrogen bond is especially suited to play a part in reactions oc-

[1] Other names, such as hydrogen bridge, have also been used.
A detailed discussion of the hydrogen bond is given in the book by G. C.
Pimentel and A. L. McClellan, *The Hydrogen Bond*, W. H. Freeman Co., San
Francisco, 1959. Many excellent review articles have been published; among
them are E. N. Lassettre, *Chem. Revs.* 20, 259 (1937); H. Hoyer, *Z. Elektrochem.*
49, 97 (1943); J. Donohue, *J. Phys. Chem.* 56, 502 (1952); A. R. Ubbelohde and
K. J. Gallagher, *Acta Cryst.* 8, 71 (1955); G. M. Badger, *Rev. Pure and App.*
Chem. (Australia) 7, 55 (1957); C. A. Coulson, *Research* (London) 10, 149 (1957);
M. Magat, *Nuovo cimento* 10, 416 (1953); D. Sokolov, *Tagungsber. der chem. Ges.*
Deutsch. Dem. Rep. 1955, 10.

449

curring at normal temperatures. It has been recognized that hydrogen bonds restrain protein molecules to their native configurations, and I believe that as the methods of structural chemistry are further applied to physiological problems it will be found that the significance of the hydrogen bond for physiology is greater than that of any other single structural feature.

The first mention of the hydrogen bond was made by Moore and Winmill,[2] who assigned to trimethylammonium hydroxide the structure

$$
\begin{array}{c}
CH_3 \\
| \\
CH_3\!-\!N\!-\!H\!-\!OH, \\
| \\
CH_3
\end{array}
$$

accounting in this way for the weakness of this substance as a base as compared with tetramethylammonium hydroxide. Recognition of the importance of the hydrogen bond and of its extensive occurrence was made by Latimer and Rodebush,[3] who used this concept in the discussion of highly associated liquids, such as water and hydrogen fluoride, with their abnormally high dielectric constant values, of the small ionization of ammonium hydroxide, and of the formation of double molecules by acetic acid. The number of molecules recognized as containing hydrogen bonds has been greatly increased by spectroscopic and crystal-structure studies and by analysis of physicochemical information.[4]

With the development of the quantum-mechanical theory of valence it was recognized[5] that a hydrogen atom, with only one stable orbital, cannot form more than one pure covalent bond[6] and that the attraction

[2] T. S. Moore and T. F. Winmill, *J. Chem. Soc.* **101**, 1635 (1912); see also P. Pfeiffer, *Ann. Chem.* **398**, 137 (1913).

[3] W. M. Latimer and W. H. Rodebush, *J.A.C.S.* **42**, 1419 (1920). G. N. Lewis (*Valence and the Structure of Atoms and Molecules*, Chemical Catalog Co., New York, 1923, p. 109) mentions that the idea was used by Huggins in an unpublished work; see also M. L. Huggins, *Phys. Rev.* **18**, 333 (1921); **19**, 346 (1922).

[4] This method was applied mainly by N. V. Sidgwick (*The Electronic Theory of Valency*, Clarendon Press, Oxford, 1927), who used it in the discussion of compounds such as the enolized β-diketones; see also Lassettre, *loc. cit.* (1).

[5] L. Pauling, *Proc. Nat. Acad. Sci. U.S.* **14**, 359 (1928).

[6] The bond-forming power of the outer orbitals of the hydrogen atom is negligibly small. It has been suggested by several authors that use may be made of an *L* orbital of hydrogen for formation of a second covalent bond. However, in case that a bond A—H with small ionic character is formed the proton is shielded almost completely by its half of the shared electron pair, and it has accordingly no power to attract an *L* electron. Only if the A—H bond were largely ionic would there occur appreciable attraction for an *L* electron, and under this circumstance the proton could use its 1*s* orbital for covalent bond for-

of two atoms observed in hydrogen-bond formation must be due largely to ionic forces. This conception of the hydrogen bond leads at once to the explanation of its important properties.

First, the hydrogen bond is a bond by hydrogen between *two* atoms; the coordination number of hydrogen does not exceed two.[7] The positive hydrogen ion is a bare proton, with no electron shell about it. This vanishingly small cation would attract one anion (which we idealize here as a rigid sphere of finite radius—see Chap. 13) to the equilibrium internuclear distance equal to the anion radius, and could then similarly attract a second anion, as shown in Figure 12-1, to form

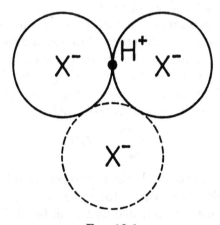

Fɪɢ. 12-1.

a stable complex. A third anion, however, would be stopped by anion-anion contacts, which prevent its close approach to the proton. From the ionic point of view the coordination number of hydrogen is thus restricted to the value two, as is observed in general.[8]

---

mation with the atom B of the group A—H—B (during the ionic phases of the A—H bond), and so would not need to call on the unstable L orbital.

[7] In some circumstances a hydrogen atom with some residual positive charge, as in the ammonium ion, is attracted by the resultant electric field of two or more negative ions. The corresponding weak interactions, although similar in nature to those involved in hydrogen-bond formation, are not conveniently included in this category.

[8] It was shown by G. A. Albrecht and R. B. Corey, *J.A.C.S.* **61**, 1087 (1939), that the crystal structure of glycine is such as to indicate strongly that one of the hydrogen atoms of the —NH_3^+ group is attracted about equally by two oxygen atoms, forming a bifurcated hydrogen bond N—H $\diagdown\!\!\!\!\genfrac{}{}{0pt}{}{O}{O}$. The structure has been refined by R. E. Marsh, *Acta Cryst.* **10**, 814 (1957), and the position of

Second, only the most electronegative atoms should form hydrogen bonds, and the strength of the bond should increase with increase in the electronegativity of the two bonded atoms. Referring to the electronegativity scale, we might expect that fluorine, oxygen, nitrogen, and chlorine would possess this ability, to an extent decreasing in this order. It is found empirically that fluorine forms very strong hydrogen bonds, oxygen weaker ones, and nitrogen still weaker ones. Although it has the same electronegativity as nitrogen, chlorine has only a very small hydrogen-bond-forming power; this may be attributed to its large size (relative to nitrogen), which causes its electrostatic interactions to be weaker than those of nitrogen.

Increasing the electronegativity of an atom increases its power of forming hydrogen bonds. The ammonium ion and its derivatives, such as $[RNH_3]^+$, form stronger hydrogen bonds than ammonia or normal amines. The phenols form stronger hydrogen bonds than aliphatic alcohols because of the increase in electronegativity of the oxygen atom resulting from resonance with structures such as

$$:\!\left\langle\!\bigcirc\!\right\rangle\!\!=\!\overset{+}{\underset{\cdot\cdot}{O}}H$$

In almost all hydrogen bonds the hydrogen atom is nearer to one of the two adjacent electronegative atoms than to the other. In ice, for example, the distance between two hydrogen-bonded oxygen atoms is 2.76 Å, and the proton has been shown by neutron diffraction to be 1.00 Å from one oxygen atom and 1.76 Å from the other (Sec. 12-4). Also in diaspore, $AlHO_2$, the oxygen-oxygen distance is 2.650 Å and the oxygen-hydrogen distances, determined by neutron diffraction, are 1.005 Å and 1.68 Å (Sec. 12-7).

The amount of partial ionic character expected for the O—H bond from the electronegativity difference of the atoms is 39 percent. Hence the $1s$ orbital of the hydrogen atom is liberated from use in covalent-bond formation with the adjacent oxygen atom to the extent of 39 percent, and hence available for formation of a fractional covalent bond with the more distant oxygen atom of the hydrogen-bonded group O—H···O; the hydrogen bond in ice can be described as involving resonance among the three structures A, B, and C:

$$
\begin{array}{lll}
A & O\!-\!H & :O \\
B & O: & H^+ \quad :O \\
C & O: & H\!-\!\!-\!\!-O
\end{array}
$$

the proton has been verified by neutron diffraction by J. H. Burns and H. A. Levy, *Am. Cryst. Ass'n Meeting* June (1958). The bifurcated hydrogen bond seems to be present also in crystals of iodic acid, HIO_3 (M. T. Rogers and L. Helmholz, *J.A.C.S.* **63**, 278 [1941]), and in nitramide, NH_2NO_2 (C. A. Beevers and A. S. Trotman-Dickenson, *Acta Cryst.* **10**, 34 [1957]).

(Here the dashes represent pure covalent bonds.)   A rough idea of the amount of covalent bonding to the more distant oxygen atom can be obtained by use of the equation relating interatomic distance to bond number for fractional bonds, Equation 7-7.   The long $H \cdots O$ distance in ice is 0.80 Å greater than the single-bond distance, corresponding to bond number 0.05.   Hence we conclude that for the hydrogen bonds in ice the three structures A, B, and C contribute about 61 percent, 34 percent, and 5 percent, respectively.[9]   The contribution of structure C for diaspore is similarly calculated from the distance 1.68 Å to be 6 percent.   The shortest reported $O—H \cdots O$ bonds have oxygen-oxygen distance 2.40 Å (Sec. 12-7).   This is only 0.06 Å greater than the distance expected for two half-bonds, 2.34 Å, and it is likely that symmetrical hydrogen bonds between oxygen atoms are present in a few substances.

In general the $A—H \cdots B$ hydrogen bond can be taken to be approximately linear; for example, in diaspore the angle between the internuclear lines A—H and A- - -B has been found by neutron diffraction (Sec. 12-7) to be 12.1°.   An estimate of the strain energy for deviation of the $O—H \cdots O$ bonds with oxygen-oxygen distance 2.76 Å (as in ice) has been made;[10] it is that the strain energy of bending the hydrogen bond is $0.003\delta^2$ kcal/mole, where δ is the deviation, in degrees, from a straight angle of the O—H and $H \cdots O$ bonds at the hydrogen atom.

The strain energy of extending or compressing the $O—H \cdots O$ bond with length 2.76 Å (as in ice) has been calculated from the compressibility of ice to be $12 (D - D_0)^2$ kcal/mole, in which $D - D_0$ is the change in length of the bond in Å (Sec. 12-8).

In all molecules and crystals containing hydrogen bonds $A—H \cdots B$ the angles between the bond A—H and other bonds formed by atom A correspond to the principles discussed in Chapter 3; for example, in alcohols (Sec. 12-5) the angle $R—O—H \cdots$ is close to 105°.   Also, in general the angles between the weak $H \cdots B$ bond and the other bonds formed by atom B are those that would be predicted for a covalent H—B bond.   There are some exceptions to this rule, however; an example is urea, in which two of the $O \cdots H—N$ bonds formed by the oxygen atom are in the plane of the molecule, as expected for the struc-

ture $\diagdown \diagup C{=}O\!:$, and the other two are out of the plane.   These hydrogen

bonds are very weak; the observed $O \cdots H—N$ distance 3.03 Å corresponds (Equation 7-7) to only 1.7 percent contribution of the covalent long-bond structure C.

In general a hydrogen bond $A—H \cdots B$ can be considered to involve

[9] L. Pauling, *J. chim. phys.* **46**, 435 (1949).
[10] L. Pauling and R. B. Corey, *Fortschr. Chem. org. Naturstoffe* **11**, 180 (1954)

an electron pair of atom B.   An exception is urea, in which the oxygen atom, with two electron pairs available, forms four hydrogen bonds. Another exception is ammonia; one unshared electron pair of the nitrogen atom is involved in the formation of three hydrogen bonds.   It will be seen in the following section that these three N—H$\cdots$N hydrogen bonds affect the physical properties of the substance to about the same extent that those of hydrogen fluoride are affected by one F—H$\cdots$F bond.

12-2. THE EFFECT OF THE HYDROGEN BOND ON THE PHYSICAL PROPERTIES OF SUBSTANCES

It is the hydrogen bond that determines in the main the magnitude and nature of the mutual interactions of water molecules and that is consequently responsible for the striking physical properties of this uniquely important substance.   In this section we shall discuss the melting point, boiling point, and dielectric constant of water and related substances; other properties of water are treated later (Sec. 12-4).

For the sequence of related substances H_2Te, H_2Se, and H_2S the melting points and boiling points show the decreasing courses expected in view of the decreasing molecular weights and van der Waals forces[11] (Fig. 12-2).   The continuation of the sequence in the way indicated by the values for the noble gases would lead to the expectation of values of about $-100°C$ and $-80°C$, respectively, for the melting point and boiling point of water.   The observed values of these quantities are very much higher; this is the result of the formation of hydrogen bonds, which have the extraordinary effect of doubling the boiling point of the substance on the Kelvin scale.

The melting points and boiling points of ammonia and hydrogen fluoride are also considerably higher than the values extrapolated from the sequences of analogous compounds, the effects being, however, somewhat smaller than for water.   This decrease for ammonia is due in part to the smaller electronegativity of nitrogen than of oxygen and in part to the presence in the ammonia molecule of only one unshared electron pair, which must serve as the source of attraction for the protons involved in all the hydrogen bonds formed with the N—H groups of other molecules.   Hydrogen fluoride can form only one-half as many hydrogen bonds as water, and, although its F—H$\cdots$F bonds are stronger than the O—H$\cdots$O bonds in water and ice, the resultant effects are smaller for this substance than for water.

[11] The van der Waals forces for these substances are due mainly to dispersion forces, which decrease with decrease in atomic number for atoms of similar structure.   London's calculations (F. London, *Z. Physik* **63**, 245 (1930) have shown the interaction of permanent dipoles to contribute only a small amount to the van der Waals forces for a substance such as hydrogen chloride.

It is worthy of note that from the existence of both the melting point effect and the boiling point effect the deduction can be made that some of the hydrogen bonds existing in crystals of hydrogen fluoride, water, and ammonia are ruptured on fusion and that others (more than one-half of the total) are retained in the liquid, even at the boiling point, and are then ruptured on vaporization.   Indeed, the very strong hydrogen bonds of hydrogen fluoride tend to hold the molecules together even in the vapor, which is partially polymerized.

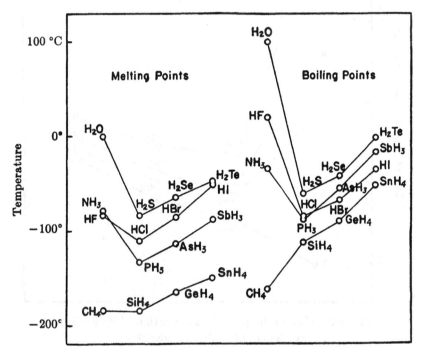

Fig. 12-2.—The melting points and boiling points of isoelectronic sequences of hydride molecules.

Methane, with no power to form hydrogen bonds, shows the expected very low boiling point.   Its melting point, however, lies about 20° higher than the expected value; the explanation of this is not known.

Properties that are related to melting point and boiling point also show the effect of hydrogen-bond formation; this is illustrated for the molal heat of vaporization in Figure 12-3.[12]

The abnormally high dielectric-constant values observed for certain liquid substances, such as water and ammonia, were attributed by Latimer and Rodebush to continued polymerization through hydrogen-

[12] Figures similar to 12-2 and 12-3 were published by F. Paneth in his volume of George Fisher Baker Lectures, *Radio-Elements as Indicators*, McGraw-Hill Book Co., New York, 1928.

bond formation. In Figure 12-4 the comparison is made of the values of the dielectric constant of liquid substances, measured[13] at 20°C, and the values of the electric dipole moments of the molecules of the sub-

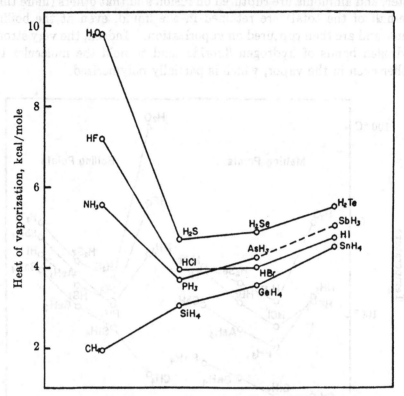

FIG. 12-3.—The enthalpies of vaporization of isoelectronic sequences of hydride molecules.

stances in the gaseous state or in solution in nonpolar solvents. It is seen that most of the points lie close to a simple curve, represented in the figure.[14] The points for methylamine, ammonia, the alcohols, water, hydrogen peroxide, hydrogen fluoride, and hydrogen cyanide, however, lie above the curve. For all of these substances except the

[13] The value shown for the dielectric constant of liquid hydrogen fluoride at 20°C, 65, is extrapolated from measurements made at 0°C and lower temperatures. The value 87 for the dielectric constant of hydrogen peroxide is obtained by linear extrapolation of the value found for a 46 percent aqueous solution of the substance and that for pure water.

[14] By consideration of the molal volume and by other refinements a still closer correlation of dielectric constant of liquids and molecular dipole moments for substances which do not form hydrogen bonds might be achieved; the simple comparison made above is, however, suitable to our purpose.

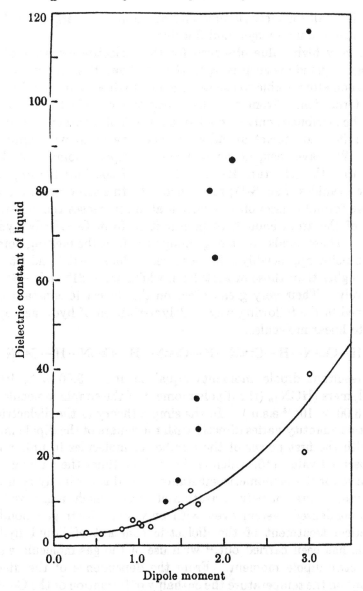

Fig. 12-4.—The dielectric constants of polar liquids plotted against the dipole moments of the gas molecules. From left to right the substances shown by open circles are AsH₃, HI, PH₃, HBr, H₂S, CHCl₃, HCl, (C₂H₅)₂O, SOCl₂, SO₂, SO₂Cl₂, (CH₃)₂CO, CH₃NO₂; by solid circles, CH₃NH₂, NH₃, CH₃OH, C₂H₅OH, H₂O, HF, H₂O₂, HCN.

last hydrogen-bond formation is expected, to an extent, indeed, which is roughly related to the magnitude of the deviation from the curve, which is small for methylamine and ammonia and larger for the substances containing oxygen and fluorine.[15]

The very high value observed for the dielectric constant of liquid hydrogen cyanide is surprising in that it shows that in this substance the carbon atom is able to use its attached hydrogen atom in hydrogen-bond formation. From the electronegativity scale the C—H bond would be attributed only a small amount of ionic character, insufficient to permit it to attract an adjacent negative atom with appreciable force. We have seen, however, from the dipole moment of the CN group that the structure R—C$^+$::N:$^-$ is of considerable importance for the cyanides (Sec. 8-1); resonance with this structure, involving a positive formal charge on the carbon atom, increases the electronegativity of the atom enough to permit it to form C—H$\cdots$N hydrogen bonds. These bonds are strong enough to affect the melting point and boiling point appreciably; the observed values, $-12°$ and $25°C$, are much higher than those of acetylene, which are $-81°$ and $-84°C$, respectively. Their very great effect on the dielectric constant can be explained in the following way. Polymerization of hydrogen cyanide leads to linear molecules,

$$H—C\equiv N\cdots H—C\equiv N\cdots H—C\equiv N\cdots H—C\equiv N\cdots H—C\equiv N,$$

with resultant dipole moments equal to about $3.00\ n \times 10^{-18}$ D for polymers $(HCN)_n$ (the dipole moment of the simple molecule HCN being 3.00×10^{-18} e.s.u.). In the simple theory of the dielectric constant this quantity varies directly with the square of the dipole moment and with the first power of the number of molecules in unit volume; the observed value 116, which is about three times the value given by the curve for the monomeric substance, would accordingly result from an average degree of polymerization of three, which might well occur in the condensed system even with only weak hydrogen bonds. A theoretical treatment of the dielectric constant of liquid hydrogen cyanide has been carried out,[16] with use of the gas-molecule value of the electric dipole moment. From the dependence of the dielectric constant on the temperature the enthalpy of formation of the C—H$\cdots$N bond was evaluated as 4.6 kcal/mole.

It has been pointed out[17] that data on the density of hydrogen cyanide gas show the presence of polymers $(HCN)_n$. The enthalpy of the

[15] A quantitative theoretical treatment of the dielectric constant of water and alcohols in terms of hydrogen-bond formation and making use of the gas-molecule values of the electric dipole moment has been published by G. Oster and J. G. Kirkwood, *J. Chem. Phys.* 11, 175 (1943). An alternative treatment of water has been made by L. Pauling and P. Pauling (unpublished).

[16] R. H. Cole, *J.A.C.S.* 77, 2012 (1955).

[17] W. F. Giauque and R. A. Ruehrwein, *J.A.C.S.* 61, 2626 (1939).

hydrogen bond in the dimer, H—C≡N···H—C≡N, was evaluated as 3.28 kcal/mole, and the sum of the enthalpies of the two bonds in the trimer, H—C≡N···H—C≡N···H—C≡N, as 8.72 kcal/mole. The increase in hydrogen-bond strength with increasing degree of polymerization is interesting, and can be given a simple interpretation in terms of resonance.

Crystals of hydrogen cyanide have been shown[18] to contain linear polymers $(HCN)_x$, with the C—H···N length 3.18 Å. It is interesting to note that the long polymers $(HCN)_n$ would not be expected to be able to change their orientation in the crystalline substance, and that in consequence solid hydrogen cyanide, unlike ice, would have a low dielectric constant. This has been found experimentally by Smyth and McNeight,[19] who reported the value of the dielectric constant of the solid to be about 3.

Evidence of intermolecular association through weak hydrogen-bond formation with use of a hydrogen atom attached to a carbon atom of a halogenated hydrocarbon molecule (chloroform and similar substances with ethers and glycols) has been reported.[20] The technique of proton magnetic resonance applied to solutions of chloroform in acetone and in triethylamine has shown that 1:1 complexes between solute and solvent are formed,[21] and the energy of the hydrogen bond has been shown to be 2.5 kcal/mole for Cl_3C—H···$OC(CH_3)_2$ and 4.0 kcal/mole for Cl_3C—H···$N(C_2H_5)_3$. The change with temperature of the second virial coefficient for the mixed vapor of chloroform and diethyl ether has been shown[22] to correspond to the formation of Cl_3C—H···$O(C_2H_5)_2$ molecules, with the hydrogen-bond energy 6.0 kcal/mole.

There is an interesting difference in properties between fluoro compounds and the corresponding hydrogen compounds that can be explained by the assumption of the formation of C—H···X bonds. For example, trifluoroacetyl chloride, F_3CCOCl, has a boiling point below 0°C, whereas that of acetyl chloride is 51°C; similarly, trifluoroacetic acid anhydride, $(F_3CCO)_2O$, boils at 20°C and acetic acid anhydride at 137°C.

The degree of polymerization of hydrogen fluoride, water, hydrogen peroxide, and the alcohols is without doubt much greater than that of

[18] W. J. Dulmage and W. N. Lipscomb, *Acta Cryst.* **4**, 330 (1951).

[19] C. P. Smyth and S. A. McNeight, *J.A.C.S.* **58**, 1723 (1936).

[20] S. Glasstone, *Trans. Faraday Soc.* **33**, 200 (1937); D. B. McLeod and F. J. Wilson, *ibid.* **31**, 596 (1935); G. F. Zellhoefer, M. J. Copley, and C. S. Marvel, *J.A.C.S.* **60**, 1337 (1938); and many later papers.

[21] C. M. Huggins, G. C. Pimentel, and J. N. Shoolery, *J. Chem. Phys.* **23**, 1244 (1955).

[22] J. H. P. Fox and J. D. Lambert, *Proc. Roy. Soc. London* **A120**, 557 (1952)

hydrogen cyanide.   The dielectric constants of these substances remain smaller than that of hydrogen cyanide, however, because for them polymerization is not accompanied by a linear increase in the magnitude of the resultant dipole moment of the molecule.   Hydrogen fluoride, for example, tends to form hydrogen bonds at about 140° angles, and a molecule $(HF)_n$, such as the following one,

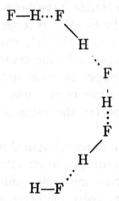

may have a very small resultant dipole moment; liquid hydrogen fluoride probably also contains ring molecules with zero moment in large numbers (Sec. 12-3).

Hydrogen-bond formation is of importance also for various other properties of substances, such as the solubility of organic liquids in water and other solvents, melting points of substances under water,[23] viscosity of liquids,[24] second virial coefficient of gases,[25] choice of crystal structure, cleavage and hardness of crystals, infrared absorption spectra, and proton magnetic resonance.   Some of these are discussed in the following sections of this chapter.

12-3. HYDROGEN BONDS INVOLVING FLUORINE ATOMS

The strongest hydrogen bond known is that in the hydrogen difluoride ion, HF_2^-.   The enthalpy of formation of HF_2^- (g) from HF (g) and F^- (g) has been evaluated as 58 ± 5 kcal/mole by Waddington[26]

[23] N. V. Sidgwick, W. J. Spurrell, and T. E. Davies, *J. Chem. Soc.* 107, 1202 (1915); W. Baker, *ibid.* 1934, 1684; H. O. Chaplin and L. Hunter, *ibid.* 1938, 375; E. D. Amstutz, J. J. Chessick, and I. M. Hunsberger, *Science* 111, 305 (1950).

[24] C. E. Kendall, *Chem. & Ind.* (London) 1944, 211.

[25] Fox and Lambert, *loc. cit.* (22).

[26] T. C. Waddington, *Trans. Faraday Soc.* 54, 25 (1958).   Several theoretical calculations for the symmetrical model of the ion have given approximately the same value: a very simple treatment (L. Pauling, *Proc. Roy. Soc. London* A114, 181 [1927]) leads to 49.5 kcal/mole, and a somewhat more refined one (M. Davies, *J. Chem. Phys.* 15, 739 [1947]) to 47.3 kcal/mole.

by the use of a calculated crystal energy and thermochemical quantities. This value is about eight times as great as that for any other hydrogen bond (Sec. 12-12).[27]

Evidence that the proton lies midway between the fluorine atoms in the crystal KHF_2 has been provided by entropy measurements,[28] study of the polarized infrared spectrum,[29] neutron diffraction,[30] and nuclear spin magnetic resonance.[31] The uncertainty in the location of the proton at the midpoint between the fluorine atoms is reported to be ± 0.10 Å for the neutron diffraction study and ± 0.06 Å for the nuclear magnetic resonance study.

The fluorine-fluorine distance in the HF_2^- ion in the crystal KHF_2 has been determined[32] to be 2.26 ± 0.01 Å. Hence the H—F half-bond in this ion has the length 1.13 Å, which is 0.21 Å greater than the H—F single-bond length in HF, rather than 0.18 Å, as given by Equation 7-7.

The hydrogen bonds in the polymers $(HF)_n$ present in gaseous hydrogen fluoride are much weaker than those in the HF_2^- ion. The average bond energy (enthalpy) was evaluated by Fredenhagen[33] as 6.02 kcal/mole. Fredenhagen found evidence for various polymers with n equal to 3 or more; the dimer seems to be less stable than the higher polymers. The enthalpy of formation of $(HF)_6$, which presumably has a ring structure with six hydrogen bonds (bond angle 120°), from 6HF was reported by Simons and Hildebrand[34] to be 40 kcal/mole, corresponding to 6.7 kcal/mole per F—H$\cdots$F hydrogen bond. The electron-diffraction study[35] of $(HF)_n$ has given the value 1.00 ± 0.06 Å for F—H and 2.55 ± 0.05 Å for F—H$\cdots$F, leading to 1.55 Å for H$\cdots$F. This value corresponds to about 9 percent covalent character of the H$\cdots$F bond. The bond angle in the gas polymers is reported to be $140° \pm 5°$.

---

[27] The enthalpy of formation of HF_2^- (aq) from HF (aq) and F^- (aq) is only about 4 kcal/mole. Hence the hydrogen bonds formed by F^- and HF with water molecules must be much stronger than those formed by HF_2^-. If we make the reasonable assumption that most of the difference is due to the four hydrogen bonds between F^- and ligated water molecules, each of these O—H$\cdots$$F^-$ bonds is to be assigned bond energy about 13 kcal/mole.

[28] E. F. Westrum, Jr., and K. S. Pitzer, *J.A.C.S.* **71**, 1940 (1949).

[29] R. Newman and R. M. Badger, *J. Chem. Phys.* **19**, 1207 (1951).

[30] S. W. Peterson and H. A. Levy, *J. Chem. Phys.* **20**, 704 (1952).

[31] J. S. Waugh, F. B. Humphrey, and D. M. Yost, *J. Phys. Chem* **57**, 486 (1953).

[32] L. Helmholz and M. T. Rogers, *J.A.C.S.* **61**, 2590 (1939).

[33] K. Fredenhagen, *Z. anorg. Chem.* **218**, 161 (1934).

[34] J. H. Simons and J. H. Hildebrand, *J.A.C.S.* **46**, 2183 (1924).

[35] S. H. Bauer, J. Y. Beach, and J. H. Simons, *J.A.C.S.* **61**, 19 (1939).

Crystalline hydrogen fluoride has been found[36] to contain infinite zigzag chains with F—H···F distance 2.49 ± 0.01 Å and bond angle 120.1°. These values are more accurate than those for the gas phase.

Cady[37] has made the crystalline substances KH_2F_3, KH_3F_4, and KH_4F_5, and Winsor and Cady[38] have made CsH_2F_3, CsH_3F_4, and CsH_6F_7. The structures of these crystals have not yet been determined. It seems likely that the $H_nF_{n+1}^-$ ions in these crystals contain zigzag hydrogen-bonded chains, but there is the possibility that the structures involve a central fluoride ion with three or more HF molecules attached by hydrogen bonds; for example, $H_4F_5^-$ might have the tetrahedral structure

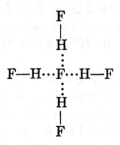

with F—H···F distance about 2.35 Å.

The crystal structure of NH_4HF_2 is of interest in that it is completely determined by hydrogen bonds.[39] In KHF_2 each potassium ion has eight equidistant fluorine neighbors. The structure of NH_4HF_2 is similar,[40] except that four of the eight fluorine atoms, surrounding the nitrogen atom tetrahedrally, are drawn in to the distance 2.80 ± 0.02 Å through the formation of N—H···F hydrogen bonds, the other four being at about 3.1 Å. The structure is shown in Figure 12-5.

The F—H—F⁻ distance in NH_4HF_2 is 2.32 ± 0.02 Å, about 0.06 Å greater than in KHF_2. This increase may be the effect of the N—N···F hydrogen bonds in partially saturating the valence of the fluorine atom, and thus decreasing the strength of the F—H—F bonds.

[36] M. Atoji and W. N. Lipscomb, *Acta Cryst.* **7**, 173 (1954). It is interesting that D. F. Horing and W. E. Osberg, *J. Chem. Phys.* **23**, 662 (1955), have obtained evidence from infrared spectra for the existence of zigzag chains in the low-temperature crystalline forms of HCl and HBr, with H···X—H bond angle about 107° for HCl and 97° for HBr. The high-temperature forms have structure with cubic closest packing of the HX molecules, either rotating or with random orientation.

[37] G. H. Cady, *J.A.C.S.* **56**, 1431 (1934).

[38] R. V. Winsor and G. H. Cady, *J.A.C.S.* **70**, 1500 (1948).

[39] L. Pauling, *Z. Krist.* **85**, 380 (1933); M. T. Rogers and L. Helmholz, *J.A.C.S.* **62**, 1533 (1940).

[40] Pauling, also Rogers and Helmholz, *loc. cit.* (39).

Ammonium azide, NH_4N_3, has the same structure[41] as NH_4HF_2; the N—H···N hydrogen bonds have the length 2.98 Å.

The ammonium fluoride crystal, NH_4F, has a structure closely resembling that of wurtzite (Fig. 7-6). Each nitrogen atom is bonded by hydrogen bonds to four tetrahedrally arranged fluorine ions, the N—H···F distance being 2.66 Å. The value of the energy of the N—H···F bond has been discussed by Sherman[42] by the comparison of the experimental value of the crystal energy of the substance with the value calculated for an ionic structure not involving hydrogen bonds,

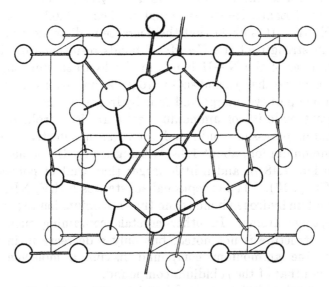

FIG. 12-5.—The atomic arrangement of the crystal NH_4HF_2. The large circles represent nitrogen atoms and the smaller circles fluorine atoms, with hydrogen bonds indicated by double lines.

use being made of thermochemical data for the other ammonium halides. The added stability of the crystal due to hydrogen-bond formation is 18.3 kcal/mole greater for ammonium fluoride than for ammonium iodide. If we assume the extra interaction energy of the ammonium ion with the surrounding iodide ions in the latter crystal[43]

---

[41] L. K. Frevel, *Z. Krist.* **94**, 197 (1936); E. W. Hughes, Ph.D. dissertation, Cornell University, 1935.

[42] J. Sherman, *Chem. Revs.* **11**, 93 (1932).

[43] The data for ammonium chloride and bromide, treated similarly, lead to values of about 6 to 8 kcal/mole for the extra energy of interaction of the ammonium ion with the surrounding chloride and bromide ions. In these crystals each ammonium ion is surrounded by eight halogenide ions at cube corners. It can form hydrogen bonds with four, at tetrahedron corners, at a time. There

to be about 2 kcal/mole, the value 5 kcal/mole is obtained for the N—H$\cdots$F bond energy in ammonium fluoride.

The structure of NH_4F is closely similar to that of ice (see the following section): the atoms are similarly arranged, and the dimensions differ by only 3.7 percent (N—H$\cdots$F, 2.66 Å; O—H$\cdots$O, 2.76 Å). It has been found[44] that the two substances form crystalline solutions containing as much as 10 percent ammonium fluoride. Ammonium fluoride is the only substance known to have appreciable solubility in ice.

In the hydrazinium difluoride crystal,[45] $N_2H_6F_2$, there are hydrogen bonds with distance N—H$\cdots$F equal to 2.62 $\pm$ 0.02 Å, 0.04 Å less than in NH_4F. It is likely that the decrease in distance represents an increase in strength of the N—H$\cdots$F bonds resulting from increased ionic character of the N—H bonds in the hydrazinium ion, $N_2H_6^{++}$, with two positive charges and six hydrogen atoms, over the ammonium ion, with one positive charge and four hydrogen atoms.

The great majority of ammonium salts are isomorphous with the corresponding potassium and rubidium salts, the effective ionic radius of the ammonium ion, about 1.48 Å, being nearly the same as that of rubidium ion, 1.48 Å, and a little larger than that of potassium ion, 1.33 Å (Chap. 13). The exceptional substances NH_4F, NH_4HF_2, and NH_4N_3 contain hydrogen bonds from the ammonium ion to surrounding electronegative atoms. In other crystals containing such bonds a change in structure is not noted, but only a decrease in interatomic distances, the ammonium compound having a molecular volume smaller than that of the rubidium compound.

It is interesting that in general fluorine atoms attached to carbon do not have significant power to act as proton acceptors in the formation of hydrogen bonds[46] in the way that would be anticipated from the large difference in electronegativity of fluorine and carbon.

12-4. ICE AND WATER; CLATHRATE COMPOUNDS

The crystal structure of ice has been shown by x-ray investigation,[47] which has led to the assignment of the oxygen atoms to positions in the

is evidence that at room temperature the ammonium ion changes freely from one orientation to another.

[44] R. F. Brill and S. Zaromb, *Nature* **173**, 316 (1954); S. Zaromb and R. F. Brill, *J. Chem. Phys.* **24**, 895 (1956); S. Zaromb, *ibid.* **25**, 350 (1956).

[45] M. L. Kronberg and D. Harker, *J. Chem. Phys.* **10**, 309 (1942).

[46] This fact was pointed out to me by V. Schomaker.

[47] D. M. Dennison, *Phys. Rev.* **17**, 20 (1921); W. H. Bragg, *Proc. Phys. Soc. London* **34**, 98 (1922); W. H. Barnes, *Proc. Roy. Soc. London* **A125**, 670 (1929); H. D. Megaw, *Nature* **134**, 900 (1934); S. Hillesund, *Ark. norske Vidensk. Acad.* No. 8 (1942).

lattice, to be similar to that of wurtzite (Fig. 7-6), each oxygen atom being surrounded tetrahedrally by four other oxygen atoms, at the distance 2.76 Å, as shown in Figure 12-6. This is a very open structure, which causes ice to have a low density; hydrogen sulfide, for example, crystallizes in a closest-packed arrangement, each sulfur atom (hydrogen sulfide molecule) having 12 equidistant neighbors. The ice

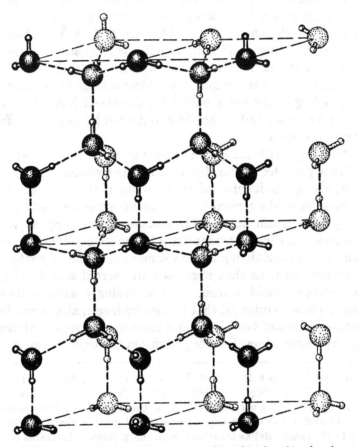

FIG. 12-6.—The arrangement of molecules in the ice crystal. The orientation of the water molecules, as represented in the drawing, is arbitrary; there is one proton along each oxygen-oxygen axis, closer to one or the other of the two oxygen atoms.

structure is, however, just that expected in case O—H···O hydrogen bonds are formed, with each bond making greater or less use of one of the four valence electron pairs of each of the two bonded oxygen atoms.[48]

[48] It has been found (H. König, *Z. Krist.* **105,** 279 [1944]) that water vapor condenses at very low temperatures to produce a cubic modification of ice, closely similar to ordinary ice, but like sphalerite (Fig. 7-5) rather than wurtzite. The

The question now arises as to whether a given hydrogen atom is midway between the two oxygen atoms it connects or closer to one than to the other. The answer to this is that it is closer to one than to the other, and that (with few exceptions) each oxygen atom has two hydrogen atoms bonded to it by strong bonds. In the gas molecule the O—H distance is 0.96 Å, and the magnitudes of the changes in properties from steam to ice are not sufficiently great to permit us to assume that this distance is increased in ice to 1.38 Å. There is, for example, only a rather small difference in the frequencies of the vibrational motions of the molecule involving stretching the O—H bonds observed for ice and water vapor; this difference has been interpreted[49] as corresponding to the value 0.99 Å for the O—H bond distance. A more accurate value, 1.01 Å, has been obtained by neutron diffraction of deuterium oxide ice.[50]

An interesting verification of the existence of discrete water molecules in ice is provided by the discussion of its residual entropy, which, moreover, also gives definite information about the orientation of the water molecules in the crystal.[51]   It is found experimentally that ice[52] and heavy ice[53] retain appreciable amounts of entropy at very low temperatures. If each water molecule in the ice crystal were oriented in a definite way, permitting the assignment of a unique configuration to the crystal, such as that suggested by Bernal and Fowler,[54] the residual entropy would vanish. We accordingly assume that each water molecule is so oriented that its two hydrogen atoms are directed approximately toward two of the four surrounding oxygen atoms, that only one hydrogen atom lies along each oxygen-oxygen line, and that

---

edge of the unit cube is 6.37 ± 0.02 Å at −190°C (F. V. Shallcross and G. B. Carpenter, *J. Chem. Phys.* **26,** 782 [1957]). Condensation below −140°C leads to a deposit that gives diffuse rings in the x-ray diffraction pattern; between −140 and −120°C the deposited material gives a sharp pattern corresponding to the sphaleritelike structure, with sharp rings. The same results are obtained for deuterium oxide. The hydrogen-bond length at −120°C is 2.751 Å for both the cubic form and the hexagonal form of both light water and deuterium oxide (M. Blackman and N. D. Lisgarten, *Proc. Roy. Soc. London* A239, 93 [1957]).

[49] P. C. Cross, J. Burnham, and P. A. Leighton, *J.A.C.S.* **59,** 1134 (1937).

[50] S. W. Peterson and H. A. Levy, *Acta Cryst.* **10,** 70 (1957).

[51] L. Pauling, *J.A.C.S.* **57,** 2680 (1935).

[52] W. F. Giauque and M. Ashley, *Phys. Rev.* **43,** 81 (1933); W. F. Giauque and J. W. Stout, *J.A.C.S.* **58,** 1144 (1936).

[53] E. A. Long and J. D. Kemp, *J.A.C.S.* **58,** 1829 (1936).

[54] J. D. Bernal and R. H. Fowler, *J. Chem. Phys.* **1,** 515 (1933); these authors also suggested that at temperatures just below the melting point, but not at lower temperatures, the molecular arrangement might be partially or largely irregular.

under ordinary conditions the interaction of nonadjacent molecules is such as not to stabilize appreciably any one of the many configurations satisfying these conditions with reference to the others. Thus we assume that an ice crystal can exist in any one of a large number of configurations, each corresponding to certain orientations of the water molecules. It can change from one configuration to another by rotation of some of the molecules or by motion of some of the hydrogen nuclei, each moving 0.76 Å from a position 1.00 Å from one oxygen atom to the similar position near the other bonded atom.[55] It is probable that both processes occur. The fact that at temperatures above about 200°K the dielectric constant of ice is of the order of magnitude of that of water shows that the molecules can reorient themselves with considerable freedom, the crystal changing in the stabilizing presence of the electric field from unpolarized to polarized configurations satisfying the above conditions.[56]

When a crystal of ice is cooled to very low temperatures it is caught in some one of the many possible configurations; but it does not assume (in a reasonable period of time) a uniquely determined configuration with no randomness of molecular orientation. It accordingly retains the residual entropy $k \ln W$, in which k is the Boltzmann constant and W is the number of configurations accessible to the crystal.

Let us now calculate W. In a mole of ice there are $2N$ hydrogen nuclei. If each had the choice of two positions along its O—O axis, one closer to one and the other closer to the second oxygen atom, there would be 2^{2N} configurations. However, many of these are ruled out by the condition that each oxygen atom have two attached hydrogen atoms. Let us consider a particular oxygen atom and the four surrounding hydrogen nuclei. There are 16 arrangements of this OH_4 group; one with all four hydrogen nuclei close to the oxygen atom, corresponding to the ion $(H_4O)^{++}$, four corresponding to $(H_3O)^+$, six to H_2O, four to $(OH)^-$, and one to O^{--}. The acceptable arrangements assigning two strongly bonded hydrogen nuclei to this oxygen atom accordingly comprise six-sixteenths or three-eighths of the total. Of these, only three-eighths are suitable with respect to the second oxygen atom, and so on; the number of configurations W is hence $2^{2N}(\frac{3}{8})^N$ or $(\frac{3}{2})^N$.

---

[55] The protons will tend to jump in this way in groups, so as to leave each oxygen atom with two protons attached; ice is so similar to water that we are assured that the concentrations of $(OH)^-$ and $(H_2O)^+$ ions present in ice are very small.

[56] At the April 1937 meeting of the American Chemical Society at Chapel Hill, North Carolina, L. Onsager reported that values of the dielectric constant calculated for this model agree approximately with experiment.

This leads to the theoretical value $k \ln (\frac{3}{2})^N = R \ln \frac{3}{2} = 0.806$ cal/mole degree for the residual entropy of ice. The experimental values are 0.82 cal/mole degree for ordinary ice and 0.77 cal/mole degree for heavy ice; the agreement with the theoretical value provides strong support for the postulated structure involving hydrogen bonds with the hydrogen nucleus unsymmetrically placed between the two bonded oxygen atoms.[57]

Verification of this disordered structure for ice has been made by neutron diffraction. The intensities of the diffracted beams of neutrons were found to correspond to a structure in which the scattering power of a half-hydrogen atom is assigned to each of the four tetrahedral sides about each oxygen atom; that is, the intensities correspond to occupancy of half of the sides by hydrogen atoms.[58] A reinvestigation of single crystals of deuterium oxide at $-50°C$ and $-150°C$ by neutron diffraction[59] has led to the determination of the O—D distance is 1.01 Å and the D—O—D angles as close to tetrahedral ($109.5° \pm 0.5°$).

Of the enthalpy of sublimation of ice, 12.20 kcal/mole, about one-fifth can be attributed to ordinary van der Waals forces (as estimated from values for other substances); the remainder, 10 kcal/mole, represents the rupture of hydrogen bonds and leads to the value 5 kcal/mole for the energy of the O—H⋯O hydrogen bond in ice. The small value 1.44 kcal/mole of the enthalpy of fusion of ice shows that on melting only about 15 percent of the hydrogen bonds are broken.

Measurements of the alternating-current conductivity of ice[60] have given results indicating that water molecules on internal surfaces of the ice crystal engage in a special kind of random walk in which the molecule is always attached to the surface by one hydrogen bond (bipedal random walk). The temperature dependence of the local conductivity corresponds to an activation energy of 5.2 kcal/mole, which may be interpreted as the energy required to break a hydrogen bond. Nuclear magnetic resonance (proton spin) studies of water have permitted the determination of the self-diffusion coefficient and the spin-lattice relaxation time. The temperature dependence of these quantities corresponds to an energy of activation for the ratio of the self-diffusion co-

[57] It was found by K. S. Pitzer and L. V. Coulter, *J.A.C.S.* 60, 1310 (1938), that sodium sulfate decahydrate has a residual entropy of 1.7 cal/mole degree, corresponding to some randomness of orientation of water molecules. In some other crystals the hydrogen bonds are ordered, so that there is no residual entropy; examples are H_2SO_4 (T. R. Rubin and W. F. Giauque, *ibid.* 74, 800 [1952]) and $ZnSO_4 \cdot 7H_2O$ (R. E. Barieau and W. F. Giauque, *ibid.* 72, 5676 [1950]; W. F. Giauque, R. E. Barieau, and J. E. Kunzler, *ibid.* 5685).
[58] E. O. Wollan, W. L. Davidson, and C. G. Shull, *Phys. Rev.* 75, 1348 (1949).
[59] Peterson and Levy, *loc. cit.* (50).
[60] E. J. Murphy, *J. Chem. Phys.* 21, 1831 (1953).

efficient to the viscosity that decreases from 5.5 kcal/mole at 2°C to 3.5 at 100°C, and for the spin-lattice relaxation time that decreases from 5.5 at 2°C to 3 at 100°C.[61] The value 5.0 kcal/mole for the energy of the hydrogen bond between two water molecules in water vapor has been obtained through the determination of the second virial coefficient of water vapor.[62]

The enthalpy of sublimation of hydrogen peroxide, 14.1 kcal/mole, leads, when corrected by subtraction of the estimated value 4 kcal/mole for the energy of van der Waals attraction, to the same value for the hydrogen-bond energy as for water.

The problem of the structure of liquid water has attracted much attention, but as yet no completely satisfactory solution to it has been found. We shall postpone the discussion of this problem until after the description of the structure of certain crystalline hydrates of simple substances.

Clathrate Compounds.—In 1811 Humphry Davy[63] showed that water is a component of the phase that had earlier been thought to be crystalline chlorine, and 12 years later Faraday[64] reported an analysis corresponding to the formula $Cl_2 \cdot 10H_2O$. Later studies have indicated that the composition is close to $Cl_2 \cdot 8H_2O$. Since Faraday's time similar crystalline hydrates of many simple substances, including the noble gases and simple hydrocarbons, have been reported. X-ray studies of hydrates of xenon, chlorine, bromine, sulfur dioxide, hydrogen sulfide, methyl bromide, methyl iodide, ethyl chloride, chloroform, and some other substances[65] showed that some of these crystals have a cubic unit of structure with edge about 12.0 Å and others have a cubic unit of structure with edge about 17.0 Å. Structures have been proposed for these two types of hydrates.[66] The results of a thorough x-ray study of one crystal, chlorine hydrate, have been reported.[67] The structure found for chlorine hydrate is shown in Figures 12-7 and 12-8. The 20 water molecules can be placed at the corners of a pentagonal dodecahedron, as shown in Figure 12-7. They form hydrogen bonds along the dodecahedron edges. The angles of a regular pentagon are 108°, closely approximating the tetrahedral angle. The expected value of the length of the edge of the dodecahedron is accordingly about

[61] J. H. Simpson and H. Y. Carr, *Phys. Rev.* 111, 1201 (1958).

[62] J. S. Rowlinson, *Trans. Faraday Soc.* 45, 974 (1949).

[63] H. Davy, *Phil. Trans. Roy. Soc. London* 101, 155 (1811).

[64] M. Faraday, *Quart. J. Sci.* 15, 71 (1823).

[65] M. von Stackelberg, O. Gotzen, J. Pietuchovsky, O. Witscher, H. Fruhbuss, and W. Meinhold, *Fortschr. Mineral.* 26, 122 (1947).

[66] W. F. Claussen, *J. Chem. Phys.* 19, 259, 662, 1425 (1951); L. Pauling and R. E. Marsh, *Proc. Nat. Acad. Sci. U.S.* 36, 112 (1952).

[67] Pauling and Marsh, *loc. cit.* (66).

2.76 Å, the hydrogen-bond length for two water molecules. In the
unit of structure, which is a cube 11.88 Å on edge (for chlorine hydrate),
there are two dodecahedra, one at the cube corner and the second,
with different orientation, at the center (Fig. 12-7). Eight of the 20
water molecules in each dodecahedron form hydrogen bonds with corresponding molecules in the eight surrounding dodecahedra. These
hydrogen bonds extend directly out from the center of the polyhedron,
along the threefold axes of the crystal. In addition, there are six
water molecules in interstices between four dodecahedra, such that
each forms four hydrogen bonds, one with a water molecule in each of

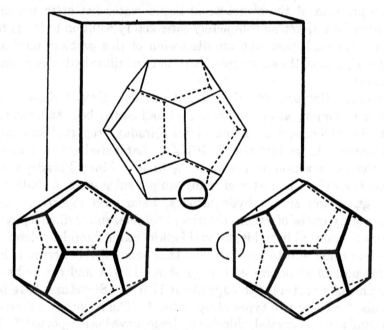

Fig. 12-7.—The arrangement of water molecules in the chlorine hydrate crystal. Some of the water molecules are at the
corners of pentagonal dodecahedra, as indicated. Some additional water molecules (circles) are needed to complete the structure. Hydrogen bonds are formed along the edges of the dodecahedra, and also between adjacent dodecahedra and between
the dodecahedra and the interstitial water molecules.

the four surrounding dodecahedra, as shown in Figure 12-8. The arrangement of the 46 water molecules in the unit cube is such as to define
six tetrakaidecahedra, in addition to the two pentagonal dodecahedra
per unit cube. The tetrakaidecahedra (Figure 12-8) are polyhedra
with two hexagonal faces and eight pentagonal faces. The values of
the parameters of the oxygen atoms, as determined from the x-ray pattern, are such as to correspond to the value 2.75 Å for the hydrogen-

bond lengths throughout the crystal. Each of the 46 water molecules in this framework forms four hydrogen bonds that are approximately tetrahedral in orientation.

Chlorine molecules, with somewhat random orientation, occupy the tetrakaidecahedra. There are accordingly six Cl_2 molecules per unit cube. It is likely also that there are two water molecules per unit cube, occupying the cavities in the dodecahedron; these cavities are too small for a chlorine molecule. The composition of the crystal thus is $Cl_2 \cdot 8H_2O$.

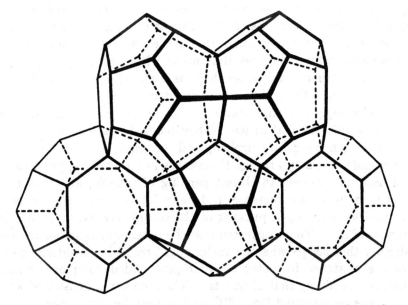

FIG. 12-8.—A portion of the hydrogen-bond framework in the chlorine hydrate crystal. The water molecules are grouped into tetrakaidecahedra as well as dodecahedra.

In xenon hydrate, methane hydrate, and other hydrates involving small molecules the eight polyhedron centers per unit cube are occupied by the xenon or methane molecules, their composition accordingly being $Xe \cdot 5\frac{3}{4}H_2O$ and $CH_4 \cdot 5\frac{3}{4}H_2O$.

Compounds of this sort, in which molecules are entrapped by a lattice formed by other molecules, are called *clathrate compounds*. Often the lattice is formed by molecules linked to one another by hydrogen bonds. The structures of many of these substances have been determined by Powell and his coworkers.[68] An especially interesting class is that formed by quinol, *p*-dihydroxybenzene.[69] The hydroxyl groups

[68] For reviews see H. M. Powell, *J. Chem. Soc.* 1948, 61; *Research* (London) 1, 353 (1948).

[69] H. M. Powell and P. Riesz, *Nature* 161, 52 (1948); H. M. Powell, *J. Chem. Soc.* 1950, 298, 300, 468.

of the quinol molecules form hydrogen bonds with one another in such a way as to form two infinite interpenetrating frameworks, not bonded to one another. (In this respect the structure resembles that of cuprite, Fig. 7–9.) There are cavities in the structure, large enough to accommodate a small molecule: one cavity per three molecules of quinol. Known substances of this sort have the composition $3C_6H_4(OH)_2 \cdot M$, with M = Ar, Kr, Xe, HCl, HBr, H_2S, SO_2, CO_2, HCN, H_2C_2, HCOOH, CH_3OH, and CH_3CN.

Water.—The problem of the structure of liquid water is an interesting one that is still far from complete solution. There is no doubt that water, like other liquids, has a structure that involves a great deal of randomness, and yet it is likely that there are certain configurations of groups of water molecules that occur with high frequency in the liquid. The structure for water that has received serious consideration for many years is the one proposed by Bernal and Fowler.[70] Bernal and Fowler suggested that water retains in part a hydrogen-bonded structure similar to that of ice. They pointed out that as more and more hydrogen bonds are broken, with increase in temperature, the oxygen molecules may arrange themselves in a manner approximating more and more closely to closest packing of spheres, and that there would be a significant increase in density for this sort of packing, as compared with the open packing of the completely hydrogen-bonded structure of ice. They suggested also the possibility that the decrease in density that occurs when ice melts might result in considerable part from the existence in water of hydrogen-bonded complexes with a structure resembling that of quartz. The increase in density of water that occurs on warming from 0°C to 4°C may be attributed to a decrease in the concentration of aggregates with the icelike structure and increase in the number of complexes with a quartzlike structure[71] or some other denser structure.

The suggestion that water contains significant numbers of aggregates with a quartzlike structure must be rejected, because there is no way in which a complex with a quartzlike structure can be stabilized relative to one with a tridymitelike or cristobalitelike structure. For each of these structures each oxygen atom is surrounded tetrahedrally by four other oxygen atoms, with which it forms hydrogen bonds. In the quartzlike structure the hydrogen bonds are bent through a considerable angle from the normal angle 180°, and the presence of these bent

---

[70] Bernal and Fowler, *loc. cit.* (54).

[71] Ordinary ice is sometimes described as resembling tridymite, a hexagonal form of silicon dioxide, with SiO_4 tetrahedra sharing corners. Cristobalite is a closely similar cubic form. Quartz is another hexagonal form in which the tetrahedra are arranged in a different way, such as to increase the density by 16 percent.

hydrogen bonds should introduce a serious destabilization.

It seems not unreasonable to discuss the structure of liquid water in terms of the structure of methane hydrate, discussed above. If each of the methane molecules in the crystal of methane hydrate were replaced by water molecules, there would be obtained a crystal with a completely hydrogen-bonded framework of 46 water molecules per cubic unit, plus 8 water molecules per cubic unit in the centers of the polyhedra, forming no hydrogen bonds. The number of hydrogen bonds is 85 percent of the number in ice, as is indicated by the enthalpy of fusion (mentioned above). The density of this crystal, assuming the dimensions to be the same as for methane hydrate, is 1.00 g/cm³, which is that of liquid water. Moreover, it is found on analysis of the structures that the pentagonal dodecahedra may be arranged relative to one another in a large number of ways, so that highly random structures for liquid water might be based upon aggregates of water molecules bonded to one another in this way. The radial distribution function for water as calculated from the x-ray diffraction pattern, the dispersion of dielectric constant, and some other properties of water are found to be compatible with a structure of this sort.[72]

12-5. ALCOHOLS AND RELATED SUBSTANCES

In crystalline alcohols the molecules are usually combined by hydrogen bonds to polymers of the type

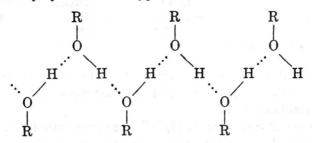

An example is the methanol crystal, which has such a structure,[73] with O—H···O distance 2.66 Å at −110°C. The hydrogen bonds form zigzag chains, as represented above.

It is not necessary that many of these bonds be broken for the crystal to melt to a liquid containing long-chain[74] or ring polymers—indeed, if the liquid contained only ring polymers, such as $(ROH)_6$,

---

[72] L. Pauling, *Trans. Internat. Conf. on the Hydrogen Bond* Ljubljana, Sept. 1957; L. Pauling and P. Pauling, unpublished research.

[73] K. J. Tauer and W. N. Lipscomb, *Acta Cryst.* **5**, 606 (1952).

[74] X-ray evidence for chain structures in liquid alcohols has been presented by W. C. Pierce and D. P. MacMillan, *J.A.C.S.* **60**, 779 (1938).

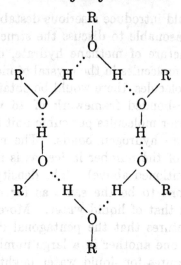

no loss of energy due to the rupture of hydrogen bonds would occur on fusion. For this reason the heats of fusion and the melting points of alcohols are only slightly abnormal, whereas the heats of vaporization and boiling points show the effect of the hydrogen bonds strongly; and in consequence the liquid state is stable over a wide temperature range. It is instructive to compare ethyl alcohol with its isomer dimethyl ether; the respective values of some physical constants of these substances are as follows:

	C_2H_5OH	$(CH_3)_2O$	*Difference*
Melting point	− 115°C	− 141°C	26°C
Boiling point	78°	− 25°	103°
Molal enthalpy of sublimation	11.3	6.3	5.0

The difference in the molal enthalpies of sublimation, 5.0 kcal/mole, may be accepted as an approximate value of the energy of the O—H$\cdots$O bond in crystalline ethanol.

The tetramer of methanol, $(CH_3OH)_4$, has been found to be present in the vapor.[75] Its structure can be assumed to involve a square of four hydrogen bonds:

[75] W. Weltner, Jr., and K. S. Pitzer, *J.A.C.S.* **73**, 2606 (1951)

The enthalpy of formation of the tetramer from four monomeric gas molecules is 24.2 kcal/mole. One-fourth of this value, 6.05 kcal/mole, may be taken to be the energy of the hydrogen bond; a small correction should be made for the van der Waals attraction. The value thus agrees with that given above for ethanol.

Studies of the association of ethanol molecules in carbon tetrachloride solution have been made by infrared spectroscopy, permitting the evaluation of the enthalpies of formation of the dimer, trimer, and tetramer.[76] The value for the tetramer, 22.56 kcal/mole, corresponds to 5.64 kcal/mole for the energy of the hydrogen bond (no correction for van der Waals attraction is made because the interaction with the solvent counteracts it). This value agrees moderately well with the value obtained above from the enthalpy of sublimation. The values for the dimer and the trimer, 5.09 and 10.18 kcal/mole, presumably correspond to one and two hydrogen bonds, respectively.

The alcohol pentaerythritol, $C(CH_2OH)_4$, forms tetragonal crystals, with the structure[77] shown in Figure 12-9. The hydrogen bonds, with the O—H$\cdots$O distance 2.69 Å, bind the molecules into layers. The crystal shows a correspondingly good basal cleavage.

In this alcohol the hydrogen bonds bind the oxygen atoms together into square groups, as shown above for the methanol tetramer.

The hydrogen atoms in the pentaerythritol crystal have been located by a neutron-diffraction study of the deuterated substance.[78] They have an ordered arrangement; each hydrogen atom is 0.94 ± 0.03 Å from the nearest oxygen atom, and the angle C—O—H has the value 110°. The angle between the directions O—H and O- - -O is 6°, so that there is a 9° bend in the hydrogen bond (deviation of the O—H$\cdots$O angle by 9° from a straight angle).

In resorcinol, *m*-dihydroxybenzene,[79] there are infinite $\cdots$ OHOH $\cdots$ spirals (α modification) and staggered chains (β modification). The O—H$\cdots$O distance is about 2.70 Å.

The hydrogen bonds in crystalline and liquid ammonia are weaker than those in ice and water for two reasons: the small ionic character of the N—H bond gives it only small hydrogen-bond-forming power, and the one unshared electron pair of the NH_3 molecule must serve for

[76] W. C. Coburn, Jr., and E. Grunwald, *J.A.C.S.* **80**, 1318 (1958).

[77] E. G. Cox, F. J. Llewellyn, and T. H. Goodwin, *J. Chem. Soc.* **1937**, 882; E. W. Hughes, unpublished investigation; I. Nitta and T. Watanabé, *Nature* **140**, 365 (1937); *Sci. Papers Inst. Phys. Chem. Res.* (Tokyo) **34**, 1669 (1938). The intramolecular interatomic distances are C—C = 1.548 ± 0.011 Å, C—O = 1.425 ± 0.014 Å: R. Shiono, D. W. J. Cruikshank, and E. G. Cox, *Acta Cryst.* **11**, 389 (1958).

[78] J. Hvoslef, *Acta Cryst.* **11**, 383 (1958).

[79] J. M. Robertson, *Proc. Roy. Soc. London* **A157**, 79 (1936); J. M. Robertson and A. R. Ubbelohde, *ibid.* **A167**, 122 (1938).

all of the bonds formed by the molecule with other N—H groups, whereas water has an electron pair for each hydrogen bond. In the ammonia crystal[80] each nitrogen atom has six neighbors[81] at 3.380 ±0.004 Å, this distance representing a weak N—H$\cdots$N bond; the stronger N—H$\cdots$N bonds in NH_4N_3 show the distance 2.94 − 2.99 Å. From the heat of sublimation, 6.5 kcal/mole, the energy of the N—H$\cdots$N

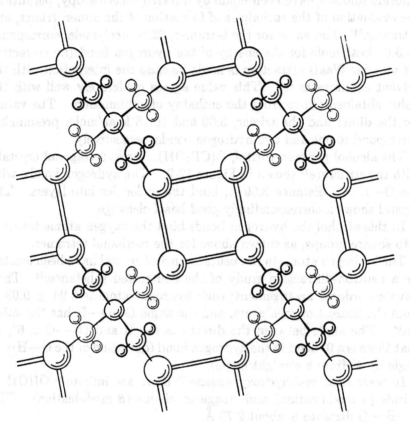

Fig. 12-9.—The structure of pentaerythritol, $C(CH_2OH)_4$. Large circles represent oxygen atoms, circles of intermediate size carbon atoms, and small circles hydrogen atoms attached to carbon atoms. Hydrogen bonds are shown by double lines.

bond in ammonia can be calculated to be about 1.3 kcal/mole, with use of the estimated value 2.6 kcal/mole for the van der Waals energy.

[80] H. Mark and E. Pohland, *Z. Krist.* **61**, 532 (1925); J. de Smedt, *Bull. Ac. Roy. de Belgique* **10**, 655 (1925); I. Olovsson and D. H. Templeton, *Acta Cryst.* in press (1959).

[81] The structure of the crystal represents a small distortion from cubic closest packing; the six next nearest neighbors are at 3.95 Å

12-6. CARBOXYLIC ACIDS

The hydrogen bonds formed by water are not sufficiently strong to lead to an appreciable concentration of polymerized molecules in the vapor phase. The oxygen atoms of carboxyl groups can, however, form stronger hydrogen bonds, leading to the formation of stable double molecules of formic acid and acetic acid. The structure of the formic acid dimer as determined by the electron-diffraction method[82] is the following:

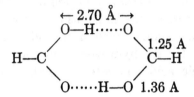

The value 2.70 Å for the O—H⋯O distance in this substance is smaller than that in ice, 2.76 Å, as expected for this stronger bond. From the enthalpy of dimerization,[83] 14.12 kcal/mole, the O—H⋯O bond energy is found to have the value 7.06 kcal/mole. The value 7.6 kcal/mole is similarly found for the hydrogen-bond energy in acetic acid.[84] These values are about 50 percent greater than those for ice.

The distance from each hydrogen atom to the nearer of the two adjacent oxygen atoms in the dimer of acetic acid has been reported[85] to be 1.075 ± 0.015 Å; this is considerably greater than the value 1.01 Å for ice, as is to be expected in consequence of the increased strength of the hydrogen bond.

The increased strength of this hydrogen bond can be accounted for in the following way. The resonance of the molecule to the structure

gives a resultant positive charge to the oxygen atom

[82] L. Pauling and L. O. Brockway, *Proc. Nat. Acad. Sci. U.S.* **20**, 336 (1934); J. Karle and L. O. Brockway, *J.A.C.S.* **66**, 574 (1944). The same structure was found for the dimers of acetic acid and trifluoroacetic acid.

[83] A. S. Coolidge, *J.A.C.S.* **50**, 2166 (1928).

[84] M. D. Taylor, *J.A.C.S.* **73**, 315 (1951).

[85] R. C. Herman and R. Hofstadter, *Phys. Rev.* **53**, 940 (1938); *J. Chem. Phys.* **6**, 534 (1938). This value is obtained by applying Badger's rule to frequencies observed in the infrared absorption spectra of light and heavy acetic acid (CH_3COOH and CH_3COOD).

which donates the proton in hydrogen-bond formation, and thus increases the ionic character of the O—H bond and the positive charge of the hydrogen atom. It also gives to the other oxygen atom, the proton acceptor, an increased negative charge. Both of these effects operate to increase the strength of the O—H$\cdots$O bond.

It is interesting to note that in general the strength of an unsymmetrical hydrogen bond A—H$\cdots$B is increased by increasing the resultant positive charge of A and the negative charge of B.

Benzoic acid and other carboxylic acids have been shown to be associated with double molecules in solution in certain solvents, such as benzene, chloroform, carbon tetrachloride, and carbon disulfide.[86] The value 4.2 kcal/mole for the hydrogen-bond energy has been found in this way for benzoic acid and *o*-toluic acid, and 4.7 kcal/mole for *m*-toluic acid.

Benzoic acid exists in the monomeric form in solution in acetone, acetic acid, ethyl ether, ethyl alcohol, ethyl acetate, and phenol; in these solutions the single molecules are stabilized by hydrogen-bond formation with the solvent.

Salicylic acid forms double molecules in solvents such as benzene and carbon tetrachloride. It has, moreover, been shown by the spectroscopic method (Sec. 12-8) that the double molecule contains no OH groups which are not involved in hydrogen-bond formation. This results from the assumption by the molecule of the following structure:

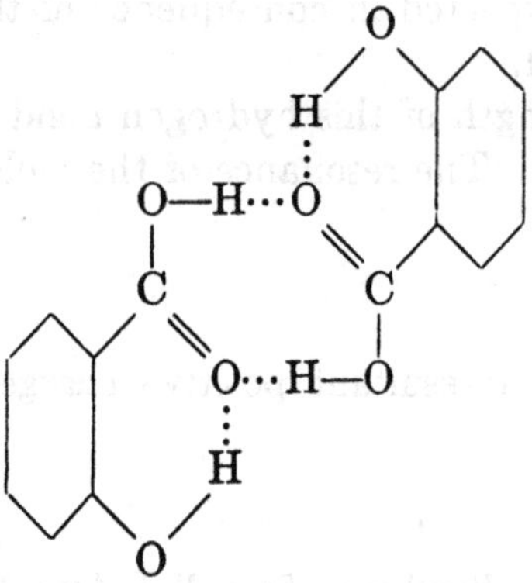

The two carboxyl groups are joined as in the dimer of formic acid, and in addition each hydroxyl group is bonded to an oxygen atom of the adjacent carboxyl group.[87] It has become customary to refer to this ring formation as *chelation* (from χηλή, a crab's claw) through hydro-

---

[86] F. T. Wall and F. W. Banes, *J.A.C.S.* **67**, 898 (1945).

[87] This structure is found also in the crystal: W. Cochran, *Acta Cryst.* **4**, 376 (1951).

gen-bond formation, the term being used also in a wider sense.[88]

The effect of chelation or internal hydrogen-bond formation on the properties of salicylic acid is striking. Branch and Yabroff[89] have pointed out that salicylic acid is a much stronger acid than its meta and para analogs because of the effect of the hydrogen bond with the hydroxyl group in saturating in part the proton attraction of the carboxylate ion. The effect is still more pronounced in 2,6-dihydroxybenzoic acid,[90] with the structure

This substance is a stronger acid than phosphoric acid or sulfurous acid; its acid constant has the value 5×10^{-2}.

In crystals of *o*, *m*, and *p*-hydroxybenzoic acids hydrogen bonds are formed between molecules (and for the ortho compound within the molecule), whereas only the ortho compound can form the chelate bond in the single molecules of the vapor. The heat of sublimation is hence expected to be smaller for salicylic acid than for its analogs, and related properties should differ accordingly. This is observed to be the case; the vapor pressures of the substances at 100°C have the relative values 1320, 5, and 1. The quantities $RT \ln 1320/5$ and $RT \ln 1320/1$ have the values 4.16 and 5.36 kcal/mole, respectively; we deduce therefore that the energy of the hydrogen bond in the *o*-hydroxybenzoic acid molecule is approximately 4.7 kcal/mole. In this argument the reasonable assumptions are made that the free-energy values of the three crystals are the same and that the free-energy values of the gases differ only by the hydrogen-bond energy of the ortho compound.

The effect of hydrogen bonds on the physical properties of crystals is shown in a striking way by oxalic acid. This substance exists in two anhydrous crystal forms.[91] One of these, the α form, contains layers of molecules held together by hydrogen bonds, the structure of a layer

[88] G. T. Morgan and H. D. K. Drew, *J. Chem. Soc.* 117, 1457 (1920).

[89] G. E. K. Branch and D. L. Yabroff, *J.A.C.S.* 56, 2568 (1934).

[90] W. Baker, *Nature* 137, 236 (1936).

[91] S. B. Hendricks, *Z. Krist.* 91, 48 (1935); E. G. Cox, M. W. Dougill, and G. A. Jeffrey, *J. Chem. Soc.* 1952, 4854.

being represented schematically by the following diagram

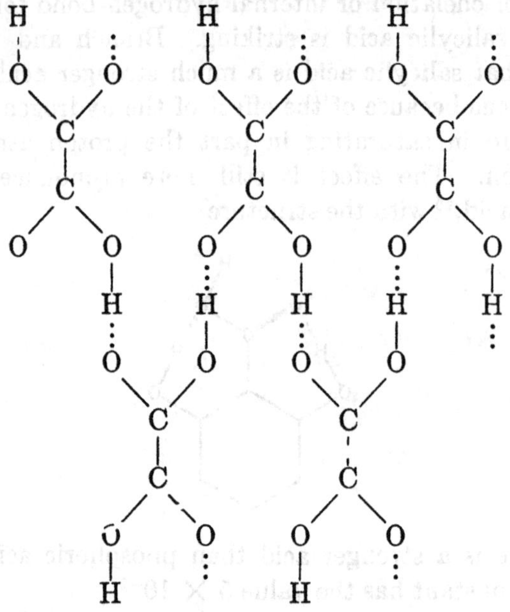

The crystal can correspondingly be easily cleaved into layers, this cleavage not breaking any of the hydrogen bonds. In the β form of the crystals there are long chains of molecules, with the structure

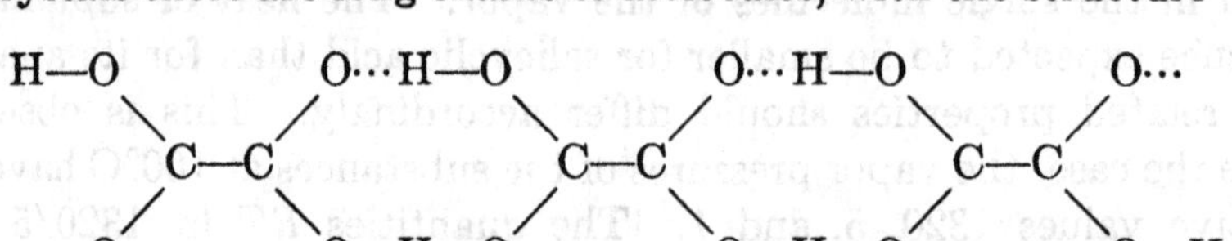

These crystals cleave along two planes parallel to the axis of these strings, breaking up into long laths. The O—H···O distances for both forms are about 2.65 Å.

Similar structures have been found for many other dicarboxylic acids, including succinic acid,[92] $COOH(CH_2)_2COOH$; glutaric acid, $COOH(CH_2)_3COOH$; adipic acid, $COOH(CH_2)_4COOH$, and sebacic acid, $COOH(CH_2)_8COOH$. Crystal structure determinations have also been made of many carboxylic acid hydrates; in all of the crystals the carboxyl groups form hydrogen bonds, usually with water molecules. An example is oxalic acid dihydrate;[93] in this crystal the O—H···O distance is 2.50 Å.

Of the many crystals of known structure containing hydrogen bonds

[92] J. D. Morrison and J. M. Robertson, *J. Chem. Soc.* 1949, 980.
[93] J. D. Dunitz and J. M. Robertson, *J. Chem. Soc.* 1947, 142.

we shall mention a few, in addition to those referred to above, in order to indicate the stereochemical properties of this bond.

Boric acid[94] contains layers of $B(OH)_3$ molecules held together by hydrogen bonds as indicated in Figure 12-10, which represents a portion of one layer. The crystal cleaves easily along the layer plane. Each oxygen atom forms two hydrogen bonds, the O—H$\cdots$O distance being 2.72 ± 0.01 Å. These bonds are coplanar with the BO_3 groups.

The substances ammonium trihydrogen paraperiodate, $(NH_4)_2H_3IO_6$, and potassium dihydrogen phosphate, KH_2PO_4, contain one hydrogen atom for every two oxygen atoms. In their crystals hydrogen bonds are formed between oxygen atoms of adjacent complex anions, each oxygen atom forming one such bond. The structure[95] of the hexagonal

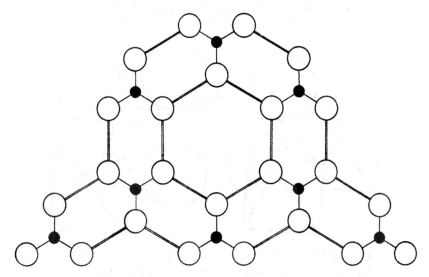

Fig. 12-10.—The arrangement of atoms in a layer of the boric acid crystal. Large circles represent oxygen atoms and small circles boron atoms. The double lines represent hydrogen bonds.

crystal $(NH_4)_2H_3IO_6$ is shown in Figure 12-11. The IO_6 groups, which lie on trigonal axes, are rotated about these axes in such a way as to bring each oxygen atom to the distance 2.60 ± 0.05 Å from an oxygen atom of an adjacent group, with which it forms a hydrogen bond. The structure[96] of the tetragonal crystal KH_2PO_4 is similar; the PO_4 groups rotate about the diagonal axes on which they lie to give the O—H$\cdots$O bond distance the value 2.487 ± 0.005 Å.

[94] W. H. Zachariasen, *Z. Krist.* **88**, 150 (1934); *Acta Cryst.* **7**, 305 (1954).

[95] L. Helmholz, *J.A.C.S.* **59**, 2036 (1937).

[96] S. B. Hendricks, *Am. J. Sci.* **15**, 269 (1927); J. West, *Z. Krist.* **74**, 306 (1930); G. E. Bacon and R. S. Pease, *Proc. Roy. Soc. London* **220A**, 397 (1953).

FIG. 12-11.—The structure of the crystal $(NH_4)_2H_3IO_6$.
Circles represent ammonium ions, octahedra $[IO_6]^{5-}$ ions,
and double lines hydrogen bonds.

In diaspore, $AlHO_2$, with the structure[97] represented in Figure 12-12, the oxygen atoms occur in pairs connected by a hydrogen bond; the crystal accordingly contains the group O—H···O. The O—H···O distance is 2.650 ± 0.003 Å. The related crystal lepidocrocite,[98] FeO(OH), contains oxygen atoms of two kinds (Fig. 12-13). Those of the first kind are bonded to iron atoms only, whereas each of those of the second kind forms two hydrogen bonds.

A valuable contribution to the understanding of the nature of the

[97] F. J. Ewing, *J. Chem. Phys.* **3**, 203 (1935). Groutite, $MnHO_2$, has the same structure: R. L. Collin and W. N. Lipscomb, *Acta Cryst.* **2**, 104 (1949).
[98] F. J. Ewing, *Acta Cryst.* **3**, 420 (1935).

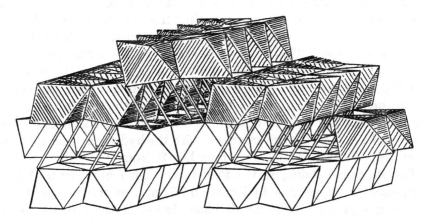

FIG. 12-12.—The structure of the diaspore crystal, AlHO$_2$. Oxygen atoms are at the corners of the octahedra and aluminum atoms at their centers. Hydrogen bonds are indicated by tubes.

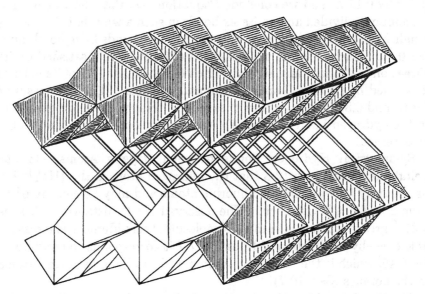

FIG. 12-13.—The structure of lepidocrocite, FeO(OH).

hydrogen bond and its contribution to the stabilization of structures has been made by Busing and Levy, in their accurate determination of the atomic positions in diaspore, including the location of the hydrogen atoms.[98a]   As shown in Figure 12-12, each aluminum atom is ligated to six oxygen atoms.   The aluminum-oxygen bond lengths are Al—O_I = 1.858 ± 0.004 Å (1) and 1.851 ± 0.002 Å (2) and Al—O_{II} = 1.980 ± 0.003 Å (1) and 1.975 ± 0.003 Å (2).   The atoms O_I are those that do not define the edges shared between adjacent octahedra, and O_{II} are those that define these edges.   The O—H···O bond distance was found to have the value given above, with the hydrogen atom 1.005 Å from O_{II}.   The bond O_{II}—H deviates by 12° from O_{II}- - -O_I.

The stronger bonding to O_{II} than to O_I (bond numbers 0.85 and 0.09, respectively, as given by Equation 7-7) may be understood by the following argument.   The Al—O bonds have a considerable amount of ionic character, such as to place a positive electric charge on the aluminum atoms.   The closest approach of two aluminum atoms is across a shared octahedral edge.   It is to be expected that the electrostatic repulsion of these atoms would cause a stretching of the O_{II}—Al bonds (as well as a decrease in the length of the shared edge, as discussed in Sec. 13-6).   The observed Al—O_I and Al—O_{II} bond lengths differ by 0.12 Å, and we conclude (Equation 7-7) that the valence 3 of aluminum is divided among its six bonds in such a way that the Al—O_I bonds have bond number 0.61 and the Al—O_{II} bonds have bond number 0.39.   Each O_I atom thus has 1.83 of its valence 2 satisfied by its three bonds to aluminum atoms.   Each O_{II} atom has 1.17 of its valence 2 satisfied by its three bonds to aluminum atoms.   The valences 0.17 for O_I and 0.83 for O_{II} remain unsatisfied.   Hence the hydrogen atom in the hydrogen bond between O_I and O_{II} is closer to O_{II} than to O_I, as observed.

Symmetrical Hydrogen Bonds between Oxygen Atoms.—It was pointed out in Section 12-3 that the hydrogen atom in the $[HF_2]^-$ ion lies midway between the two fluorine atoms and may be considered to form a half-bond with each.   The observed F—H distance, 1.13 Å, is 0.21 Å greater than it is in the HF molecule; this difference is a reasonable one—by application of Equation 7-7 it corresponds to bond number 0.45, which is the value usually found for bridging hydrogen atoms in the boranes (Sec. 10-7).

The O—H bond length in water is 0.96 Å.   The O—H distance for a symmetrical O—H—O hydrogen bond is 1.17 Å; the two oxygen atoms would then be 2.34 Å apart.

For most hydrogen bonds between oxygen atoms the O- - -O distance

[98a] W. Busing and H. Levy, *Acta Cryst.* 11, 798 (1958).

lies between 2.50 Å and 2.80 Å. The values 2.44 ± 0.02 Å for nickel dimethylglyoxime[99] and 2.40 ± 0.02 Å for acetamide hemihydrochloride,[100] $(NH_2COCH_3)_2 \cdot HCl$ are exceptions. The location of the hydrogen atom midway between the two atoms in the hydrogen diacetamide cation has been verified by a neutron-diffraction study of the crystal.[101]

The structure of this ion is

$$\left[\begin{array}{c} \text{CH}_3 \\ \text{H}_2\text{N}-\text{C} \\ \text{O}-\text{H}-\text{O} \\ \text{C}-\text{NH}_2 \\ \text{H}_3\text{C} \end{array} \right]^{+}$$

The N—C distance is 1.303 ± 0.013 Å and the C—O distance is 1.244 ± 0.012 Å. These bond lengths indicate that the two structures N—C=O and N=C—O make nearly equal contributions: N—C=O about 57 percent and N=C—O about 43 percent. Hence the oxygen atom has about 0.43 of its valence unsatisfied.

The difference 0.24 Å between the O—H distance in this compound and in water corresponds to bond number 0.40 (Equation 7-7). Accordingly the formation of a symmetrical hydrogen bond permits the satisfaction of the valence of the oxygen atom.

12-7. THE SPECTROSCOPIC STUDY OF THE HYDROGEN BOND

A very important method of studying the hydrogen bond was developed by Wulf, Hendricks, Hilbert, and Liddel[102] and applied by them in the study of a large number of compounds. Some of the results of this investigation are described below. The experimental method used is the study of the infrared absorption spectrum of substances in carbon tetrachloride solution, the spectral region of interest being that corresponding to the frequencies characteristic of the stretching of O—H or N—H bonds. Similar studies have been made

[99] L. E. Godycki, R. E. Rundle, R. C. Voter, and C. B. Banks, *J. Chem. Phys.* 19, 1205 (1951); L. Godycki and R. E. Rundle, *Acta Cryst.* 6, 487 (1953).

[100] W. J. Takei and E. W. Hughes, *Acta Cryst.*, in press (1959).

[101] E. W. Peterson and H. Levy, unpublished communication to E. W. Hughes.

[102] U. Liddel and O. R. Wulf, *J.A.C.S.* 55, 3574 (1933); O. R. Wulf and U. Liddel, *ibid.* 57, 1464 (1935); G. E. Hilbert, O. R. Wulf, S. B. Hendricks, and U. Liddel, *Nature* 135, 147 (1935); *J.A.C.S.* 58, 548 (1936); S. B. Hendricks, O. R. Wulf, G. E. Hilbert, and U. Liddel, *ibid.* 1991; O. R. Wulf, U. Liddel, and S. B. Hendricks, *ibid.* 2287; O. R. Wulf and L. S. Deming, *J. Chem. Phys.* 6, 702 (1938).

by many other investigators. A detailed account of the work in this field is given in a book by Pimental and McClellan.[103]

Compounds Showing Strong Hydrogen-Bond Formation.—The frequency of the vibration corresponding essentially to the stretching of the O—H bond in molecules containing this group lies in the neighborhood of 3500 cm⁻¹ (in wave-number units), its first overtone being at about 7000 cm⁻¹. The absorption spectrum in this infrared region of a solution of methanol in carbon tetrachloride solution[104] is shown in Figure 12-14; it consists of a well-defined peak at about 7115 cm⁻¹.

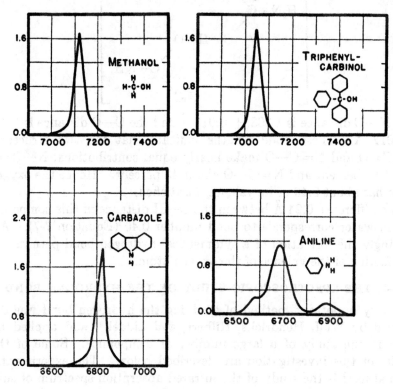

Fig. 12-14.—Infrared absorption spectra of methanol, triphenylcarbinol, carbazole, and aniline in carbon tetrachloride solution (Hilbert, Wulf, Hendricks, and Liddell). Ordinates represent the molal absorption coefficient and abscissas the wave number, in cm⁻¹.

Other alcohols show a similar absorption spectrum; for triphenylcarbinol, for example, there is no noticeable difference from methanol except for a shift in frequency to 7050 cm⁻¹. The N—H group gives similar spectra, in the region near 6850 cm⁻¹, as is seen from the curve

[103] Pimentel and McClellan, *op. cit.* (1).
[104] See the articles listed in footnote 102.

reproduced in Figure 12–14 for carbazole, [structure]. A more com-

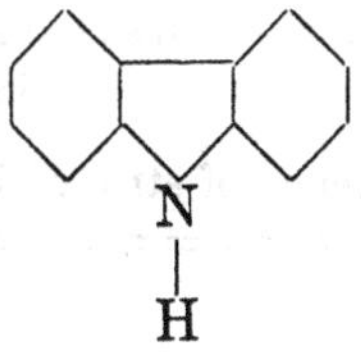

plex spectrum would be expected for the amino group, because of the
presence of two interacting N—H bonds; the curve shown for aniline
is characteristic of this group.

The striking observation was made by Hilbert, Wulf, Hendricks, and
Liddel that OH and NH groups that are involved in strong hydrogen
bond formation do not absorb radiation in this way in the 7000 cm^{-1}
region, nor in the region of the other overtones of the O—H and N—H
oscillations.   Instead of a sharp peak, the spectrum of these substances
shows only a weak and diffuse absorption band in these regions.   This
effect was observed for all the substances investigated which were
known from other evidence to contain hydrogen bonds, including, for

example, *o*-nitrophenol (Fig. 12–15) and salicylaldehyde, [structure] ,

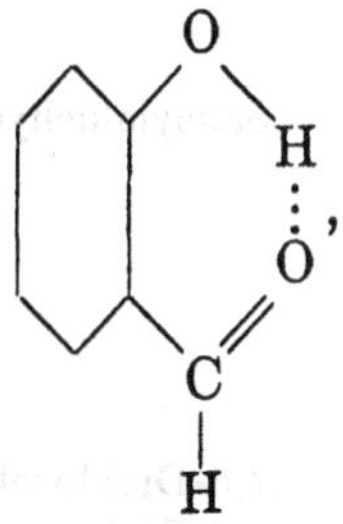

the physical properties of which indicate hydrogen-bond formation between the hydroxyl group and an adjacent oxygen atom.

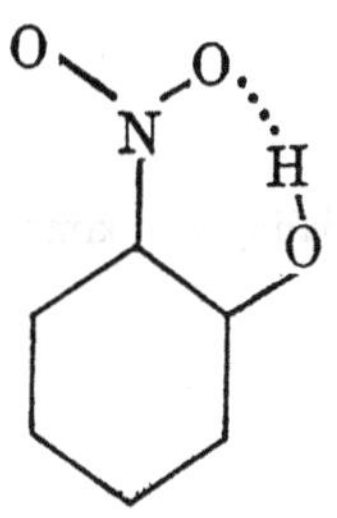

FIG. 12-15.—The atomic arrangement in the
ortho-nitrophenol molecule.   The interatomic
distances and bond angles are given their correct
values; it is seen that the C—H bond is directed
toward an oxygen atom of the nitro group.

It is probable that the change in the nature of the infrared absorption
bands on hydrogen-bond formation is connected with the interaction
between the vibration of the hydrogen atom and the vibration of the
heavier atoms, such that the infrared radiation is absorbed over a
wide range of frequencies, representing the combinations of the

TABLE 12-1.—Substances Forming Strong Intramolecular
Hydrogen Bonds

(Absence of strong absorption in the 7000 cm⁻¹ region)

---

o-Nitrophenol
2,6-Dinitrophenol
1-Nitronaphthol-2
2-Nitroresorcinol
Methyl salicylate, *o*-$C_6H_4OHCOOCH_3$
o-Hydroxyacetophenone, *o*-$C_6H_4OHCOCH_3$
1,4-Dihydroxy-5,8-naphthoquinone
1,5-Dihydroxyanthraquinone
4,6-Diacetylresorcinol
2,4-Dinitroresorcinol
4,6-Dinitroresorcinol
2,2′-Dihydroxybenzophenone
1,8-Dihydroxyanthraquinone
2,5-Dihydroxydiethylterephthalate

Acetylacetone, $CH_3{-}C{=}CH{-}C{-}CH_3$
                        | ‖
                        OH   O

Salicylaldehyde,

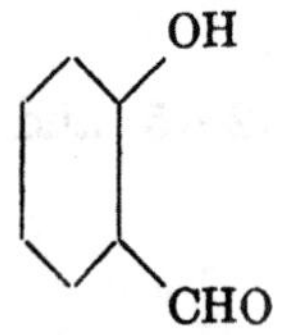

2,5-Dichlorobenzeneazo-1-naphthol-2
2,5-Dichloro-2′-hydroxy-4-methyl-5′-chloroazobenzene
Phenylazo-1-naphthol-2
Salicylaldehyde-anil, 2-$OHC_6H_4CH{=}NC_6H_5$
2-Hydroxy-5-methylbenzophenoneoxime acetate,

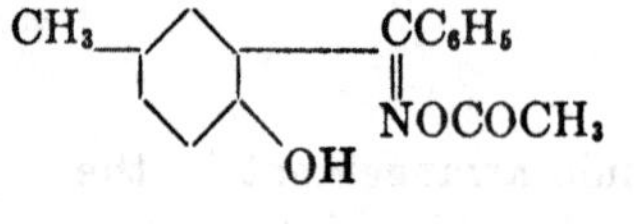

Salicylaldoxime acetate,

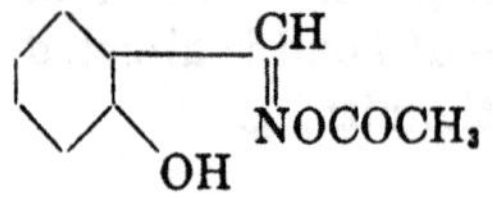

Salicylaldehyde-α-methyl-α-phenylhydrazone,

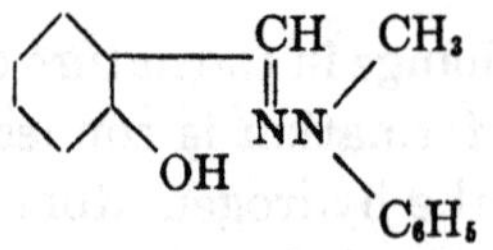

(*See next page*)

TABLE 12-1.—(*continued*)

Salicylaldehydedimethylhydrazone,

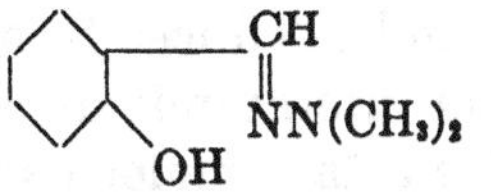

$$\text{CH}$$
$$\|$$
$$\text{NN(CH}_3)_2$$
OH

stretching frequency of the principal bond formed by the hydrogen atom and the many low vibrational frequencies of the heavier atoms of the groups connected by the hydrogen bond.[105]

This method of investigation, applied to nearly one hundred substances, has provided valuable information regarding the conditions that favor the formation of strong hydrogen bonds. In Table 12-1 a list is given of some of the substances found not to absorb strongly in the 7000 cm^{-1} region, and hence inferred to contain strong hydrogen

TABLE 12-2.—SUBSTANCES NOT FORMING STRONG INTRAMOLECULAR HYDROGEN BONDS

(Presence of strong absorption in the 7000 cm^{-1} region)

m-Nitrophenol
p-Nitrophenol
o-Cresol, *o*-C$_6$H$_4$CH$_3$OH
o-Chlorophenol
Catechol, *o*-C$_6$H$_4$(OH)$_2$
Resorcinol, *m*-C$_6$H$_4$(OH)$_2$
Hydroquinone, *p*-C$_6$H$_4$(OH)$_2$
Benzoin, C$_6$H$_5$COCHOHC$_6$H$_5$
Ethyl lactate, CH$_3$CHOHCOOC$_2$H$_5$
o-Hydroxybenzonitrile
o-Phenylphenol
3,6-Dibromo-2,5-dihydroxydiethylterephthalate
m-Hydroxybenzaldehyde
p-Hydroxybenzaldehyde
p-Hydroxyazobenzene

bonds between the OH or NH group and adjacent electronegative atoms in the molecule. A complementary list is given in Table 12-2 of molecules that do absorb strongly in this region; it is inferred that these substances do not form intramolecular hydrogen bonds at all or form only very weak bonds of the type discussed in the following section.

From these results it can be concluded that in *o*-nitrophenol and similar molecules the steric conditions for forming strong hydrogen

[105] R. M. Badger and S. H. Bauer, *J. Chem. Phys.* **5**, 369 (1937); M. Davies and G. B. B. M. Sutherland, *ibid.* **6**, 755 (1938); S. Bratoz, D. Hadzi, and N. Sheppard, *Spectrochim. Acta* **8**, 249 (1956); G. C. Pimentel, *J.A.C.S.* **79**, 3323 (1957).

bonds are satisfied, whereas in other molecules, such as *m*-nitrophenol and *o*-hydroxybenzonitrile, they are not.   The evidence provided by this spectroscopic method agrees in general with that found in other ways, and the rules that can be deduced from it can be interpreted in a reasonable way in terms of interatomic distances and bond angles.

The Formation of Weak Intramolecular Hydrogen Bonds.—The spectra of many of the substances containing hydroxyl groups studied by Wulf and his coworkers consist of a single sharp peak in the neighborhood of 7050 cm^{-1}, as illustrated in Figure 12-14.   Other substances which absorb strongly in this region (and are shown in this way not to be forming strong hydrogen bonds with use of the hydroxyl and amino groups) give curves of different types, involving pronounced frequency shifts and often splitting of the peak into two components, as shown in Figure 12-16.   It has been suggested[106] that this complexity of the observed spectra is due to the presence in the solution of two or more types of hydroxyl or amino groups with different characteristic frequencies—the groups of different type being either in different molecular species, as in the case of *o*-chlorophenol discussed below, or in the same molecule, as in catechol.   This suggestion has received strong support, through the experimental verification[107] of predictions based on it.

The substances resorcinol, hydroquinone, *m*-nitrophenol, and 2,6-dimethylphenol, as well as many others, show a single absorption peak resembling that of phenol very closely, not only in shape but also in position, the maxima for these five substances appearing at 7050, 7065, 7035, 7060, and 7050 cm^{-1}, respectively.   This shows that there is very little interaction between a phenolic hydroxyl group and another group substituted in the meta or para position (or, in the case of alkyl groups, in the ortho position) in the benzene ring; the interaction through the ring produces only small frequency shifts of the order of magnitude of 20 cm^{-1}.

In phenol and substituted phenols the C—O bond has some double-bond character, as discussed in Section 8-3.   This tends to cause the hydrogen atom to lie in the plane of the benzene ring.   The phenol molecule can thus assume either of the two configurations

O—H   H—O

⬡   and   ⬡   , which, however, are equivalent, so that we expect

[106] L. Pauling, *J.A.C.S.* **58**, 94 (1936).

[107] Wulf, Liddel, and Hendricks, *loc. cit.* (102); O. R. Wulf and E. J. Jones, *J. Chem. Phys.* **8**, 745 (1940); O. R. Wulf, E. J. Jones, and L. S. Deming, *ibid.* 753.

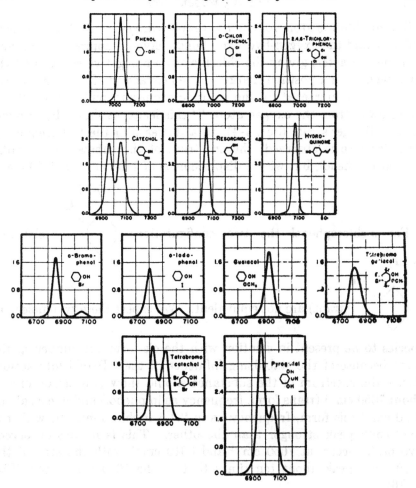

FIG. 12-16.—Infrared absorption spectra of phenol and related substances in carbon tetrachloride solution (Wulf and collaborators).

for phenol only one molecular species and a sharp OH absorption peak, as observed at 7050 cm^{-1}.

Similarly, the two configurations

and

for 2,4,6-trichlorophenol are equivalent, and we expect a single peak for this substance also.   However, we can predict that it will occur at a lower frequency than that for phenol, because of the attraction of the adjacent chlorine atom for the hydroxyl hydrogen.   Both the carbon-chlorine and the oxygen-hydrogen bonds have an appreciable amount of ionic character, giving the chlorine atom a negative charge and the hydrogen atom a positive charge.   The interaction of these causes the proton to be attracted by the chlorine atom and pulled a small distance away from the oxygen atom;[108] this leads, as shown by application of Badger's rule (Sec. 7-4), to a decrease in the OH frequency. The decrease is observed; the spectrum of the substance is similar to that of phenol, but with a displacement of 160 cm^{-1}, to 6890 cm^{-1}.

For *o*-chlorophenol   the   two   configurations   (cis)   and (trans) are not equivalent.   We expect these two molecular species to be present in solution with the cis form outnumbering the trans because of the stabilizing influence of the OH···Cl interaction. Hence the spectrum of the substance should show two peaks; one at about 7050 cm^{-1} (trans form, frequency as in phenol) and one at about 6890 cm^{-1} (cis form, frequency as in 2,4,6-trichlorophenol), with the 6890 cm^{-1} peak stronger than the other.   This is in fact observed; two peaks occur, at 7050 cm^{-1} and 6910 cm^{-1}, with the area of the 6910 cm^{-1} peak about ten times that of the 7050 cm^{-1} peak (Fig. 12-16).

The infrared absorption spectrum thus shows that *o*-chlorophenol in solution in carbon tetrachloride consists of about 91 percent cis molecules and 9 percent trans molecules.   The cis molecules are more stable than the trans molecules by a standard free-energy difference of about 1.4 kcal/mole (calculated from the ratio of the areas of the peaks). This is presumably the difference in free energy of the cis molecule with its intramolecular hydrogen bond and the trans molecule with a weaker hydrogen bond with a solvent molecule.

The weak hydrogen bond in *o*-chlorophenol stabilizes the gas molecule relative to those of the meta and para isomers, whereas the crystalline and liquid phases of the three substances, in which hydrogen bonds can be formed between adjacent molecules, have about the same stabil-

[108] The equilibrium O—H distance is increased by about 0.01 Å by interaction with the chlorine atom.

ity.   In consequence, the boiling point of the ortho isomer, 176°C, is lower than those of the others, 214° and 217°C, respectively.   The effect is shown also by the melting points, which have the values 7°, 0°, and −4° for three crystalline modifications of o-chlorophenol, 29° for m-chlorophenol, and 41° for p-chlorophenol.

An absorption peak at 6620 cm^{-1} has been observed for liquid o-chlorophenol by Errera and Mollet.[109]   The further decrease in frequency below the value 6910 cm^{-1} for the trans form of the molecule can be explained by assuming that the liquid contains double molecules with the structure

These double molecules would be stabilized by the energy of the strong O—H⋯O hydrogen bond.   The formation of the bond would increase the electronegativity of the oxygen atom on the right, causing its O—H bond to have increased ionic character; this would increase the positive charge on the attached hydrogen atom, and lead to a stronger O—H⋯Cl hydrogen bond, with its resultant decrease in OH vibrational frequency.

A spectrum of the same type as that of the solution has been observed for the vapor of o-chlorophenol also.[110]

The absorption curves for o-bromophenol and o-iodophenol are similar to those for o-chlorophenol, the shifted peaks lying at 6860 and 6800 cm^{-1}, respectively.   Guaiacol, o-methoxyphenol, shows a single peak at 6930 cm^{-1}, corresponding to the cis configuration

[109] J. Errera and P. Mollet, *J. phys. Radium* 6, 281 (1935).

[110] R. M. Badger and S. H. Bauer, *J. Chem. Phys.* 4, 711 (1936).   The effect of change in temperature in changing the relative amounts of the two molecular forms has been studied by L. R. Zumwalt and R. M. Badger (*ibid.* 7, 87 [1939]; *J.A.C.S.* 62, 305 [1940]), who found for the energy of the hydrogen bond in the gas molecule the value 3.9 ± 0.7 kcal/mole, and for the free energy of the bond 2.8 ± 0.5 kcal/mole.

O—H···O—CH$_3$; there seems to be no appreciable amount of the trans-

form present. In this molecule the O—H···O hydrogen bond that is formed is weak compared with other O—H···O bonds because of the unfavorable steric conditions.

The broad peak observed for tetrabromoguaiacol has its maximum close to 6810 cm^{-1}, showing that the proton attraction of the O—H···Br bond is greater than that of the O—H···O bond under the steric conditions present in this molecule.

Catechol shows two nearly equal peaks, at 6970 and 7060 cm^{-1}. Of the three configurations

for this molecule, the third is the most stable, inasmuch as it is stabilized relative to the second by the O—H···O interaction, and the first is made unstable by repulsion of the similarly charged hydrogen atoms. The third configuration accounts satisfactorily for the observed spectrum of two equal peaks.

The effect of the weak hydrogen bond on the boiling point of catechol is noticeable. The substance boils at 245°C, whereas the boiling point of resorcinol is 277° and that of hydroquinone 285°.

Pyrogallol shows a peak at 7050 cm^{-1} and another with doubled area at 6960 cm^{-1}; this spectrum corresponds to the structure

The two equal peaks observed for tetrabromocatechol at 6820 and 6920 cm^{-1} similarly correspond to the structure

The weak hydrogen bonds shown by this spectroscopic method to be present in the molecules discussed above and in many others are not of great significance in affecting the melting and boiling points and other physical properties of substances, nor do they lead to isomeric forms of substances of sufficient stability to permit their separation. It is, however, possible that these bonds are strong enough to influence the chemical properties of substances, and especially the rates of chemical reactions.

Factors Affecting Hydrogen-Bond Formation.—It is seen by reference to Tables 12-1 and 12-2 that a phenolic hydroxyl group forms a strong hydrogen bond with an oxygen atom of an adjacent nitro group. The conditions here are favorable to the formation of this bond. The conjugation of the groups with the benzene ring causes the planar configuration

to be stable. This places the hydroxyl oxygen atom 2.6 Å from a nitro oxygen atom, with the hydrogen atom directed approximately toward it.

The nitro group is able to form hydrogen bonds with two hydroxyl groups, as in 2-nitroresorcinol, with the structure

Carboxyl oxygen is also effective in hydrogen-bond formation, as in methyl salicylate, and in the dimer of salicylic acid and in 2,6-dihydroxybenzoic acid, mentioned in Section 12-6.

In 1,8-dihydroxyanthraquinone, with the structure

the carbonyl oxygen acts as receptor for two hydrogen bonds.

In most of these substances hydrogen-bond formation involves clos ing a six-membered ring (counting the hydrogen atom), the values of interatomic distances and bond angles being such as to favor the forma- tion of a strong hydrogen bond. On the other hand, a strong hydrogen bond is not formed with completion of a five-membered ring, the con- ditions being unfavorable. Ethyl lactate, for which the structure

can be written, shows a large infrared absorption peak at 6900 cm^{-1}, representing a weak hydrogen bond, and a small peak at 7050 cm^{-1}, cor- responding to a small number of molecules with configurations that do not permit this weak bond to be formed. The bond is weak in this substance for two reasons: the longer hydrogen-oxygen distance is large (2.02 Å, which is 0.22 Å greater than in ice and hence corresponds to a bond less than half as strong), and the hydrogen atom is not well di- rected toward the outer part of the oxygen atom, where the unshared electron pairs are located.

The possibility of forming a six-membered ring does not insure that a strong hydrogen bond will be formed, for other steric effects may operate unfavorably. In *o*-hydroxybenzonitrile, with the structure

the 180° C—C≡N bond angle causes the O—H···N distance to have a large value, about 3.5 Å, and, moreover, the hydrogen atom is not directed toward the unshared electron pair of the nitrogen atom; in consequence only a weak attraction occurs.

Another interesting example has been discussed by Hilbert, Wulf, Hendricks, and Liddel, that of 3,6-dibromo-2,5-dihydroxydiethyl-terephthalate. For this substance the configuration

might be expected, involving strong hydrogen bonds. The spectroscopic study shows, however, that the hydrogen bonds formed are weak, an absorption peak being observed at 6810 cm^{-1}. This is interpreted in a reasonable way as resulting from the steric repulsion of the bromine atoms for the ethoxy groups, which causes rotation about the C—COOC$_2$H$_5$ bonds and thus increases the O—H···O distance by several tenths of an Ångström. The effect is shown to be due to the bromine atoms by the fact that 2,5-dihydroxydiethylterephthalate contains strong hydrogen bonds, as shown by the absence of infrared absorption in the 7000 cm^{-1} region.

In general the usual rules of stereochemistry (planarity for conjugated systems, tetrahedral values of bond angles) apply to both of the atoms connected by hydrogen bonds. Many examples have been quoted of planar aggregates involving hydrogen bonds (boric acid, oxalic acid, etc.) and approximately tetrahedral bond angles for the hydrogen-bonded atoms. As mentioned earlier, it is not surprising that these rules apply with smaller force to the atom B than to the atom A of a group A—H···B, as illustrated by the urea crystal.

Other examples of structures involving strong hydrogen bonds are given in the following sections.

12-8. HYDROGEN BONDS IN PROTEINS

The polypeptide chains of protein molecules are coiled in a precise way. Hydrogen bonds play an important part in determining the configurations of these molecules. A great deal has been learned in recent years about the N—H···O hydrogen bonds formed by the peptide groups of the polypeptide chains; little is as yet known about the hydrogen bonds formed by the side chains of the amino-acid residues.

From the determination of the structure of crystals of amides and simple peptides the structure shown in Figure 12-17 has been assigned to the amide group in polypeptide chains. The N—C bond has about

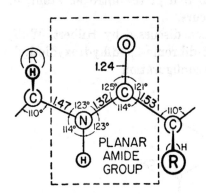

FIG. 12-17.—Fundamental dimensions of polypeptide chains as derived from x-ray crystal-structure determinations of amino acids and simple peptides.

40 percent double-bond character (bond length 1.32 Å). The group is planar, and it has been found to have the trans configuration in all substances studied except the cyclic peptides (diketopiperazine).

There is essential freedom of rotation about the single bonds between the amide groups and the α carbon atoms, permitting the polypeptide chain to assume many configurations. Certain configurations are stabilized by the formation of N—H···O hydrogen bonds.[111]

Structure determinations of crystals of amino acids and simple peptides have shown that in general the N—H···O bond is linear (to within about 10°), and that the nitrogen-oxygen distance is equal to 2.79 ±0.12 Å. The oxygen atom lies on the extension of the N—H bond axis. The energy of the hydrogen bond seems not to depend greatly on the angle at the oxygen atom, but there is some evidence that maximum stability results from having all four atoms N—H···O=C' on the same axis.

Estimates have been made of the amount of strain energy associated with the deviation from the optimum values of the structural parameters caused by steric factors.[112] These estimates can be expressed by

[111] For a summary of work in this field see Pauling and Corey, *loc. cit.* (10).
[112] L. Pauling and R. B. Corey, *Proc. Nat. Acad. Sci. U.S.* **37**, 251, 729 (1951).

giving the amount of change in the parameters that corresponds to a strain energy of 0.1 kcal/mole. This amount of strain is found from known force constants to correspond to stretching or compressing the single bonds to the α carbon atom by 0.02 Å or the conjugated bonds C'—N and C'—O by 0.01 Å, or to changing a bond angle by 3°, or to rotating the two ends of the amide group by 3° from the planar configuration. The value of the compressibility of ice, 12×10^{-6} cm²kg⁻¹, corresponds to the strain energy 0.1 kcal/mole for stretching or compressing the O—H···O hydrogen bonds (with length 2.76 Å) by 0.09 Å; this value presumably applies also to the N—H···O hydrogen bonds, with length 2.79 Å. A bend of the N—H···O bond by 6° from a straight angle at the hydrogen atom has been estimated to produce 0.1 kcal/mole of strain energy.

Two helical configurations of polypeptide chains have been found that satisfy the structural requirements for maximum stability of the amide groups and the N—H···O bonds.[113] One of these, called the γ helix, is a rather large helix, with a hole along its axis. It is probably made unstable, relative to other structures, by its small van der Waals attraction energy, and it has not been recognized in nature. The other structure, the α helix, is a compact arrangement of the polypeptide chain about the helical axis. It has been verified by x-ray diffraction and infrared birefringence as the configuration of many synthetic polypeptides and proteins, especially the fibrous proteins of the α-keratin class (hair, horn, fingernail, muscle). There is evidence also that the α helix is a principal structural feature of many globular proteins, such as hemoglobin.

The α helix is represented in Figure 12-18. Each amide group is attached by a hydrogen bond to the third one from it in either direction along the polypeptide chain. There are 3.60 amino-acid residues per turn of the helix. The total rise of the helix per turn—the pitch of the helix—is about 5.38 Å, which corresponds to 1.49 Å per residue. The amino-acid side chains extend away from the helix axis, as shown in Figure 12–18.

There are two stable arrangements of nearly completely extended polypeptide chains forming hydrogen bonds with neighboring chains.[114] They are the parallel-chain pleated sheet (Fig. 12–19) and the antiparallel-chain pleated sheet (Fig. 12–20). The identity distance in the direction of the chains is found to be different for the two structures when the requirement that the N—H···O bonds be linear is imposed;

[113] L. Pauling, R. B. Corey, and H. R. Branson, *Proc. Nat. Acad. Sci. U.S.* **37**, 205 (1951).

[114] L. Pauling and R. B. Corey, *Proc. Nat. Acad. Sci. U.S.* **37**, 729 (1951); **39**, 253 (1953).

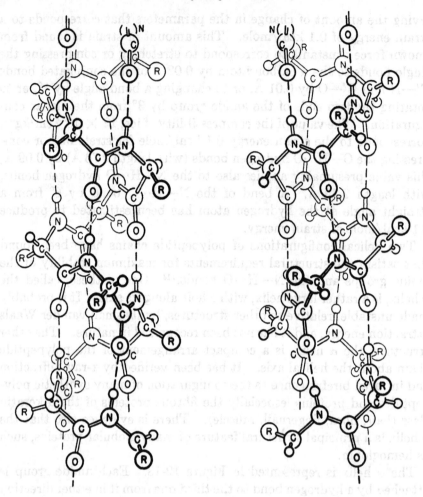

FIG. 12-18.—A drawing showing two possible forms of the alpha helix; the one on the left is a left-handed helix, and the one on the right is a right-handed helix. The amino acid residues have the L-configuration in each case.

it is 6.5 Å for the parallel-chain pleated sheet and 7.0 Å for the anti-parallel-chain pleated sheet. Silk fibroin and synthetic poly-*L*-alanine have been found to have the antiparallel-chain pleated-sheet structure.[115] It is likely that the β-keratin structure (assumed by the α-keratin proteins when they are stretched) is that of the parallel-chain pleated sheet.

[115] R. E. Marsh, R. B. Corey, and L. Pauling, *Biochim. et Biophys. Acta* **16**, 1 (1955); *Acta Cryst.* **8**, 710 (1955).

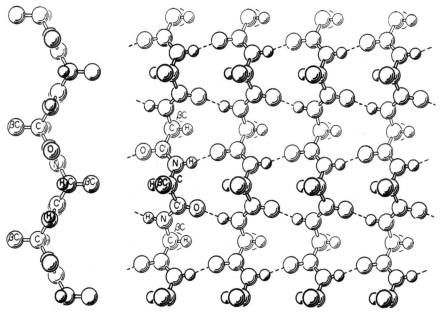

Fig. 12-19.—A drawing of the parallel-chain pleated sheet. The hydrogen bonds are approximately at right angles to the chain direction.

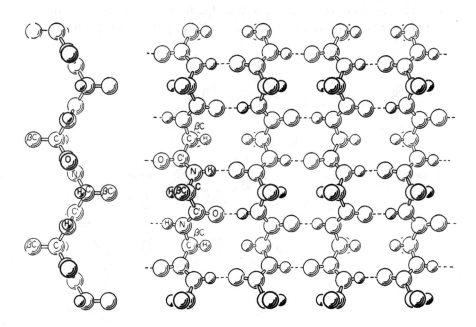

Fig. 12-20.—A drawing of the antiparallel-chain pleated sheet. Residues in Fig. 12-19 and 12-20 are shown with the D configuration. Pleated sheets with L amino-acid residues have structures that are the mirror images of those shown in the drawings.

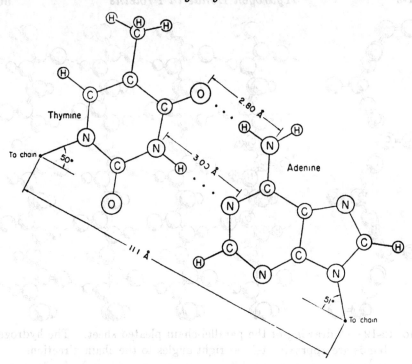

FIG. 12-21.—A drawing showing how molecules of adenine and thymine may form a complementary pair held together by two hydrogen bonds.

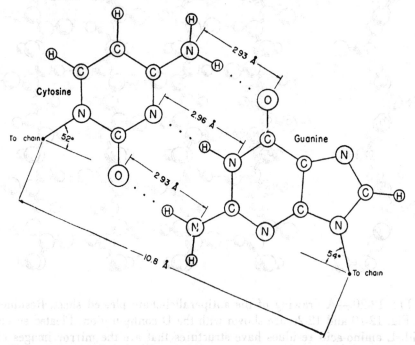

FIG. 12-22.—A drawing showing how cytosine and guanine may form a complementary pair held together by three hydrogen bonds.

12-9. HYDROGEN BONDS IN NUCLEIC ACIDS

Nucleic acids are of great interest because they are the units of heredity, the genes, and because they control the manufacture of proteins and the functions of the cells of living organisms. Hydrogen bonds play an important part in the novel structure proposed for deoxyribonucleic acid by Watson and Crick.[116] This structure involves a detailed complementariness of two intertwined polynucleotide chains, which form a double helix.[117] The complementariness in structure of the two chains was attributed by Watson and Crick to the formation of hydrogen bonds between a pyrimidine residue in one chain and a purine residue in the other, for each pair of nucleotides in the chains.

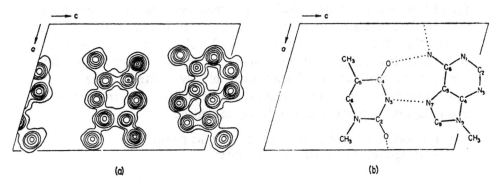

(a)   (b)

Fig. 12-23.—The arrangement of atoms in the crystal containing equal numbers of molecules of 1-methylthymine and 9-methyladenine, as determined by Hoogsteen. The two molecules with their hydrogen bonds are indicated by the diagram at the right. The diagram at the left shows the contour lines representing levels of electron density, as determined by the intensities of the x-ray diffraction maxima.

The pyrimidines found in the deoxyribonucleic acids are thymine and cytosine, and the purines are adenine and guanine. Their structures have been discussed in Section 8-8.

It is to be expected that these molecules would form with one another N—H$\cdots$O hydrogen bonds with length about 2.8 Å and N—H$\cdots$N hydrogen bonds with length about 3.0 Å. A reasonable way in which this might be done[118] is shown in Figures 12-21 and 12-22. This is

[116] J. D. Watson and F. H. Crick, *Nature* 171, 737, 964 (1953); *Cold Spring Harbor Symposia Quant. Biol.* 18, 123 (1953).

[117] The possibility that the duplication of genes occurs by a two-stage mechanism, with molecule A serving as a template for the synthesis of the complementary molecule A⁻¹, and then A⁻¹ serving as the template for the synthesis of a molecule complementary to it and identical with the original molecule A, was suggested by L. Pauling and M. Delbrück, *Science* 92, 77 (1940).

[118] L. Pauling and R. B. Corey, *Arch. Biochem. Biophys.* 65, 164 (1956).

essentially the arrangement proposed by Watson and Crick; it differs only in that three hydrogen bonds are formed between cytosine and guanine, whereas Watson and Crick had suggested, on the basis of some chemical evidence, that only two are formed.

A number of other types of hydrogen-bond formation between pyrimidines and purines have been discussed by Donohue.[119]

A strong indication that an arrangement differing from that of Watson and Crick is involved in nucleic acids has been provided by the study by Hoogsteen[120] of the crystal structure of the 1:1 compound of 1-methylthymine and 9-methyladenosine. In each of these nitrogen bases a methyl group has been attached in the position of attachment of the sugar (ribose or deoxyribose) in the nucleic acids. The structure found is shown in Figure 12-23. It is seen that one of the hydrogen bonds between the two molecules is not that assumed by Watson and Crick (Fig. 12-21): it involves the atom N_7 of the five-membered ring of adenine, instead of N_1 of the six-membered ring. Further complete structure determinations of substances closely related to the nucleic acids may be expected to provide a deeper insight into the nature of these important constituents of living organisms.

[119] J. Donohue, *Proc. Nat. Acad. Sci. U.S.* **42**, 60 (1956).
[120] K. Hoogsteen, *Acta Cryst.*, in press (1959).

CHAPTER 13

The Sizes of Ions and the Structure

of Ionic Crystals

OF all the different types of atomic aggregates, ionic crystals have been found to be most suited to simple theoretical treatment. The theory of the structure of ionic crystals described briefly in the following sections was developed about 40 years ago by Born, Haber, Landé, Madelung, Ewald, Fajans, and other investigators. The simplicity of the theory is due in part to the importance in the interionic interactions of the well-understood Coulomb terms and in part to the spherical symmetry of the electron distributions of the ions with noble-gas configurations.

13-1. INTERIONIC FORCES AND CRYSTAL ENERGY

The electron distribution function, as given by quantum-mechanical calculations, for an ion with the electronic configuration of a noble gas or with a completed 18 shell (such as Zn^{++}, with 18 outer electrons occupying the $3s$ orbital, the three $3p$ orbitals, and the five $3d$ orbitals in pairs) is spherically symmetrical,[1] showing that the interaction of the ion with other ions is independent of direction. The nature of the electron distribution functions for the alkali and halogenide ions is indicated by the drawings in Figure 13-1, which represent the results of theoretical calculations. It is seen that the successive $K, L, M, \cdots$ shells of electrons in an ion become evident as successive regions of large electron density. The electron distributions of isoelectronic ions, such as F^- and Na^+, are similar, and show the operation of increased effective nuclear charge from halogenide ion to the corresponding alkali ion in holding the electrons closer to the nucleus.

The interaction of two ions i and j with electric charges $z_i e$ and $z_j e$

[1] A. Unsöld, *Ann. Physik* **82**, 355 (1927).

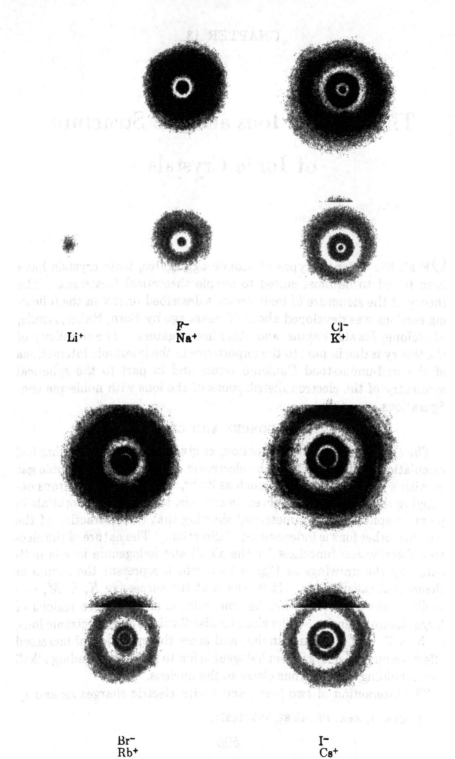

FIG 13-1.—Drawing showing the distribution of electrons
in the alkali and halogenide ions.

can be described in the following way. At large distances the ions attract or repel one another by Coulomb interaction of their charges, the potential function for this interaction being $z_i z_j e^2 / r_{ij}$, with r_{ij} the distance between the ions. In addition, some attraction results from the polarization of each ion in the field of the other; we shall neglect this attraction, which except at very small distances is negligible in comparison with the Coulomb attraction or repulsion. As the ions are brought closer together, so that their outer electron shells begin to overlap, an additional characteristic repulsive force becomes operative, resulting from the overlapping of the ions. It is this repulsive force that opposes the Coulomb attractive force between a positive and a negative ion and causes them to come to equilibrium at a finite value of the internuclear distance.[2]

The characteristic repulsive potential falls off very rapidly in value with increase in r_{ij}. It was suggested by Born that it be approximated by an inverse power of r_{ij}, so that the mutual potential energy of two ions would be written as

$$V_{ij} = \frac{z_i z_j e^2}{r_{ij}} + \frac{b_{ij} e^2}{r_{ij}{}^n} \tag{13-1}$$

The total potential energy of an ionic crystal MX with the sodium chloride arrangement can be obtained by summing the terms V_{ij} over all the pairs of ions in the crystal, and the quotient of this quantity by the number of stoichiometric molecules MX in the crystal is the potential energy of the crystal per molecule MX. Since in the crystal all of the interionic distances are related to the smallest interionic distance R by geometrical factors, the potential energy of the crystal can be written as

$$V = -\frac{A e^2 z^2}{R} + \frac{B e^2}{R^n} \tag{13-2}$$

The constant A in this expression is called the *Madelung constant*. It can be evaluated by straightforward mathematical methods.[3] Values of

---

[2] In addition to these interactions the van der Waals interactions (dispersion forces) between the ions in an ionic molecule or crystal should be considered. This effect has been discussed by M. Born and J. E. Mayer, *Z. Physik* **75**, 1 (1932), and by J. E. Mayer, *J. Chem. Phys.* **1**, 270 (1933). Multipole polarization of ions in alkali halogenide crystals has been discussed on the basis of a simple quantum-mechanical theory by H. Lévy, thesis, Calif. Inst. Tech., 1938.

[3] P. Appell, *Acta Math.* **4**, 313 (1884); E. Madelung, *Physik. Z.* **19**, 524 (1918); P. P. Ewald, *Ann. Physik* **64**, 253 (1921); M. Born, *Z. Physik* **7**, 124 (1921); O. Emersleben, *Physik. Z.* **24**, 73, 79 (1923); Y. Sakamoto, *J. Chem. Phys.* **28**, 164 (1958). For a brief description of these methods see J. Sherman, *Chem. Revs.* **11**, 93 (1932). Very simple methods of evaluating Madelung constants have been proposed by H. M. Evjen, *Phys. Rev.* **39**, 680 (1932), and K. Højendahl,

TABLE 13-1.—VALUES OF MADELUNG CONSTANTS[a]

Structure	A_{R_0}	A_{δ_0}	A_{α_0}
Sodium chloride, M^+X^-	1.74756	2.20179	3.49513
Cesium chloride, M^+X^-	1.76267	2.03536	2.03536
Sphalerite, M^+X^-	1.63806	2.38309	3.78292
Wurtzite, M^+X^-	1.64132	2.386	
Fluorite, $M^{++}X_2^-$	5.03878	7.33058	11.63656
Cuprite, $M_2^+X^{--}$	4.11552	6.54364	9.50438
Rutile, $M^{++}X_2^-$	4.816	7.70	
Anatase, $M^{++}X_2^-$	4.800	8.04	
CdI_2, $M^{++}X_2^-$	4.71	6.21	
β-Quartz, $M^{++}X_2^-$	4.4394	9.5915	
Corundum, $M_2^{+++}X_3^{--}$	25.0312	45.825	
Perovskite, $M^+M^{++}X_3^-$		12.37747	12.37747

[a] Values of A_{R_0} are such that the Coulomb energy per stoichiometric molecule is $- A_{R_0}e^2/R_0$ with R_0 the smallest anion-cation distance; those for A_{δ_0} and A_{α_0} have similar meanings, with δ_0 the cube root of the molecular volume and α_0 the edge of the cubic unit of structure (for cubic crystals). The parameter values used for structures containing parameters are those found experimentally.

A for the more important ionic crystals are given in Table 13-1. The magnitudes of the A values are seen to be reasonable on comparison with those for finite molecules; A for an isolated molecule Na^+Cl^- is 1, the Coulomb energy being $-1 \cdot e^2/R_0$, whereas for a sodium chloride crystal with the same interionic distance the crystal energy is about 75 percent greater, the value of A_{R_0} being 1.74756.

The factor z^2 is introduced in the Coulomb term to permit the application of the equation to crystals containing multivalent ions; with $z = 1$ it applies to substances Na^+Cl^-, $Mg^{++}F_2^-$, etc., and with $z = 2$ to $Mg^{++}O^{--}$, $Ti^{++++}O_2^{--}$, etc.

At equilibrium the attractive forces and the repulsive forces are balanced. The value of R (called R_0) at which this occurs can be found by differentiating V of Equation 13-2 with respect to R, equating to zero, and solving for R_0:

$$\frac{dV}{dR} = \frac{Ae^2z^2}{R^2} - \frac{nBe^2}{R^{n+1}}$$

$$\frac{Ae^2z^2}{R_0^2} - \frac{nBe^2}{R_0^{n+1}} = 0$$

$$R_0 = \left(\frac{nB}{Az^2}\right)^{1/(n-1)} \tag{13-3}$$

Kgl. Danske Videnskab. Selskab. **16,** 135 (1938). Values reliable to about 1 percent can be obtained by an expression involving the ligancies of the ions: D. H. Templeton, *J. Chem. Phys.* **21,** 2097 (1953); **23,** 1826 (1955).

With B and n known, this equation can be used to calculate R_0. Actually it is R_0 that can be easily determined experimentally; from its value the repulsive coefficient B can then be found by the equation

$$B = \frac{R_0{}^{n-1}Az^2}{n} \qquad (13\text{-}4)$$

provided that the value of n is known.

The Born exponent n can be evaluated from the results of experimental measurements of the compressibility of the crystal, which depends on the second derivative, d^2V/dR^2. It is found that for all crystals n lies in the neighborhood of 9. A somewhat better approximation to the experimental values is shown in Table 13-2; for a crystal of mixed-ion type an average of values of this table is to be used (6 for LiF, for example).

TABLE 13-2.—VALUES OF THE BORN EXPONENT n

Ion Type	n
He	5
Ne	7
Ar, Cu^+	9
Kr, Ag^+	10
Xe, Au^+	12

It is convenient to introduce the symbol $U_0 = -NV_0$ (with N Avogadro's number) to represent the *crystal energy*. U_0 is a positive quantity, representing the heat of formation per mole of MX from $M^+(g)$ and $X^-(g)$.

The substitution in Equation 13-2 of expression 13-4 for B leads to the following equation for the crystal energy U_0:

$$U_0 = \frac{NAe^2z^2}{R_0}\left(1 - \frac{1}{n}\right) \qquad (13\text{-}5)$$

It is seen that the crystal energy is smaller in magnitude than the Coulomb energy (with changed sign) by the fractional amount $1/n$, which is approximately 10 percent. With $R_0 = 2.814$ Å for sodium chloride and $n = 8$, U_0 is given the value 179.2 kcal/mole by this equation. This value, representing the heat of formation of NaCl(c) from $Na^+(g) + Cl^-(g)$, may be considered to be uncertain by about 2 percent, that is, about 4 kcal/mole, because of uncertainty in the form of the energy function V. A more refined treatment,[4] involving consider-

---

[4] Born and Mayer, *loc. cit.* (2); J. E. Mayer and L. Helmholz, *Z. Physik* **75**, 19 (1932).

ation of van der Waals forces and use of an exponential repulsive potential, has been found to give the value 183.1 kcal/mole, and a direct thermochemical measurement, mentioned below,[5] has provided the value 181.3 kcal/mole, substantiating the estimate of 2 percent as the order of reliability of the Born expression.

The Born-Haber Thermochemical Cycle.—The following cycle was devised by Born and Haber[6] to relate the crystal energy to other thermochemical quantities:

$$
\begin{array}{ccc}
& U & \\
MX(c) \longrightarrow & & M^+(g) + X^-(g) \\
\Big\uparrow{-Q} & & \Big\downarrow{-I+E} \\
& -S - \tfrac{1}{2}D & \\
M(c) + \tfrac{1}{2}X_2(g) \longleftarrow & & M(g) + X(g)
\end{array}
$$

(For convenience the cycle is given for the special case of an alkali halide.)   Here U is the crystal energy, I the ionization energy of the metal $M(g)$, E the electron affinity of $X(g)$, S the heat of sublimation of the metal, D the heat of dissociation of the halogen molecule, and Q the heat of formation of $MX(c)$ from the elements $M(c)$ and $\tfrac{1}{2}X_2(g)$. All of the quantities represent enthalpy change, $-\Delta H°$, for the corresponding reactions at 25°C.   The condition that the total change in enthalpy for the cycle is zero leads to the equation

$$ U = Q + S + I + \tfrac{1}{2}D - E \tag{13-6} $$

A few years ago experimental values were available for Q, S, I, and D, but not for E; the procedure adopted in testing the equation was to use the equation with calculated values of U_0 (Equation 13-5) to find E, and as a test of the method to examine the constancy of E for a series of alkali halogenides containing the same halogen.   The values obtained in this way were found to be constant to within about ± 3 kcal/mole.   However, later experimental determinations of the values of the electron affinities of the halogen atoms by direct methods have shown that Equation 13-5 for the crystal energy is in general reliable only to about 2 percent.

Work on the direct determination of the electron affinities of halogen atoms was begun by Mayer,[7] who, with his students, measured directly the equilibrium constant for dissociation of alkali halogenide gas molecules into ions or of gas halogenide ions into atoms and electrons. Other methods have also been used, especially some involving mass

[5] L. Helmholz and J. E. Mayer, *J. Chem. Phys.* **2**, 245 (1934).
[6] M. Born, *Verhandl. deut. physik. Ges.* **21**, 13 (1919); F. Haber, *ibid.* 750.
[7] J. E. Mayer, *Z. Physik* **61**, 789 (1930).

spectrometry. The values obtained in these ways[8] for $-\Delta H°$ of addition of an electron to a halogen atom at 25°C are F, 83.5; Cl, 87.3; Br, 82.0; and I, 75.7 kcal/mole. These values are good to about ± 1.5 kcal/mole.

We may illustrate the reliability of the Born equation by an example. For NaF the values $Q = 136.0$, $S = 26.0$, $I = 120.0$, $\frac{1}{2}D = 18.3$, and $E = 83.5$ kcal/mole (all for 25°C) lead, with Equation 13-6, to 216.8 kcal/mole for U, the crystal enthalpy at 25°C. The value of U_0 calculated by Equation 13-5 with $R_0 = 2.307$ Å and $n = 7$ is 215.5 kcal/mole. This value, with the pV correction 1.2 kcal/mole, leads to 216.7 kcal/mole for U, in excellent agreement with the experimental value. The mean deviation found for the alkali halogenides is 3 kcal/mole.

In general the results of recent investigations support the thesis that the forces operative in ionic crystals are those, described above, that underlie the Born equation for the crystal energy; and we may feel justified in investigating the further consequences of this postulate. The question of the sizes of ions is studied from this point of view in the following section.

13-2. THE SIZES OF IONS: UNIVALENT RADII AND CRYSTAL RADII[9]

It is possible to make an approximate quantum-mechanical calculation of the forces operating between ions in a crystal and to predict values for the equilibrium interionic distance, the crystal energy, the compressibility, and other properties of the crystal. This calculation has been made in a straightforward manner for lithium hydride (Li^+H^-, with the sodium chloride structure) by Hylleraas, with results in good agreement with experiment.[10] A thorough theoretical treatment of

---

[8] For F: N. I. Yonov, *J. Exptl. Theoret. Phys. U.S.S.R.* **18**, 174 (1948); G. Kimball and M. Metlay, *J. Chem. Phys.* **16**, 779 (1948); J. L. Margrave, *ibid.* **22**, 636 (1954); I. N. Bakulina and N. I. Yonov, *Doklady Akad. Nauk S.S.S.R.* **105**, 680 (1955); T. L. Bailey, *J. Chem. Phys.* **28**, 792 (1958). For the other halogens: Helmholz and Mayer, *loc. cit.* (5); P. P. Sutton and J. E. Mayer, *J. Chem. Phys.* **3**, 20 (1935); J. J. Mitchell and J. E. Mayer, *ibid.* **8**, 282 (1940); K. J. McCallum and J. E. Mayer, *ibid.* **11**, 56 (1943); P. M. Doty and J. E. Mayer, *ibid.* 323; D. T. Vier and J. E. Mayer, *ibid.* **12**, 28 (1944); J. P. Blewett, *Phys. Rev.* **49**, 900 (1936); G. Glockler and M. Calvin, *J. Chem. Phys.* **3**, 771 (1935); **4**, 492 (1936); Bakulina and Yonov, *loc. cit.*; Bailey, *loc. cit.*

[9] The treatment given in this section and in some later sections of the chapter was published in 1927 (L. Pauling, *J.A.C.S.* **49**, 765 [1927]).

[10] E. A. Hylleraas, *Z. Physik* **63**, 771 (1930). The calculated value of the crystal energy is 219 kcal/mole, and the Born-Haber cycle value is 218 kcal/mole, using for the electron affinity of hydrogen the reliable quantum-mechanical value 16.480 kcal/mole (see *Introduction to Quantum Mechanics*, Sec. 29c). The calculated value for the lattice constant, 4.42 Å, is less reliable than the value

ionic crystals has been carried out by Löwdin.[11]   The theoretical treatment is complex and laborious, however, and for chemical considerations it is desirable to have a set of empirical or semiempirical values of ionic radii that reproduce experimental lattice constants to within 1 or 2 percent.

It has been found possible to formulate a semiempirical set of ionic radii by using as the starting point only five experimental values of interionic distances, namely, the observed cation-anion distances in NaF, KCl, RbBr, CsI, and Li_2O.   The way in which this was done is described below.

Since the electron distribution function for an ion extends indefinitely, it is evident that no single characteristic size can be assigned to it.   Instead, the apparent ionic radius will depend upon the physical property under discussion and will differ for different properties.   We are interested in ionic radii such that the sum of two radii (with certain corrections when necessary) is equal to the equilibrium distance between the corresponding ions in contact in a crystal.   It will be shown later that the equilibrium interionic distance for two ions is determined not only by the nature of the electron distributions for the ions, as shown in Figure 13-1, but also by the structure of the crystal and the ratio of radii of cation and anion.   We take as our standard crystals those with the sodium chloride arrangement, with the ratio of radii of cation and anion about 0.75 and with the amount of ionic character of the bonds about the same as in the alkali halogenides, and calculate crystal radii of ions such that the sum of two radii gives the equilibrium interionic distance in a standard crystal.

The crystals NaF, KCl, RbBr, and CsI, with observed interionic distances 2.31, 3.14, 3.43, and 3.85 Å respectively, are standard crystals—it will be seen later that their radius ratios are about 0.75.   (The value 3.85 Å for the Cs^+—I^- distance in the modification of cesium iodide with the sodium chloride arrangement was obtained by subtracting 2.7 percent from the observed value for the crystal with the cesium chloride arrangement.)   The size of an ion is determined by the distribution of the outermost electrons, which varies in a simple way for isoelectronic ions, being inversely proportional to the effective nuclear charge operative on these electrons.   The effective nuclear charge is equal to the actual nuclear charge Ze minus the screening effect Se of the other electrons in the ion, and we write for a sequence of isoelectronic ions the

---

for the energy, and the poor agreement with the experimental value 4.08 Å is not significant.

[11] P.-O. Löwdin, *A Theoretical Investigation into Some Properties of Ionic Crystals*, thesis, Uppsala, 1948; *Phil. Mag. Suppl.* **5**, 1 (1956).

equation

$$R_1 = \frac{C_n}{Z - S} \qquad (13\text{-}7)$$

in which C_n is a constant determined by the total quantum number of the outermost electrons in the ions. A complete set of values of the screening constant S has been obtained, partially by theoretical calculation[12] and partially by the interpretation of observed values of mole refraction[12] and x-ray term values[13] of atoms. For ions with the neon structure, for example, S for the outermost electrons has the value 4.52, and the effective nuclear charges for Na^+ and F^- are thus 6.48 e and 4.48 e, respectively. By division of the Na^+—F^- distance 2.31 Å in the inverse ratio of these values for the effective nuclear charge the values 0.95 Å for the crystal radius of sodium ion and 1.36 Å for that of fluoride ion were obtained.

In a similar way the values K^+ 1.33, Cl^- 1.81, Rb^+ 1.48, Br^- 1.95, Cs^+ 1.69, and I^- 2.16 Å were obtained. The value 0.60 Å was selected for Li^+ in order to give agreement, when combined with the oxygen radius 1.40 Å discussed below, with the observed Li^+O^{--} distance 2.00 Å in Li_2O.

For the alkali and halogenide ions these radii represent the relative extension in space of the outer electron shells; that is, they may be considered to be a measure of the relative sizes of the ions; and in addition they have such absolute values as to cause their sums to be equal to interionic distances in standard crystals. By the use of Equation 13-7 and the values for the constants C_n given by the alkali and halogenide ions radii were obtained for all ions with the helium, neon, argon, krypton, and xenon structures. These radii represent correctly the relative sizes of the outer electron shells of the ions, compared with those for the alkali and halogen ions; *they do not, however, have absolute values such that their sums are equal to equilibrium interionic distances.* The significance of the radii is the following: if the Coulomb attractive and repulsive forces in a standard crystal (with the sodium chloride arrangement) containing one of these cations, with charge $+ze$, and one of these anions, with charge $-ze$, were to have the magnitude corresponding to charges $+e$ and $-e$, respectively (as though the ions were univalent), and the characteristic repulsive forces were to retain their actual magnitude, the equilibrium interionic distance would be equal to the sum of these radii. That is, these radii are the radii the multivalent ions would possess if they were to retain their electron distributions but to enter into Coulomb interaction as if they were univalent.

[12] L. Pauling, *Proc. Roy. Soc. London* A114, 181 (1927).
[13] L. Pauling and J. Sherman, *Z. Krist.* 81, 1 (1932).

TABLE 13-3.—CRYSTAL RADII AND UNIVALENT RADII OF IONS

+1	++	+3	+4	+5	+6	+7	(inert gas)	−	−−	3−	4−
			C⁴⁺ 0.15 (0.29)	N⁵⁺ 0.11 (0.25)	O⁶⁺ 0.09 (0.22)	F⁷⁺ 0.07 (0.19)	He (0.93)	H⁻ 2.08 (2.08)			
Li⁺ 0.60 (0.60)	Be⁺⁺ 0.31 (0.44)	B³⁺ 0.20 (0.35)							O⁻⁻ 1.40 (1.76)	N³⁻ 1.71 (2.47)	C⁴⁻ 2.60 (4.14)
Na⁺ 0.95 (0.95)	Mg⁺⁺ 0.65 (0.82)	Al³⁺ 0.50 (0.72)	Si⁴⁺ 0.41 (0.65)	P⁵⁺ 0.34 (0.59)	S⁶⁺ 0.29 (0.53)	Cl⁷⁺ 0.26 (0.49)	Ne (1.12)	F⁻ 1.36 (1.36)	S⁻⁻ 1.84 (2.19)	P³⁻ 2.12 (2.79)	Si⁴⁻ 2.71 (3.84)
K⁺ 1.33 (1.33)	Ca⁺⁺ 0.99 (1.18)	Sc³⁺ 0.81 (1.06)	Ti⁴⁺ 0.68 (0.96)	V⁵⁺ 0.59 (0.88)	Cr⁶⁺ 0.52 (0.81)	Mn⁷⁺ 0.46 (0.75)	Ar (1.54)	Cl⁻ 1.81 (1.81)			
Cu⁺ 0.96 (0.96)	Zn⁺⁺ 0.74 (0.88)	Ga³⁺ 0.62 (0.81)	Ge⁴⁺ 0.53 (0.76)	As⁵⁺ 0.47 (0.71)	Se⁶⁺ 0.42 (0.66)	Br⁷⁺ 0.39 (0.62)			Se⁻⁻ 1.98 (2.32)	As³⁻ 2.22 (2.85)	Ge⁴⁻ 2.72 (3.71)
Rb⁺ 1.48 (1.48)	Sr⁺⁺ 1.13 (1.32)	Y³⁺ 0.93 (1.20)	Zr⁴⁺ 0.80 (1.09)	Cb⁵⁺ 0.70 (1.00)	Mo⁶⁺ 0.62 (0.93)		Kr (1.69)	Br⁻ 1.95 (1.95)			
Ag⁺ 1.26 (1.26)	Cd⁺⁺ 0.97 (1.14)	In³⁺ 0.81 (1.04)	Sn⁴⁺ 0.71 (0.96)	Sb⁵⁺ 0.62 (0.89)	Te⁶⁺ 0.56 (0.82)	I⁷⁺ 0.50 (0.77)			Te⁻⁻ 2.21 (2.50)	Sb³⁻ 2.45 (2.95)	Sn⁴⁻ 2.94 (3.70)
Cs⁺ 1.69 (1.69)	Ba⁺⁺ 1.35 (1.53)	La³⁺ 1.15 (1.39)	Ce⁴⁺ 1.01 (1.27)				Xe (1.90)	I⁻ 2.16 (2.16)			
Au⁺ 1.37 (1.37)	Hg⁺⁺ 1.10 (1.25)	Tl³⁺ 0.95 (1.15)	Pb⁴⁺ 0.84 (1.06)	Bi⁵⁺ 0.74 (0.98)							

These radii are called the *univalent radii* of the ions. Values of univalent radii are given in parentheses in Table 13-3.

The *crystal radii* of multivalent ions, such that the sum of two crystal radii is equal to the actual equilibrium interionic distance in a crystal containing the ions, can be calculated from their univalent radii by multiplication by a factor obtained from consideration of Equation 13-3. From this equation it is seen that the equilibrium interionic distance in a crystal containing ions with valence z is

$$R_z = \left(\frac{nB}{Az^2}\right)^{1/(n-1)}$$

If the Coulomb forces were to correspond to $z = 1$ (univalent ions), with the characteristic repulsive coefficient B unchanged, the equilibrium interionic distance would be

$$R_1 = \left(\frac{nB}{A}\right)^{1/(n-1)}$$

From these expressions it is seen that crystal radii R_z and univalent radii R_1 are related by the equation

$$R_z = R_1 z^{-2/(n-1)} \tag{13-8}$$

This equation, with the values of n given by Table 13-2, has been used in calculating the values of the crystal radii given in Table 13-3.

Values are given in the table for univalent and crystal radii of ions with outer 18-shells (Cu^+, Ag^+, Au^+, etc.) also. These were calculated by using the same values for C_n as for argon, krypton, and xenonlike ions, respectively, with the appropriate screening constants. It might seem at first that the 18-shell radii should be larger than the values calculated in this way, since there are ten electrons in the outermost subshell (nd) and only six (np) for the ions of noble-gas type. However, the nd orbitals have their maxima somewhat nearer the nucleus (for given effective nuclear charge) than the corresponding np orbitals; in consequence, the density of the ten d electrons is, in the outer part of the ion, about equal to that for six p electrons, and this effect permits the simple calculation to be made without correction.

Univalent radii and crystal radii are represented graphically in Figure 13-2 as functions of the atomic number Z, and crystal radii are shown also in Figure 13-3. It is seen that there is great regularity in the univalent-radii sequences. The crystal radii deviate from the univalent radii in an understandable way, which, however, introduces such apparent lack of system in the crystal radii as to have prevented

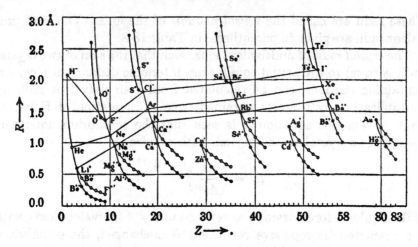

FIG. 13-2.—The crystal radii (solid circles) and
univalent radii (open circles) of ions.

an early satisfactory interpretation of the empirical information on interionic distances.

This valence effect shows up strikingly in the comparison of interatomic distances for isoelectronic sequences. In crystals containing
covalent bonds the interatomic distance remains almost constant

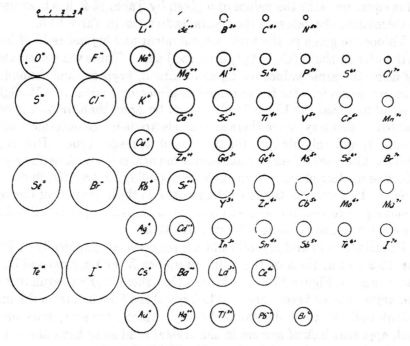

FIG. 13-3.—The crystal radii of ions.

through an isoelectronic sequence, such as Ge—Ge, 2.44 Å; Ga—As, 2.44 Å; Zn—Se, 2.45 Å; Cu—Br, 2.46 Å. Here the effect of decrease in nuclear charge of one atom is balanced by that of increase for the other. In ionic crystals, however, a decrease of about 10 percent is uniformly observed for the change from a crystal M^+X^- to its isoelectronic $M^{++}X^{--}$, as in the examples K^+—Br^-, 3.29; $Ca^{++}Se^{--}$, 2.96 Å; and Na^+—Cl^-, 2.81; Mg^{++}—S^{--}, 2.54 Å. This decrease is not due to lack of compensation of the changes in the electron distributions (the univalent radius sums remain nearly constant), but results instead from the effect of doubling the electric charges on the ions.

With increase in valence there are corresponding great increases in hardness, melting point, and other properties of ionic crystals of an isoelectronic sequence.

The comparison of the radii with experimental values of interionic distances and the discussion of various corrections are treated in the following sections.

The first roughly correct values assigned to ionic radii were those obtained by Landé[14] from the assumption that in the lithium halogenide crystals the halogen ions are in mutual contact (see Sec. 13-3). More accurate values were then given in 1923 by Wasastjerna,[15] who divided the observed interionic distances in crystals in ratios determined by the mole refraction values of the ions, the mole refraction being roughly proportional to ionic volume. The following values were given by Wasastjerna:

O^{--} 1.32 Å	F^- 1.33 Å	Na^+ 1.01 Å	Mg^{++} 0.75 Å
S^{--} 1.69	Cl^- 1.72	K^+ 1.30	Ca^{++} 1.02
Se^{--} 1.77	Br^- 1.92	Rb^+ 1.50	Sr^{++} 1.20
Te^{--} 1.91	I^- 2.19	Cs^+ 1.75	Ba^{++} 1.40

These agree with the values of Table 13-3 to within about 0.10 Å in general.

Wasastjerna's table of radii was then revised and greatly extended by Goldschmidt by the use of empirical data.[16] Goldschmidt based his values on Wasastjerna's values 1.33 Å for F^- and 1.32 Å for O^{--}, and, using data obtained from crystals that he considered to be essentially ionic in nature, he deduced from this starting point empirical values of the crystal radius for over 80 ions. His values (indicated by G) are compared with those from Table 13-3 in Table 13-4.

The agreement is good in general, and would be better if Goldschmidt had selected 1.40 Å instead of 1.32 Å for O^{--} as the basis for

[14] A. Landé, *Z. Physik* **1**, 191 (1920).

[15] J. A. Wasastjerna, *Soc. Sci. Fenn. Comm. Phys. Math.* **38**, 1 (1923).

[16] V. M. Goldschmidt, "Geochemische Verteilungsgesetze der Elemente," *Skrifter Norske Videnskaps-Akad. Oslo, I, Mat.-Naturv. Kl.*, **1926**.

TABLE 13-4.—COMPARISON OF CRYSTAL RADII FROM TABLE 13-3
WITH GOLDSCHMIDT'S VALUES

		Li	Be^{++}		
		0.60	0.31		
	G	.78	.34		
O^{--}	F$^-$	Na$^+$	Mg^{++}	Al^{3+}	Si^{4+}
1.40	1.36	0.95	0.65	0.50	0.41
G 1.32	1.33	.98	.78	.57	.39
S^{--}	Cl$^-$	K$^+$	Ca^{++}	Sc^{3+}	Ti^{4+}
1.84	1.81	1.33	0.99	0.81	0.68
G 1.74	1.81	1.33	1.06	.83	.64
Se^{--}	Br$^-$	Rb$^+$	Sr^{++}	Y^{3+}	Zr^{4+}
1.98	1.95	1.48	1.13	0.93	0.80
G 1.91	1.96	1.49	1.27	1.06	.87
Te^{--}	I$^-$	Cs$^+$	Ba^{++}	La^{3+}	Ce^{4+}
2.21	2.16	1.69	1.35	1.15	1.01
G 2.11	2.20	1.65	1.43	1.22	1.02

the values for bivalent ions. W. L. Bragg and his collaborators in their early important investigations[17] on the structure of silicates and related crystals selected 1.35 Å for the radius of O^{--} as well as F$^-$, this value having been indicated by the average O—O distance 2.7 Å observed in crystals showing anion contact (Sec. 13-5).

TABLE 13-5.—EMPIRICAL CRYSTAL RADII

Fr$^+$	1.76 Å	Ra^{++}	1.40 Å			Ac^{+++}	1.18 Å
NH$_4^+$	1.48	Yb^{++}	1.13	Ce^{+++}	1.11 Å	Th^{+++}	1.14
Ga$^+$	1.13	Ge^{++}	0.93	Pr^{+++}	1.09	Pa^{+++}	1.12
In$^+$	1.32	Sn^{++}	1.12	Nd^{+++}	1.08	U^{+++}	1.11
Tl$^+$	1.40	Pb^{++}	1.20	Pm^{+++}	1.06	Np^{+++}	1.09
Hf^{++++}	0.81	Pr^{++++}	0.92	Sm^{+++}	1.04	Pu^{+++}	1.07
Pr^{++++}	.92	Eu^{++}	1.12	Eu^{+++}	1.03	Am^{+++}	1.06
Ti^{++}	.90	Ti^{+++}	0.76	Gd^{+++}	1.02	Pa^{++++}	0.98
V^{++}	.88	V^{+++}	.74	Tb^{+++}	1.00	U^{++++}	.97
Cr^{++}	.84	Cr^{+++}	.69	Dy^{+++}	0.99	Np^{++++}	.95
Mn^{++}	.80	Mn^{+++}	.66	Ho^{+++}	.97	Pu^{++++}	.93
Fe^{++}	.76	Fe^{+++}	.64	Er^{+++}	.96	Am^{++++}	.92
Co^{++}	.74	Co^{+++}	.63	Tm^{+++}	.95		
Ni^{++}	.72	Ni^{+++}	.62	Yb^{+++}	.94		
Pd^{++}	.86	V^{++++}	.60	Lu^{+++}	.93		
		Cr^{++++}	.56				
		Mn^{++++}	.54				

[17] W. L. Bragg and J. West, *Proc. Roy. Soc. London*, A114, 450 (1927); W. L. Bragg, *The Atomic Structure of Minerals*, Cornell University Press, 1937.

Empirical crystal radius values, based on $O^{--} = 1.40$ Å and designed for application to the same standard crystals, are given in Table 13-5. These are in part obtained from Goldschmidt's set with suitable small corrections.[18]

13-3. THE ALKALI HALOGENIDE CRYSTALS

The alkali halogenides all crystallize with the sodium chloride arrangement (Figs. 1-1, 13-4) except cesium chloride, bromide, and

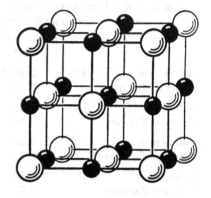

FIG. 13-4.—The arrangement of sodium ions and chloride ions in the sodium chloride crystal (see also Fig. 1-1).

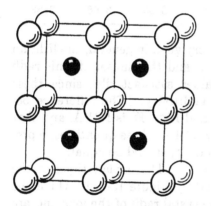

FIG. 13-5.—The arrangement of cesium ions and chloride ions in the cesium chloride crystal.

iodide,[19] which have the cesium chloride arrangement, shown in Figure 13-5. (The chloride, bromide, and iodide of rubidium have also been found to assume this structure under high pressure,[20] the transitions

---

[18] The values for the ions La^{+++} to Lu^{+++} are from D. H. Templeton and C. H. Dauben, *J.A.C.S.* **76**, 5237 (1954).

[19] These salts have been found to have the sodium chloride arrangement when deposited from the vapor onto cleavage surfaces of mica or certain other crystals: L. G. Schulz, *J. Chem. Phys.* **18**, 996 (1950). The observed interionic distances $Cs^+—Cl^- = 3.47$, $Cs^+—Br^- = 3.62$, $Cs^+—I^- = 3.83$ Å.

[20] J. C. Slater, *Phys. Rev.* **23**, 488 (1924); P. W. Bridgman, *Z. Krist.* **67**, 363 (1927); L. Pauling, *ibid.* **69**, 35 (1928); R. B. Jacobs, *Phys. Rev.* **53**, 930 (1938); **54**, 468 (1938). The potassium halides other than the fluoride are reported to show similar transitions at about 20,000 kg/cm²; see also P. W. Bridgman, *ibid.* **57**, 237 (1940).

occurring at about 5000 kg/cm²; and cesium chloride has a high-temperature modification with the sodium chloride arrangement, stable above 460°C.[21] Crystals of rubidium bromide grown from aqueous solution onto oriented silver films have the cesium chloride arrangement,[22] with Rb^+—Br^- = 3.53 Å. The observed values of interionic distances for the crystals with the sodium chloride structure are compared with the radius sums in Table 13-6.

TABLE 13-6.—INTERIONIC DISTANCES FOR ALKALI HALOGENIDE CRYSTALS WITH THE SODIUM CHLORIDE STRUCTURE

		Li^+	Na^+	K^+	Rb^+	Cs^+
Radius sum	F^-	1.96 Å	2.31 Å	2.69 Å	2.84 Å	3.05 Å
Observed distance		2.01	2.31	2.67	2.82	3.01
Radius sum	Cl^-	2.41	2.76	3.14	3.29	3.50
Observed distance		2.57	2.81	3.14	3.29	3.47
Radius sum	Br^-	2.55	2.90	3.28	3.43	3.64
Observed distance		2.75	2.98	3.29	3.43	3.62
Radius sum	I^-	2.76	3.11	3.49	3.64	3.85
Observed distance		3.02	3.23	3.53	3.66	3.83

It is seen that the agreement is not very good in general, the lithium salts showing especially large deviations, and that no set of ionic radii could reproduce the experimental values satisfactorily, since these values do not satisfy the criterion of additivity. The difference between the observed values for Li^+—I^- and Li^+—F^- is 1.01 Å, and that between Rb^+—I^- and Rb^+—F^- is only 0.84 Å; these quantities, representing the difference in radius of I^- and F^-, should be equal.

Anion Contact and Double Repulsion.[23]—The explanations of the deviations from additivity are indicated by Figure 13-6, in which the circles have radii corresponding to the crystal radii of the ions and are drawn with the observed interionic distances. It is seen that for LiCl, LiBr, and LiI *the anions are in mutual contact*, as suggested in 1920 by Landé.[14] A simple calculation shows that if the ratio $\rho = r_+/r_-$ of the radii of cation and anion falls below $\sqrt{2} - 1 = 0.414$ anion-anion contact will occur rather than cation-anion contact (the ions being considered as rigid spheres). A comparison of apparent anion radii in these crystals and crystal radii from Table 13-3 is given in Table 13-7.

The radius ratio for lithium fluoride is 0.44. In this crystal each anion is approaching contact not only with the surrounding cations but

[21] C. D. West, *Z. Krist.* **88**, 94 (1934).

[22] L. G. Schulz, *J. Chem. Phys.* **19**, 504 (1952).

[23] Pauling, *loc. cit.* (9).

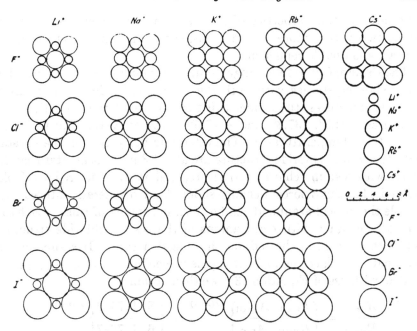

FIG. 13-6.—The arrangement of ions in cube-face layers of alkali halogenide crystals with the sodium chloride structure.

also with other anions.   In consequence, the repulsive forces are larger than they would be for either anion-cation or anion-anion contact alone, and equilibrium with the attractive Coulomb forces is reached with a lattice constant such that the cation-anion distance is larger than the sum of the radii and the anion-anion distance is also larger than twice the anion radius.   This phenomenon of *double repulsion* is shown by sodium iodide, bromide, and chloride.

It is seen that the radius ratio is an important quantity in influencing the properties of ionic crystals.   Its significance in the chemistry of ionic substances was first pointed out by Magnus[24] and was emphasized by Goldschmidt[25] in the field of crystal chemistry.   The effect of the ra-

TABLE 13-7.—HALOGENIDE ION RADII IN THE LITHIUM HALOGENIDES

	Apparent radius in Li^+X^-	Crystal radius
Cl^-	1.82 Å	1.81 Å
Br^-	1.95	1.95
I^-	2.12	2.16

[24] A. Magnus, *Z. anorg. Chem.* 124, 288 (1922)
[25] Goldschmidt, *loc. cit.* (16).

dius ratio on the properties of ionic substances is discussed below in connection with a more refined treatment of the phenomena of anion contact and double repulsion.

The Cesium Chloride Arrangement.—Ammonium chloride, bromide, and iodide crystallize with both the sodium chloride and the cesium chloride structures, the former being stable above the transition temperatures (184.3°, 137.8°, and −17.6°C., respectively), and the latter below these temperatures.[26] The interionic distances for the crystals with the cesium chloride structure are about 3 percent greater than for those with sodium chloride structure, and it was inferred by Goldschmidt that this 3 percent change should hold in general. Moreover, a simple theoretical argument can be given in support of this. Each cation is in contact with eight anions in the cesium chloride structure, but with only six in the sodium chloride structure. It seems not unreasonable to place the ratio of repulsive coefficients B_{CsCl}/B_{NaCl} equal to 8/6; with use of Equation 13-3 we then obtain

$$\frac{R_{CsCl}}{R_{NaCl}} = \left\{ \frac{B_{CsCl}}{B_{NaCl}} \cdot \frac{A_{NaCl}}{A_{CsCl}} \right\}^{1/(n-1)} = \left\{ \frac{8}{6} \cdot \frac{1.7476}{1.7627} \right\}^{1/(n-1)}$$

which with $n = 9$ gives $R_{CsCl}/R_{NaCl} = 1.036$ and with $n = 12$ gives 1.027. For this reason the Cs^+—I^- distance used in the derivation of radii in Section 13-2 was taken to be 2.7 percent less than the observed value.[27]

The observed interionic distances for the cesium and rubidium halogenides (the latter being at high pressure) with the cesium chloride structure compared with the crystal radius sums in Table 13-8.

The increase in interionic distance of approximately 3 percent is of especial interest in relation to the question of the relative thermodynamic stability of the two structures. It is seen from Equation 13-5 that if the Born postulate were valid the cesium chloride and sodium chloride modifications of a substance would have the same energy (and presumably nearly the same free energy) when the equilibrium interionic distances have the same ratio as the Madelung constants, that is, at the value $R_{CsCl}/R_{NaCl} = A_{CsCl}/A_{NaCl} = 1.0135$. Actually the transition occurs at about 1.030 for the rubidium halogenides, and the stability of the cesium chloride modification of the cesium halogenides shows that the equilibrium ratio is greater than 1.022 to 1.029 for them

[26] At still lower temperatures the substances undergo further transitions to forms characterized by decreased freedom of rotational motion of the ammonium ions.

[27] A more detailed discussion of the change in interionic distance accompanying this transition for the cesium and rubidium halides has been published: Pauling, *loc. cit.* (20).

TABLE 13-8.—INTERIONIC DISTANCES FOR CRYSTALS WITH THE
CESIUM CHLORIDE STRUCTURE

	Observed distance	Radius sum	Ratio
CsCl	3.56 Å	3.50 Å	1.027
CsBr	3.72	3.64	1.022
CsI	3.96	3.85	1.029
RbCl	3.41[a]	3.29	1.036
RbBr	3.53	3.43	1.028
RbI	3.75[b]	3.64	1.030

[a] Calculated from the density changes reported by Bridgman as accompanying the transitions. This value is probably a little too large.

[b] R. B. Jacobs, *Phys. Rev.* **54**, 468 (1938).

also. This requires the crystal energy for the cesium chloride structure to be greater than that given by the Born equation (relative to that for the sodium chloride structure) by about 2 percent, which is about 3 kcal/mole. Various suggestions have been made as to the source of this extra stability (van der Waals forces,[28] multipolar deformation,[29] etc.), but the question must be considered as still unsettled.

A Detailed Discussion of the Effect of Relative Ionic Sizes on the Properties of the Alkali Halogenides.—A simple detailed representation of interionic forces in terms of ionic radii has been formulated that leads to complete agreement with the observed values of interionic distances for alkali halogenide crystals and provides a quantitative theory of the anion-contact and double-repulsion effects.[30]

Let us assume that the mutual potential energy of two ions A and B at the distance r_{AB} can be expressed approximately by the equation

$$u_{AB} = \frac{z_A z_B e^2}{r_{AB}} + \beta_{AB} B_0 e^2 \frac{(r_A + r_B)^{n-1}}{r_{AB}^n} \qquad (13\text{-}9)$$

in which $z_A e$ and $z_B e$ are the electric charges of the ions, r_A and r_B constants representing their radii (which we shall call *standard radii*), B_0 is a characteristic repulsive coefficient, and β_{AB} is a constant with the value 1 for univalent cation-anion interaction, 1.25 for cation-cation interaction, and 0.75 for anion-anion interaction.[31] This form for the

[28] Born and Mayer, *loc. cit.* (2).

[29] Lévy, *op. cit.* (2).

[30] L. Pauling, *J.A.C.S.* **50**, 1036 (1928); *Z. Krist.* **67**, 377 (1928).

[31] The values for β_{AB} are obtained from a quantum-mechanical discussion: L. Pauling, *Z. Krist.* **67**, 377 (1928). Equation 13-9 and the following equations differ slightly from those originally published in that $(r_A + r_B)^{n-1}$ replaces $(r_A + r_B)^n$.

repulsive term, with inclusion of the factor $(r_A + r_B)^{n-1}$, is reasonable in that it makes the repulsive forces increase in magnitude with increase in the sizes of the ions.

For a crystal with the sodium chloride structure containing univalent cations and anions with radii r_+ and r_-, respectively, the total energy per molecule then becomes

$$V = -\frac{Ae^2}{R} + 6B_0e^2 \frac{(r_+ + r_-)^{n-1}}{R^n} + 6 \cdot 1.25 B_0 e^2 \frac{(2r_+)^{n-1}}{(\sqrt{2}\,R)^n}$$

$$+ 6 \cdot 0.75 B_0 e^2 \frac{(2r_-)^{n-1}}{(\sqrt{2}\,R)^n} \qquad (13\text{-}10)$$

in which the first term on the right, containing the Madelung constant, results from summing the Coulomb terms (the first term of 13-9), R being the minimum cation-anion distance in the crystal, the second term represents the repulsion between each cation and its six anion neighbors, the third term the repulsion of each cation and its nearest cation neighbors, at the distance $\sqrt{2}\,R$, and the fourth term the repulsion of anion-anion neighbors at the distance $\sqrt{2}\,R$. The repulsions of more distant ions are neglected. This equation can be rewritten in the form

$$V = -\frac{Ae^2}{R} + \frac{6B_0e^2}{R^n}\left\{(r_+ + r_-)^{n-1} + \frac{1.25(2r_+)^{n-1}}{(\sqrt{2})^n}\right.$$

$$\left. + 0.75\frac{(2r_-)^{n-1}}{(\sqrt{2})^n}\right\} \qquad (13\text{-}11)$$

which is analogous to Equation 13-2; the equilibrium value of R is accordingly found from Equation 13-3 to be

$$R_0 = (r_+ + r_-)F(\rho), \qquad (13\text{-}12)$$

in which $F(\rho)$ is a function of the radius ratio $\rho = r_+/r$, with the form

$$F(\rho) = \left(\frac{6nB_0}{A}\right)^{1/(n-1)}\left\{1 + \frac{1.25}{(\sqrt{2})^n}\left(\frac{2\rho}{\rho+1}\right)^{n-1}\right.$$

$$\left. + \frac{0.75}{(\sqrt{2})^n}\left(\frac{2}{\rho+1}\right)^{n-1}\right\}^{1/(n-1)} \qquad (13\text{-}13)$$

It is convenient to give B_0 a value ($nB_0 = 0.262$) such as to make $F(\rho)$ equal to unity for $\rho = 0.75$; this causes R_0 to be equal to the sum of the standard radii of cation and anion for crystals with this radius ratio, which was selected as standard in Section 13-2 because it is approximated by isoelectronic alkali and halogenide ion pairs. More-over, the exponent n is assumed to have the value 9 for all the alkali halogenides, for the sake of simplicity.

The form of the correction factor $F(\rho)$ as a function of ρ is shown in Figure 13-7. The broken line represents the same function for rigid spheres ($n = \infty$). It is seen that for ρ less than about 0.35 anion contact is effective in determining the equilibrium interionic distances, and between 0.35 and 0.60 the curve for $n = 9$ rises above the rigid sphere

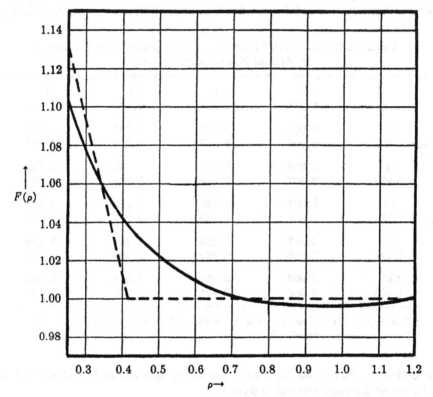

FIG. 13-7.—The function $F(\rho)$ showing the effect of radius ratio on equilibrium interionic distance of crystals with the sodium chloride arrangement.

curve because of the operation of the phenomenon of double repulsion. It is interesting to note that at $\rho = 0.28$, corresponding to LiI, $F(\rho)$ falls about 1 percent below the rigid anion contact curve; this explains the low value for the I^-—I^- distance in this crystal (Table 13-7).

It is possible by assigning suitable values to r_+ and r_- for the nine alkali and halogenide ions to calculate, with Equations 13-12 and 13-13, values of R_0 for the 17 alkali halogenides with the sodium chloride structure at room temperature that agree with the experimental values to within 0.001 Å on the average. The comparison of calculated and observed values is given in Table 13-9. The agreement is striking, especially when it is considered that the radius-ratio effect for lithium iodide amounts to 0.247 Å, nearly 10 percent of R_0; and there is accordingly little doubt that the deviations from additivity in the interionic distances in the alkali halogenide crystals are to be attributed to this effect.[32]

The standard radii r_+ and r_- for the ions have the same values as the corresponding crystal radii (to within 0.008 Å) except for F⁻ and Cs⁺, which show somewhat larger deviations, −0.019 and −0.034 Å, re-

TABLE 13-9.—CALCULATED AND OBSERVED INTERIONIC DISTANCES IN ALKALI HALOGENIDE CRYSTALS

	F⁻ $r_- = 1.341$	Cl⁻ 1.806	Br⁻ 1.951	I⁻ 2.168
Li⁺	2.009[a]	2.566	2.747	3.022
$r_+ = 0.607$	2.009	2.566	2.747	3.025
Na⁺	2.303	2.814	2.980	3.233
0.958	2.307	2.814	2.981	3.231
K⁺	2.664	3.139	3.293	3.529
1.331	2.664	3.139	3.293	3.526
Rb⁺	2.817	3.283	3.434	3.664
1.484	2.815	3.285	3.434	3.663
Cs⁺	3.005	3.451	3.598	3.823
1.656	3.005	3.47[b]	3.62[b]	3.83[b]

[a] The upper value of each pair is calculated, the lower observed.
[b] These values are unreliable.

spectively. It is possible that these deviations are to be attributed to the use of the constant value 9 for n.

[32] Similar calculations with use of an exponential form for the repulsive potential have been made by M. L. Huggins and J. E. Mayer, *J. Chem. Phys.* 1, 643 (1933), and M. L. Huggins, *ibid.* 5, 143 (1937). The problem has been treated also by J. A. Wasastjerna, *Soc. Sci. Fenn. Comm. Phys. Math.* VIII, 21 (1935).

It is possible to extend the treatment of this section, based on Equation 13-9, to crystals other than the alkali halogenides, with suitable choice of standard radii. It is found that the standard radii differ somewhat in general from the univalent radii of Table 13-3, because of the different choice of values of n. An approximate value for the standard radius of a multivalent ion can be obtained by multiplying its crystal radius by $z^{1/4}$, z being the magnitude of the valence of the ion; this is the correction factor from crystal radius to univalent radius corresponding to $n = 9$.

The deviations in additivity of interionic distances in the alkali halogenides resulting from the radius-ratio effect may be expected to be associated with irregularities in other properties of the substances. For some properties the radius ratio is unimportant; thus the interatomic distance in a gaseous diatomic salt molecule is not a function of it (for only the radius sum enters in the equation expressing the potential energy of two ions), nor is the energy of formation of such a molecule from free ions. In order to separate the effect of the radius ratio from other effects let us define for each substance a corresponding hypothetical standard substance, namely, one with the same radius sum $r_+ + r_-$ and the same ionic properties otherwise, but with the standard radius ratio $\rho = 0.75$. The properties attributed to this hypothetical substance will be designated as corrected for the radius effect or, briefly, corrected.

The properties to be expected for the hypothetical alkali halogenides with $\rho = 0.75$ are the following: The equilibrium interionic distances would be additive, being equal to $r_+ + r_-$. The crystal energy, which is inversely proportional to the interionic distances, would show a corresponding regularity. A large number of properties of salts depend essentially on the crystal energy—the heat of fusion, heat of sublimation, melting point, boiling point, solubility, etc. All of these properties would exhibit for the hypothetical alkali halogenides a regular dependence on the interionic distance, and hence the values of any one of these properties should vary monotonically for a sequence LiX, NaX, KX, RbX, CsX, or MF, MCl, MBr, MI. The properties of the actual alkali halogenides deviate greatly from this expected regularity, as is seen from Figures 13-8 and 13-9, which show on the left the experimental values of their melting points and boiling points.

The explanation of these irregularities is given by the radius-ratio effect. The crystal energy of the corrected and that of the actual alkali halogenide crystal can be calculated with Equation 13-5. Values of the difference ΔU_0 between the corrected and the actual energy obtained in this way are given in Table 13-10. This energy quantity is required to correct the heat of sublimation, as the energy of a gaseous molecule is not a function of the radius ratio.

The heat of sublimation at room temperature is equal to the sum of the heat of fusion at the melting point, the heat of vaporization at the boiling point, and the difference between the heat capacity of the solid and liquid and that of the vapor, integrated from room temperature to the boiling point; so that the energy correction ΔU_0 is to be divided among these quantities. For potassium chloride the heat of fusion amounts to 10 percent of the heat of sublimation, the integrated heat capacity difference to 30 percent, and the heat of vaporization to 60

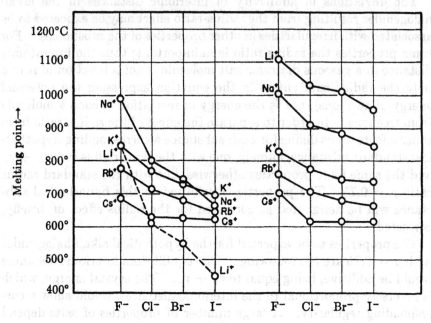

FIG. 13.8.—The observed melting points of the alkali halogenides (left) and values corrected for the radius-ratio effect (right.)

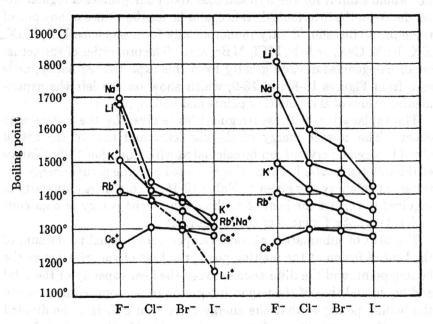

FIG. 13-9.—The observed boiling points of the alkali halogenides (left) and values corrected for the radius-ratio effect (right).

percent. It might be reasonable to apportion the correction ΔU_0 in these ratios; however, it is to be expected that the transition from crystal to liquid would destroy in some part the octahedral coordination of the ionic aggregate, causing the heat of fusion to assume a larger share of the radius-ratio correction, and that furthermore the coordination surviving in the liquid would decrease rapidly with increasing temperature, causing the heat content of the liquid also to assume more than its proportionate share of ΔU_0. The following calculations were made by apportioning 20 percent of ΔU_0 to the heat of fusion and 40 percent

TABLE 13-10.—THE EFFECT OF RADIUS RATIO ON CRYSTAL ENERGY, BOILING POINTS, AND MELTING POINTS OF THE ALKALI HALOGENIDES

		F⁻	Cl⁻	Br⁻	I⁻
Li⁺	$\Delta U_0 =$	7.9	12.7	13.8	15.1 kcal/mole
	$\Delta T_{BP} =$	132°	212°	230°	252°
	$\Delta T_{MP} =$	264°	424°	460°	504°
Na⁺		0.4	3.3	4.2	5.4
		7°	55°	70°	90°
		14°	110°	140°	180°
K⁺		− 0.6	0.1	0.5	1.2
		− 10°	2°	8°	20°
		− 20°	4°	16°	40°
Rb⁺		− 0.5	− 0.3	− 0.1	0.5
		− 8°	− 5°	− 2°	8°
		− 16°	− 10°	− 4°	16°
Cs⁺		0.5	− 0.4	− 0.3	− 0.1
		8°	− 7°	− 5°	− 2°
		16°	− 13°	− 10°	− 4°

to the heat of vaporization, these proportions being chosen partly in order to give satisfactory results in the consideration below of melting points and boiling points.

The heat of sublimation and related heat quantities are not known accurately enough throughout the series to permit a direct test of the effect. The boiling point of a substance is, however, related to its heat of vaporization by Trouton's rule, according to which the entropy of vaporization is a constant. For the alkali halogenides this constant is found experimentally to be about 25 cal/mole degree. If the relation is assumed to hold for the corrected alkali halogenides, the boiling point correction, in degrees, is $\Delta T_{BP} = 0.40 \Delta U_0 / 0.024$. If we similarly assume the entropy of fusion to be constant (Richard's rule) with the value 6.0 cal/mole degree, the melting point correction is $\Delta T_{MP} = 0.20$ $\Delta U_0/0.0060$. The values calculated for ΔT_{BP} and ΔT_{MP} are included in Table 13-10.

The observed melting points and boiling points of the alkali halogenides (Figs. 13-8 and 13-9, left side) show large irregularities; thus the boiling points and melting points of all the lithium salts lie below those of the corresponding sodium salts. It has been suggested[33] that these irregularities are due to deformation of the ions. Our calculations show, however, that they result mainly from the radius-ratio effect. The corrected melting points and boiling points vary in a regular manner throughout each sequence and correspond closely in qualitative behavior to the interionic distances, except for a small deviation shown by the cesium salts.

Alkali Halogenide Gas Molecules.—The amount of partial ionic character of the single bond in an alkali halogenide gas molecule is estimated[34] from the electronegativity differences to lie between 43 percent for LiI(g) and 94 percent for CsF(g). (The bonds in the crystals have roughly the same fraction of ionic character: the ionic structures are favored relative to the gas molecule by the increase in the Madelung constant, and the covalent structures are favored by having six or eight positions, rather than one; these two effects roughly counteract one another.) It might accordingly be expected that the structure and properties of the gas molecules could be discussed with use of an equation analogous to Equation 13-12, and with the same values of the standard radii and of the repulsion coefficient B_0.

It might be thought that this treatment would provide a poor approximation because of the neglect of polarization of each of the two ions in the electric field of the other.[35] However, there is reason to think that the neglect of polarization does not introduce great error. First, the effect of multipole polarization as well as of the partial covalent character of the bonds is taken into account in the treatment of the crystals by the evaluation of the Born exponent n from the observed compressibility and of the repulsion factor from the observed interionic distance. Second, in the gas molecule, in which there is dipole polarization mainly of the anion, its effect in causing increased attraction of the ions may be largely neutralized by the increased repulsion caused

[33] K. Fajans, *Z. Krist.* **61**, 18 (1925).

[34] These values should be decreased by a few percent to take account of the structures involving a bond formed with a pair of π electrons of the halogenide ion and a π orbital of the alkali ion.

[35] For a treatment with consideration of polarization, see E. S. Rittner, *J. Chem. Phys.* **19**, 1030 (1951); E. J. W. Verwey and J. H. deBoer, *Rec. trav. chim.* **59**, 633 (1940).

by the increase in electron density of the anion on the side adjacent to the cation.

During the last few years the application of the techniques of microwave spectroscopy to these gas molecules has provided much precise information about their properties.[36] It has been found[37] that the observed values of the internuclear distances can be closely approximated by use of an equation analogous to Equation 13-12. The potential energy of the molecule, analogous to 13-10 for the crystal, is

$$ V = -\frac{z^2 e^2}{r} + \frac{B_0 e^2 (r_+ + r_-)^{n-1}}{r^n} \qquad (13\text{-}14) $$

By differentiating V with respect to r and equating the derivative to zero, the equilibrium internuclear distance r_e is found to be

$$ r_e = (r_+ + r_-)(nB_0)^{1/(n-1)} \qquad (13\text{-}15) $$

The calculated values of r_e obtained by giving nB_0 its crystal value 0.262, r_+ and r_- their crystal values (Table 13-9), and n the crystal-compressibility values of Table 13-2 agree roughly with the observed values. The calculation is, however, sensitive to n; a change in n by 1 (this is about the reliability of the values in Table 13-2) corresponds to a change in r_e of about 0.050 Å. We may accordingly make use of the 15 accurate experimental values for r_e given in Table 13-11 to evaluate n. The values of n obtained in this way are given just below the experimental values of r_e in the table. They differ from those of Table 13-2 by an average of 1.2. They can be expressed with an average deviation of ± 0.06 as the sum of values n_+ and n_- for the ions, as shown in Table 13-11 just below the symbols of the ions. The corresponding deviation for r_e is ± 0.003 Å.

These values of n_+ and n_- may be used to predict values of the equilibrium interionic distance for the five molecules for which experimental values have not yet been obtained. The predicted values are given in parentheses.

The calculated values of the vibrational frequency and of the heat of formation agree well with the available experimental values. It seems likely that this simple model, with neglect of polarization, may

[36] This information has been summarized by A. Honig, M. Mandel, M. L. Stitch, and C. H. Townes, *Phys. Rev.* **96**, 629 (1954). Values of r_e have been reported also for some other halogenides from high-temperature microwave studies by A. H. Barrett and M. Mandel, *ibid.* **109**, 1572 (1958); GaCl, 2.2017; GaBr, 2.3525; GaI, 2.5747; InCl, 2.4011; InBr, 2.5432; InI, 2.754; TlF, 2.0844; TlCl, 2.4848; TlBr, 2.6181; TlI, 2.8135.

[37] L. Pauling, *Proc. Nat. Acad. Sci. India* **A25**, 1 (1956). The discussion in the text corrects some numerical errors in this paper.

be used in general for the prediction of the properties of ionic molecules.

The model has been applied[38] to the dimers of alkali halogenide molecules, $M_2X_2(g)$. The molecule M_2X_2 has the form of a rhombus, with the edge M^+—X^- about 0.17 Å greater than in MX(g) and with a difference in the M^+—M^+ and X^-—X^- diagonals as expected from the radii (extreme values of the difference are —0.51 Å for LiI and +0.17

TABLE 13-11.—INTERIONIC DISTANCES IN ALKALI
HALOGENIDE GAS MOLECULES

	F− $n_- = 1.9$	Cl− 4.2	Br− 4.8	I− 5.6
Li+ $n_+ = 4.5$	(1.520 Å) (6.4)	(2.029 Å) (8.7)	2.170 9.20	2.392 Å 10.10
Na+ $n_+ = 5.2$	(1.846 Å) (7.1)	2.361 Å 9.62	2.502 Å 10.01	2.712 Å 10.66
K+ $n_+ = 5.1$	(2.139 Å) (7.0)	2.667 Å 9.30	2.821 Å 9.87	3.048 Å 10.69
Rb+ $n_+ = 4.9$	(2.242 Å) (6.8)	2.787 Å 9.08	2.945 Å 9.70	3.177 Å 10.53
Cs+ $n_+ = 4.6$	2.345 Å 6.46	2.906 Å 8.64	3.072 Å 9.37	3.315 Å 10.35

Å for CsF). The calculated values of $-\Delta H°$ for the reaction $2MX(g) \rightarrow M_2X_2(g)$ range from about 41 kcal/mole for Cs_2I_2 to 59 kcal/mole for Li_2F_2. They agree reasonably well with the experimental values.[39]

Other complexes have also been observed:[40] $MM'X_2$, M_3X_3, M_4X_4, M_2X^+, $M_3X_2^+$, $M_4X_3^+$. Similar hydroxide and water complexes ($[KOH_2]^+$) have also been reported.[41]

[38] T. A. Milne and D. Cubicciotti, *J. Chem. Phys.* **29**, 846 (1958). Similar calculations with polarization have also been made: C. T. O'Konski and W. I. Higuchi, *ibid.* **23**, 1174 (1955).

[39] N. A. Yonov, *Doklady Akad. Nauk S.S.S.R.* **59**, 467 (1948); R. C. Miller and P. Kusch, *J. Chem. Phys.* **25**, 860 (1956); R. F. Porter and R. C. Schoonmaker, *ibid.* **29**, 1070 (1958); J. Berkowitz and W. A. Chupka, *ibid.* 653; P. Kusch, *ibid.* **28**, 981 (1958); A. C. Pugh and R. F. Barrow, *Trans. Faraday Soc.* **54**, 671 (1958); S. H. Bauer, R. M. Diner, and R. F. Porter, *J. Chem. Phys.* **29**, 991 (1958); R. C. Schoonmaker and R. F. Porter, *ibid.* **30**, 991 (1959); M. Eisenstadt, V. S. Rao, and G. M. Rothberg, *ibid.* 604.

[40] R. F. Porter and R. C. Schoonmaker, *J. Chem. Phys.* **28**, 168 (1958); *ibid.* **62**, 486 (1958).

[41] Porter and Schoonmaker, *loc. cit.* (40); W. A. Chupka, *J. Chem. Phys.* **30**, 458 (1959).

13-4. THE STRUCTURE OF OTHER SIMPLE IONIC CRYSTALS

The Alkaline-Earth Oxides, Sulfides, Selenides, and Tellurides.—
All the alkaline-earth compounds with oxygen, sulfur, selenium, and
tellurium crystallize with the sodium chloride arrangement except
beryllium oxide and magnesium telluride, which have the wurtzite
structure, and beryllium sulfide, selenide, and telluride, which have the
sphalerite structure. On comparison of observed interionic distances
with the sums of the crystal radii, shown in Table 13-12, excellent agreement is found except for the magnesium compounds. This agreement
provides a striking verification of the arguments used in the derivation
of the table of crystal radii, inasmuch as the experimental values for
these substances were in no way involved in the formulation of the
table.

TABLE 13-12.—INTERIONIC DISTANCES IN CRYSTALS $M^{++}X^{--}$ WITH
THE SODIUM CHLORIDE ARRANGEMENT

		Mg^{++}	Ca^{++}	Sr^{++}	Ba^{++}
Radius sum	O^{--}	2.05 Å	2.39 Å	2.53 Å	2.75 Å
Observed distance		2.10	2.40	2.54	2.75
Radius sum	S^{--}	2.49	2.83	2.97	3.19
Observed distance		2 54	2.83	3.00	3.18
Radius sum	Se^{--}	2.63	2.97	3.11	3.33
Observed distance		2.72	2.96	3.11	3.31
Radius sum	Te^{--}		3.20	3.34	3.56
Observed distance			3.17	3.33	3.50

In magnesium sulfide and selenide the anions are in contact; the
radii deduced on this assumption (1.80 Å for S^{--}, 1.93 Å for Se^{--}) are
slightly smaller than the crystal radii. The ratio $R_{Mg}{}^{++}/R_0{}^{--}$ is 0.46,
which lies in the region in which double repulsion is operative; this accounts for the high value observed for magnesium oxide.

**Crystals with the Rutile and the Fluorite Structures; Interionic Dis-
tances for Substances of Unsymmetrical Valence Type.**—In a crystal
of a substance of unsymmetrical valence type, such as fluorite, CaF_2
(Fig. 13-10), the equilibrium cation-anion interionic distance cannot be
expected necessarily to be given by the sum of the crystal radii of the bivalent calcium ion and the univalent fluoride ion. The sum of the univalent radii of calcium and fluoride, 2.54 Å, would give the equilibrium
interionic distance in a hypothetical crystal with attractive and repulsive forces corresponding to the sodium chloride arrangement.

This can now be corrected for the valence effect in the following way.[42]

According to Equation 13-3 the ratio of equilibrium distances for the two structures is

$$\frac{R_{CaF2}}{R_{NaCl}} = \left\{ \frac{B_{CaF_2}}{B_{NaCl}} \cdot \frac{A_{NaCl}}{A_{CaF_2}} \right\}^{1/(n-1)}$$

Now in fluorite there are eight cation-anion contacts per stoichiometric molecule, and in NaCl six; we accordingly assume the ratio B_{CaF_2}/B_{NaCl}

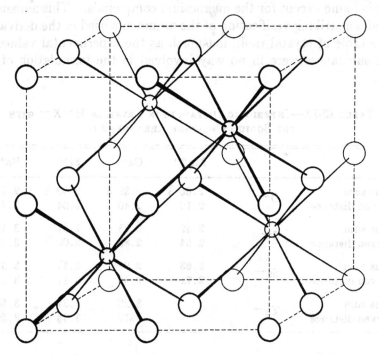

Fɪɢ. 13-10.—The structure of the cubic crystal fluorite, CaF_2. Small circles represent calcium ions and large circles fluoride ions.

to have the value $\frac{8}{6}$. Introducing this and the values of A_{NaCl} (1.7476) and A_{CaF_2} (5.0388) in this equation and placing n equal to 8 (the mean of the values of Ca^{++} and F^-), we obtain $R_{CaF_2}/R_{NaCl} = 0.894$, which on multiplication by the sum of the univalent radii gives for R_{CaF_2} the value 2.27 Å.

This value is somewhat smaller than the observed Ca^{++}—F^- distance for fluorite, 2.36 Å. It is probable that the anion-anion contacts

---

[42] Pauling, *loc. cit.* (9); a more detailed discussion of methods in calculating interionic distances from univalent radii has been given by W. H. Zachariasen, *Z. Krist.* 80, 137 (1931).

in fluorite make the ratio B_{CaF_2}/B_{NaCl} somewhat greater than $\frac{2}{3}$, perhaps about $\frac{5}{6}$ (that is, proportional to the number of ions); this leads to a calculated value of 2.32 Å.

It is found empirically, indeed, that these complicated and unreliable calculations need not be made; in general even for substances of unsymmetrical valence type the interionic distances are very closely approximated by the sum of the crystal radii. For fluorite this sum is 2.35 Å, which agrees very well with the observed value. The reason for this is apparent; in the crystal radius of Ca^{++} a correction for bivalence of cation and anion is made, and this has nearly the same magnitude as the correction for bivalence of cation alone made for the sum of the univalent radii of calcium and fluorine.

A comparison of observed interionic distances and crystal radius sums for some crystals with the fluorite structure and some with the rutile structure is given in Table 13-13. It is seen that in general the agreement is excellent. Similarly good agreement is found for other ionic crystals of unsymmetrical valence type, the available data being so extensive as to prevent their reproduction here. For these crystals, as for those of symmetrical valence type also, greater refinement can be attained in the discussion of interionic distances by considering the effect of change of ligancy. This question is treated in the following section.

It is of interest in connection with the theory of interionic forces discussed in the preceding section to consider the rutile crystal in somewhat greater detail. The structure of this crystal, aside from the absolute dimensions, depends on two parameters, the axial ratio c_0/a_0 and another parameter determining the positions of the oxygen atoms (Fig. 3-2). This parameter can be given a value, for any value of c_0/a_0, such as to make the distances from the cation to the six surrounding anions equal, as is reasonable for an ionic crystal. The Madelung constant is then found on calculation to be given by the expression

$$A_{R_0} = 4.816 - 4.11\left(0.721 - \frac{c_0}{a_0}\right)^2 \qquad (13\text{-}16)$$

It is seen that A has its maximum value, corresponding to maximum stability of the crystal, if R_0 is constant, at $c_0/a_0 = 0.721$. This is greater than the values observed, which lie near 0.66 (0.660 for MgF_2, 0.644 for TiO_2, etc.). The explanation of this is seen when the F^-—F^- and O^{--}—O^{--} distances are considered; the axial ratio 0.721 corresponds to a very small value of the minimum anion-anion distance—about 2.40 Å, as compared with 2.72 and 2.80 Å, twice the crystal

TABLE 13-13.—INTERIONIC DISTANCES FOR SUBSTANCES OF
UNSYMMETRICAL VALENCE TYPE

	Radius sum	Observed distance		Radius sum	Observed distance
Crystals with the fluorite structure					
CaF_2	2.35 Å	2.36 Å	Na_2O	2.35 Å	2.40 Å
SrF_2	2.49	2.50	K_2O	2.69	2.79
BaF_2	2.71	2.68	Rb_2O	2.88	2.92
RaF_2	2.76	2.76	Li_2S	2.44	2.47
$SrCl_2$	2.94	3.02	Na_2S	2.79	2.83
$BaCl_2$	3.16	3.18	K_2S	3.17	3.20
CdF_2	2.33	2.34	Rb_2S	3.32	3.31
HgF_2	2.46	2.40	Li_2Se	2.58	2.59
EuF_2	2.48	2.51	Na_2Se	2.93	2.95
PbF_2	2.56	2.57	K_2Se	3.31	3.32
$LaOF$	2.53*	2.49	Li_2Te	2.81	2.82
$AcOF$	2.56*	2.57	Na_2Te	3.16	3.17
$PuOF$	2.45*	2.47	K_2Te	3.54	3.53
ZrO_2	2.20	2.20	PaO_2	2.38	2.36
HfO_2	2.21	2.21	UO_2	2.37	2.37
ThO_2	2.42	2.42	NpO_2	2.35	2.35
CeO_2	2.41	2.34	PuO_2	2.33	2.34
PrO_2	2.32	2.32	AmO_2	2.32	2.33
Li_2O	2.00	2.00			
Crystals with the rutile structure					
MgF_2	2.01	2.02	TiO_2	2.08	1.96
MnF_2	2.16	2.17	SnO_2	2.11	2.10
FeF_2	2.12	2.14	PbO_2	2.24	2.22
CoF_2	2.10	2.10	VO_2	2.00	1.96
NiF_2	2.08	2.08	CrO_2	1.96	1.93
ZnF_2	2.10	2.12	MnO_2	1.94	1.95
PdF_2	2.22	2.22			

* Average for the two anions.

radius of F^- and O^{--}, respectively. It is accordingly anion-anion repulsion that increases the value of a_0 for crystals of the rutile type. A quantitative treatment[43] with use of a potential function similar to that of Equation 13-10 leads to $c_0/a_0 \cong 0.66$, in agreement with experiment. The F^-—F^- distance then becomes about 2.60 Å and the O^{--}—O^{--} distance about 2.50 Å; these values are less than twice the crystal radii of the anions, showing that a compromise is reached in the attempt of the anion-anion repulsion to decrease the axial ratio and that of the Madelung constant to increase it.

[43] L. Pauling, *Z. Krist.* **67**, 377 (1928).

The same phenomenon is shown also by anatase and brookite, two additional crystal modifications of titanium dioxide. In these crystals also the minimum O^{--}—O^{--} distances have the low value 2.50 Å. A theoretical treatment like that described above has been carried out for anatase, with results in good agreement with experiment.

It might be expected that the repulsion of cations along the c axis would cause the four M—X bond lengths to the shared edges to be greater than the other two. This effect is found[44] for MgF_2, Mg—F = 1.997(4), 1.928(2); MnF_2, Mn—F = 2.132(4), 2.102(2); and ZnF_2, Zn—F = 2.043(4), 2.015(2). It is, however, not found for rutile itself:[45] Ti—O = 1.946(4), 1.984(2); the reason for this deviation is not clear. In SnO_2 the distances are equal (2.052(4), 2.056(2)).

Manganese ion, Mn^{++}, with five electrons with parallel spin in $3d$ orbitals ($^6S_0$), and zinc ion, Zn^{++}, with a completed $3d$ subshell ($^1S_0$), have spherical symmetry, as has also Mg^{++}, with the argon structure ($^1S_0$). Other bipositive ions of the transition metals do not have spherical symmetry and might be expected to orient their $3d$ electrons in such a way as to permit stabilization of the MF_2 crystals by increase in the length of the M—F bonds to the shared edges and decrease of the others. This effect is shown in a pronounced way[46] in FeF_2 (Fe—F = 2.122(4), 1.993(2)), and also[47] in CrO_2 (Cr—O = 1.92(4), 1.87(2)). CoF_2 (2.046(4), 2.032(2)) and NiF_2 (2.018(4), 1.986(2)) show nearly the same differences as for the spherically symmetrical ions.[46] These differences in properties may be correlated with the fact that octahedral symmetry can result from electrons occupying either a set of three d orbitals (xy, yz, and zx—Chap. 5) or a set of two d orbitals (the remaining two). For Fe^{++}, with four odd d electrons, octahedral symmetry cannot be achieved, whereas it can be for Co^{++} and Ni^{++}.

VO_2 has a distorted rutile structure involving V—V bonds (Chap. 11).

The Effect of Ligancy on Interionic Distance.—It was pointed out in the preceding section that an approximate value for the ratio of interionic distances for two modifications of a substance can be obtained by use of the equation

$$\frac{R_{II}}{R_I} = \left\{\frac{A_I B_{II}}{A_{II} B_I}\right\}^{1/(n-1)} \tag{13-17}$$

[44] W. H. Baur, *Acta Cryst.* **9**, 515 (1956); **11**, 488 (1958).

[45] D. T. Cromer and K. Herrington, *J.A.C.S.* **77**, 4708 (1955); Baur, *loc. cit* (44).

[46] Baur, *loc. cit.* (44).

[47] O. Glemser, U. Hauschild, and F. Trupel, *Z. anorg. Chem.* **277**, 113 (1954).

in which the repulsive coefficients B_I and B_{II} are assumed to be proportional to the numbers of cation-anion contacts for the two structures. With $B_{CsCl}/B_{NaCl} = \frac{8}{6}$, this equation leads to $R_{CsCl}/R_{NaCl} = 1.036$ for $n = 9$, in approximate agreement with the experimental results.

Goldschmidt[48] emphasized the necessity of making corrections of this type for the effect of ligancy, and suggested the factor 1.03 for changing from ligancy 6 to 8, and 0.93 to 0.95 for changing from 6 to 4.

We see from application of Equation 13-17 that the correction depends in the main on the ligancy of the cation, that is, the number of anions grouped about each cation. With $n = 9$ Equation 13-17 leads to the following ratios:

$$6 \rightarrow 8: \quad \frac{R_{CsCl}}{R_{NaCl}} = 1.036, \quad \frac{R_{fluorite}}{R_{rutile}} = 1.031, \quad \left(\frac{8}{6}\right)^{1/8} = 1.036;$$

$$6 \rightarrow 4: \quad \frac{R_{sphalerite\ or\ wurtzite}}{R_{NaCl}} = 0.957, \quad \frac{R_{\beta-quartz}}{R_{rutile}} = 0.960,$$

$$\left(\frac{4}{6}\right)^{1/8} = 0.950.$$

It is seen that the changes from the standard sodium chloride and rutile arrangements, with ligancy 6, to cesium chloride and fluorite, respectively, with ligancy 8, are nearly the same, as are also those to sphalerite or wurtzite and β-quartz, respectively, with ligancy 4. Moreover, the values are nearly the same as those calculated by ignoring the differences in Madelung constants, that is, by placing $A_{II}/A_I = 1$.

Values of the ratio $\{B_{II}/B_I\}^{1/n-1}$ with B_I equal to 6 (ligancy 6 having been chosen as the standard in the derivation of the table of ionic radii) and with B_{II} equal to the ligancy for the second structure are given in Table 13-14 for various values of the exponent n.

TABLE 13-14.—CORRECTION FACTOR FOR CHANGE OF LIGANCY
FROM THE STANDARD VALUE 6

$n =$	6	7	8	9	10	11	12
Ligancy							
12	1.149	1.122	1.104	1.091	1.080	1.072	1.065
9	1.085	1.070	1.060	1.052	1.046	1.041	1.038
8	1.059	1.049	1.042	1.037	1.032	1.029	1.026
7	1.031	1.026	1.022	1.019	1.017	1.016	1.014
6	1.000	1.000	1.000	1.000	1.000	1.000	1.000
5	0.964	0.970	0.974	0.978	0.980	0.982	0.984
4	0.922	0.935	0.944	0.951	0.956	0.960	0.964

48 Goldschmidt. *loc. cit.* (16).

TABLE 13-15.—INTERATOMIC DISTANCES M—X FOR LIGANCY 12

	Crystal radius sum	Corrected sum	Observed distance
K—F	2.69 Å	2.97 Å	2.90 Å
Rb—F	2.84	3.12	3.01
Cs—F	3.05	3.31	3.20
K—Cl	3.14	3.43	3.44–3.50
Rb—Cl	3.29	3.57	3.50–3.60
Cs—Cl	3.50	3.77	3.60–3.70
K—Br	3.28	3.56	3.64–3.68

As an example of the application of this table we may consider the Al—O distances for octahedral and tetrahedral coordination of oxygen ions about aluminum. In corundum (α-Al_2O_3), topaz ($Al_2SiO_4F_2$), diaspore (AlOOH, Fig. 12-12), and many other crystals with aluminum octahedra the observed values for the Al—O distances lie close to 1.90 Å, the sum of the crystal radii of the ions. The value observed for tetrahedral coordination in crystals such as sodalite ($NaAl_3Si_3O_{12}Cl$), zunyite ($Al_{13}Si_5O_{20}(OH)_{18}Cl$), natrolite ($Na_2Al_2Si_3O_{10} \cdot 2H_2O$), the feldspars, and other aluminosilicates is 1.78 $\pm$ 0.02 Å. The sum of the radii corrected by the appropriate factor from the table is 1.78 Å, in good agreement with the experimental value. The Si—O distance in the SiO_4 tetrahedron has been discussed in Section 9-6.

In cubic crystals M_2RX_6, such as potassium fluosilicate, with the structure shown in Figure 5-1, each ion M is surrounded by 12 X ions. Many of these compounds have been studied, the observed values for M—X being given in Table 13-15. In general these agree well with the radius sums corrected to ligancy 12. It is interesting that this agreement is found even though in many of the substances (M_2SeCl_6, M_2PtCl_6, etc.) the R—X bonds are essentially covalent; this reflects the fact that van der Waals radii and ionic radii are nearly equal.

Some of the complex fluorides crystallize with a hexagonal structure, which has been studied by Hoard and Vincent[49] for potassium hexafluogermanate and ammonium hexafluogermanate and by Gossner and Kraus[50] for the hexagonal modification of ammonium fluosilicate. The structure is based on hexagonal closest packing of the M and X ions (Sec. 13-5), with, however, considerable distortion, leading to greater compactness than for the cubic structure described above. Thus the density of hexagonal ammonium fluosilicate is 7 percent greater than that of the cubic form of the substance. The distortion is such as to

[49] J. L. Hoard and W. B. Vincent, *J.A.C.S.* **61**, 2849 (1939).
[50] B. Gossner and O. Kraus, *Z. Krist.* **88**, 223 (1934).

bring nine of its twelve fluorine neighbors somewhat closer to each univalent cation than the other three, the K—F distances in K_2GeF_6 being 2.84 Å (six), 2.86 Å (three), and 3.01 Å (three). The value 2.85 Å predicted for the K^+—F^- distance for ligancy 9 is hence in excellent agreement with experiment.

It is to be emphasized that equilibrium interionic distances are less well defined than covalent bond lengths; their values depend not only on ligancy, but also on radius ratio (anion contact, double repulsion), amount of covalent bond character, and other factors, and a simple discussion of all the corrections that have been suggested and applied cannot be given. On the other hand, we have a reliable picture of the forces operating between ions, and it is usually possible to make a reliable prediction about interionic distances for particular structures.

The Effect of Radius Ratio in Determining the Relative Stability of Different Structures.—It is seen from Table 13-14 that the transition from the rutile structure, with ligancy 6 for the cations, to the fluorite structure, with ligancy 8, is accompanied by an increase in equilibrium interionic distance R_0 of about 3.7 percent for $n = 9$. The Madelung constant A has for the two structures the values 4.816 and 5.039, respectively, with ratio 1.046, and accordingly (Equation 13-5) the fluorite structure will be the more stable of the two so long as the increase in R_0 from rutile to fluorite remains less than 4.6 percent. The discussion of the radius-ratio effect in the preceding section indicates the conditions under which the rutile structure becomes stable. Let us consider a cation M^{++} in the center of a cube of anions X^- in the fluorite structure (Fig. 13-10). If the repulsive forces from cations to anions are stronger than those between anions, they will determine the equilibrium cation-anion interionic distance, which will be equal to the sum of the crystal radii with correction for ligancy, whereas if the anion-anion forces are the larger (anion contact) or of about the same magnitude (double repulsion) the value of R_0 will be larger than the corrected crystal radius sum, and in consequence the structure will be unstable relative to the rutile structure. The value of the radius ratio ρ at which this effect sets in can be calculated in the following way: If r_+ and r_- represent the univalent radii of the ions (the use of univalent radii is pertinent here because the discussion depends on the relative magnitudes of cation-anion and anion-anion repulsions), double repulsion will be effective when $2r_-$ and $r_+ + r_-$ are in the ratio $1 : \sqrt{3}/2$, that is, the ratio of edge and half body diagonal of a cube. From the equation $(r_+ + r_-)/2r_- = \sqrt{3}/2$ we obtain

$$\rho = \sqrt{3} - 1 = 0.732 \qquad (13\text{-}18)$$

as the limiting radius ratio for stability of the fluorite structure, or in general for structures with cubic coordination; for values of ρ less than 0.732 the rutile structure is expected to occur for ionic crystals MX_2.

TABLE 13-16.—RADIUS RATIO VALUES FOR CRYSTALS WITH RUTILE
AND FLUORITE STRUCTURES

Rutile structure		Fluorite structure	
	ρ		ρ
MgF_2	0.60	CaF_2	0.87
ZnF_2	.65	SrF_2	.97
TiO_2	.55	BaF_2	1.12
GeO_2	.43	CdF_2	0.84
SnO_2	.55	HgF_2	.92
PbO_2	.60	$SrCl_2$	.73
		ZrO_2	.62
		CeO_2	.72

The comparison with experiment is shown in Table 13-16; it is seen that with two exceptions (ZrO_2 and CeO_2) the conditions $\rho < 0.73$ and $\rho > 0.73$ for stability of the rutile and fluorite structures, respectively, are satisfied.

TABLE 13-17.—RADIUS RATIO VALUES FOR CRYSTALS WITH
OCTAHEDRAL AND TETRAHEDRAL COORDINATION

Octahedral coordination		Tetrahedral coordination	
	ρ		ρ
PbO_2	0.60	GeO_2	0.43
SnO_2	.55	SiO_2	.37
GeO_2	.43	BeF_2	.32
MgF_2	.60		

A similar calculation leads to the limiting value $\rho = \sqrt{2} - 1 = 0.414$ for transition from octahedral to tetrahedral coordination. The extent to which this is verified by experiment is shown in Table 13-17.

It is interesting that GeO_2, with $\rho = 0.43$, crystallizes with both the rutile and the quartz structure.

The discussion of the relation between radius ratio and ligancy is continued in Section 13-6.

13-5. THE CLOSEST PACKING OF LARGE IONS IN IONIC CRYSTALS

In many of the structures that are assumed by crystals of ionic substances the large ions are arranged in closest packing (Sec. 11-5). As pointed out by Landé in 1920, this is the case for the sodium chloride structure; in the lithium halogenides it is the closest-packed anion lattices that essentially determine the values of the lattice constant. This observation about the sodium chloride arrangement should, indeed, be attributed to Barlow, who in 1898, in discussing this structure for spheres with radius ratio 0.414, as shown in Figure 1-1 (taken from his paper), pointed out that the large spheres are in cubic closest packing.

The sphalerite, wurtzite, antifluorite (Li_2S), cadmium iodide, cadmium chloride, and many other arrangements also involve closest packing of the large ions.

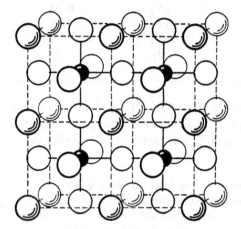

FIG. 13-11.—The structure of the cubic crystal $KMgF_3$. Potassium ions are represented by large shaded circles. They are at the corners of the unit cube. The fluoride ions, represented by large open circles, are at the face-centered positions, and the magnesium ions, represented by small circles, are at the center of the cubes. This structure is often called the perovskite structure; perovskite is the mineral $CaTiO_3$.

It was pointed out by Bragg and West[51] in 1927 that in many silicate crystals and other minerals the volume per oxygen atom lies between 14 and 20 Å³, indicating that in these crystals oxygen ions (with volume per ion 15.5 Å³ for crystal radius 1.40 Å) are in closest packing, the small metal ions being inserted in the interstices. This idea was of great use to Bragg and his coworkers in their successful attack on the structures of many silicate minerals.

Double hexagonal closest packing (Sec. 11-5) was first found[52] for the oxygen ions in brookite, the orthorhombic form of titanium dioxide, and for the oxygen and fluoride ions in topaz,[53] $Al_2SiO_4F_2$. It has since been reported for the halogens in one modification of cadmium iodide,[54] in mercuric bromide,[55] and in mercuric chloride[56] and for chloride and

[51] Bragg and West, *loc. cit.* (17).
[52] L. Pauling and J. H. Sturdivant, *Z. Krist.* **68**, 239 (1928).
[53] L. Pauling, *Proc. Nat. Acad. Sci. U.S.* **14**, 603 (1928).
[54] O. Hassel, *Z. physik. Chem.* **B22**, 333 (1933).
[55] H. J. Verweel and J. M. Bijvoet, *Z. Krist.* **77**, 122 (1931).
[56] H. Braekken and L. Harang, *Z. Krist.* **68**, 123 (1928).

hydroxide in CdOHCl.[57]   It has been found, moreover, that modifications of cadmium bromide and nickel bromide exist for which the sequence of layers *A*, *B*, and *C* of bromine atoms is largely a random one.[58]

It often occurs that in crystals containing the larger cations (K^+, Rb^+, Cs^+, Ba^{++}, NH_4^+, etc.) these large cations and the anions together form a closest-packed array.   An example of this is the $KMgF_3$ arrangement (the so-called perovskite structure), shown in Figure 13-11.   It is seen that K^+, with radius 1.33 Å, and $3F^-$, with radius 1.36 Å, together are in cubic closest packing, with the small ions Mg^{++} in the centers of fluoride octahedra.   A similar arrangement of alkali and halogenide ions is also shown by the potassium chlorostannate structure (Fig. 5-1), and by the structures reported for $Cs_3Tl_2Cl_9$, $Cs_3As_2Cl_9$, and similar substances.[59]

13-6. THE PRINCIPLES DETERMINING THE STRUCTURE OF COMPLEX IONIC CRYSTALS

Simple ionic substances, such as the alkali halogenides, have little choice of structure; only a very few relatively stable ionic arrangements corresponding to the formula M^+X^- exist, and the various factors that influence the stability of the crystal are pitted against one another, with no one factor necessarily finding clear expression in the decision between the sodium chloride and the cesium chloride arrangement. For a complex substance, such as mica, $KAl_3Si_3O_{10}(OH)_2$, or zunyite, $Al_{13}Si_5O_{20}(OH)_{18}Cl$, on the other hand, many conceivable structures differing only slightly in nature and stability can be suggested, and it might be expected that the most stable of these possible structures, the one actually assumed by the substance, will reflect in its various features the different factors that are of significance in determining the structure of ionic crystals.   It has been found possible to formulate a set of rules about the stability of complex ionic crystals, as described in the following paragraphs.   These rules were obtained[60] in part by induction from the structures known in 1928 and in part by deduction from the equations for crystal energy.   They are not rigorous in their derivation or universal in their application, but they have been found useful as a criterion for the probable correctness of reported structures for complex crystals and as an aid in the x-ray investigation of crystals

[57] J. L. Hoard and J. D. Grenko, *Z. Krist.* **87**, 110 (1934).

[58] J. M. Bijvoet and W. Nieuwenkamp, *Z. Krist.* **86**, 466 (1933); J. A. A. Ketelaar, *ibid.* **88**, 26 (1934).

[59] J. L. Hoard and L. Goldstein, *J. Chem. Phys.* **3**, 117, 199 (1935).

[60] L. Pauling, in *Sommerfeld Festschrift*, S. Hirzel, Liepzig, 1928; *J.A.C.S.* **51**, 1010 (1929).

by making possible the suggestion of reasonable structures for experimental test. The rules are, moreover, of some significance for molecules and complex ions.

The substances to which the rules apply are those in which the bonds are largely ionic in character rather than largely covalent, and in which all or most of the cations are small (with radius less than 0.8 Å) and multivalent, the anions being large (greater than 1.35 Å in radius) and univalent or bivalent. The anions that are most important are those of oxygen and fluorine.

The differentiation between cations and anions in regard to size and charge is reflected in the rules; markedly different roles are attributed to cations and anions in a crystal. The rules are based upon the concept of the coordination of anions at the corners of a tetrahedron, octahedron, or other polyhedron about each cation, as assumed in the early work of W. L. Bragg on the silicate minerals, and they relate to the nature and interrelations of these polyhedra.

The Nature of the Coordinated Polyhedra.—The first rule, relating to the nature of the coordinated polyhedron of anions about a cation, is the following: *A coordinated polyhedron of anions is formed about each cation, the cation-anion distance being determined by the radius sum and the ligancy of the cation by the radius ratio.*

In crystals containing highly charged cations the most important terms in the expression for the crystal energy are those representing the interaction of each cation and the adjacent anions. The negative Coulomb energy of the cation-anion interactions causes each cation to attract a number of anions, which approach to the equilibrium distance from it. This distance is given with some accuracy by the sum of the crystal radii of cation and anion, as discussed in earlier sections of this chapter.

If too many anions are grouped around one cation, the anion-anion repulsion becomes strong enough to prevent the anions from approaching this closely to the cation. The increase in Coulomb energy resulting from increase in the cation-anion distance then makes the structure less stable than another structure with fewer anions about each cation. This phenomenon has been discussed in Section 13-4, where it was shown that the transition from cubic to octahedral coordination would occur at about the value 0.732 for ρ, the ratio of univalent radii of anion and cation, and that the transition from octahedral to tetrahedral coordination would occur at about $\rho = 0.414$ It may be mentioned also that the square antiprism, the polyhedron with 16 equal edges shown in Figure 13-12, is a more satisfactory ionic coordination polyhedron than the cube, and that the transition from the antiprism to the octahedron would occur at about $\rho = 0.645$[61] (Table 13-18).

[61] The coordination described in Chap. 5 for the $[Mo(CN)_8]^{----}$ ion corre-

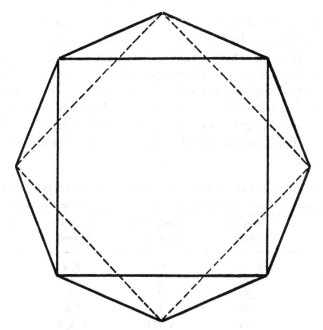

FIG. 13-12.—The square antiprism.

TABLE 13-18.—VALUES OF THE MINIMUM RADIUS RATIO FOR STABILITY
OF VARIOUS COORDINATION POLYHEDRA

Polyhedron	Ligancy	Minimum radius ratio
Cubo-octahedron	12	1.000
	9[a]	0.732
Cube	8	.732
Square antiprism	8	.645
	7[b]	.592
Octahedron	6	.414
Tetrahedron	4	.225

[a] This polyhedron, with 18 equal edges, is obtained by adding three atoms at the centers of the vertical faces of a right triangular prism.

[b] This polyhedron is obtained by adding an atom at the center of a face of an octahedron.

Values of the radius ratio for various cations relative to oxygen are given in Table 13-19, together with the predicted ligancies, from Table

---

sponds to a polyhedron with 12 triangular faces (not equilateral). Transition from it to the cube would occur at about $\rho = 0.667$. This polyhedron is found for ZrO_8 groups in zircon, $ZrSiO_4$.

13-18. The observed ligancies that are given in the fourth column of Table 13-19 are seen to lie close to the predicted values. The values in boldface type are those usually found for the cation. The other values are observed only in a few crystals. In the cases where the observed ligancy deviates greatly from the expected value, such as 12 for K^+ in crystals such as mica, $KAl_3Si_3O_{10}(OH)_2$, it is probable that the other ions present play the most important part in determining the configuration. Tetrahedral coordination has been observed about Si^{4+} in

TABLE 13-19.—VALUES OF THE LIGANCY FOR CATIONS WITH OXYGEN ION

Ion	Radius ratio	Predicted ligancy	Observed ligancy	Strength of bond
B^{3+}	0.20	3 or 4	3, 4	1 or ¾
Be^{++}	.25	4	4	½
Li^+	.34	4	4	¼
Si^{4+}	.37	4	4, 6	1
Al^{3+}	.41	4 or 6	4, 5, 6	¾ or ½
Ge^{4+}	.43	4 or 6	4, 6	1 or ⅔
Mg^{++}	.47	6	6	⅓
Na^+	.54	6	6, 8	⅙
Ti^{4+}	.55	6	6	⅔
Sc^{3+}	.60	6	6	½
Zr^{4+}	.62	6 or 8	6, 8	⅔ or ½
Ca^{++}	.67	8	7, 8, 9	¼
Ce^{4+}	.72	8	8	½
K^+	.75	9	6, 7, 8, 9, 10, 12	⅑
Cs^+	.96	12	12	1/12

scores of crystals, and octahedral coordination in only one, SiP_2O_7 (which does not occur in nature); it is clear that this crystal is to be considered as exceptional.

The ions with transition values of the radius ratio are especially interesting. Boron is tricoordinate in H_3BO_3, Be_2BO_3OH (hambergite), CaB_2O_4, and many other crystals and tetracoordinate in $CaB_2Si_2O_8$ (danburite) and duodecitungstoboric acid; in $KH_8B_5O_{12}$ and several other complex borates some boron atoms are tetracoordinate and some are tricoordinate. Aluminum ion forms oxygen tetrahedra in many aluminosilicates and octahedra in others; it has ligancies 4 and 6 in sillimanite, 5 and 6 in andalusite, and 6 alone in cyanite, all three stable minerals having the composition Al_2SiO_5. Germanium dioxide is dimorphous, showing both a quartzlike (ligancy 4) and a rutilelike modification (ligancy 6).   Zirconium is octahedrally coordinated in several crystals, but has ligancy 8 in zircon, $ZrSiO_4$.[62]

The calcite-aragonite transition may be discussed as an illustration of the significance of the radius ratio in determining the choice among alternative structures for a substance.[63]   Calcium carbonate crystallizes with the well-known rhombohedral calcite structure, in which calcium ion has ligancy 6, and with the pseudo-hexagonal orthorhombic aragonite structure, in which calcium ion has ligancy 9.   The choice of these structures by univalent nitrates, bivalent carbonates, and tervalent borates is shown in Table 13-20, together with the ratios of univalent radii of the cations and oxygen.

TABLE 13-20.—VALUES OF RADIUS RATIO FOR NITRATES, CARBONATES, AND BORATES

		ρ		ρ		ρ
Calcite structure	$LiNO_3$	0.34	$MgCO_3$	0.47	$ScBO_3$	0.60
	$NaNO_3$	.54	$ZnCO_3$	.50	$InBO_3$	.59
			$CdCO_3$	.65	YBO_3	.68
			$CaCO_3$	.67		
Aragonite structure	KNO_3	0.76	$CaCO_3$	0.67	$LaBO_3$	0.79
			$SrCO_3$	.75		
			$BaCO_3$	.87		
$RbNO_3$ structures	$RbNO_3$	0.84				
	$CsNO_3$	.96				

It is seen that the transition occurs at about $\rho = 0.67$, and that this is the value of ρ for the dimorphous substance calcium carbonate.   At $\rho \geqq 0.85$ transition occurs to other structures, shown by rubidium nitrate and cesium nitrate, in which the univalent cation probably has ligancy twelve.

The Number of Polyhedra with a Common Corner: The Electrostatic Valence Rule.—In silica crystals, SiO_2, each silicon ion is surrounded by four oxygen ions[64] at tetrahedron corners.   In order for the stoichio-

---

[62] The relation of radius ratio to the formulas of oxygen acids has been discussed by L. Pauling, *J.A.C.S.* **55**, 1895 (1933); see also E. Zintl and W. Morawietz, *Z. anorg. Chem.* **236**, 372 (1938).

[63] V. M. Goldschmidt, *loc. cit.* (16); V. M. Goldschmidt and H. Hauptmann, *Nachr. Ges. Wiss. Göttingen*, **1932**, 53.

[64] Here, as elsewhere in this chapter, the use of the word ion is to be interpreted as meaning that the bonds are largely ionic but not necessarily of the extreme

metric ratio $1Si:2O$ to be retained it is accordingly necessary that on the average each oxygen ion act as a corner of two tetrahedra. This might be achieved by having alternate oxygen ions serve as corners of one tetrahedron and three tetrahedra, or in other such way; the following *electrostatic valence rule* requires, however, that each oxygen ion serve as a corner of two tetrahedra: Let ze be the electric charge of a cation and ν its coordination number; we then define the *strength of the electrostatic bond* to each coordinated anion as

$$s = \frac{z}{\nu}$$

and make the postulate that *in a stable ionic structure the valence of each anion, with changed sign, is exactly or nearly equal to the sum of the strengths of the electrostatic bonds to it from the adjacent cations*; that is, that

$$\zeta = \sum_i s_i = \sum_i \frac{z_i}{\nu_i}, \qquad (13\text{-}19)$$

in which $-\zeta e$ is the electric charge of the anion and the summation is taken over the cations at the centers of all the polyhedra of which the anion forms a corner.

In justification of the rule it may be pointed out that it leads to stability of the crystal by placing the anions with large negative charges in positions with large positive potentials, inasmuch as the bond strength for a cation is an approximate measure of the contribution of the cation to the positive potential at the polyhedron corner (the factor $1/\nu$ corresponding to the larger cation-anion distance and the greater number of adjacent anions in the case of cations with larger coordination number). It has been shown by Bragg[65] that the rule can be given a simple interpretation and justification in terms of lines of force. Lines of force start from cations in numbers proportional to their valence, and end on anions. We divide these lines of force for each cation equally among the bonds to the corners of its coordinated polyhedron; the rule then states that each anion receives from the cations to which it is coordinated enough lines of force to satisfy its valence. It is not

---

ionic type. The bonds in these crystals may have a large amount (50 percent or even more) of covalent character. If the bonds resonate among the alternative positions, the valence of the metal atom will tend to be divided equally among the bonds to the coordinated atoms, and a rule equivalent to the electrostatic valence rule would express the satisfaction of the valences of the nonmetal atoms.

[65] W. L. Bragg, *Z. Krist.* **74**, 237 (1930); *op. cit.* (17).

necessary for lines of force to connect distant ions, and in consequence the crystal is stable.

This simple rule restricts greatly the acceptable structures for a substance, and it has been found useful in the determination of the structures of complex ionic crystals, including especially the silicate minerals. The rule is satisfied nearly completely by most of the structures that have been reported for the silicate minerals, deviations by as much as $\pm \frac{1}{8}$ being rare. Somewhat larger deviations from the rule are occasionally found for substances prepared in the laboratory, for which stability as great as for minerals is not expected.

Values of electrostatic bond strengths are given in Table 13-19. It is seen that an oxygen ion ($\zeta = 2$) may be satisfied by two silicon bonds, one silicon bond plus two octahedral aluminum bonds, one silicon bond plus three octahedral magnesium bonds, four octahedral aluminum bonds, three titanium bonds, and in various other ways. These are exemplified by many crystals:[66] 2Si in the various forms of silica and in the disilicates, metasilicates, and other silicates in which silicon tetrahedra share corners; Si + 2Al(6) in topaz ($Al_2SiO_4F_2$), muscovite ($KAl_3Si_3O_{10}(OH)_2$), cyanite (Al_2SiO_5), etc.; Si + 3Mg in phlogopite ($KMg_3AlSi_3O_{10}(OH)_2$), olivine (Mg_2SiO_4), etc.; 4Al(6) in corundum (Al_2O_3), cyanite, etc.; 3Ti in rutile, anatase, and brookite (TiO_2); Si + 2Be in phenacite (Be_2SiO_4); Si + Al(6) + 2Ca(8) in garnet ($Ca_3Al_2Si_3O_{12}$); Si + 2Zr(8) in zircon ($ZrSiO_4$); Si + Al(6) + Be(4) in beryl ($Be_3Al_2Si_6O_{18}$).

Fluoride and hydroxide ions are saturated by bonds of total strength 1. This is achieved by two aluminum octahedral bonds, as in hydrargillite ($Al(OH)_3$), with the structure shown in Figure 13-17, topaz ($Al_2SiO_4F_2$), zunyite, described below, and many other crystals, and also by three magnesium octahedra in brucite, $Mg(OH)_2$, and other crystals.

Many aluminosilicates are based on a complete framework of linked tetrahedra similar to those of the various forms of silica, but involving aluminum ions with coordination number 4 as well as silicon ions. The oxygen ions common to an aluminum and a silicon tetrahedron are then reached by bonds of total strength $\frac{7}{4}$, and require a bond of strength $\frac{1}{4}$ for saturation. Such a bond is not provided by a cation with large charge and small radius; it is therefore necessary that large univalent or bivalent cations, namely, alkali or alkaline-earth ions, be present to the extent of one alkali ion or one-half an alkaline-earth ion for every quadricoordinate aluminum ion. This requirement of the electrostatic valence rule is thoroughly substantiated by the formulas of the

[66] See Bragg, *op. cit.* (17).

zeolites, feldspars, and other aluminosilicates with tetrahedral frame
works, as is shown by the following small list:

Orthoclase, $KAlSi_3O_8$	Edingtonite, $BaAl_2Si_3O_{10} \cdot 4H_2O$
Celsian, $BaAl_2Si_2O_8$	Chabazite, $NaAlSi_2O_6 \cdot 3H_2O$
Albite, $NaAlSi_3O_8$	Marialite, $Na_4Al_3Si_9O_{24}Cl$
Anorthite, $CaAl_2Si_2O_8$	Meionite, $Ca_4Al_6Si_6O_{24}(SO_4,CO_3)$
Analcite, $NaAlSi_2O_6 \cdot H_2O$	Nepheline, $Na_3KAl_4Si_4O_{16}$
Natrolite, $Na_2Al_2Si_3O_{10} \cdot 2H_2O$	Kaliophilite, $KAlSiO_4$
Scolecite, $CaAl_2Si_3O_{10} \cdot 3H_2O$	Leucite, $KAlSi_2O_6$
Thomsonite, $NaCa_2Al_5Si_5O_{20} \cdot 6H_2O$	Sodalite, $Na_4Al_3Si_3O_{12}Cl$

In all of these crystals the ratio of number of oxygen atoms to number of aluminum atoms and silicon atoms is 2:1, as required for a complete tetrahedral framework, and the number of alkali and alkaline-
earth atoms is that required by the argument given above. (In a few
cases, such as sodalite, described below, more alkali ion is present, balanced by halogenide ion or a similar anion.)

Two adjacent aluminum tetrahedra, sharing a corner, contribute
two bonds of strength $\frac{3}{4}$ to the shared oxygen atom. The total of $\frac{3}{2}$
(which might be increased by a small amount by bonds from alkali or
alkaline-earth ions) represents a deviation from the electrostatic valence rule such that in general in aluminosilicates of the tetrahedral
framework type the Al/Si ratio does not exceed 1, and when it equals 1
there is good ordering, with alternation of the aluminum and silicon
tetrahedra.

The tetrahedral framework crystals have interesting properties. It
is sometimes possible for the alkali and alkaline-earth ions to be inter-
changed with others in solution; this property permits the zeolites to
be used for softening water. The water molecules also can be removed
and replaced by other molecules without the destruction of the crystal.
The presence of definite available positions for occupancy by large cations or water molecules is clearly indicated by the formulas of such isomorphous pairs as natrolite and scolecite, differing in the replacement
of $2Na^+ + 2H_2O$ by $Ca^{++} + 3H_2O$.

The structure of sodalite, $Na_4Al_3Si_3O_{12}Cl$, a representative crystal of
the framework class, is shown in Figure 13-13. It is interesting that
the same framework is present in ultramarine (lapis lazuli).[67] In the
ultramarines sulfur complexes, S_x^{--}, to which the blue color is attributed, are present in place of chlorine. The selenium and tellurium
analogues are blood-red and yellow, respectively.

The feldspars, such as albite, $NaAlSi_3O_8$, have a compact aluminosili-

[67] F. M. Jaeger, *Spatial Arrangements of Atomic Systems and Optical Activity*,
etc., McGraw-Hill Book Co., New York, 1930; E. Podschus, U. Hofmann, and
K. Leschewski, *Z. anorg. Chem.* **228**, 305 (1936)

Fig. 13-13.—A model showing the structure of the cubic crystal sodalite, $Na_4Al_3Si_3O_{12}Cl$. SiO_4 and AlO_4 tetrahedra alternate, with shared corners. The large spheres represent chloride ions. The sodium ions are not shown.

cate tetrahedral framework with the alkali and alkaline-earth ions in the interstices.[68]

The mineral zunyite, $Al_{13}Si_5O_{20}(OH)_{18}Cl$, may be described as an example of a complex silicate.[69]   Of the thirteen aluminum ions twelve show octahedral coordination, the twelve octahedra forming the group

[68] W. H. Taylor, *Z. Krist.* **85**, 425 (1933); W. H. Taylor, J. A. Darbyshire, and H. Strunz, *ibid.* **87**, 464 (1934); F. Laves and U. Chaisson, *J. Geol.* **58**, 584 (1950); J. R. Goldsmith and F. Laves, *Geochim. et Cosmochim. Acta* **6**, 100 (1954); S. W. Bailey and W. H. Taylor, *Acta Cryst.* **8**, 621 (1955); R. B. Ferguson, R. J. Traill, and W. H. Taylor, *ibid.* **11**, 331 (1958).

[69] L. Pauling, *Z. Krist.* **84**, 442 (1933); B. Kamb, *Acta Cryst.* **12**, in press (1959).

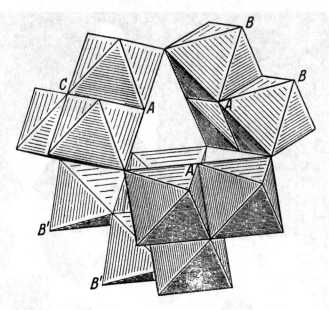

Fig. 13-14.—The group of 12 aluminum octahedra in zunyite. Groups of this type are attached to one another by corners B and B', to silicon tetrahedra by corners A, and to the aluminum tetrahedron by the shared corner C.

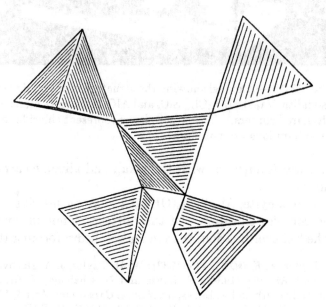

Fig. 13-15.—The group of five silicon tetrahedra in zunyite

shown in Figure 13-14. The five silicon ions are present as the tetrahedral complex Si_5O_{16} (Fig. 13-15). These two complexes are combined together, along with an aluminum tetrahedron, in the way shown in Figure 13-16. Of the anions in the formula 4 O^{--} are held by two silicon tetrahedra, 12 O^{--} by a silicon tetrahedron and two aluminum octahedra, and 18 OH^- by two aluminum octahedra; these all satisfy the electrostatic valence rule. The four remaining oxygen ions are common to an aluminum tetrahedron and three aluminum octahedra, with total bond strength $2\frac{1}{4}$. The chloride ion is introduced to balance this excess of bond strength.

The clay minerals, micas, and chlorites make an interesting group.[70] In hydrargillite,[71] $Al(OH)_3$, there are pseudo-hexagonal layers of octahedra, as shown in Figure 13-17, and linked tetrahedral layers with almost the same dimensions occur as part of the framework of tridymite and cristobalite (Fig. 13-18). If all of the tetrahedra of such a layer are turned in the same direction their unshared hydroxide ions can be condensed with three-fourths of the hydroxide ions on one side of the hydrargillite layer, with elimination of water. This gives a double layer, as shown on the right of Figure 13-19; it represents a layer of kaolin, with formula $Al_2Si_2O_5(OH)_4$. The complete crystal contains these neutral layers loosely piled together; the layers are easily separated, making the substance soft and giving it pronounced basal cleavage. The modifications of kaolin (kaolinite, dickite, and nacrite) differ in the way in which the layers are superimposed.[72] In halloysite the very thin crystal layers occur as minute cylinders; the curving of the layers might be expected from the lack of equivalence of the two sides.

If layers of silicon tetrahedra are condensed on both sides of a hydrargillite layer, a substance is obtained having the composition $Al_2Si_4O_{10}(OH)_2$. This is the clay mineral pyrophyllite. The substance $Mg_3Si_4O_{10}(OH)_2$ obtained similarly from a brucite layer (Fig. 7-10) is the mineral talc. Both of these substances, involving the loose superposition of neutral layers, are very soft, with extreme basal cleavage.[73]

By the replacement of one-fourth of the silicon ions in a talc or pyrophyllite layer by aluminum ions a layer is obtained with a negative elec-

[70] L. Pauling, *Proc. Nat. Acad. Sci. U.S.* **16**, 123, 578 (1930).

[71] H. D. Megaw, *Z. Krist.* **87**, 185 (1934).

[72] J. W. Gruner, *Z. Krist.* **83**, 75, 394 (1932); **85**, 345 (1933); S. B. Hendricks, *Nature* **142**, 38 (1938); *Am. Mineralogist* **23**, 295 (1938); *Z. Krist.* **100**, 509 (1939); G. W. Brindley and K. Robinson, *Mineral Mag.* **27**, 242 (1946); **28**, 393 (1948).

[73] The structures have been described in detail by J. W. Gruner, *Z. Krist* **88**, 412 (1934); see also S. B. Hendricks, *ibid.* **99**, 264 (1938), who has reported that there is some randomness of superposition of the layers, and B. B. Zvjagin and Z. G. Pinsker, *Doklady Akad. Nauk S.S.S.R.* **68**, 505 (1949).

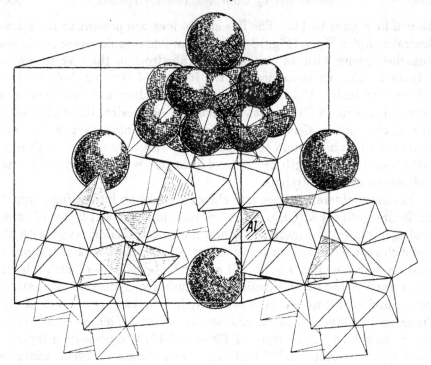

FIG. 13-16.—The structure of the cubic crystal zunyite, $Al_{13}Si_5O_{20}$
$(OH)_{18}Cl$, showing the arrangement of aluminum octahedra and aluminum and silicon tetrahedra. The large spheres represent the chloride
ions.

FIG. 13-17.—Layers of aluminum octahedra with
shared edges as in hydrargillite, $Al(OH)_3$.

tric charge, corresponding to the composition $[Mg_3AlSi_3O_{10}(OH)_2]^-$ or $[Al_2AlSi_3O_{10}(OH)_2]^-$. A neutral crystal can be built up by alternating these layers with layers of potassium ions or other alkali ions, which fit into the pockets formed by rings of six oxygen ions of each of the adjacent layers (Figs. 13-18 and 13-19). The mica crystals obtained in this way have the composition $KMg_3AlSi_3O_{10}(OH)_2$ (phlogopite) and $KAl_2AlSi_3O_{10}(OH)_2$ (muscovite). The general formula of the mica minerals can be written as $(K, Na) X_nAlSi_3O_{10}(OH, F)_2$, with $X = Al^{3+}$, Mg^{++}, Fe^{3+}, Fe^{++}, Mn^{3+}, Mn^{++}, Ti^{4+}, Li^+ (ions with ligancy 6) and n lying between 2 and 3. It is interesting that in lithium micas (lepidolite, zinnwaldite) the lithium ions are in the octahedral layers and not in the positions occupied by potassium ion.

In the margarites or brittle micas the potassium ions are largely replaced by calcium ions, the ideal composition of margarite being $CaX_nAl_2Si_2O_{10}(OH)_2$. In talc and pyrophyllite the layers are electrically neutral and are held together only by van der Waals forces; these crystals are therefore very soft and feel soapy to the touch. To separate the layers in mica it is necessary to break the electrostatic bonds of the univalent potassium ions, so that the micas are not so soft, and thin plates are sufficiently elastic to straighten out after being bent. Separation of layers in the brittle micas involves breaking bonds of bipositive calcium ions; these minerals are hence harder and are brittle, but still show perfect basal cleavage. The sequence of hardness of the minerals on the Mohs scale is the following: talc and pyrophyllite, 1 to 2; the micas, 2 to 3, the brittle micas, $3\frac{1}{2}$ to 5.

By replacement of one-third of the magnesium ions in a brucite layer by aluminum ions a positively charged octahedral layer of composition $[Mg_2Al(OH)_6]^+$ is obtained. Layers of this type can be alternated with negatively charged mica layers to give substances with the structure shown in Figure 13-20. Their general formula is $X_mY_4O_{10}(OH)_8$, with m between 4 and 6; X represents cations with octahedral coordination and Y cations (Al^{3+} and Si^{4+}) with tetrahedral coordination. These are called the chlorite minerals[74] (chlorite, penninite, clinochlore, amesite, etc.).

There are many ways in which the electrostatic valence rule can be used other than those relating directly to the structure of crystals. Some of these are discussed in the following paragraphs.

Although the metasilicates, disilicates, and other silicates in which tetrahedron corners are shared are very stable, the corresponding compounds of phosphorus and sulfur are unstable. The explanation of

[74] Pauling, *loc. cit.* (70), 578. The methods of superposition of layers in the micas and chlorites have been discussed by W. W. Jackson and J. West, *Z. Krist.* **76**, 211 (1931), and by R. C. McMurchy, *ibid.* **88**, 420 (1934).

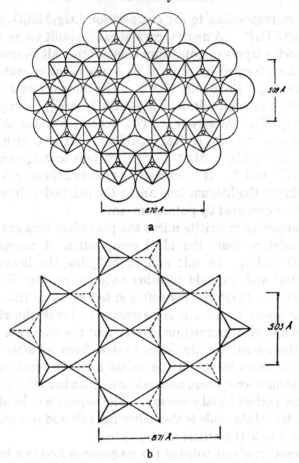

FIG. 13-18.—The fundamental layers of the clays, micas, and chlorites. (a) A hydrargillite layer of octahedra. The light circles indicate oxygen atoms, the heavier ones hydroxide ions in mica. (b) A tetrahedral layer from β-cristobalite or β-tridymite. A silicon *(continued on opposite page)*

this is as follows: An oxygen ion shared by two silicon tetrahedra satisfies the electrostatic valence rule, whereas there is an infraction by $\frac{1}{2}$ for the common corner of two phosphorus tetrahedra and by 1 for two sulfur tetrahedra. In consequence, the pyrophosphates and metaphosphates are unstable—they do not occur at all as minerals and in solution they hydrolyze easily to orthophosphates—and the pyrosulfates are exceedingly unstable. For the same reason silicon dioxide is stable but phosphorus pentoxide and sulfur trioxide combine with water with great avidity.

The electrostatic valence rule can be satisfied for sulfuric acid by the formation of OHO hydrogen bonds between molecules, the strength of the electrostatic bond of a shared proton being taken as $\frac{1}{2}$. This situa-

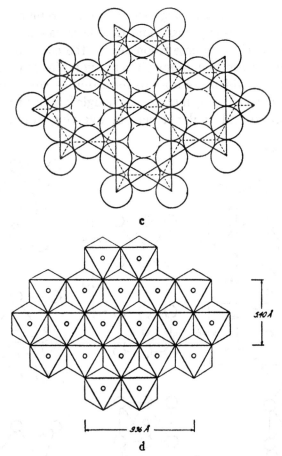

c

d

atom is located at the center of each tetrahedron and an oxygen atom at each corner. (c) A tetrahedral layer in which all the tetrahedra point in the same direction. (d) A complete layer of octahedra (brucite layer).

tion occurs in many crystals, in which for each of two adjacent oxygen ions, presumably bonded together by a proton, the sum of bond strengths from other cations is $\frac{5}{6}$. It has been mentioned in Chapter 12 that the proton between two oxygen ions is usually closer to one than to the other; it seems wise[75] in some cases to divide the total bond strength unequally, in about the ratio 5/6:1/6.

The acid strengths in the sequence $Si(OH)_4$, $PO(OH)_3$, $SO_2(OH)_2$, $ClO_3(OH)$ can be discussed qualitatively in a simple way. The bonds from the central atoms have the strengths 1, $\frac{5}{4}$, $\frac{3}{2}$, and $\frac{7}{4}$, respectively, leaving each oxygen atom with further bonding power of 1, $\frac{3}{4}$, $\frac{1}{2}$, and $\frac{1}{4}$, respectively, for hydrogen. In consequence, silicic acid is a very weak acid, phosphoric acid is a weak acid, sulfuric acid is a strong acid, and

[75] C. A. Beevers and C. M. Schwartz, *Z. Krist.* 91, 157 (1935).

perchloric acid is a very strong acid.[76] (This argument is valid even
for molecules containing single and double covalent bonds provided
that there is nearly complete resonance of the covalent bonds among
all the coordinated oxygen atoms; it is not valid for acids of such atoms
as molybdenum, which are able to form double covalent bonds with
some of the adjacent oxygen atoms and single bonds with others.)

In hydrargillite and similar structures involving aluminum octahedra

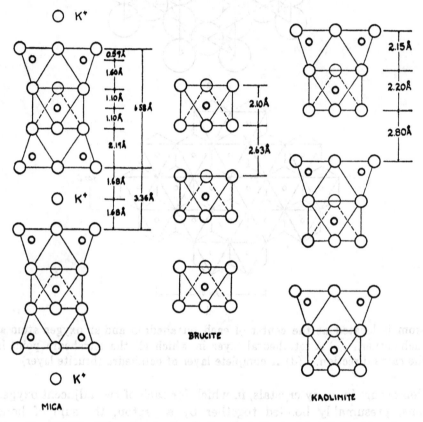

Fig. 13-19.—The structures of mica, brucite, and kaolinite, showing
the sequence of layers normal to the cleavage planes. Large circles
represent O^{--} or OH^- (or K^+ when so marked), small ones Si^{4+} or
Al^{3+} at tetrahedron centers and Mg^{++} or Al^{3+} at octahedron centers.

with shared hydroxide ions the electrostatic valence rule is satisfied as
for silicic acid; $Al(OH)_3$ is hence expected to be about as strong an acid
as $Si(OH)_4$. An aluminum tetrahedron with corners shared with silicon tetrahedra is, however, similar to the perchlorate ion, and the acid
obtained by replacing the potassium ion of mica by hydrogen ion

[76] See A. Kossiakoff and D. Harker, *J.A.C.S.* **60**, 2047 (1938).

should be very strong. This has been verified by experiment for the acid obtained from clays by replacement of their alkali ions by hydrogen ion. The alkali aluminates $MAlO_2$ are salts of acids obtained by the polymerization of tetrahedral $Al(OH)_4$ groups.[77]

The Sharing of Polyhedron Corners, Edges, and Faces.—Characteristic features of the structure of complex ionic crystals other than those included within the field of application of the electrostatic valence rule can be summarized in further rules dealing with the sharing of polyhedron corners, edges, and faces. The electrostatic valence rule indicates the number of polyhedra with a common corner, but makes no prediction as to the number of corners common to two polyhedra, that is, as to whether they share one corner only, two corners defining an edge, or three or more corners defining a face. In rutile, brookite, and anatase, for example, each oxygen ion is common to three titanium octahedra, but the number of edges shared by each octahedron with adjoining octahedra is two in rutile, three in brookite, and four in anatase. The significance of this difference in structure is contained in the following rule: *The presence of shared edges and especially of shared faces in a coordinated structure decreases its stability; this effect is large for cations with large valence and small ligancy.*

The decrease in stability arises from the cation-cation Coulomb terms. The sharing of an edge between two regular tetrahedra brings the cations at their centers to a distance from each other only 0.58 times that obtaining when the tetrahedra share a corner only; and the sharing of a face decreases this distance to 0.33 times that for a shared corner (Fig. 13-21). The corresponding

[77] T. F. W. Barth, *J. Chem. Phys.* **3**, 323 (1935).

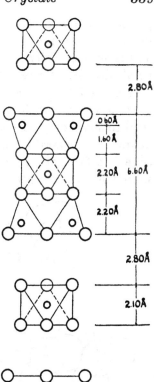

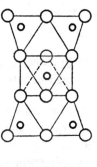

CHLORITE

Fig. 13-20.—The sequence of layers normal to the cleavage plane of the chlorite minerals, showing the alternation of mica layers such as $[Mg_3AlSi_2O_{10}(OH)_2]^-$ with charged brucitelike layers $[Mg_2Al(OH)_6]^+$.

positive Coulomb terms cause a large increase in crystal energy and decrease in stability of the crystal to accompany the sharing of edges and faces, especially for highly charged cations. The effect is less for octahedra than for tetrahedra, the interatomic distance ratios[78] being 0.71 and 0.58 in place of 0.58 and 0.33.

In agreement with this rule, it is observed that silicon tetrahedra tend to share only corners with other silicon tetrahedra or other polyhedra. No crystal is known in which two silicon tetrahedra share an edge or a face, and in most of the silicate structures only corners are shared between silicate tetrahedra and other polyhedra also. This rule

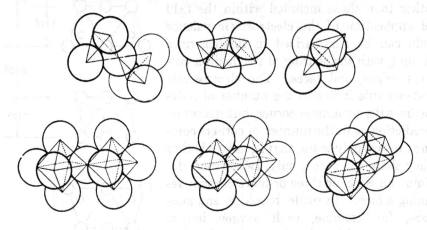

Fig. 13-21.—The sharing of a corner, an edge, and a face by a pair of tetrahedra and by a pair of octahedra.

leads to the formation of framework structures such as that shown in Figure 13-13, rather than the more compact structures that could be built by the sharing of edges and faces. It requires that metasilicates (and also metaphosphates and related substances) contain rings of three or more tetrahedra with shared corners ($[Si_3O_9]^{6-}$ in benitoite,[79] $BaTiSi_3O_9$, $[Si_6O_{18}]^{12-}$ in beryl,[80] $Be_3Al_2Si_6O_{18}$, etc.) or infinite chains, as in diopside,[81] $CaMg(SiO_3)_2$, rather than $[Si_2O_6]^{4-}$ groups formed by the sharing of an edge by two tetrahedra.

This rule, like the others mentioned in this chapter, can be used as a criterion for the largely ionic character of the bonds in a substance. The rule is obeyed by all of the forms of silica, in which the bonds have as much ionic character as covalent character. In SiS_2, on the

[78] These values are for undistorted polyhedra. Some compensating distortion always occurs, as discussed in a later paragraph.

[79] W. H. Zachariasen, *Z. Krist.* **74**, 139 (1930).

[80] W. L. Bragg and J. West, *Proc. Roy. Soc. London* A111, 691 (1926).

[81] B. E. Warren and W. L. Bragg, *Z. Krist.* **69**, 168 (1928).

other hand, the SiS_4 tetrahedra share edges with one another, forming

infinite strings, $\cdots Si \diagdown \diagup Si \diagdown \diagup Si \diagdown \diagup Si \cdots$ with S atoms at top ($S \quad S \quad S \quad S$) and bottom ($S \quad S \quad S \quad S$) (Fig. 11-20); this sub-

stantiates the idea that the Si—S bonds are largely covalent.

It is interesting that rutile, with only two shared edges per octahedron, is reported to be more stable than brookite and anatase, and, moreover, that many substances MX_2 have the rutile structure, whereas only titanium dioxide has been reported to have the brookite and anatase structures.

Another rule relating to the sharing of corners, edges, and faces is the following: *In a crystal containing different cations those with large valence and small coordination number tend not to share polyhedron elements with each other.* This rule expresses the fact that cations with large electric charges tend to be as far apart from each other as possible in order to reduce their contribution to the Coulomb energy of the crystal.

The rule requires that in silicates the silicon tetrahedra share no elements with each other if the oxygen-silicon ratio is equal to or greater than four. This is found to be true in general (topaz, zircon, olivine, other orthosilicates). Most of the few exceptional substances now known contain extra oxygen atoms in the form of hydroxide ions. These include the clay minerals, micas, and chlorites, discussed above, and also the mineral hemimorphite, $Zn_4Si_2O_7(OH)_2 \cdot H_2O$, in which there is a disilicate group.[82] There is one crystal, danburite, $CaB_2Si_2O_8$, that contains a framework of silicon and boron tetrahedra in which the silicon tetrahedra do not alternate with the boron tetrahedra, as indicated by the rule, but instead occur in Si_2O_7 pairs.

When coordinated polyhedra about cations with large charge do share edges or faces with one another, it is to be expected that the repulsion of the cations will lead to such a deformation of the polyhedra as to increase the cation-cation distance. This can occur without change in cation-anion distance by decrease in length of the shared edges. It will continue until the effect of the cation-cation repulsion is balanced by the characteristic repulsion of the anions defining the shared edges. Theoretical calculations for rutile and anatase, described in Section 13-4, have shown that this effect for edges shared between titanium octahedra leads to an oxygen-oxygen distance of 2.50 Å, in place of the normal distance 2.80 Å. The edges bounding the face shared between two aluminum octahedra in corundum, Al_2O_3, are

[82] T. Ito and J. West, *Z. Krist.* **83**, 1 (1932).

also 2.50 Å long, and the same edge length has been found for edges shared by aluminum octahedra in other crystals (diaspore, hydrargillite) and for edges shared by other polyhedra.

The shortening of shared edges can be used as another test of the amount of ionic character of the bonds in a substance. In essentially covalent crystals such as niccolite, NiAs, and marcasite, FeS_2, MX_4 octahedra that share faces occur; the edges bounding these faces are, however, longer than unshared edges, instead of shorter, as required by the rule for essentially ionic crystals. In the gas molecules Al_2Cl_6, Al_2Br_6, and Al_2I_6 the edge shared by the two AlX_4 tetrahedra is shorter than the other edges;[83] this shows that the Al—X bonds have appreciable ionic character.

Minerals, including the silicate minerals, many of which have been thoroughly investigated by x-ray methods, provide excellent illustrations for the rules given in the preceding paragraphs. The sulfide minerals, on the other hand, show general lack of agreement with the rules; their bonding is largely covalent (Sec. 11-16).

[83] K. J. Palmer and N. Elliott, *J.A.C.S.* **60**, 1852 (1938).

CHAPTER 14

A Summarizing Discussion of Resonance
and Its Significance for Chemistry

14-1. THE NATURE OF RESONANCE

NOW that we have considered some of the ways in which the idea
of resonance has brought clarity and unity into modern structural
chemistry, has led to the solution of many problems of valence theory,
and has assisted in the correlation of the chemical properties of substances with the information obtained about the structure of their
molecules by physical methods, we may well inquire again into the
nature of the phenomenon of resonance.[1]

The goal of a structural investigation of a system is the description of
the system in terms of simpler entities. This description may be divided into two parts, the first relating to the material particles or bodies
of which the system is considered to be composed, and the second to the
ways in which these particles or bodies are interrelated, that is, to their
interactions and interconnections. In describing a system it is usually
convenient to resolve it first into the next simpler parts, rather than
into its ultimate constituents, and then to carry the resolution further
and further in steps. We are thoroughly accustomed to this way of describing the material constitution of substances. The use of the concept of resonance permits the extension of the procedure to include the
discussion not only of the next simpler constituent bodies but also of
their interactions. Thus the material description of the benzene molecule as containing carbon and hydrogen atoms, which themselves contain electrons and nuclei, is amplified by use of the resonance concept
in the following way: The structure of the normal benzene molecule
corresponds to resonance between the two Kekulé structures, with

[1] A thorough discussion of this question is given by G. W. Wheland, *Resonance
in Organic Chemistry*, John Wiley and Sons, New York, 1955.

smaller contributions by other valence-bond structures, and the molecule is stabilized and its other properties are changed somewhat by this resonance from those expected for either Kekulé structure alone; each Kekulé structure consists of a certain distribution of single and double bonds, with essentially the properties associated with these bonds in other molecules; each bond represents a type of interaction between atoms that can be discussed in terms of the resonance between structures differing in the interchange of electrons between atomic orbitals.

It was pointed out in Section 1-3 and again in Section 6-5 that the selection of the primary structures for the discussion of any particular case of quantum-mechanical resonance is arbitrary, but that this arbitrariness (which has an analogue in the classical resonance phenomenon) does not impair the value of the concept of resonance.

14-2. THE RELATION BETWEEN RESONANCE AND TAUTOMERISM

No sharp distinction can be made between tautomerism and resonance; but it is convenient in practice to make a distinction between the two that is applicable to all except the borderline cases.

Tautomers are defined as isomers that are readily interconvertible. It is clear that the distinction between tautomerism and ordinary isomerism is very vague indeed, and that it depends on the interpretation of the adverb "readily." It is customary to designate as tautomers those isomers whose half-lives (with respect to interconversion) are, under ordinary circumstances, less than the time required for laboratory operations to be carried out (some minutes or hours), so that the separation of the isomers from the equilibrium mixtures is difficult. The distinction between tautomers and ordinary isomers has no molecular significance whatever, since it is dependent on the accidental rate of ordinary human activity.

It is possible, on the other hand, to define tautomerism and electronic resonance in a way that gives structural significance to them individually.

Let us consider as a definite example a benzene molecule, which may have different substituent groups in the 1, 2, · · · , 6 positions. The nuclei of the molecule vibrate relative to one another in a manner determined by the electronic energy function for nuclear configurations.[2] For most molecules this electronic energy function is such that there is one most stable nuclear configuration, about which the nuclei carry out small vibrations, with amplitudes of about 0.1 Å. If the molecule can be assigned a single valence-bond structure, the nature of this equilibrium configuration can be predicted by the rules of stereochemistry. Thus for tetramethylethylene the expected configuration is

[2] *Introduction to Quantum Mechanics.*

$$
\begin{array}{ccc}
CH_3 & & CH_3 \\
\diagdown & & \diagup \\
\alpha & C=C & \alpha \\
\diagup & & \diagdown \\
CH_3 & & CH_3
\end{array}
$$

with the angle α equal to about 110° (close to the tetrahedral angle 109°28′). This has been verified experimentally. We may describe benzene, however, as involving resonance between the two valence-bond structures ⬡ (I) and ⬡ (II). This resonance is so rapid that its frequency[3] (the resonance energy divided by Planck's constant h) is a thousandfold greater than the frequency of nuclear vibration, so that resonance between the Kekulé structures occurs in the time required for the nuclei to move an inappreciable distance (0.0001 Å). Hence the effective electronic energy function determining the nuclear configuration is not that for either Kekulé structure, but instead that corresponding to the Kekulé resonance. Since the predicted stable configurations for the two Kekulé structures do not differ greatly, there is an intermediate configuration that is the stable equilibrium configuration for the actual resonating molecule. This is the hexagonal planar configuration with 120° bond angles.

The magnitude of the resonance integral, which determines the resonance energy and the resonance frequency, depends on the nature of the structures involved. In benzene it is large (about 36 kcal/mole), but it might have been much smaller. Let us consider what the benzene molecule would be like if the value of the resonance integral were very small, so that the resonance frequency would be less than the frequency of nuclear oscillation. For each nuclear configuration there would be more or less electronic resonance of the Kekulé type. We may discuss three nuclear configurations, a, b, and c.

a b c

[3] An effort to determine the resonance frequency greatly disturbs the molecule, in such a way that after the experiment has been carried out the molecule may not be in its original state. Hence some care must be used in the interpretation of the expression resonance frequency. The argument in the text can be carried out by quantum-mechanical methods without use of this expression.

In *a* and *c* the bond angles to the substituent groups approach 110° and 125° in alternate pairs, corresponding to the tetrahedral model for alternating single and double bonds in the ring, and in *b* the bond angles are all 120°. Now for configuration *a* the valence-bond structure I is stable, whereas structure II is unstable because of the strain involved in the bond angles. Since the resonance integral is assumed to be small in value, this energy difference would cause structure II to be unimportant. The normal electronic state of the molecule would be represented for this nuclear configuration essentially by Kekulé structure I alone, with only a negligible amount of resonance with structure II.

Similarly, for configuration *c* structure II alone would be of significance.

The intermediate configuration *b* would involve complete resonance between I and II. Since the resonance energy is assumed to be very small, and this configuration corresponds to bond-angle strain for both I and II, the configuration would be less stable than *a* or *c*.

This hypothetical benzene molecule would accordingly oscillate for some time about the configuration *a*, with essentially the valence-bond structure I; it might then pass through the configuration *b*, with resonance to structure II becoming complete, and then oscillate for some time about configuration *c*, with essentially the valence-bond structure II.

The chemical properties of this hypothetical benzene would be just those expected for the valence-bond structures I and II, and, indeed, the substance would be correctly described as a mixture of these two isomers or tautomers.

It is evident that we may define tautomerism and resonance in the following reasonable way: *When the magnitudes of the electronic resonance integral (or integrals) and of the other factors determining the electronic energy function of a molecule are such that there are two or more well-defined stable nuclear equilibrium configurations, we refer to the molecule as capable of existing in tautomeric forms; when there is only one well-defined stable nuclear equilibrium configuration, and the electronic state is not satisfactorily represented by a single valence-bond structure, we refer to the molecule as a resonating molecule.*

This may be expressed somewhat more loosely by saying that, whereas a tautomeric substance is a mixture of two types of molecules, differing in configuration, in general the molecules of a substance showing electronic resonance are all alike in configuration and structure.

Each of the tautomeric forms of a substance may show electronic resonance; tautomerism and resonance are not mutually exclusive. Let us discuss 5-methylpyrazole as an example. This substance exists in two tautomeric forms, A and B, differing in the position of the N-hydrogen atom.

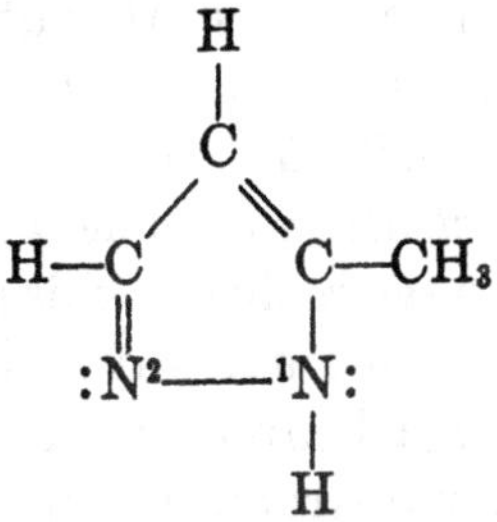

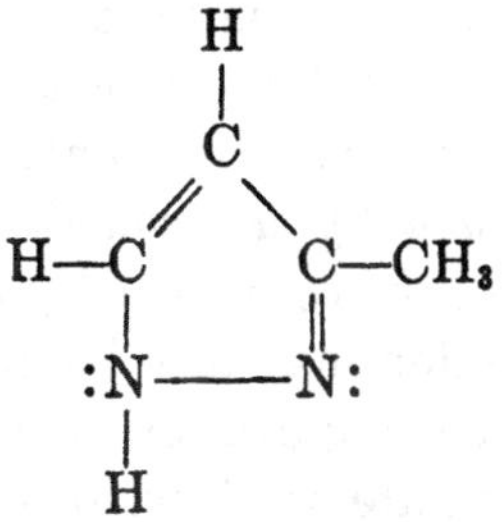

Each of these tautomers in its normal state is represented, not by the conventional valence-bond structure shown above, but by a resonance hybrid of this structure and others.   For tautomer A, with the hydrogen atom attached to the nitrogen atom 1, the principal resonance is between structures A I and A II, with A I the more important; smaller contributions are made also by other structures such as A III.   Similar resonance occurs for tautomer B.   Thus for both tautomers the principal resonance

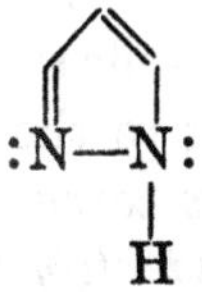

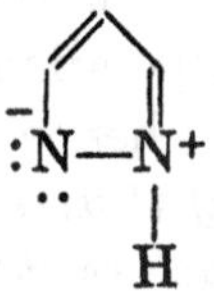

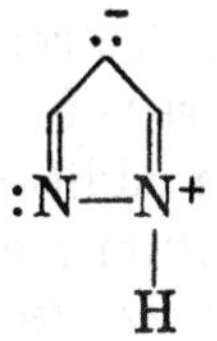

is between valence-bond structures I  and II  , with I more important for A and II for B; but it is not correct (according to our conventional nomenclature for electronic resonance) to say that methylpyrazole resonates between the structures

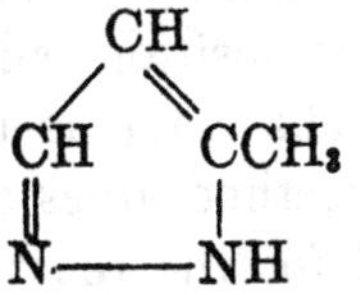

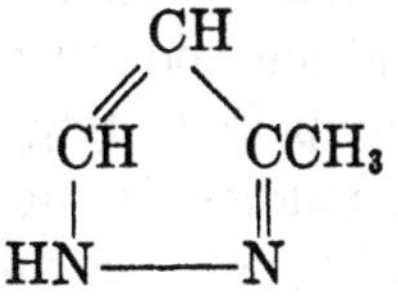

14-3. THE REALITY OF THE CONSTITUENT STRUCTURES OF A RESONATING SYSTEM

It is often asked whether or not the constituent structures of a resonating system, such as the Kekulé structures for the benzene molecule, are to be considered as having reality.   There is one sense in which this question may be answered in the affirmative, but the answer is definitely negative if the usual chemical significance is attributed to the structures.   *A substance showing resonance between two or more valence-bond structures does not contain molecules with the configurations and properties usually associated with these structures.*   The constituent structures of the resonance hybrid do not have reality in this sense.

The question may also be discussed in a different way.   The stable equilibrium configuration of the nuclei of a benzene molecule is not that appropriate to either of the two Kekulé structures, but is the intermediate hexagonal configuration.   The valence-bond structures I and II are hence to be interpreted as being

somewhat different from those for nonresonating molecules.   They mean that the electronic motion is that corresponding to alternating single and double bonds, but with the equilibrium internuclear distances constant (1.40 Å), rather than alternating between 1.54 Å and 1.33 Å.   The electronic wave function for the normal benzene molecule can be composed of terms corresponding to the Kekulé structures I and II, plus some additional terms; hence, according to the fundamental ideas of quantum mechanics, *if it were possible to carry out an experimental test of the electronic structure that would identify structure I or structure II, each structure would be found for the molecule to the extent determined by the wave function.*   The difficulty for benzene and for other molecules showing electronic resonance is to devise an experimental test that could be carried out quickly enough and that would distinguish among the structures under discussion.   In benzene the frequency of Kekulé resonance is only a little less than the frequency of the bonding resonance of electron pairs, so that the time required for the experiment is closely limited.

Most methods of testing bond type involve the motion of nuclei. The chemical method, such as substitution at positions adjacent to a hydroxyl group in testing for double-bond character as used in the Mills-Nixon studies, is one of these.   This method gives only the resultant bond type over the period required for the reaction to take place.   Since this period is much longer than that of ordinary electronic resonance, the chemical method cannot be used in general to identify the constituent structures of a resonating molecule.   Only if the resonance frequency is very small (less than the frequencies of nuclear vibration) can the usual methods be applied to identify the constituent structures; and in this case the boundary between resonance and tautomerism is approached or passed.

The foregoing statement is not to be construed as meaning that chemical methods, as well as physical methods, cannot be used as the basis for inference about the nature of resonating structures. This inference is based on the resultant bond type, and not on the direct identification of individual structures.

14-4. THE FUTURE DEVELOPMENT AND APPLICATION OF THE CONCEPT OF RESONANCE

When we compare our present knowledge of structural chemistry with that of 30 years ago and become cognizant of the extent to which clarity has been brought into this field of knowledge by the extensive application of the concept of resonance, we are tempted to speculate about the future development of this concept and the nature of the further applications of it that may be made.

The applications of the idea of resonance that have been made during the last thirty years are in the main qualitative in nature. This represents only the first step, which should be followed by more refined treatments with quantitative significance. Some rough quantitative considerations, such as those about interatomic distances, the partial ionic character of bonds, and the energy of resonance of molecules among several valence-bond structures, have been described in the preceding chapters of this book; these, however, deal with only small portions of the broad field of structural chemistry. The ultimate goal, a theory permitting the quantitative prediction of the structure and properties of molecules, is still far away.

In this book the discussion has been restricted almost entirely to the structure of the normal states of molecules, with little reference to the great part of chemistry dealing with the mechanisms and rates of chemical reactions. It seems probable that the concept of resonance can be applied very effectively in this field. The "activated complexes" that represent intermediate stages in chemical reactions are, almost without exception, unstable molecules that can be described as resonating among several valence-bond structures. Thus, according to the theory of Lewis, Olson, and Polanyi, Walden inversion occurs in the hydrolysis of an alkyl halogenide by the following mechanism:

$$
\begin{array}{ccccc}
& H & & H & & H \\
& \backslash & & | & & / \\
HO^- + R_1{-}C{-}I & \longrightarrow & HO\text{---}C\text{---}I & \longrightarrow & HO{-}C{-}R_1 + I^- \\
& / & & / \quad \backslash & & \backslash \\
& R_2 & & R_1 \quad R_2 & & R_2
\end{array}
$$

The activated complex can be described as involving resonance of the fourth bond of carbon between the hydroxide and iodide ions. Some very interesting quantum-mechanical calculations bearing on the theory of chemical reactions have been made by Eyring and Polanyi and their collaborators and by other investigators. It is to be hoped that the quantitative treatments can be made more precise and more reliable; but before this can be done effectively, an extensive development of the qualitative theory of chemical reactions must take place, probably in terms of resonance.

Among the most interesting problems of science are those of the structure and properties of substances of biological importance. I have little doubt that in this field resonance and the hydrogen bond are of great significance, and that these two structural features will be found to play an important part in such physiological phenomena as the contraction of muscle and the transmission of impulses along nerves and in the brain. A conjugated system provides the only way of transmitting an effect from one end to another of a long molecule; and the hydrogen bond is the only strong and directed intermolecular interaction which can come into operation quickly. It will be many years before our understanding of molecular structure becomes great enough to encompass in detail such substances as the proteins, with highly specific properties (such as those shown by antibodies) which must be attributed to their possession of well-defined and complex molecular structures; but the attack on these substances by the methods of modern structural chemistry can be begun now, and it is my belief that this attack will ultimately be successful.

The foregoing paragraph has been reproduced without change from the first edition of this book (1939). The discoveries about the structure of polypeptide chains in proteins and polynucleotide chains in nucleic acids that have been made during the last decade have been largely based on considerations of resonance (planarity of the amide group, purines, pyrimidines) and hydrogen-bond formation. We may ask what the next step in the search for an understanding of the nature of life will be. I think that it will be the elucidation of the nature of the electromagnetic phenomena involved in mental activity in relation to the molecular structure of brain tissue. I believe that thinking, both conscious and unconscious, and short-term memory involve electromagnetic phenomena in the brain, interacting with the molecular (material) patterns of the long-term memory, obtained from inheritance or experience. What is the nature of the electromagnetic phenomena? What is the nature of the molecular patterns? What is the mechanism of their interaction? These are problems of structural chemistry that we may now strive to solve.

APPENDICES AND INDICES

Values of Physical Constants

(Chemists' Scale)

Velocity of light c	$= 2.99793 \times 10^{10}$ cm/sec
Electronic charge e	$= 4.80286 \times 10^{-10}$ statcoulomb
Mass of electron m	$= 9.1083 \times 10^{-28}$ g
Mass of proton M_p	$= 1.67239 \times 10^{-24}$ g
Mass of neutron M_n	$= 1.67470 \times 10^{-24}$ g
Planck's constant h	$= 6.62517 \times 10^{-27}$ erg sec
Avogadro's number N	$= 0.60232 \times 10^{24}$ mole^{-1}
Faraday F	$= 96,495.7$ coulomb/mole
Boltzmann's constant k	$= 1.38044 \times 10^{-16}$ erg/deg
Gas constant R	$= 1.9872$ cal deg^{-1} mole^{-1}
Bohr magneton μ_B	$= 0.92731 \times 10^{-20}$ erg/gauss
Ratio of physical to chemical atomic weights	$= 1.000272$
Energy of 1 ev	$= 1.60206 \times 10^{-12}$ erg
Energy of 1 ev	$= 23.063$ kcal/mole
Wave length of 1-ev quantum	$= 12,397.67$ Å
Wave number of 1-ev quantum	$= 8,066.03$ cm^{-1}
Energy of 1 g of mass	$= 5.6100 \times 10^{32}$ ev

The Bohr Atom

IN his first paper on atomic structure Niels Bohr[1] discussed quantized circular orbits of the electron (mass m, charge $-e$) about the nucleus in the hydrogen atom and hydrogenlike ions (mass M, charge $+Ze$). A possible state of motion of the electron is in a circular orbit. According to classical mechanics, the orbit might have any radius. Bohr derived a set of quantized orbits by making the assumption that the angular momentum of the atom should be an integral multiple of $h/2\pi$, where h is Planck's constant.

The relation between the speed v of the electron in a circular orbit about the nucleus and the radius r of the orbit can be derived by use of Newton's laws of motion. A geometrical construction shows that the acceleration of the electron toward the center of the orbit is v^2/r, and hence the force required to produce this acceleration is mv^2/r. This force is the force of attraction Ze^2/r^2 of the electron and the nucleus; hence we write the equation

$$mv^2/r = Ze^2/r^2$$

or, multiplying by r,

$$mv^2 = Ze^2/r \qquad \text{(II-1)}$$

It may be noted that this equation corresponds to the virial theorem (Sec. 1-4). The term on the left is twice the kinetic energy, and that on the right is the potential energy with changed sign.

The angular momentum for the electron in its orbit is mrv. Bohr s postulate for quantizing the circular orbits is represented by the equation

$$mrv = nh/2\pi \qquad \text{(II-2)}$$

---

[1] N. Bohr, *Phil. Mag.* **26**, 1 (1913).

In this equation n is the quantum number for the hydrogen atom, assumed to have the values 1 (for the normal state of the atom), 2 (for the first excited state), 3, 4, 5, and so on.

These two equations are easily solved. It is found that the radius of the circular Bohr orbit for quantum number n is equal to $n^2h^2/4\pi^2Zme^2$. This can be written as n^2a_0/Z, in which a_0 has the value 0.530 Å. The speed of the electron in its orbit is found to be $v = 2\pi Ze^2/nh$. For the normal hydrogen atom, with $Z = 1$ and $n = 1$, this speed is 2.18×10^8 cm/sec, about 0.7 percent that of the speed of light.

The energy of the atom, the sum of the kinetic energy and the potential energy, is

$$E_n = -2\pi^2Z^2e^4m/n^2h^2 \qquad \text{(II-3)}$$

In the above calculation the system has been treated as though the nucleus were stationary and the electron moved in a circular orbit about the nucleus. The correct application of Newton's laws of motion to the problem of two particles with inverse-square force of attraction leads to the result that both particles move about their center of mass. The center of mass is the point on the line between the centers of the two particles such that the two radii are inversely proportional to the masses of the two particles. The equations for the Bohr orbits with consideration of motion of the nucleus are the same as those given above, except that the mass of the electron, m, is to be replaced by the reduced mass of the two particles, μ, defined by the expression $1/\mu = 1/m + 1/M$, where M is the mass of the nucleus.

Hydrogenlike Orbitals

THE wave functions for a state of a hydrogenlike atom described by the quantum numbers n (total quantum number), l (azimuthal quantum number), and m (magnetic quantum number) are usually expressed in terms of the polar coordinates r, θ, and ϕ. The orbital wave function is a product of three functions, each depending on one of the coordinates:

$$\psi_{nlm}(r, \theta, \phi) = R_{nl}(r)\Theta_{lm}(\theta)\Phi_m(\phi) \qquad \text{(III-1)}$$

In this equation the functions Φ, Θ, and R have the forms

$$\Phi_m(\phi) = \frac{1}{\sqrt{2\pi}} e^{im\phi} \qquad \text{(III-2)}$$

$$\Theta_{lm}(\theta) = \left\{ \frac{(2l+1)(l-|m|)!}{2(l+|m|)!} \right\}^{1/2} P_l^{|m|}(\cos\theta) \qquad \text{(III-3)}$$

and

$$R_{nl}(r) = -\left[\left(\frac{2Z}{na_0}\right)^3 \frac{(n-l-1)!}{2n\{(n+l)!\}^3} \right]^{1/2} e^{-\rho/2} \rho^l L_{n+l}^{2l+1}(\rho) \qquad \text{(III-4)}$$

in which

$$\rho = \frac{2Z}{na_0} r \qquad \text{(III-5)}$$

and

$$a_0 = \frac{h^2}{4\pi^2\mu e^2} \qquad \text{(III-6)}$$

TABLE III-1.—THE FUNCTIONS $\Phi_m(\phi)$

$$\Phi_0(\phi) = \frac{1}{\sqrt{2\pi}} \qquad \text{or} \qquad \Phi_0(\phi) = \frac{1}{\sqrt{2\pi}}$$

$$\Phi_1(\phi) = \frac{1}{\sqrt{2\pi}} e^{i\phi} \qquad \text{or} \qquad \Phi_{1\ \cos}(\phi) = \frac{1}{\sqrt{\pi}} \cos\phi$$

$$\Phi_{-1}(\phi) = \frac{2}{\sqrt{2\pi}} e^{-i\phi} \qquad \text{or} \qquad \Phi_{1\ \sin}(\phi) = \frac{1}{\sqrt{\pi}} \sin\phi$$

$$\Phi_2(\phi) = \frac{1}{\sqrt{2\pi}} e^{i2\phi} \qquad \text{or} \qquad \Phi_{2\ \cos}(\phi) = \frac{1}{\sqrt{\pi}} \cos 2\phi$$

$$\Phi_{-2}(\phi) = \frac{1}{\sqrt{2\pi}} e^{-i2\phi} \qquad \text{or} \qquad \Phi_{2\ \sin}(\phi) = \frac{1}{\sqrt{\pi}} \sin 2\phi$$

TABLE III-2.—THE WAVE FUNCTIONS $\theta_{lm}(\theta)$

$l = 0$, s orbitals:

$$\Theta_{00}(\theta) = \frac{\sqrt{2}}{2}$$

$= 1$ p orbitals:

$$\Theta_{10}(\theta) = \frac{\sqrt{6}}{2} \cos\theta$$

$$\Theta_{1\pm1}(\theta) = \frac{\sqrt{3}}{2} \sin\theta$$

$= 2$, d orbitals:

$$\Theta_{20}(\theta) = \frac{\sqrt{10}}{4} (3\cos^2\theta - 1)$$

$$\Theta_{2\pm1}(\theta) = \frac{\sqrt{15}}{2} \sin\theta \cos\theta$$

$$\Theta_{2\pm2}(\theta) = \frac{\sqrt{15}}{4} \sin^2\theta$$

$= 3$ f orbitals:

$$\Theta_{30}(\theta) = \frac{3\sqrt{14}}{4} \left(\frac{5}{3}\cos^3\theta - \cos\theta \right)$$

$$\Theta_{3\pm1}(\theta) = \frac{\sqrt{42}}{8} \sin\theta(5\cos^2\theta - 1)$$

$$\Theta_{3\pm2}(\theta) = \frac{\sqrt{105}}{4} \sin^2\theta \cos\theta$$

$$\Theta_{3\pm3}(\theta) = \frac{\sqrt{70}}{8} \sin^3\theta$$

TABLE III-3.—THE HYDROGEN RADIAL WAVE FUNCTIONS

$n = 1$, K shell:

$l = 0$, $1s$　$R_{10}(r) = (Z/a_0)^{3/2} \cdot 2e^{-\rho/2}$

$n = 2$, L shell:

$l = 0$, $2s$　$R_{20}(r) = \dfrac{(Z/a_0)^{3/2}}{2\sqrt{2}}\,(2 - \rho)e^{-\rho/2}$

$l = 1$, $2p$　$R_{21}(r) = \dfrac{(Z/a_0)^{3/2}}{2\sqrt{6}}\,\rho e^{-\rho/2}$

$n = 3$, M shell:

$l = 0$, $3s$　$R_{30}(r) = \dfrac{(Z/a_0)^{3/2}}{9\sqrt{3}}\,(6 - 6\rho + \rho^2)e^{-\rho/2}$

$l = 1$, $3p$　$R_{31}(r) = \dfrac{(Z/a_0)^{3/2}}{9\sqrt{6}}\,(4 - \rho)\rho e^{-\rho/2}$

$l = 2$, $3d$　$R_{32}(r) = \dfrac{(Z/a_0)^{3/2}}{9\sqrt{30}}\,\rho^2 e^{-\rho/2}$

$n = 4$, N shell:

$l = 0$, $4s$　$R_{40}(r) = \dfrac{(Z/a_0)^{3/2}}{96}\,(24 - 36\rho + 12\rho^2 - \rho^3)e^{-\rho/2}$

$l = 1$, $4p$　$R_{41}(r) = \dfrac{(Z/a_0)^{3/2}}{32\sqrt{15}}\,(20 - 10\rho + \rho^2)\rho e^{-\rho/2}$

$l = 2$, $4d$　$R_{42}(r) = \dfrac{(Z/a_0)^{3/2}}{96\sqrt{5}}\,(6 - \rho)\rho^2 e^{-\rho/2}$

$l = 3$, $4f$　$R_{43}(r) = \dfrac{(Z/a_0)^{3/2}}{96\sqrt{35}}\,\rho^3 e^{-\rho/2}$

$n = 5$, O shell:

$l = 0$, $5s$　$R_{50}(r) = \dfrac{(Z/a_0)^{3/2}}{300\sqrt{5}}\,(120 - 240\rho + 120\rho^2 - 20\rho^3 + \rho^4)e^{-\rho/2}$

$l = 1$, $5p$　$R_{51}(r) = \dfrac{(Z/a_0)^{3/2}}{150\sqrt{30}}\,(120 - 90\rho + 18\rho^2 - \rho^3)\rho e^{-\rho/2}$

$l = 2$, $5d$　$R_{52}(r) = \dfrac{(Z/a_0)^{3/2}}{150\sqrt{70}}\,(42 - 14\rho + \rho^2)\rho^2 e^{-\rho/2}$

$l = 3$, $5f$　$R_{53}(r) = \dfrac{(Z/a_0)^{3/2}}{300\sqrt{70}}\,(8 - \rho)\rho^3 e^{-\rho/2}$

$l = 4$, $5g$　$R_{54}(r) = \dfrac{(Z/a_0)^{3/2}}{900\sqrt{70}}\,\rho^4 e^{-\rho/2}$

$n = 6$, P shell:

$l = 0$, $6s$　$R_{60}(r) = \dfrac{(Z/a_0)^{3/2}}{2160\sqrt{6}}\,(720 - 1800\rho + 1200\rho^2 - 300\rho^3 + 30\rho^4 - \rho^5)e^{-\rho/2}$

$l = 1$, $6p$　$R_{61}(r) = \dfrac{(Z/a_0)^{3/2}}{432\sqrt{210}}\,(840 - 840\rho + 252\rho^2 - 28\rho^3 + \rho^4)\rho e^{-\rho/2}$

The quantity a_0 is interpreted in the Bohr theory as the radius of the smallest orbit in the hydrogen atom; its value is 0.530 Å.

The functions $P_l^{|m|}(\cos \theta)$ are the associated Legendre functions, and the functions $L_{n+l}^{2l+1}(\rho)$ are the associated Laguerre polynomials.

The wave functions are normalized, so that

$$\int_0^\infty \int_0^\pi \int_0^{2\pi} \psi_{nlm}^*(r, \theta, \phi)\psi_{nlm}(r, \theta, \phi)r^2 \sin \theta d\phi d\theta dr = 1 \quad \text{(III-7)}$$

ψ^* is the complex conjugate of ψ. The functions in r, θ, and ϕ are separately normalized to unity:

$$\int_0^{2\pi} \Phi_m^*(\phi) \Phi_m(\phi)d\phi = 1 \quad \text{(III-8)}$$

$$\int_0^\pi \{\Theta_{lm}(\theta)\}^2 \sin \theta d\theta = 1 \quad \text{(III-9)}$$

$$\int_0^\infty \{R_{nl}(r)\}^2 r^2 dr = 1 \quad \text{(III-10)}$$

In Tables III-1, III-2, and III-3 there are given the expressions for the three component parts of the hydrogenlike wave functions for all values of the quantum numbers that relate to the normal states of atoms. The expressions for $\Phi_m(\phi)$ are given in both the complex form and the real form.

Russell-Saunders States of Atoms Allowed
by the Pauli Exclusion Principle

IN Section 2-7 it was pointed out that the allowed Russell-Saunders states of an atom with two electrons with different total quantum numbers can be found by combining the electron spins to produce a resultant spin, corresponding to the total spin quantum number S (in this case 0 and 1), combining the orbital angular momenta of the electrons to produce the values of the total angular momentum quantum number L that are permitted by the magnitudes of the individual orbital angular momenta of the electrons, and then combining the total spin angular momentum vector and the total orbital angular momentum vector in all of the ways permitted by the magnitudes of the vectors that correspond to the total angular momentum quantum number J, with J having integral values when S is integral (an even number of electron spins) and half-integral values ($\frac{1}{2}$, $\frac{3}{2}$, $\cdots$) when S has half-integral values (corresponding to an odd number of electrons). Then, in Section 2-8, it was mentioned that the Pauli exclusion principle introduces a restriction in case that the two electrons have the same value of the total quantum number. In particular, the normal state of the helium atom corresponds to the electron configuration $1s^2$, with each electron having $n = 1$, $l = 0$, $m_l = 0$, and, of course, $s = \frac{1}{2}$; the Pauli exclusion principle requires that one of the electrons have $m_s = +\frac{1}{2}$ and the other have $m_s = -\frac{1}{2}$, so that the resultant spin angular momentum is zero, and the state must be a singlet state, $^1S_0$. The corresponding triplet state, $^3S_1$, is excluded by the exclusion principle, and in fact does not exist.

The application of the Pauli exclusion principle is necessary for the understanding of the normal states of atoms. There is a simple way of

determining the allowed Russell-Saunders states for an atom with two
or more electrons in the same subgroup (same values of n and l).

Sometimes the allowed states can be discovered by a simple argument. For example, let us discuss the normal state of the nitrogen
atom. The nitrogen atom, with seven electrons, has $1s^2 2s^2 2p^3$ as its
most stable electron configuration. By the argument given above, the
two $1s$ electrons contribute nothing to the spin angular momentum or
the orbital angular momentum of the atom, as do also the two $2s$ electrons. The values of the quantum numbers S, L, and J for the normal
state of the atom can accordingly be found by consideration of the
three $2p$ electrons. The three electrons might give rise to one or more
quartet states, with the spin quantum number $S = \frac{3}{2}$, and doublet
states, with the spin quantum number $S = \frac{1}{2}$. By Hund's first rule,
the quartet states will be more stable than the doublet states, and accordingly we may discuss the quartet states, in the search for the normal state. Each of the $2p$ electrons has $l = 1$, and the possible
values of the resultant angular momentum quantum number are hence
$L = 0$, 1, 2, and 3. The quartet state might accordingly be ^4S, ^4P,
^4D, or ^4F. In order to obtain a quartet state, with $S = \frac{3}{2}$, the spins of
the three $2p$ electrons must be parallel. These three electrons have accordingly the same values of the quantum numbers n, l, s, and m_s, equal
respectively to 2, 1, $\frac{1}{2}$, and $+\frac{1}{2}$ (for orientation of the resultant spins
in the positive direction). The Pauli exclusion principle requires that
they differ from one another; accordingly the remaining quantum number, m_l, must have the values $+1$, 0, and -1 for the three electrons,
respectively, and hence the resultant orbital angular momentum must
be zero $(L = 0)$. Accordingly the one quartet state allowed by the
exclusion principle for the configuration $2p^3$ is $^4S_{3/2}$. The normal state
of the nitrogen atom is thus found, in agreement with experiment, to
be $1s^2 2s^2 2p^3\ ^4S_{3/2}$, as given in Table 2-6.

A somewhat more extended argument is needed to show that the al-
lowed doublet states for this configuration are $^2D_{3/2}$, $^2D_{1/2}$, and $^2S_{1/2}$.
The method used to reach this conclusion will be illustrated by application to a simpler case, that of two equivalent p electrons (two electrons
with the same value of n and with $l = 1$).

The Zeeman Effect.—It was discovered by the Dutch physicist
P. Zeeman that spectral lines may be split into two or more components
when a magnetic field is applied to the atoms that are emitting or absorbing the radiation. This effect is called the Zeeman effect. The
splitting of the spectral lines is due to a splitting of the energy levels
into two or more components as a result of the interaction of the magnetic moments associated with the spin of the electrons and their orbital motion with the magnetic field.

We may use the configuration $2p3p$ as an example. The Russell-Saunders states corresponding to this configuration, beginning with the most stable one, are $^3D_1$, $^3D_2$, $^3D_3$, $^3P_0$, $^3P_1$, $^3P_2$, $^3S_1$, $^1D_2$, $^1P_1$, and $^1S_0$. There are accordingly ten energy levels. However, when a magnetic field is applied, all of these levels except those corresponding to $J = 0$ are split into several levels by the interaction of the magnetic moments with the magnetic field. For example, the states with $J = 1$ are split into three levels, corresponding to the total magnetic quantum number $M_J = -1, 0$, and $+1$, and those with $J = 2$ are split into five levels,

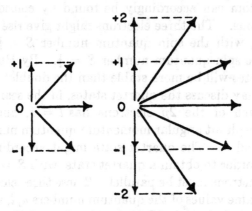

Fig. IV-1.—A diagram showing the orientation in a vertical magnetic field of the total angular momentum vector corresponding to the value 1 for the angular momentum quantum number J, and also to the value 2 for this quantum number. For $J = 1$ there are three orientations, corresponding to the values $-1, 0$, and $+1$ for the total magnetic quantum number M_J, and for $J = 2$ there are five orientations. This diagram also represents the orientation of the total spin angular momentum and the total orbital angular momentum for the Paschen-Back effect for the ^3D states, with quantum numbers $S = 1$ and $L = 2$. The diagram at the left represents the orientation of the spin vector and that at the right the independent orientation of the orbital vector in the vertical magnetic field.

corresponding to $M_J = -2, -1, 0, +1$, and $+2$ (Fig. IV-1). In general a state with given value of J is split into $2J + 1$ levels. No further splitting can be obtained. The energy level in the absence of the magnetic field is said to be *degenerate*, the degeneracy of the state being $2J + 1$; the Russell-Saunders state $^3D_1$ is in fact three states, which in the absence of a magnetic field happen to coincide in energy. The application of the magnetic field is said to remove the degeneracy.

By adding the values of $2J + 1$ for the ten Russell-Saunders states listed above we see that there are in fact 36 states based upon the configuration $2p3p$ and that application of a magnetic field gives rise to 36 energy levels.

The change in energy due to the magnetic field is equal to

$$\Delta E = M_J g \mu_B H \qquad \text{(IV-1)}$$

In this equation M_J is the total magnetic quantum number, g is a factor that will be discussed later, μ_B is the Bohr magneton equal to $eh/4\pi mc$, and H is the strength of the magnetic field. The splitting of energy levels into equally separated components is illustrated in Figure IV-2.

The Paschen-Back Effect.—It was discovered by Paschen and Back[1] that when the magnetic field becomes so strong that the Zeeman splitting of the energy levels of a Russell-Saunders state becomes approximately as great as the separation of the states with different values of J, such as $^3D_3$, $^3D_2$, and $^3D_1$, the nature of the pattern of energy levels

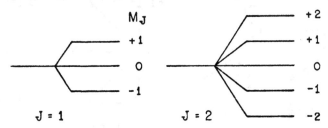

Fig. IV-2.—Energy levels for states with total angular momentum quantum number $J = 1$ (left) and 2 (right) in the Zeeman effect. The degenerate energy level is split into three or five components by the magnetic field, corresponding to different values of the magnetic quantum number M_J.

changes. In a strong magnetic field the coupling between the orbital angular momentum and the spin angular momentum to give a resultant angular momentum represented by J is broken; instead, the orbital angular momentum, represented by L, and the spin angular momentum, represented by S, orient independently in the magnetic field, in the ways determined by the orbital magnetic quantum number M_L and the spin magnetic quantum number M_S. This situation is illustrated for the multiplet $^3D_1$, $^3D_2$, and $^3D_3$ in Figure IV-1. In this figure the spin angular momentum is shown to be oriented in three ways in the magnetic field, corresponding to $M_S = -1$, 0, and $+1$, and the orbital angular momentum in five ways, corresponding to $M_L = -2$, -1, 0, $+1$, and $+2$. The orientations of the spin and the orbital angular momenta are independent of one another, and accordingly 15 quantum states are represented. Similarly, the Paschen-Back effect for $^3P_0$, $^3P_1$, and $^3P_2$ gives rise to nine quantum states; the other Russell-Saunders states, with either $S = 0$ or $P = 0$, do not show a Paschen-

[1] F. Paschen and E. Back, *Physica* 1, 261 (1921).

Back effect; they correspond to a total of 12 quantum states, so that the total for the configuration remains 36, as discussed for the Zeeman case, above. The application of a magnetic field of gradually increasing strength cannot lead to the destruction of quantum states or the formation of new ones, but only to the change in their energy values.

The Extreme Paschen-Back Effect.—If the magnetic field is very strong, the interactions that cause the spins of the electrons to combine to a resultant spin and the orbital moments to combine to a resultant orbital moment are broken. Then each electron orients its spin independently in the magnetic field, having two possible values, $+\frac{1}{2}$ and

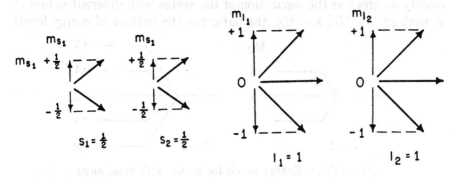

Fig. IV-3.—Diagram representing orientations of the spin vectors of the two electrons and the orbital angular momentum vectors of the two electrons in the extreme Paschen-Back effect for an atom with two $2p$ electrons. The two spins orient themselves separately in the vertical magnetic field, as do also the two orbital angular momentum vectors. Each electron spin can assume orientations such that the component of angular momentum along the field direction is represented by the quantum number $m_s = +\frac{1}{2}$ or $-\frac{1}{2}$, and each orbital angular momentum may orient itself in such a way that the component of the orbital angular momentum along the field direction is represented by the quantum number $m_l = +1$, 0, or -1.

$-\frac{1}{2}$. Similarly, each orbital moment orients itself independently in the magnetic field, there being only one orientation ($m_l = 0$) for an s electron, three ($m_l = -1, 0, +1$) for a p electron, and so on. For the configuration $2p3p$ there are two orientations of the spin for each electron and three orientations for the orbital moment, as shown in Figure IV-3. These orientations are independent of one another; accordingly the extreme Paschen-Back effect for this configuration gives rise to $2 \times 2 \times 3 \times 3 = 36$ quantum states. These quantum states are equal in number to those for the ten Russell-Saunders states listed above, and also to those for the Paschen-Back states.

Two Equivalent p Electrons.—If the two electrons do not have different values of the total quantum numbers, then some of the extreme

Paschen-Back states indicated in Figure IV-3 are excluded by the exclusion principle. For example, the electrons cannot both have $m_s = +\frac{1}{2}$ and $m_l = +1$; this is an excluded state. The allowed states for two equivalent p electrons are seen by inspection to be 15 in number; they are listed in Table IV-1. Note that these allowed states require that the quantum numbers m_{s_1} and m_{l_1} for the first electron be not identical with m_{s_2} and m_{l_2} for the second electron; moreover, two assignments of quantum numbers that differ only in the interchange of the two electrons are not counted as two states, but only as one.

TABLE IV-1.—ALLOWED STATES FOR TWO EQUIVALENT
p ELECTRONS

m_{s1}	m_{s2}	m_{l1}	m_{l2}	$M_S = m_{s1} + m_{s2}$	$M_L = m_{l1} + m_{l2}$
$+1/2$	$+1/2$	$+1$	0	$+1$	$+1$
		$+1$	-1	$+1$	0
		0	-1	$+1$	-1
$+1/2$	$-1/2$	$+1$	$+1$	0	$+2$
		$+1$	0	0	$+1$
		$+1$	-1	0	0
		0	$+1$	0	$+1$
		0	0	0	0
		0	-1	0	-1
		-1	$+1$	0	0
		-1	0	0	-1
		-1	-1	0	-2
$-1/2$	$-1/2$	$+1$	0	-1	$+1$
		$+1$	-1	-1	0
		0	-1	-1	-1

The correlation of the extreme Paschen-Back states with Russell-Saunders states can be made by considering the Paschen-Back effect. By adding the electron magnetic spin quantum numbers and electron magnetic orbital quantum numbers values of the resultant spin and orbital quantum numbers M_S and M_L are obtained. These may be interpreted at once in terms of Russell-Saunders states. The presence of $M_S = +1$ and -1 (as well as 0) requires that there be some triplet states, with $S = 1$. The values $M_S = +1$ or -1 are correlated with $M_L = +1, 0$, and -1, but not $+2$ or -2; hence there are no ^3D states, but only ^3P. When the nine values of M_S and M_L corresponding to ^3P are crossed out, there remain only the values $M_S = 0$ together with $M_L = +2, +1, 0, 0, -1$, and -2; it is seen that these correspond to the states ^1D and ^1S. Accordingly the Russell-Saunders states $^3P_0$,

$^3P_1$, $^3P_2$, $^1D_2$, and $^1S_0$ are allowed for two equivalent p electrons.

In Table IV-2 are listed the allowed Russell-Saunders states for equivalent s, p, d, and some cases of f electrons.

The Landé g-Factor.—The magnetic moment of an atom can be expressed in a simple way in terms of its angular momentum. The Bohr unit of angular momentum is $h/2\pi$, and the Bohr unit of magnetic moment (the Bohr magneton) is $he/4\pi mc$. An electron moving in an orbit with x units of angular momentum has orbital magnetic moment equal to x Bohr magnetons.

However, the magnetic moment of electron spin bears a different relation to the spin angular momentum; it is approximately twice as great. This can be expressed by saying that the Landé g-factor for orbital motion of the electron is 1 and that for spin of the electron is 2.

TABLE IV-2

Equivalent s Electrons

$$s - {}^2S$$
$$s^2 - {}^1S$$

Equivalent p Electrons

$p^1 -$		^2P				
$p^2 - {}^1S$			^1D		^3P	
$p^3 -$		^2P		^2D		^4S
$p^4 - {}^1S$			^1D		^3P	
$p^5 -$		^2P				
$p^6 - {}^1S$						

Equivalent d Electrons

d^1	^2D					
$d^2 - {}^1(SDG)$		$^3(PF)$				
$d^3 -$	^2D		$^2(PDFGH)$	$^4(PF)$		
$d^4 - {}^1(SDG)$		$^3(PF)$	$^1(SDFGI)$	$^3(PDFGH)$	^5D	
$d^5 -$	^2D		$^2(PDFGH)$	$^4(PF)$	$^4(SDFGI)$	$^4(DG)$ ^6S
$d^6 - {}^1(SDG)$		$^3(PF)$	$^1(SDFGI)$	$^3(PDFGH)$	^5D	
$d^7 -$	^2D		$^2(PDFGH)$	$^4(PF)$		
$d^8 - {}^1(SDG)$		$^3(PF)$				
$d^9 -$	^2D					
$d^{10} - {}^1S$						

Equivalent f Electrons

f^1	^2F	
f^2	$^1(SDGI)$	$^3(PFH)$
f^{12}	$^1(SDGI)$	$^3(PFH)$
f^{13}	^2F	
f^{14}	^1S	

TABLE IV-3.—THE LANDÉ g-FACTOR FOR RUSSELL-SAUNDERS COUPLING

Singlets, $S = 0$

Term	L	J = 0	1	2	3	4	5
^1S	0	0/0					
^1P	1	..	1				
^1D	2	..	..	1			
^1F	3	..	..	..	1		
^1G	4	..	..	..	..	1	
^1H	5	..	..	..	..	..	1

Doublets, $S = 1/2$

Term	L	J = 1/2	3/2	5/2	7/2	9/2	11/2
^2S	0	2					
^2P	1	2/3	4/3				
^2D	2	..	4/5	6/5			
^2F	3	..	..	6/7	8/7		
^2G	4	..	..	..	8/9	10/9	
^2H	5	..	..	..	..	10/11	12/11

Triplets, $S = 1$

Term	L	J = 0	1	2	3	4	5	6
^3S	0	..	2					
^3P	1	0/0	3/2	3/2				
^3D	2	..	1/2	7/6	4/3			
^3F	3	..	..	2/3	13/12	5/4		
^3G	4	..	..	..	3/4	21/20	6/5	
^3H	5	..	..	..	..	4/5	31/30	7/6

Quartets, $S = 3/2$

Term	L	J = 1/2	3/2	5/2	7/2	9/2	11/2	13/2
^4S	0		2					
^4P	1	8/3	26/15	8/5				
^4D	2	0	6/5	48/35	10/7			
^4F	3	..	2/5	36/35	26/21	4/3		
^4G	4	..	..	4/7	62/63	116/99	14/11	
^4H	5	..	..	..	2/3	32/33	162/143	16/13

Quintets, $S = 2$

Term	L	J = 0	1	2	3	4	5	6	7
^5S	0	..	..	2					
^5P	1	..	5/2	11/6	5/3				
^5D	2	0/0	3/2	3/2	3/2				
^5F	3	..	0	1	5/4	27/20	7/5		
^5G	4	..	..	1/3	11/12	23/20	19/15	4/3	
^5H	5	..	..	..	1/2	9/10	11/10	17/14	9/7

Sextets, $S = 5/2$

Term	L	J = 1/2	3/2	5/2	7/2	9/2	11/2	13/2	15/2
^6S	0			2					
^6P	1		12/5	66/35	12/7				
^6D	2	10/3	28/15	58/35	100/63	14/9			
^6F	3	-2/3	16/15	46/35	88/63	142/99	16/11		
^6G	4		0	6/7	8/7	14/11	192/143	18/13	
^6H	5	..	..	2/7	52/63	106/99	172/143	50/39	4/3

Septets, $S = 3$

Term	L	J = 0	1	2	3	4	5	6	7	8
^7S	0	..	..	2						
^7P	1	..	..	7/3	23/12	7/4				
^7D	2	..	3	2	7/4	33/20	8/5			
^7F	3	0/0	3/2	3/2	3/2	3/2	3/2	3/2		
^7G	4	..	-1/2	5/6	7/6	13/10	41/30	59/42	10/7	
^7H	5	..	..	0	3/4	21/20	6/5	9/7	75/56	11/8

Octets, $S = 7/2$

Term	L	J = 1/2	3/2	5/2	7/2	9/2	11/2	13/2	15/2	17/2
^8S	0	..	..	16/7	2	16/9				
^8P	1	..	..	16/7	122/63	16/9				
^8D	2	..	14/5	72/35	38/21	56/33	18/11			
^8F	3	4	2	12/7	34/21	52/33	222/143	20/13		
^8G	4	-4/3	14/15	44/35	86/63	140/99	206/143	284/195	22/15	
^8H	5	..	-2/5	24/35	22/21	40/33	186/143	88/65	118/85	24/17

The Landé g-factor for an atom is the ratio of the magnetic moment of the atom in Bohr magnetons and the angular momentum of the atom in Bohr units $h/2\pi$.

It is possible to calculate the g-factor for an atom in a Russell-Saunders state by considering the angles between the vectors S and L and the vector J. The total angular momentum, in units $h/2\pi$, is $\sqrt{J(J+1)}$. The magnetic moment in the direction of the angular momentum vector (the component perpendicular to the angular momentum vector cancels out) is equal to the components of the magnetic moment along S and that along L in the direction of J. The value can be calculated by trigonometry, the magnitudes of the vectors S and L being taken to be proportionally equal to $\sqrt{S(S+1)}$ and $\sqrt{L(L+1)}$, respectively. The equation obtained in this way is

$$g(J) = \frac{3J(J+1) + S(S+1) - L(L+1)}{2J(J+1)}$$

Values of the Landé g-factor calculated with this equation are given in Table IV-3.

Resonance Energy

A DETAILED discussion of resonance energy can be found in books on quantum mechanics. A simple problem, the dependence of the energy of resonance between two structures on the difference in energy of the structures, is discussed in the following paragraphs.

The value of the energy of a system corresponding to a wave function ψ (normalized to unity) is

$$E = \int \psi^* H \psi d\tau \qquad \text{(V-1)}$$

Here τ represents the coordinates for the system (x, y, z for each electron and each nucleus) and H is the Hamiltonian operator corresponding to the total energy of the system. The integral is taken over the whole of configuration space for the system. ψ^* is the complex conjugate of ψ.

Let us now consider a normalized function ψ_I corresponding to a reasonable structure for the system. The corresponding value of the energy, as given by Equation V-1, is H_{II}, defined by the equation

$$H_{ij} = \int \psi_i^* H \psi_j d\tau \qquad \text{(V-2)}$$

Similarly, the energy for another function ψ_{II}, corresponding to another structure for the system, is $H_{II\ II}$.

Now let us consider resonance between the two structures. We may write for the resonance structure the wave function

$$\psi = a\psi_I + b\psi_{II} \qquad \text{(V-3)}$$

In order for this function to be normalized ($\int \psi^* \psi d\tau = 1$) the coefficients must satisfy the condition

$$a^2 + 2ab\Delta_{I\ II} + b^2 = 1 \qquad (V\text{-}4)$$

(Note that $\Delta_{II\ I} = \Delta_{I\ II}$ if the functions are real, as we shall assume.) Here Δ is the overlap integral

$$\Delta_{ij} = \int \psi_i^* \psi_j d\tau \qquad (V\text{-}5)$$

The variation principle of quantum mechanics states that the true wave function for the normal state of a system is the one that minimizes the energy. We may accordingly find the best function ψ by finding the ratio a/b that minimizes E (Equation V-1).

A simple way of achieving this end is by application of Lagrange's method of undetermined multipliers. Let us consider the function F, such that

$$F = \int \psi^* H \psi d\tau - \lambda \int \psi^* \psi d\tau \qquad (V\text{-}6)$$

The second integral is constant, and hence for any value of λ the function F has its minimum at the same place as E.

The expression for F is

$$F = a^2 H_I + 2ab H_{I\ II} + b^2 H_{II} - \lambda(a^2 + 2ab\Delta_{I\ II} + b^2) \qquad (V\text{-}7)$$

Here and in the following equations H_I is written for $H_{I\ I}$ and H_{II} for $H_{II\ II}$. To find the minimum we differentiate F with respect to a and to b and equate to zero:

$$\left.\begin{aligned}
\frac{\partial F}{\partial a} &= 2a(H_I - \lambda) + 2b(H_{I\ II} - \lambda\Delta_{I\ II}) = 0 \\
\frac{\partial F}{\partial b} &= 2a(H_{I\ II} - \lambda\Delta_{I\ II}) + 2b H_{II} = 0
\end{aligned}\right\} \qquad (V\text{-}8)$$

These are two homogeneous linear equations in the two unknowns a and b; they have a solution only if the determinant formed by their coefficients vanishes:

$$\begin{vmatrix} H_I - \lambda & H_{I\ II} - \lambda\Delta_{I\ II} \\ H_{I\ II} - \lambda\Delta_{I\ II} & H_{II} - \lambda \end{vmatrix} = 0 \qquad (V\text{-}9)$$

This equation can be solved for the two values of λ that satisfy it. Each of them may then be substituted in Equations V-8 and V-4 to evaluate a and b. When this is done it is found that the quantity λ is equal to the energy E.

Often the approximation of neglecting the overlap integral $\Delta_{I\,II}$ is made. Equation V-9, which is called the secular equation, then becomes

$$\begin{vmatrix} H_I - E & H_{I\,II} \\ H_{I\,II} & H_{II} - E \end{vmatrix} = 0 \qquad (V\text{-}10)$$

The roots of this equation are

$$E = (H_I + H_{II})/2 \pm \{H^2_{I\,II} + (H_{II} - H_I)^2/4\}^{1/2}$$

The lower of these roots (with the negative sign) lies below $H_{I\,II}$, the energy of the more stable structure, by the amount

Effective resonance energy =

$$- (H_{II} - H_I)/2 + \{H^2_{I\,II} + (H_{II} - H_I)^2/4\}^{1/2} \qquad (V\text{-}11)$$

The effective resonance energy, given in this equation, is the amount of stabilization of the normal state of the system relative to the more stable of the two structures, structure I. It is shown in Figure 1-6, as a function of the difference in energy of the two structures, $H_{II} - H_I$.

The secular equation minimizing the energy for a more general wave function can easily be set up in the same way. Let the assumed wave function be

$$\psi = c_1\psi_1 + c_2\psi_2 + \cdots + c_m\psi_m \qquad (V\text{-}12)$$

Application of the Lagrange method, used above, leads to the following set of homogeneous simultaneous linear equations as the condition for minimum energy:

$$\sum_{k=1}^{m} c_k(H_{nk} - \Delta_{nk}E) = 0, \qquad n = 1, 2, \cdots m \qquad (V\text{-}13)$$

The condition that the set have a solution is

$$\begin{vmatrix} H_{11} - \Delta_{11}E & H_{12} - \Delta_{12}E & \cdots & H_{1m} - \Delta_{1m}E \\ H_{21} - \Delta_{21}E & H_{22} - \Delta_{22}E & \cdots & H_{2m} - \Delta_{2m}E \\ \cdots & \cdots & \cdots & \cdots \\ H_{m1} - \Delta_{m1}E & H_{m2} - \Delta_{m2}E & \cdots & H_{mm} - \Delta_{mm}E \end{vmatrix} = 0 \qquad (V\text{-}14)$$

The lowest root of this equation gives the best approximation to the value of the energy E provided by the wave functions of the assumed form V-12. The values of the coefficients c_k can then be found by use of Equations V-13 with this value of E inserted.

Wave Functions for Valence-

Bond Structures

IN his valuable paper "Molecular Energy Levels and Valence Bonds" Slater developed a method of formulating approximate wave functions for molecules and constructing the corresponding secular equations.[1] Let $a, b, \cdots$ represent atomic orbitals, each occupied by one valence electron, and α and β represent the electron spin functions for spin orientation $+\frac{1}{2}$ and $-\frac{1}{2}$, respectively. Slater showed that the following function corresponds to a valence-bond structure with bonds $a\text{——}b, c\text{——}d$, and so forth:

$$\frac{1}{2^{n/2}} \sum_R (-1)^R R \left\{ \frac{1}{((2n)!)^{1/2}} \right.$$
$$\left. \cdot \sum_P (-1)^P a(1)\beta(1)b(2)\alpha(2)c(3)\beta(3)d(4)\alpha(4) \cdots \right\} \quad \text{(VI-1)}$$

Here $1, 2, \cdots$ represent the electrons. P is the operation of permuting the electrons among the spin-orbit functions, for example, interchanging 1 and 2 between $a\beta$ and $b\alpha$. There are $(2n)!$ of these operations in the permutation group; $2n$ is the number of electrons for n bonds. The symbol $(-1)^P$ is 1 if P involves an even number of interchanges of pairs of electrons and -1 if it involves an odd number. The function in the brackets satisfies the Pauli exclusion principle. R represents the 2^n operations of interchanging the spin functions α and β for orbitals (such as a and b) that are bonded together.

The Slater valence-bond function leads to an energy expression that contains the single exchange integrals between bonded orbitals, such as a and b, with the coefficient $+1$. These integrals are usually negative,

[1] J. C. Slater, *Phys. Rev.* **38**, 1190 (1931).

and hence, with coefficient $+1$, stabilize the system, corresponding to attraction and bond formation. The single exchange integrals between orbitals not bonded to one another, such as a and c, occur with the coefficient $-\frac{1}{2}$, which corresponds to repulsion.

A simple graphical method of formulating the independent valence-bond structures for a molecule was discovered by Rumer.[2] This method has been extended to permit the secular equation for a set of resonating valence-bond structures to be written without difficulty.[3] Quantum-mechanical treatments of aromatic and conjugated molecules have been carried out by many investigators. The subject of molecular quantum mechanics is too extensive to be reviewed in this book.

[2] G. Rumer, *Nachr. Ges. Wiss. Göttingen* 1932, 337.
[3] L. Pauling, *J. Chem. Phys.* 1, 280 (1933).

Molecular Spectroscopy

A GREAT deal of information about the structure of molecules has been obtained by analysis of their spectra, which are called *molecular spectra* or *band spectra*. The spectrum of a molecule can be interpreted in terms of an energy-level diagram, similar to those described for atoms in Chapter 2. The transition from one energy level to another is associated with the emission or absorption of a quantum of light, with frequency related to the energy difference of the two energy levels by the Bohr frequency rule; that is, the energy $h\nu$ of the absorbed or emitted light quantum is equal to the difference in energy of the two states.

It has been found that the total energy of a molecule in its various quantized states can be represented approximately as the sum of three terms, the electronic energy, the vibrational energy, and the rotational energy:

$$W_{\text{total}} = W_{\text{electronic}} + W_{\text{vibration}} + W_{\text{rotation}} \qquad \text{(VII-1)}$$

The values of each of the three terms are determined by quantum numbers, called the electronic quantum numbers, the vibrational quantum numbers, and the rotational quantum numbers, respectively. Usually the different electronic states of molecules differ by a large amount in energy, so that transition from one electronic state to another involves emission or absorption of a light quantum with frequency in the visible or ultraviolet range; sometimes electronic transitions occur with much smaller frequencies, corresponding to the infrared or microwave range. The vibrational energy levels usually lie moderately close together, so that transitions between them correspond to absorption or emission of radiation in the near infrared region. Usually the rotational energy levels for one electronic and vibrational state lie very close to one another. For molecules containing light atoms the transition from one rotational state to another corresponds to radiation in the far infrared, and for molecules containing heavier atoms to radiation in the microwave region.

Electronic Energy Curves; The Morse Function.—Born and Oppenheimer[1] carried out a quantum-mechanical treatment of molecules, making use of the fact that the nuclei in a molecule are several thou-

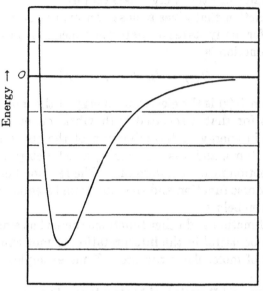

Fɪɢ. VII-1.—A curve representing the electronic energy of a diatomic molecule as a function of the distance between the nuclei. The zero for energy is the energy of the separated atoms. The minimum of the curve corresponds to the equilibrium value of the internuclear distance. The curve shown, which approximates closely the observed electronic energy curves for many states of diatomic molecules, corresponds to the Morse function.

sand times heavier than the electrons, and were able to show that an approximate solution of the wave equation for a molecule can be obtained by solving the wave equation for the electrons alone, with the nuclei held in a fixed configuration. The energy values obtained in this way, as a function of the configuration of the nuclei, can then be used as the potential energy function determining the modes of vibration of the nuclei. For a diatomic molecule, for example, the electronic energy of the molecule (including the energy of repulsion of the two nuclei) is found on approximate solution of the wave equation to be a curve with the general shape shown in Figure VII-1. At large distances between the two nuclei the energy has a value equal to the sum of the energies

[1] M. Born and J. R. Oppenheimer, *Ann. Physik* **84**, 457 (1927).

of the two atoms. As the atoms approach one another there is attraction and the energy curve falls below the value 0, corresponding to separated atoms. At a certain internuclear distance, usually represented by the symbol r_e, the curve has a minimum; that is, at $r = r_e$ the electronic energy of the molecule is a minimum. With further decrease in the value of r the energy rises rapidly.

A simple function that gives a close approximation to the electronic energy curve for many states of diatomic molecules is the Morse function.[2] This function is

$$U(r) = D_e \{1 - e^{-a(r-r_e)}\}^2 \qquad \text{(VII-2)}$$

In this equation $U(r)$ is the electronic energy of the molecule (the total energy except for that associated with vibration or rotation), D_e is the difference in energy of the minimum of the curve and the value for separated atoms, and a is a constant, which determines the curvature of the function near its minimum. The relation between the constants of the Morse function and the vibrational frequency of the molecule will be given below.

The Morse function and other functions somewhat similar to it have been found to be useful in the interpretation of molecular spectra and the discussion of molecular structure. Some examples are mentioned in Chapter 3.

The Vibration and Rotation of Molecules.—The nature of the vibrational motion and the values of the vibrational energy levels of a molecule are determined by the electronic energy function, such as that shown in Figure VII-1. The simplest discussion of the vibrational motion of a diatomic molecule is based upon the approximation of the energy curve in the neighborhood of its minimum by a parabola; that is, it is assumed that the force between the atoms of the molecule is proportional to the displacement of the internuclear distance from its equilibrium value r_e. This corresponds to the approximate potential function

$$U(r) = \tfrac{1}{2}k(r - r_e)^2 \qquad \text{(VII-3)}$$

A potential energy function of this type is called a Hooke's-law potential energy function.

Solution of the Schrödinger wave equation for the motion of the nuclei for this potential energy function leads to the following expression for the vibrational energy of the molecule:

$$W_{\text{vibration}} = (v + \tfrac{1}{2})h\nu_e \qquad \text{(VII-4)}$$

[2] P. M. Morse, *Phys. Rev.* **34**, 57 (1929).

In this equation v, the vibrational quantum number, may assume the values 0, 1, 2, $\cdots$. The frequency ν_e is the classical frequency of motion corresponding to this potential function; it is related to the Hooke's-law constant k by the equation

$$\nu_e = \tfrac{1}{2}\pi\sqrt{k/\mu} \qquad \text{(VII-5)}$$

Here μ is the reduced mass of the two nuclei, related to the masses μ_1 and μ_2 of the nuclei by the equation

$$1/\mu = 1/\mu_1 + 1/\mu_2 \qquad \text{(VII-6)}$$

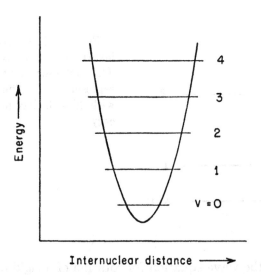

Fig. VII-2.—Some vibrational energy levels for an idealized diatomic molecule. The electronic energy curve has been approximated by a parabola, corresponding to a Hooke's-law interaction between the two atoms. The first five vibrational states are represented. They are separated by the energy difference $h\nu$. The lowest vibrational state, with $v = 0$, has the zero-point vibrational energy $\tfrac{1}{2}h\nu$.

It is seen that the vibrational energy levels are equally spaced, being separated by the energy value $h\nu_e$. The vibrational energy of the lowest vibrational state, with $v = 0$, is $\tfrac{1}{2}h\nu_e$; even in the lowest state the molecule has this amount of vibrational energy. This quantity is called the zero-point vibrational energy of the molecule (Fig. VII-2).

Many molecules are found by experiment to have vibrational energy levels that get closer together as the quantum number v increases. This behavior is represented by the Morse function, Equation VII-2.

The vibrational energy corresponding to the Morse function is given by the expression

$$W_{\text{vibration}} = (v + \tfrac{1}{2})h\nu_e - (v + \tfrac{1}{2})^2 \frac{h^2\nu_e{}^2}{4D_e} \qquad \text{(VII-7)}$$

$$\nu_e = \frac{a}{2\pi} \sqrt{2D_e/\mu} \qquad \text{(VII-8)}$$

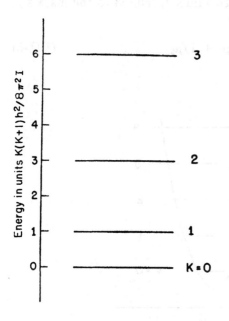

Fig. VII-3.—Rotational energy levels for a diatomic molecule. The first four rotational states are represented. The lowest state, with $K = 0$, has no rotational energy.

Solution of the wave equation for rotation of a rigid diatomic molecule leads to the following expression for rotational energy:

$$W_{\text{rotation}} = K(K + 1)\frac{h^2}{8\pi^2 I_e} \qquad \text{(VII-9)}$$

Here K is the rotational quantum number, with values 0, 1, 2, $\cdots$. I_e is the moment of inertia of the molecule, equal to μr_e^2.

An energy-level diagram of the rotational states for the lowest electronic and vibrational quantum numbers of a molecule is shown in Figure VII-3. It is seen that the rotational energy levels are not equally spaced, but show increasingly larger separations. From the experimental values of the energy levels the moment of inertia of the molecule can be calculated by the use of Equation VII-9. Because the molecule is not rigid the value of the moment of inertia, and accordingly of the average internuclear distance, depends somewhat on the vibrational quantum number v, and also on the rotational state given by the rotational quantum number K, as well as, of course, the electronic state. The symbol r_0 is usually used to represent the average internuclear distance for the lowest state of the molecule, that is, for $v = 0$ and $K = 0$. Usually it differs by no more than about 0.001 Å from r_e, the distance corresponding to the minimum of the electronic energy.

Similarly, the symbol D_0 is used for the difference in energy of the lowest state, with $v = 0$ and $K = 0$, and the separated atoms; it is called the dissociation energy of the molecule. It is smaller than D_e by the amount $\frac{1}{2}h\nu_0$, the zero-point vibrational energy.

Microwave Spectroscopy.—Since 1945 a great deal of information about the structure of molecules has been obtained by the methods of microwave spectroscopy. Apparatus has been developed for the convenient generation and study of microwaves with wavelengths in the range from 1 mm to 3 cm. Whereas only the simplest molecules, with hydrogen as one of the two atoms, have moments of inertia small enough to permit the pure rotation spectrum to be studied by the methods of infrared spectroscopy, many others have pure rotational transitions corresponding to frequencies in the microwave region. For example, values of internuclear distance in many gas molecules of alkali halogenides, such as NaCl, have been measured by microwave spectroscopy with an accuracy of 0.0001 Å. Values of the electric dipole moment and other properties of the molecules can also be obtained by the microwave technique.[3]

As an example of the determination of interatomic distances by study of the microwave spectrum we may refer to the investigation of chloroacetylene.[4] This molecule is a linear molecule, H—C≡C—Cl, and its moment of inertia depends upon three parameters, which may be taken as the interatomic distances H—C, C≡C, and C—Cl. Microwaves with wavelengths about 0.76 cm are found to be strongly absorbed by the gas. This absorption corresponds to the transition from the rotational state $K = 1$ to the state $K = 2$. The frequency of the line is 22736.97 megacycles, that is, 2.273697 sec^{-1}. The expression for the energy levels of a rigid rotator given in Equation VII-9 can be used to obtain the value of the moment of inertia of the molecule. This value alone does not permit the evaluation of the three interatomic distances. The frequency given above is, however, for the isotopic molecular species HCCCl^{35}; the species HCCCl^{37} absorbs the frequency 22289.51 mc, DCCCl^{35} absorbs the frequency 20748.05 mc, and DCCCl^{37} absorbs the frequency 20336.94 mc. Any three of these values can be

---

[3] For a discussion of the methods of microwave spectroscopy and a summary of the results during the last few years see the book by C. H. Townes and A. L. Schawlow, *Microwave Spectroscopy*, McGraw-Hill Book Co., New York, 1955.

[4] A. A. Westenberg, J. H. Goldstein, and E. B. Wilson, Jr., *J. Chem. Phys.* **17**, 1319 (1949).

used to calculate the three interatomic distances on the assumption that they are constant for the molecules. It is found that the four values are consistent with this assumption, and the values obtained are H—C = 1.052 ± 0.001 Å, C≡C = 1.211 ± 0.001 Å, and C—Cl = 1.632 ± 0.001 Å.

Electronic Molecular Spectra.—In general the absorption and emission spectra of molecules involve change in the electronic quantum numbers as well as in the vibrational and rotational quantum numbers. These molecular spectra are complex, and their interpretation is diffi-

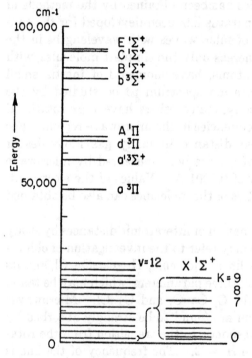

Fɪɢ. VII-4.—Some observed energy levels for the carbon monoxide molecule, as determined by the analysis of spectra. The first thirteen vibrational levels are shown for the normal electronic state of the molecule, and the rotational levels are shown for the lowest vibrational state.

cult. Much of the information that has been gathered about the properties of diatomic molecules and the simpler polyatomic molecules has been obtained by analysis of molecular spectra. Details of the methods are given in books dealing with the subject.[5]

A part of the energy-level diagram for carbon monoxide is shown in Figure VII-4. The energy levels have been obtained by analysis of the observed frequencies of the lines in the emission and absorption spectra of the molecule.

[5] G. Herzberg, *Molecular Spectra and Molecular Structure*. I. *Diatomic Molecules*, Prentice-Hall, New York, 1939; *Infrared and Raman Spectra of Polyatomic Molecules*, D. Van Nostrand Co., New York, 1945; E. B. Wilson, Jr., J. C. Decius, and P. C. Cross, *Molecular Vibrations; The Theory of Infrared and Raman Vibrational Spectra*, McGraw-Hill Book Co., New York, 1955.

The reference point for energy in the energy scale is the minimum of the electronic energy curve for the lowest electronic state.

The first 13 vibrational states ($v = 0$ to $v = 12$) for the lowest electronic state are represented by the 13 levels at the lower left corner of the diagram. Ten rotational states (for $K = 0$ to $K = 9$) are shown to the right, with a changed energy scale.

The other energy levels, from about 50,000 cm^{-1} up, represent some of the excited electronic states of the molecule. For each electronic state only the level with $v = 0$ and $K = 0$ is shown.

The lowest electronic state has the symbol $X\ ^1\Sigma^+$. X is commonly used to designate the normal electronic state, and other letters are used for other states. The left superscript 1 means that the molecule is in a singlet state, with no unpaired electrons ($S = 0$). The superscript 3 means that there are two unpaired electrons (electron-spin quantum number $S = 1$). This part of the term symbol is the same as in the Russell-Saunders symbols for atoms. The symbols Σ, Π, Δ, and so forth are used to designate the component of total orbital angular momentum of the electrons along the line passing through the two nuclei. They correspond to the values 0, 1, 2, $\cdots$, respectively. A molecule in the state $^1\Sigma$ has accordingly neither spin nor orbital angular momentum of the electrons.

Raman Spectra.—A valuable method of spectroscopic investigation was discovered by Raman and Krishnan and independently by Landsberg and Mandelstam. The method involves an effect, called the Raman effect, that occurs when light is scattered by gases, liquids, or solids. It was discovered by these investigators that when monochromatic light, with definite wavelength, is scattered by a substance, some of the scattered light has the same frequency as the incident light, but some of it has a changed frequency, either larger or smaller than the frequency of the incident light. The pattern of lines produced by a substance by this effect is called the Raman spectrum of the substance. It is found that the difference in energy of the scattered quantum and the incident quantum is equal to the difference in energy of two quantized states of the molecule involved in the scattering process. For example, a scattered line shifted by 2886.0 cm^{-1} from the incident line has been observed on scattering by hydrogen chloride; this shift agrees excellently with the center of the fundamental vibrational band in the infrared, which is at 2885.9 cm^{-1}. A great deal of information about vibrational energy levels and about the symmetry of molecules has been obtained by the study of Raman spectra.[6]

---

[6] See the books mentioned in the preceding footnote.

APPENDIX VIII

The Boltzmann Distribution Law

IN the discussion of many properties of substances it is necessary to
know the distribution of atoms or molecules among their various quantum states. An example is the theory of the dielectric constant of a
gas of molecules with permanent electric dipole moments, as discussed
in Appendix IX. The theory of this distribution constitutes the subject of statistical mechanics, which is presented in many good books.[1]
In the following paragraphs a brief statement is made about the Boltzmann distribution law, which is a basic theorem in statistical mechanics.

It is convenient and useful to express the Boltzmann distribution law
in two forms: a quantum form and a classical form. The quantum
form of the law, in its application to atoms and molecules, may be expressed as follows: The relative probabilities of various quantum states
of a system in equilibrium with its environment at absolute temperature T, each state being represented by a complete set of values of the
quantum numbers, are proportional to the Boltzmann factor $e^{-W_n/kT}$,
in which n represents the set of quantum numbers, W_n is the energy of
the quantized state, and k is the Boltzmann constant, with value
1.3804×10^{-16} erg deg^{-1}. The Boltzmann constant k is the gas-law
constant R divided by Avogadro's number; that is, it is the gas-law
constant per molecule.

We see that the Boltzmann factor has the value e^{-1}, which is equal to
0.368, when W_n is equal to kT. Hence when two states differ in energy

[1] J. Mayer and M. Mayer, *Statistical Mechanics*, John Wiley and Sons, New
York, 1944; R. C. Tolman, *Principles of Statistical Mechanics*, Oxford University
Press, 1938; R. H. Fowler, *Statistical Mechanics*, Cambridge University Press,
1936; T. L. Hill, *Statistical Mechanics*, McGraw-Hill Book Co., New York,
1956.

by kT, the probability of the state with higher energy is less than that of the state with lower energy, as given by the factor 0.368.

As an example, let us calculate the relative number of molecules of hydrogen chloride with rotational quantum number $K = 0$ and $K = 1$ in hydrogen chloride gas in thermal equilibrium at 25°C. We may take the energy of the normal state of the molecule, with $K = 0$ (also $v = 0$, as for the other states that we shall consider), as 0; the nature of the Boltzmann factor is such as to permit the zero point for energy to be chosen arbitrarily. The energy for $K = 1$ is found, by use of the equation for rotational energy of the molecule (Equation VII-9) and the value 1.275 Å for the internuclear distance, to be 4.20×10^{-15} erg. At 25°C the value of kT is 4.12×10^{-14} erg. The ratio of these values is 0.102; accordingly the Boltzmann factor for $K = 1$ has the value $e^{-0.102}$, which is 0.905, the Boltzmann factor for $K = 0$ being 1.000. We must remember, however, that the rotational level $K = 1$ consists of three states, corresponding to the three orientations of the angular momentum in space, the quantum number M_K having the values -1, 0, and $+1$. The relative total weight of the three states with $K = 1$ is accordingly $3 \times 0.905 = 2.72$, that for the state $K = 0$, which is a nondegenerate state (M_K having only the single value 0), being 1.

The center of the first vibration-rotation band for hydrogen chloride lies at 3.467μ (34,670 Å), which corresponds to the wave number 2886 cm^{-1}. Hence the first excited vibrational level, with $v = 1$ and $K = 0$, lies 2886 cm^{-1} above the normal state, with $v = 0$ and $K = 0$. Both of these levels are nondegenerate. At room temperature, 25°C, the ratio of molecules of HCl in this first vibrational excited state to the number in the normal state is found by use of the Boltzmann factor to be only 1×10^{-6}. Note that, because of the close approximation of the total vibration-rotation energy to the sum of the vibrational energy and the rotational energy, the latter being essentially the same, for given K, for the normal vibrational state and the first excited vibrational state, this value of the Boltzmann factor gives the ratio of all of the molecules with $v = 1$ and various values of K to all of those with $v = 0$ and various values of K.

The Boltzmann Distribution Law in Classical Mechanics.—The Boltzmann distribution law in classical mechanics involves the Boltzmann factor $e^{-W/kT}$ in the same way as that for quantum mechanics. In classical mechanics the state of a system can be described by giving the values of the coordinates and the momenta, for example, for a single particle the values of the three coordinates x, y, and z, and of the three linear momenta p_x, p_y, and p_z, which are equal to the mass of the particle multiplied by the components of velocity in the x, y, and z directions, respectively. The probability that the particle has coordinates between x and $x + dx$, y and $y + dy$, and z and $z + dz$ and momenta between p_x and $p_x + dp_x$, p_y and $p_y + dp_y$, and p_z and $p_z + dp_z$ is given by the expression

$$e^{-W/kT} dx\, dy\, dz\, dp_x\, dp_y\, dp_z$$

In case that the energy can be expressed as the sum of two terms, one depending only on the coordinates (the potential energy) and the other only on the momenta (the kinetic energy), the Boltzmann distribution law for coordinates can be discussed separately from that for momenta, because the Boltzmann factor can be split into the product of two exponential terms, one involving only the coordinates and the other only the momenta.

For example, for a single particle, which may be subject to the action of forces corresponding to the potential energy $V(x, y, z)$, the total energy is the sum of the potential energy and the kinetic energy $(p_x^2 + p_y^2 + p_z^2)/2m$, where m is the mass of the particle. The probability that the particle will have momenta lying between p_x and $p_x + dp_x$, p_y and $p_y + dp_y$, p_z and $p_z + dp_z$ is $e^{-mv^2/2kT} dp_x\, dp_y\, dp_z$. Here the kinetic energy $(p_x^2 + p_y^2 + p_z^2)/2m$ has been replaced by the equal quantity $mv^2/2$, where v is the velocity of the particle. The probability that the particle will have velocities lying between v and $v + dv$ can be found by integrating over a spherical shell in momentum space; it is found to be $e^{-mv^2/2kT} v^2 dv$.

This expression is the Maxwell distribution law for velocities.

Electric Polarizabilities and Electric Dipole Moments of Atoms, Ions, and Molecules

A GREAT deal of information has been obtained about the structure of molecules by the study of the electric properties of substances. A sample of a substance placed in an electric field undergoes a change in structure that is described as *electric polarization*. In general this change in structure involves motion of the electrons relative to the adjacent atomic nuclei, and also motion of the atomic nuclei relative to one another. Theoretical treatments have been developed that permit the observed polarization to be related to the properties of the atoms, ions, or molecules that compose the substance.

Electric Polarization and Dielectric Constant.—Under the influence of an electric field E acting upon a gas, liquid, or cubic crystal (the restriction to cubic crystals is made because of the complication introduced for other crystals by the dependence of their properties upon direction relative to the crystal axes), the positively and negatively charged particles that make up the substance undergo some relative motion, producing an induced average electric moment. Let P be the induced average electric moment per unit volume. The electric moment is defined as the product of the charge, positive and negative, by the distance of separation, for example, the electric moment of a pair of ions, with charge $+e$ and $-e$, the distance d apart being de. In electromagnetic theory the *electric induction D* is defined as

$$D = E + 4\pi P \qquad \text{(IX-1)}$$

and the dielectric constant ϵ is defined as

$$\epsilon = D/E = 1 + 4\pi P/E \qquad \text{(IX-2)}$$

The dielectric constant of a substance may be measured by determining the ratio of the capacity of a condenser filled with the substance and the capacity of the empty condenser. The electrical apparatus involves the condenser whose capacity is to be determined in parallel with a calibrated variable condenser, in a tuned resonant circuit; the determination is made by adjusting the variable condenser to keep the resonance frequency constant, and this requires that the sum of the capacities of the two condensers be constant.[1]

Let us first consider the dielectric constant of a gas. We assume that the molecules are far enough apart for them to contribute independently to the polarization and that the electric field E induces an electric dipole moment αE in each molecule. The quantity α is called the electric polarizability of the molecule. The number of moles per unit volume of gas is the density ρ divided by the molecular weight M, and the number of molecules in unit volume is this ratio multiplied by Avogadro's number N. Hence the polarization of the gas (the induced dipole moment per unit volume) is given by the following equation:

$$P = N \frac{\rho}{M} \alpha E \qquad (IX\text{-}3)$$

Combining this equation with Equation IX-2, we obtain

$$(\epsilon - 1) \frac{M}{\rho} = 4\pi N \alpha \qquad (IX\text{-}4)$$

This equation is not valid for liquids or crystals, but only for substances for which the dielectric constant is very close to unity, as for gases. For other substances an equation derived by consideration of the effect of the induced moments of neighboring molecules upon the molecule undergoing polarization must be considered. In a polarized medium each molecule is affected by the electric field in the region occupied by the molecule, called the local field. For many substances the local field is satisfactorily represented by the Clausius-Mossotti expression, derived in 1850. Each molecule is considered to occupy a spherical cavity. The part of the substance outside the spherical cavity undergoes polarization in the applied field. A simple calcula-

---

[1] For discussion of experimental methods and more detailed discussion of the theory see C. P. Smyth, *Dielectric Constant and Molecular Structure*, McGraw-Hill Book Co., New York, 1955; or J. W. Smith, *Electric Dipole Moments*, Butterworths, London, 1955. A more detailed theoretical treatment is given by J. H. Van Vleck, *The Theory of Electric and Magnetic Susceptibilities*, Oxford University Press, 1932.

tion shows that the shift of positive charges and negative charges corresponding to the polarization P produces inside the cavity the field $(4\pi/3)P$, in addition to the applied field E; hence the local field is given by the equation

$$E_{\text{local}} = E + \frac{4\pi}{3} P \tag{IX-5}$$

The polarization in unit volume is accordingly given by the expression

$$P = N \frac{\rho}{M} \alpha E_{\text{local}} = N\rho \frac{\alpha}{M} \left(E + \frac{4\pi}{3} P \right) \tag{IX-6}$$

This equation, together with the definition of the dielectric constant (Equation IX-2), leads at once to the equation

$$\frac{\epsilon - 1}{\epsilon + 2} \frac{M}{\rho} = \frac{4\pi}{3} N\alpha \tag{IX-7}$$

This equation, called the Lorenz-Lorentz equation, was derived in 1880 by combining the Clausius-Mossotti expression for the local field with the idea of molecular polarizability.

The principal interaction of an electromagnetic wave, such as visible light, with a substance is that of the electric field of the wave and the electric charges of the substance. The dielectric constant of the substance determines the magnitude of this interaction; in fact, it is equal to the square of the dielectric constant:

$$\epsilon = n^2 \tag{IX-8}$$

The amount of polarization of the medium by the electric field of the electromagnetic wave is a function of the frequency; for example, the dielectric constant of water is 81 when the frequency is very low or zero (static field) and falls to 1.78 for visible light. The reason for the difference is that in a static electric field or the field of an electromagnetic wave with very low frequency the molecules of water, which have a permanent electric dipole moment, are able to orient themselves in the field, and thus to produce a great increase in polarization of the liquid; whereas in the electric field of high frequency of visible light the molecular orientation cannot occur, and the only polarization that contributes to the dielectric constant is electronic polarization. A detailed discussion of the contribution of orientation of permanent molecular electric dipoles to the dielectric constant is given in a following section.

The Lorenz-Lorentz equation for the index of refraction is

$$R = \frac{n^2 - 1}{n^2 + 2} \frac{MW}{\rho} = \frac{4\pi}{3} N\alpha \tag{IX-9}$$

The quantity R is called the mole refraction.

Electronic Polarizability.—When an atom is placed in an electric field the charge distribution is changed to some extent as the electrostatic forces caused by the field operate in one direction on the nucleus and in the opposite direction on the electrons. A dipole moment is induced in the atom, with magnitude

$$\mu = \alpha E \tag{IX-10}$$

The dimensions of the polarizability α are those of volume. The polarizability of a metallic sphere is equal to the volume of the sphere, and we may anticipate that the polarizabilities of atoms and ions will be roughly equal to the atomic or molecular volumes. The polarizability of the normal hydrogen atom is found by an accurate quantum-mechanical calculation to be $4.5\ a_0^3$, that is, very nearly the volume of a sphere with radius equal to the Bohr-orbit radius a_0 ($4.19\ a_0^3$).

The Debye Equation for Dielectric Constant.—The Debye equation for the dielectric constant of a gas whose molecules have a permanent electric moment μ_0 is

$$P = \frac{\epsilon - 1}{\epsilon + 2} \frac{M}{\rho} = \frac{4\pi N}{3} \left(\frac{\mu_0^2}{3kT} + \alpha \right) \tag{IX-11}$$

This equation can be derived from Equation IX-4 by including in the expression for the polarization the contribution due to preferential orientation of the permanent dipole moments μ_0 in the field direction. The component of the dipole moment of a molecule in the field direction is $\mu_0 \cos \theta$, where θ is the polar angle between the dipole-moment vector and the field direction, and the energy of interaction is $-\mu_0 E \cos \theta$. The relative probability of orientation in volume element $\sin \theta d\theta d\phi$ (in polar coordinates) is given by the Boltzmann principle as $e^{\mu_0 E \cos \theta / kT} \sin \theta d\theta d\phi$. The average value of the component is hence given by the expression

$$\bar{u} = \frac{\displaystyle\int_0^{2\pi}\int_0^{\pi} \mu_0 \cos \theta e^{\mu_0 E \cos \theta / kT} \sin \theta d\theta d\phi}{\displaystyle\int_0^{2\pi}\int_0^{\pi} e^{\mu_0 E \cos \theta / kT} \sin \theta d\theta d\phi} \tag{IX-12}$$

(The integral in the denominator normalizes the probability.) The integrals are easily evaluated by expanding the exponential functions and retaining the first nonvanishing term:

$$\bar{\mu} = \frac{\mu_0^2 E}{kT} \frac{\displaystyle\int_0^{2\pi} \int_0^{\pi} \cos^2 \theta \sin \theta \, d\theta \, d\phi}{4\pi} \qquad \text{(IX-13)}$$

The integral (with the divisor 4π) is just the mean value of $\cos^2 \theta$ over the surface of a sphere. Its value is $\frac{1}{3}$. (The same value is found in quantum mechanics, as the average of $M_J^2/J(J+1)$, with $M_J = J$,

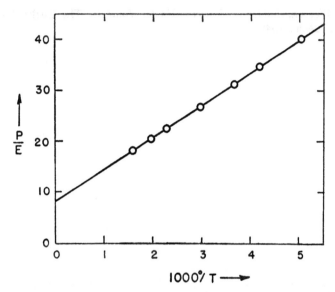

FIG. IX-1.—Values of the ratio of polarization P to field strength E for hydrogen chloride gas, as a function of the reciprocal of the absolute temperature. The slope of the line is a measure of the permanent electric dipole moment of the molecules, and the intercept of the line is a measure of the temperature-independent polarizability of the molecules.

$J - 1, \cdots -J$, for J either integral or half-integral.) Hence we obtain $\bar{\mu} = \mu_0^2 E/3kT$. This expression for the contribution of the permanent dipole moments of the molecule leads to the first term on the right in Equation IX-11. The second term, α, includes the electronic polarizability of the molecule and also the so-called atom polarization, the small change in relative positions of the nuclei caused by the field. This term is independent of the temperature.

Hydrogen chloride, for example, has dielectric constant that decreases from 1.0055 at 200°K to 1.0028 at 500°K, for constant density, corresponding to 1 atm at 0°C. Values of the polarization P (proportional to $\epsilon - 1$) are shown plotted against $1/T$ in Figure IX-1. The slope leads to the value 1.03 D for μ_0, as given in Chapter 3. (The unit 1 D, 1 debye, is equal to 1×10^{-18} statcoulomb centimeter.) The intercept of the extrapolated straight line on the P axis gives the value of the temperature-independent polarizability.

An extensive table of values of dipole moments for gases and solute molecules has been published.[2] During recent years some very accurate values have been determined by microwave spectroscopy and molecular-beam techniques.

[2] L. G. Wesson, *Tables of Electric Dipole Moments*, The Technology Press, Mass. Inst. Tech., 1948.

The Magnetic Properties of Substances

THE principal types of interaction of a substance with a magnetic field are called diamagnetism, paramagnetism, ferromagnetism, antiferromagnetism, and ferrimagnetism. They are useful in providing information about the electronic structures of the substances, especially as discussed in Chapters 5 and 11.

Diamagnetism.—It was discovered by Faraday that most substances when placed in a magnetic field develop a magnetic moment opposed to the field. Such a substance is said to be diamagnetic. (Substances that develop a moment parallel to the field are called paramagnetic substances.)[1]

A sample of a diamagnetic substance placed in an inhomogeneous magnetic field is acted on by a force that tends to push it away from the strongfield region. This force is proportional to the diamagnetic susceptibility of the substance, which is defined as the ratio of the induced moment, μ, to the field strength, H:

$$\mu = \chi H \tag{X-1}$$

The common methods of determining the magnetic susceptibility involve measuring this force.[2]

Let us consider a metal wire in the form of a circle. If a magnetic field is applied perpendicularly to the plane of the circle a current is induced in the wire. Corresponding to this current there is a magnetic field, resembling that of a magnetic dipole with orientation opposed to the field (Lenz's law).

---

[1] There is a common misapprehension that a bar of a paramagnetic substance in a uniform magnetic field sets itself parallel to the lines of force of the field and that a bar of diamagnetic substance sets itself perpendicular to the lines of force; in fact, a bar of substance either paramagnetic or diamagnetic sets itself parallel to the lines of force in a uniform field.

[2] For a description of these methods see the books in the following footnote.

The effect of the application of a magnetic field to an atom or monatomic ion is to cause the electrons to assume an added rotation about an axis parallel to the field direction and passing through the nucleus. This rotation, called the Larmor precession, has the angular velocity $eH/2mc$. The angular momentum of an electron with cylindrical radius ρ about the field axis and the angular velocity $eH/2mc$ is $eH\,\rho^2/2c$, and its magnetic moment is related to its angular momentum by the factor $-e/2mc$ and therefore has the value $-e^2\rho^2H/4mc^2$. Hence the molar diamagnetic susceptibility is

$$\chi_{\text{molar}} = -\frac{Ne^2}{4mc^2} \sum_i \overline{\rho_i^2} \qquad\qquad \text{(X-2)}$$

Here $\overline{\rho^2}$ is the average value of ρ^2 for the ith electron, and the sum is to be taken over all the electrons in the atom. For spherically symmetrical atoms, with $\rho^2 = x^2 + y^2$ and $r^2 = x^2 + y^2 + z^2$, where r is the distance of the electron from the nucleus, $\overline{\rho^2} = \frac{2}{3}\,\overline{r^2}$, and hence we may write

$$\chi_{\text{molar}} = -\frac{Ne^2}{6mc^2} \sum_i \overline{r_i^2} \qquad\qquad \text{(X-3)}$$

Measured values of the diamagnetic susceptibility of the noble gases correspond to reasonable values of $\Sigma\overline{r^2}$. For polyatomic molecules the interpretation of the diamagnetic susceptibility in terms of structural features is in general uncertain, and this property has not been found to be valuable in structural chemistry.

Some diamagnetic crystals (graphite, bismuth, naphthalene and other aromatic substances) show pronounced diamagnetic anisotropy. The observed anisotropy of crystals of benzene derivatives correspond to the molar diamagnetic susceptibility -54×10^{-6} with the field direction perpendicular to the plane of the benzene ring and -37×10^{-6} with it in the plane. This molecular anisotropy has been found to be of some use in determining the orientation of the planes of aromatic molecules in crystals.[3]

Diamagnetic susceptibility (per mole or per gram) is in general independent of the temperature.

Paramagnetism.—It is customary to restrict the use of the word paramagnetism to substances that in a magnetic field of ordinary strength develop a magnetic moment in the field direction that is pro-

---

[3] The diamagnetism of aromatic molecules has been discussed by L. Pauling, *J. Chem. Phys.* **4**, 673 (1936); K. Lonsdale, *Proc. Roy. Soc. London* A159, 149 (1937); *J. Chem. Soc.* **1938**, 364; F. London, *Compt. rend.* 205, 28 (1937); *J. phys. radium* **8**, 397 (1937).

portional to the strength of the field. (This usage excludes ferromagnetic substances.) Most paramagnetic substances have susceptibilities a hundred or a thousand times as great as the customary diamagnetic susceptibilities, and with opposite sign (mass susceptibility [per g] $+10^{-4}$ or 10^{-3}, as compared with about -1×10^{-6} for diamagnetic substances). They also have, of course, a diamagnetic contribution to the total susceptibility.

It was shown by Pierre Curie in 1895 that paramagnetic susceptibility is strongly dependent on temperature, and for many substances is inversely proportional to the absolute temperature. The equation

$$\chi_{molar} = \frac{C_{molar}}{T} + D \qquad (X\text{-}4)$$

is called Curie's law, and the constant C_{molar} called the molar Curie constant. D represents the diamagnetic contribution (it is negative).

Weber in 1854 had attributed paramagnetism to the orientation of little permanent magnets in the substance (and diamagnetism to induced currents, as discussed above). A quantitative treatment was developed by Paul Langevin in 1895, by application of the Boltzmann principle. The theory is the same as for the orientation of electric dipoles (see App. IX). It leads to the equation

$$C_{molar} = \frac{N\mu^2}{3k} \qquad (X\text{-}5)$$

in which μ is the value of the magnetic dipole moment per atom or molecule.

The value of the Bohr magneton is 0.927×10^{-20} erg gauss^{-1}. The magnetic moment μ is hence related to the molar Curie constant by the equation

$$\mu \text{ (in Bohr magnetons)} = 2.824 C_{molar}^{1/2} \qquad (X\text{-}6)$$

Curie's equation applies to gases, solutions, and some crystals. For other crystals a more general equation, the Weiss equation, may be used (derived by P. Weiss in 1907). Weiss assumed that the local magnetic field orienting the dipoles is equal to the applied field plus an added field proportional to the magnetic volume polarization M:

$$H_{local} = H + aM \qquad (X\text{-}7)$$

Application of the Boltzmann distribution law leads to the equation

$$M = \frac{N\rho\mu^2}{3kTW} (H + aM) \qquad (X\text{-}8)$$

in which ρ is the density and W the molecular weight. The molar susceptibility is defined as

$$\chi_{molar} = WM/\rho H \tag{X-9}$$

These equations lead to the Weiss equation:

$$\chi_{molal} = C_{molal}/(T - \Theta) \tag{X-10}$$

with Θ, the Curie temperature, given by the expression

$$\Theta = N\rho\mu^2 a/3kW \tag{X-11}$$

and C_{molar} by Equation X-5.

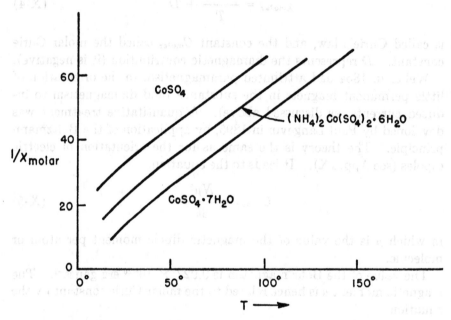

F$_{IG}$. X-1.—Curves showing the reciprocal of molar magnetic susceptibility of compounds of cobalt(II) as a function of the absolute temperature.

In a graph of $1/\chi_{molar}$ against T, the points lie on a straight line if the Weiss equation is valid. Measurements for three salts of cobalt(II) are shown in Figure X-1. It is seen that the curves are straight lines except at very low temperatures. Their slopes are the same; the slope is the reciprocal of the Curie constant, and accordingly the cobalt(II) atom has the same magnetic moment in the three substances.

Ferromagnetism.—Ferromagnetic substances assume a large magnetic polarization in weak fields, approaching a constant value (saturation) as the field strength increases. Many of them, including steel and magnetite (Fe_3O_4), retain their magnetization after the field is removed. The substances consist of domains, about 0.01 mm in diameter, that have their atomic moments parallel. In the absence of an external field different domains orient their moments in different directions (these directions are along cube edges for iron and along cube body diagonals for nickel). On application of a magnetic field the atomic moments of a domain reorient themselves. For a single crystal of pure iron saturation (magnetic moment 2.2 magnetons per atom) is achieved for an applied field of about 20 oersteds along a cube edge. With the field of 20 oersteds along a face diagonal of the cube the saturation moment is $2.2/\sqrt{2}$; it increases to 2.2 as the field is increased to about 400 oersteds (the domain is then oriented in the face-diagonal direction).

Values of the low-temperature saturation magnetic moment of ferromagnetic substances represent the maximum component of the atomic magnetic moment in the field direction; for example, for spin alone the value in Bohr magnetons is $2S$, whereas the magnetic moment obtained from the paramagnetic susceptibility is $2\sqrt{S(S+1)}$.

At higher temperatures thermal agitation diorients some of the atomic moments; at the ferromagnetic Curie temperature the substance becomes paramagnetic. The paramagnetic susceptibilities of nickel, palladium, and platinum are shown in Figure X-2. For all three substances the magnetic moment, as given by the slope of the lines, has approximately the value expected from the saturation moment for nickel in its ferromagnetic range, below 680° A (Chap. 11). Palladium and platinum are not ferromagnetic.

The nature of the local field in the ferromagnetic metals is probably that proposed by Zener.[4] It can be described as involving the interaction between the unpaired spins of atomic electrons and those of some electrons involved in forming one-electron bonds between the metal atoms.

Antiferromagnetism.—Antiferromagnetic substances are paramagnetic substances with a characteristic temperature at which the magnetic susceptibility shows a pronounced maximum. This temperature is called the antiferromagnetic transition temperature or Néel temperature (after Néel,[5] who first discussed the phenomenon). Above the Néel temperature the susceptibility depends on temperature in accordance with the Weiss equation (Equation X-10) with a negative value of the Curie temperature Θ. Below the Néel temperature the susceptibility decreases toward zero with decreasing temperature.

All of these properties can be accounted for by the assumption

[4] C. Zener, *Phys. Rev.* 81, 440 (1951); L. Pauling, *Proc. Nat. Acad. Sci. U.S.* 39, 551 (1953).

[5] L. Néel, *Ann. phys.* 18, 5 (1932); 5, 232 (1936).

(made first by Néel) that the magnetic moments of adjacent atoms are related by a resonance integral such that maximum stability is associated with alternating orientations of the moments, ↑ ↓ ↑ ↓ ↑ · · ·, rather than parallel orientation, ↑ ↑ ↑ ↑ ↑ · · ·, as in ferromagnetic substances. This interaction would make the Curie temperature Θ negative, rather than positive. Moreover, at low temperatures the interaction could become cooperative, in such a way that nearly all the atomic magnetic moments would be held in the regular antiparallel

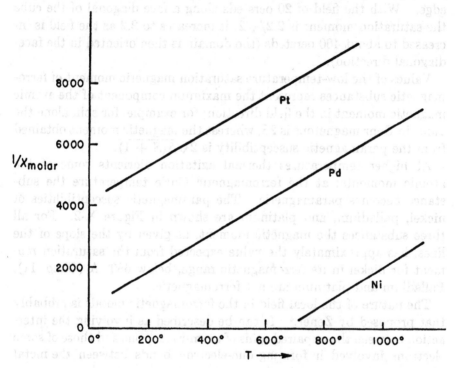

FIG. X-2.—Curves showing the reciprocal of molar paramagnetic susceptibility of nickel, palladium, and platinum as a function of the absolute temperature.

arrangement, and the susceptibility would decrease rapidly toward zero.

It is possible to determine the arrangement of positive and negative spins in an antiferromagnetic crystal by neutron diffraction; their interaction with the magnetic moment of the neutron causes their scattering powers to be different. For example, in MnF_2, which has the rutile structure (Fig. 3-2), the moments of the manganese atoms in a string of octahedra with shared edges have positive orientation, and those in the adjacent strings have negative orientation. The Néel transition temperature for MnF_2 is 72°K, and the Curie temperature Θ is −113°.

Ferrimagnetism.—Ferrimagnetic substances[6] are substances in which there is an interaction between atomic magnetic moments such as to cause them to align themselves in antiparallel orientations, as in antiferromagnetic substances, but with a difference in the total moments in the two directions, so that the resultant moment is not zero. The properties of ferrimagnetic substances are similar qualitatively to those of ferromagnetic substances: there is a Curie transition temperature, above which the substance is paramagnetic and below which it is ferromagnetic. However, the total magnetic moment indicated in the paramagnetic region is much greater than that given by saturation in the ferromagnetic region.

For example, magnetite, the first ferromagnetic substance discovered, is in fact ferrimagnetic. The crystal has composition Fe_3O_4, with 8 iron atoms in one set of equivalent positions in the unit cube and 16 iron atoms in another set. The observed paramagnetic susceptibility above the Curie temperature is compatible with the magnetic moments 5.2 for iron(II) and 5.9 for iron(III), as found in hypoligated complexes (Tables 5-2 and 5-3). The sum of the maximum components for one iron(II) moment and two iron(III) moments is 14 Bohr magnetons (spin moment only). The observed value of the ferromagnetic saturation moment is however, only 4.2 Bohr magnetons per Fe_2O_3. Néel has interpreted this fact as showing that 8Fe(II) plus 8 Fe(III) moments align themselves in parallel fashion and that the other 8Fe(III) moments in the unit cube have antiparallel orientation. The saturation moment of $MnFe_2O_4$, in which iron(II) is replaced by manganese(II), is 5.0 Bohr magnetons per $MnFe_2O_4$, and that of $NiFe_2O_4$ is 2.2 per $NiFe_2O_4$; these are the values expected for ferrimagnetism such that the iron(II) and iron(III) moments cancel. Substituted magnetites (spinels) of this sort, especially those with some zinc as well as manganese or nickel replacing iron(II), have great practical value in magnetic tape and other applications. They are called ferrites.[7]

[6] L. Néel, *Ann. phys.* **3**, 137 (1948).

[7] For reference books on magnetism see C. Kittel, *Solid State Physics*, John Wiley and Sons, New York, 1956; P. W. Selwood, *Magnetochemistry*. Interscience Publishers, New York, 1956.

The Strengths of the
Hydrohalogenic Acids

IT might be thought that hydrofluoric acid, which contains the most strongly electronegative element, would be a stronger acid than the other hydrohalogenic acids. In fact, the ionization constant of hydrofluoric acid is only 6.7×10^{-4}, whereas those of the other hydrohalogenic acids are greater than unity.

The strength of an acid in aqueous solution depends upon the difference in free energies of the hydrated ions and the undissociated molecules. Each of these free-energy terms is affected by the electronegativity of the halogen atom, and analysis of the problem shows that it is not unreasonable for hydrofluoric acid to be weaker than the other hydrohalogenic acids.[1]

The second column of Table XI-1 gives values of the free energy of formation from $H_2(g)$ and $X_2(g)$ of hydrogen ion, H^+, and halogenide ion, X^-, in aqueous solution at unit activity. These values have been obtained from those given by Latimer[2] by correction from the standard state to the gaseous state of the halogen.

The values of the free energy of formation of the negative halogenide ions (plus hydrogen ion) in aqueous solution might be expected to depend in a simple way on the electronegativity of the atoms. It is found (Fig. XI-1) that there is a linear relation, represented by the equation

$$\Delta F^\circ = -34.7(x - 2.1) \text{ kcal/mole} \qquad \text{(XI-1)}$$

[1] L. Pauling, *J. Chem. Ed.* **33**, 16 (1956). A similar discussion has been published by J. C. McCoubrey, *Trans. Faraday Soc.* **51**, 743 (1955).

[2] W. M. Latimer, *Oxidation Potentials*, 2nd edition, Prentice Hall, Inc., New York, 1952.

TABLE XI-1.—STANDARD FREE ENERGIES OF FORMATION AT 25°C OF HYDROGEN
IONS PLUS HALOGENIDE IONS AND OF HYDROGEN HALOGENIDE
MOLECULES IN AQUEOUS SOLUTION

	$\Delta F°(H^+ + X^-)$	$\Delta F°(HX)$
Hydrogen fluoride	− 66.08 kcal/mole	− 70.41 kcal/mole
Hydrogen chloride	− 31.35	− 22.8
Hydrogen bromide	− 24.95	− 13.1
Hydrogen iodide	− 14.67	− 2.0

The third column of the table gives values of the free energy of formation of the hydrogen halogenide molecules, HF, HCl, HBr, and HI, in aqueous solution.   The value for hydrogen fluoride is an experimental one.[2]   The values for the other three hydrogen halogenides are the values of the free energy of formation of the gaseous molecules with an estimated correction for the free energy of solution of the molecules. It seems likely that the correction for the free energy of solution of hydrogen chloride, hydrogen bromide, and hydrogen iodide to form the unionized molecules in aqueous solution is very close to zero.   The standard free energy of solution of phosphine is 2.6 kcal/mole and that of hydrogen sulfide is 1.4 kcal/mole; these values may be extrapolated to a value close to zero for hydrogen chloride.   Similarly, the free

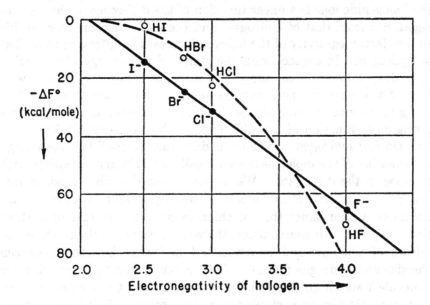

FIG. XI-1.—The standard free-energy change for the reaction of gaseous hydrogen and halogen molecules to form hydrogen halogenide molecules in aqueous solution (open circles) and hydrogen ion plus halogenide ion in aqueous solution (filled circles).

energies of solution of arsine and hydrogen selenide are 2.8 and 1.4 kcal/mole, respectively, and also suggest extrapolation to the value zero for hydrogen bromide. The value for stibine is 2.8, the same as that for arsine, and we accordingly accept zero for hydrogen iodide also.

It is seen that the values for hydrogen iodide, hydrogen bromide, and hydrogen chloride lie above those for the corresponding ions, whereas that for hydrogen fluoride lies below the value for the ions. This means that the ions in aqueous solution are more stable than the undissociated molecules for the heavier halogenides, and less stable for hydrogen fluoride.

An expression that approximates roughly the values of the free energy of the undissociated molecules is the quadratic expression

$$\Delta F^\circ = -\,23(x - 2.1)^2 \text{ kcal/mole} \qquad (XI\text{-}2)$$

The value 2.1 that appears subtracted from x, the electronegativity of the halogen, is the electronegativity of hydrogen. This equation is similar to Equation 3-12 in Chapter 3, and it represents the expected dependence of the free energy of formation of the hydrogen halogenides in relation to the electronegativity of the halogen atoms.

It is thus possible to understand why hydrofluoric acid is a weaker acid than the other hydrohalogenic acids. The stabilization energy of the halogenide ions is a linear function of the difference of the electronegativity from that of hydrogen, and these ions become more stable as the electronegativity of the halogen increases in difference from that of hydrogen. In consequence, the standard free energy of formation of hydrogen ion and fluoride ion in aqueous solution is about twice that of hydrogen ion and chloride ion. On the other hand, the free energy of formation of a hydrogen halogenide molecule is approximately a quadratic function of the difference in the electronegativity of the halogen and hydrogen. When this difference is small the free energy of formation of the molecule is very small, and it increases rapidly with increase in the difference. We would expect that the standard free energy of formation of undissociated hydrogen fluoride would be approximately four times that of chloride, and it is in fact over three times as great. In consequence, there is a reversal in the relative stabilities of the ions and undissociated molecules between hydrogen chloride and hydrogen fluoride. The undissociated hydrogen chloride molecule is stabilized by its partial ionic character to a smaller amount than the chloride ion is stabilized by its electron affinity, heat of hydration, and the like, whereas the undissociated hydrogen fluoride molecule is stabilized by its partial ionic character to a greater amount than the fluoride ion is stabilized by its electron affinity, and the like.

From the values given in the table the equilibrium constants of the hydrogen halogenides can be calculated by use of the equation $\Delta F° = -RT\ln K$. The calculated values are somewhat uncertain because of uncertainty in the estimate of the standard free energy of solution of the dissociated molecules. The values obtained in this way are 2×10^6 for HCl, 5×10^8 for HBr, and 2×10^9 for HI. These acids are accordingly very strong acids.

It is not surprising that the acid strengths increase rapidly in the sequence HF, HCl, HBr, HI. The same increase is observed also for the sequence H_2O, H_2S, H_2Se, and H_2Te. If the activity of water is taken as its molal concentration its first acid constant is 2×10^{-16}. The first acid constants of H_2S, H_2Se, and H_2Te are 1.1×10^{-7}, 1.7×10^{-4}, and 2.3×10^{-3}, respectively. Thus these four very weak acids have a total span of their first acid constants of 10^{13}. Similarly, the hydrogen halogenides have a span of 10^{13}, hydrogen iodide being 10^{13} times as strong as hydrogen fluoride. The intermediate acids in each series occupy similar positions.

The energy of stabilization of covalent bonds by partial ionic character is accordingly so great for the bonds between hydrogen and the most electronegative atoms, fluorine and oxygen, as to overcome the anion-forming tendency of these atoms and to cause hydrogen fluoride and water to be much weaker acids than the hydrogen compounds of their heavier congeners.

Bond Energy and Bond-Dissociation Energy

IN Chapter 3 and other chapters of this book much use is made of bond-energy values. These values are chosen in such a way that their sum over all of the bonds of a molecule which can be satisfactorily represented by a single valence-bond structure is equal to the enthalpy of formation of the molecule from its constituent atoms in their normal states. For example, the value of the O—H bond energy, 110 kcal /mole, is one-half the enthalpy of formation of $H_2O(g)$ from $2H(g)$ and $O(g)$.

Another quantity of much interest is the bond-dissociation energy.[1] The bond-dissociation energy of a bond in a molecule is the energy required to break that bond alone, that is, to split the molecule into the two parts that were previously connected by the bond under consideration.

The bond energy and bond-dissociation energy are the same for the bond in a diatomic molecule but are different for a bond in a polyatomic molecule. For example, the bond-dissociation energy for the O—H bond in the water molecule (splitting H_2O into $H + OH$) is 119.9 kcal/mole and that for the O—H bond in the OH radical is 101.2 kcal/mole. Their average, 110.6 kcal/mole, is the O—H bond energy.

The difference between the O—H bond-dissociation energies for H_2O and OH can be ascribed to the stabilization energy of the normal state, ^3P, of the oxygen atom. When one O—H bond is broken in the water molecule there is produced, in addition to a hydrogen atom, an OH radical, with structure :Ö—H. The radical has an unpaired elec-

---

[1] See the discussion by M. Szwarc and M. G. Evans, *J. Chem. Phys.* **18**, 618 (1950).

tron on the oxygen atom, which interacts only with electron pairs. When the second O—H bond is broken, however, an oxygen atom, :$\dot{\text{O}}\cdot$, with two unpaired electrons, configuration $1s^2 2s^2 2p^4$, is produced. There are three Russell-Saunders states corresponding to this configuration: ^1S, ^1D, and ^3P. The normal state, ^3P, involves a considerable amount of stabilization, resulting from the resonance energy of the two odd electrons (Hund's first rule); the stabilization energy has been estimated[2] to be 17.1 kcal/mole. Hence the bond-dissociation energy of OH to an oxygen atom in its valence state, rather than the more stable ^3P state, would be 118.3 kcal/mole, essentially equal to the O—H bond-dissociation energy of the first O—H bond in H_2O.

Many of the differences between values of bond-dissociation energy and bond energy are to be attributed to this effect, the resonance stabilization of Russell-Saunders atomic states with high multiplicity. In addition, the energy of resonance among two or more valence-bond structures makes an important contribution in many cases. For example, the C—H bond-dissociation energy is about 101 kcal/mole for methane, ethane, and other alkanes, but is only 77 kcal/mole for toluene, as determined by Szwarc[3] from its rate of pyrolysis and by Schissler and Stevenson[4] by electron impact. The difference, 24 kcal/mole, can be attributed to the resonance stabilization of the benzyl radical that is produced by removing one hydrogen atom from the methyl group of toluene; this radical resonates among the several structures

(see Sec. 6-4).

About the same value, 25 kcal/mole, is found for the resonance energy of the allyl radical;[5] the resonance energy stabilizing the product is that for the structures $H_2C{=}CH{-}\dot{C}H_2$ and $H_2\dot{C}{-}CH{=}CH_2$.

[2] L. Pauling, *Proc. Nat. Acad. Sci. U. S.* **35**, 229 (1949).
[3] M. Szwarc, *J. Chem. Phys.* **16**, 128 (1948).
[4] D. O. Schissler and D. P. Stevenson, *J. Chem. Phys.* **22**, 151 (1954).
[5] A. H. Sehon and M. Szwarc, *Proc. Roy. Soc. London* **A202**, 263 (1950).

Author Index

Subject Index